Vitamins and Hormones

VOLUME 65

Editorial Board

TADHG P. BEGLEY

ANTHONY R. MEANS

BERT W. O'MALLEY

LYNN RIDDIFORD

ARMEN H. TASHJIAN, JR.

VITAMINS AND HORMONES

ADVANCES IN RESEARCH AND APPLICATIONS

Editor-in-Chief

GERALD LITWACK

Department of Biochemistry and Molecular Pharmacology
Jefferson Medical College
Thomas Jefferson University
Philadelphia, Pennsylvania

VOLUME 65

ACADEMIC PRESS

An imprint of Elsevier Science

Amsterdam Boston London New York Oxford Paris
San Diego San Francisco Singapore Sydney Tokyo

This book is printed on acid-free paper.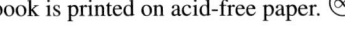

Copyright © 2002, Elsevier Science (USA).

All Rights Reserved.
No part of this publication may be reproduced or transmitted in any form or by any means, electronic or mechanical, including photocopy, recording, or any information storage and retrieval system, without permission in writing from the Publisher.

The appearance of the code at the bottom of the first page of a chapter in this book indicates the Publisher's consent that copies of the chapter may be made for personal or internal use of specific clients. This consent is given on the condition, however, that the copier pay the stated per copy fee through the Copyright Clearance Center, Inc. (www.copyright.com), for copying beyond that permitted by Sections 107 or 108 of the U.S. Copyright Law. This consent does not extend to other kinds of copying, such as copying for general distribution, for advertising or promotional purposes, for creating new collective works, or for resale. Copy fees for pre-2002 chapters are as shown on the title pages. If no fee code appears on the title page, the copy fee is the same as for current chapters.
0083-6729/2002 $35.00

Explicit permission from Academic Press is not required to reproduce a maximum of two figures or tables from an Academic Press chapter in another scientific or research publication provided that the material has not been credited to another source and that full credit to the Academic Press chapter is given.

Academic Press
An imprint of Elsevier Science.
525 B Street, Suite 1900, San Diego, California 92101-4495, USA
http://www.academicpress.com

Academic Press
84 Theobald's Road, London WC1X 8RR, UK
http://www.academicpress.com

International Standard Book Number: 0-12-709865-8

PRINTED IN THE UNITED STATES OF AMERICA
02 03 04 05 06 07 MM 9 8 7 6 5 4 3 2 1

Former Editors

ROBERT S. HARRIS
Newton, Massachusetts

JOHN A. LORRAINE
University of Edinburgh
Edinburgh, Scotland

PAUL L. MUNSON
University of North Carolina
Chapel Hill, North Carolina

JOHN GLOVER
University of Liverpool
Liverpool, England

GERALD D. AURBACH
Metabolic Diseases Branch
National Institute of Diabetes
and Digestive and Kidney Diseases
National Institutes of Health
Bethesda, Maryland

KENNETH V. THIMANN
University of California
Santa Cruz, California

IRA G. WOOL
University of Chicago
Chicago, Illinois

EGON DICZFALUSY
Karolinska Sjukhuset
Stockholm, Sweden

ROBERT OLSON
School of Medicine
State University of New York
at Stony Brook
Stony Brook, New York

DONALD B. MCCORMICK
Department of Biochemistry
Emory University School of Medicine
Atlanta, Georgia

Contents

Preface   xiii
Contributors   xv

Bone Growth Stimulators: New Tools for Treating Bone Loss and Mending Fractures
James F. Whitfield, Paul Morley, and Gordon E. Willick

I. Introduction   2
II. What is Osteoporosis?   3
III. BMUs: The Bone Repair/Remodeling Crews   4
IV. Menopause and Bone Loss   12
V. The Bone Anabolic PTHs   20
VI. The Statins   50
VII. Conclusion   58
     References   59

Using GFP-Ligand Fusions to Measure Receptor-Mediated Endocytosis in Living Cells
Lali K. Medina-Kauwe and Xinhua Chen

I. Introduction   82
II. Background   83

III. Applications   85
IV. Summary and Conclusions   93
   References   93

Mitochondrial Energy Dissipation by Fatty Acids: Mechanisms and Implications for Cell Death

Paolo Bernardi, Daniele Penzo, and Lech Wojtczak

I. Introduction   98
II. Mitochondrial Effects of Fatty Acids and Related Molecules   99
III. Cellular Effects of Fatty Acids   112
IV. Conclusions and Perspectives   116
   References   117

Trinucleotide Repeat Disease: The Androgen Receptor in Spinal and Bulbar Muscular Atrophy

Jessica L. Walcott and Diane E. Merry

I. SBMA: A Clinical and Pathological View of the Disease   128
II. The CAG/Polyglutamine Diseases   129
III. Androgen Receptor: Overview and Domain Organization   130
IV. Effects of CAG Expansion on AR Function   131
V. Molecular Pathology   133
VI. Mouse Models of SBMA   138
VII. Polyglutamine Expansion and Neuronal Dysfunction   140
   References   140

Human Prostacyclin Receptor

Emer M. Smyth and Garret A. FitzGerald

I. Introduction   150
II. Expression and Distribution of IP   151
III. Ligand Binding and Signal Transduction   152
IV. IP Regulation   153
V. Mutation Studies   155
VI. Knockout and Transgene Studies   157
VII. Multiple PGI_2 Receptors?   158

VIII. Conclusions   159
     References   159

Sterol Regulatory Element-Binding Protein Family as Global Regulators of Lipid Synthetic Genes in Energy Metabolism

Hitoshi Shimano

I. Introduction   169
II. Molecular Aspects of SREBPs   169
III. Members of the SREBP Family   174
IV. *In Vivo* Functions of SREBPs   175
V. Transcriptional Regulation of Lipid Synthesis by SREBP-2 and -1c in the Liver   178
VI. Promoter Analysis of SREBP-1c   179
VII. ADD1/SREBP-1c in Adipogenesis   181
VIII. Clinical Aspects of SREBPs   181
IX. Cross-Talk of Transcription Factors for Lipid Metabolism   183
X. Future Aspects of SREBPs   186
    References   187

Brassinosteroids: Plant Counterparts to Animal Steroid Hormones?

Steven D. Clouse

I. Introduction   196
II. Biosynthesis   198
III. Physiological Responses   207
IV. Signal Transduction   210
V. Conclusion and Future Prospects   217
    References   218

Endocannabinoids and Their Actions

Mauro Maccarrone and Alessandro Finazzi-Agrò

I. Introduction   226
II. The Endocannabinoids   226
III. Actions in the Central Nervous System   240
IV. Actions in the Periphery   243

V. Summary and Conclusions   247
   References   247

Endomorphins and Related Opioid Peptides
Yoshio Okada, Yuko Tsuda, Sharon D. Bryant, and Lawrence H. Lazarus

I. Introduction   258
II. Discovery of Endomorphins   258
III. Structure-Activity Relationship   267
IV. Conclusions and Future Directions   270
   References   271

Leptin and Melanocortin Signaling in the Hypothalamus
Christian Bjørbæk and Anthony N. Hollenberg

I. Introduction   282
II. The Leptin System   283
III. Discovery of Leptin Receptors   284
IV. Hypothalamic Signal Transduction by the Leptin Receptor   286
V. Hypothalamic Gene Regulation by Leptin   288
VI. Human Mutations in the Leptin Signaling System   291
VII. The Melanocortin System   292
VIII. The Melanocortin Receptors   293
IX. Proopiomelanocortin   296
X. Agouti-Related Peptide   297
XI. Melanocortin Coreceptors   298
XII. Targets of Melanocortin Signaling   299
XIII. Human Mutations in the Melanocortin System   300
XIV. Future Directions   301
   References   301

Function and Regulation of Cytosolic Molecular Chaperone CCT
Hiroshi Kubota

I. Introduction   314
II. Function of CCT   315

III. Structure of the CCT Complex   319
   IV. Evolution of CCT and its Eight Different Subunits   320
    V. Expression of CCT and its Subunits   321
   VI. Possible Roles of the Different Subunits   324
  VII. Conclusions   326
       References   326

Herbal Factors in The Treatment of Autoimmunity-Related Habitual Abortion
Tomoyuki Fujii

    I. Introduction   334
   II. Sairei-to and Tokishakuyaku-san and Their Constituents   334
  III. Habitual Abortion and Its Causes   335
   IV. Clinical Effects of Sai and Toki on Habitual Abortion   337
    V. Effects of Sai and Toki on the Immune Reaction of Mothers   338
   VI. Conclusions   342
       References   342

Vitamin D Receptor and Retinoid X Receptor Interactions in Motion
J. Barsony and K. Prufer

     I. General Features of VDR and RXR   346
    II. Regulation of Constitutive Activities of VDR by RXR   348
   III. Regulation of Calciotropic Effects of Calcitriol by RXR   348
    IV. Regulation of Antiproliferative Effects of Calcitriol by RXR   349
     V. RXR as a Dimerization Partner   350
    VI. RXR Effects on the Physicochemical Properties of VDR   350
   VII. RXR Effects on DNA Binding and Transcriptional Activities   352
  VIII. Dimerization Interfaces   354
    IX. Regulation of VDR/RXR Dimerization   355
     X. Subcellular Localization of VDR/RXR Heterodimers   356
    XI. Trafficking of VDR and RXR between Cytoplasm and Nucleus   357
   XII. Regulation of Unliganded VDR and RXR Shuttling   360
  XIII. Regulation of Liganded VDR and RXR Shuttling   361
   XIV. Intranuclear Trafficking and Target Site Binding of VDR and RXR   362
    XV. Outlook   363
        References   364

Index   377

Preface

Volume 65 brings some changes in the Editorial Board. We have added the following members to the board: Tadhg P. Begley and Lynn Riddiford. Previous members who are now departed from the board are Mary F. Dallman, Jenny P. Glusker, Vern L. Schramm, and Michael Sporn.

In the Editorial Board, Dr. Jenny Glusker is retiring as Professor at the Fox Chase Cancer Center and also wishes to surrender her responsibilities to *Vitamins and Hormones*. Dr. Vern Schramm finds that his schedule is already overcrowded and has made a recommendation for his replacement. I wish to thank Drs. Glusker and Schramm for their contributions to previous volumes. Some of the other members of the Editorial Board will be replaced. As a start, Dr. Tadhg Begley of Cornell University, who was the Guest editor for volume 61 of this Serial, has agreed to join the Editorial Board. Also, Dr. Lynn Riddiford of the University of Washington has agreed to join the Editorial Board. In the future, other names will be added to the roster of the Editorial Board members.

This volume contains a wide variety of subjects. The first contribution is by J. F. Whitfield, P. Morley, and G. E. Willick entitled "Bone Growth Stimulators: New Tools for Treating Bone Loss and Mending Fractures." This is followed by a treatise entitled "Using GFP-Ligand Fusions to Measure Receptor-Mediated Endocytosis in Living Cells" by L. K. Medina-Kauwe and X. Chen. P. Bernardi, D. Penzo, and L. Wojtczak provide a discussion entitled "Mitochondrial Energy Dissipation by Fatty Acids: Mechanisms and Implications for Cell Death." "Trinucleotide Repeat Diseases: the Androgen Receptor in Spinal and Bulbar Muscular Atrophy" is next reviewed by J. L. Walcott and D. E. Merry. "Human Prostacyclin Receptor" is the subject discussed by E. M. Smyth and G. A. FitzGerald. Vitamin D Receptor and Retinoid X Receptor Interactions in Motion is the subject of a contribution by J. Barsony and K. Prufer. H. Shimano reviews "Sterol Regulatory Element-Binding Protein (SREBP) Family as Global Regulators of Lipid Synthetic Genes in Energy Metabolism." The interesting subject of brassinosteroids is reviewed by S. D. Clouse in a contribution entitled "Brassinosteroids: Plant Counterparts to Animal Steroid Hormones?" "Endocannabinoids and Their Actions" is the subject of a chapter by M. Maccarrone and A. Finazzi-Agrò. Akin to this subject is an article by Y. Okada, Y. Tsuda, S. D. Bryant, and L. H. Lazarus on "Endomorphins and Related Opioid Peptides." "Leptin and Melanocortin Signaling in the Hypothalamus" is described by C. Bjørbaek and A. N. Hollenberg. H. Kubota discusses "Function and Regulation of Cytosolic Molecular Chaperone CCT." In the penultimate contribution T. Fujii discourses on the subject of "Herbal Factors in the Treatment of Autoimmunity-Related Habitual Abortion." Finally, J. Barsony and K. Prufer examine "Vitamin D Receptor and Retinoid X Receptor Interactions in Motion."

As before, I appreciate the continued interest and cooperation of the Publisher.

GERALD LITWACK

Contributors

Numbers in parentheses indicate the pages on which the authors' contributions begin.

Julia Barsony (345) Laboratory of Cell Biochemistry and Biology, National Institute of Diabetes, Digestive and Kidney Diseases, National Institutes of Health, Bethesda, Maryland 20892.

Paolo Bernardi (97) Department of Biomedical Sciences and the Venetian Institute of Molecular Medicine, University of Padova, I-35131 Padova, Italy.

Christian Bjørbæk (281) Division of Endocrinology, Beth Israel Deaconess Medical Center and Harvard Medical School, Boston, Massachusetts 02215.

Sharon D. Bryant (257) Peptide Neurochemistry, LCBRA, National Institute of Environmental Health Sciences, Research Triangle Park, North Carolina 27709.

Xinhua Chen (81) Department of Biochemistry, Institute for Genetic Medicine, University of Southern California Keck School of Medicine, Los Angeles, California 90033.

Steven D. Clouse (195) Department of Horticultural Science, North Carolina State University, Raleigh, North Carolina 27695.

Alessandro Finazzi-Agrò (225) Department of Experimental Medicine and Biochemical Sciences, University of Rome, "Tor Vergata," I-00133 Rome, Italy.

Garret A. FitzGerald (149) Center for Experimental Therapeutics, University of Pennsylvania, Philadelphia, Pennsylvania 19104.

Tomoyuki Fujii (333) Department of Obstetrics and Gynecology, Faculty of Medicine, University of Tokyo, Tokyo 113-8655, Japan.

Anthony N. Hollenberg (281) Division of Endocrinology, Beth Israel Deaconess Medical Center and Harvard Medical School, Boston, Massachusetts 02215.

Hiroshi Kubota (313) Department of Molecular and Cellular Biology and CREST/JST, Institute for Frontier Medical Sciences, Kyoto University, Kyoto 606-8397, Japan.

Lawrence H. Lazarus (257) Peptide Neurochemistry, LCBRA, National Institute of Environmental Health Sciences, Research Triangle Park, North Carolina 27709.

Mauro Maccarrone (225) Department of Experimental Medicine and Biochemical Sciences, University of Rome, "Tor Vergata," I-00133 Rome, Italy.

Lali K. Medina-Kauwe (81) Department of Biochemistry, Institute for Genetic Medicine, University of Southern California Keck School of Medicine, Los Angeles, California 90033.

Diane E. Merry (127) Department of Biochemistry and Molecular Pharmacology, Thomas Jefferson University, Philadelphia, Pennsylvania 19107.

Paul Morley (1) ParaTecH Therapeutics Inc., Ottawa, Ontario K1P 5MZ, Canada.

Yoshio Okada (257) Faculty of Pharmaceutical Sciences and High Technology Research Center, Kobe Gakuin University, Nishi-ku, Kobe 651-2180, Japan.

Daniele Penzo (97) Department of Biomedical Sciences and Venetian Institute of Molecular Medicine, University of Padova, I-35131 Padova, Italy.

Kirsten Prufer (345) Laboratory of Cell Biochemistry and Biology, National Institute of Diabetes, Digestive and Kidney Diseases, National Institutes of Health, Bethesda, Maryland 20892.

Hitoshi Shimano (167) Department of Internal Medicine, Institute of Clinical Medicine, University of Tsukuba, Tsukuba, Ibaraki 305-8575, Japan.

Emer M. Smyth (149) Center for Experimental Therapeutics, University of Pennsylvania, Philadelphia, Pennsylvania 19104.

Yuko Tsuda (257) Faculty of Pharmaceutical Sciences and High Technology Research Center, Kobe Gakuin University, Nishi-ku, Kobe 651-2180, Japan.

Jessica L. Walcott (127) Department of Biochemistry and Molecular Pharmacology, Thomas Jefferson University, Philadelphia, Pennsylvania 19107 and Graduate Group in Pharmacological Sciences, University of Pennsylvania School of Medicine, Philadelphia, Pennsylvania 19104.

James F. Whitfield (1) Institute for Biological Sciences, National Research Council of Canada, Ottawa, Ontario K1A 0R6, Canada.

Gordon E. Willick (1) Institute for Biological Sciences, National Research Council of Canada, Ottawa, Ontario K1A 0R6, Canada.

Lech Wojtczak (97) Department of Cellular Biochemistry, Nencki Institute of Experimental Biology, 02-093 Warsaw, Poland.

Bone Growth Stimulators

New Tools for Treating Bone Loss and Mending Fractures

James F. Whitfield,[*] Paul Morley,[†] and Gordon E. Willick[*]

[*]Institute for Biological Sciences, National Research Council of Canada, Ottawa, Ontario, Canada K1A 0R6, and [†]ParaTecH Therapeutics Inc., Ottawa, Ontario, K1P 5M2, Canada

I. Introduction
II. What is Osteoporosis?
III. BMUs: The Bone Repair/Remodeling Crews
    A. The Call to Action
    B. The Diggers
    C. The Fillers
    D. The Decline and Fall of BMU Efficiency
IV. Menopause and Bone Loss
    A. Estrogen, Fat, Brains, and Bones
    B. How Does an Estrogen Loss Cause an Osteoclast Population Explosion?
    C. How to Stop Menopausal Bone Loss
V. The Bone Anabolic PTHs
    A. Anabolics
    B. PTHs-Induced Bone Growth and Fracture Mending in Animals
    C. PTHs-Induced Bone Growth in Humans
    D. The Mechanism of PTHs' Anabolic Action
    E. The Clinical Prospects of the PTHs

VI. The Statins
  A. Animal Studies
  B. Osteogenic Mechanism
  C. Human Studies
  D. Statins' Clinical Prospects
VII. Conclusion
  References

In the new millennium, humans will be traveling to Mars and eventually beyond with skeletons that respond to microgravity by self-destructing. Meanwhile in Earth's aging populations growing numbers of men and many more women are suffering from crippling bone loss. During the first decade after menopause all women suffer an accelerating loss of bone, which in some of them is severe enough to result in "spontaneous" crushing of vertebrae and fracturing of hips by ordinary body movements. This is osteoporosis, which all too often requires prolonged and expensive care, the physical and mental stress of which may even kill the patient. Osteoporosis in postmenopausal women is caused by the loss of estrogen. The slower development of osteoporosis in aging men is also due at least in part to a loss of the estrogen made in ever smaller amounts in bone cells from the declining level of circulating testosterone and is needed for bone maintenance as it is in women. The loss of estrogen increases the generation, longevity, and activity of bone-resorbing osteoclasts. The destructive osteoclast surge can be blocked by estrogens and selective estrogen receptor modulators (SERMs) as well as antiosteoclast agents such as bisphosphonates and calcitonin. But these agents stimulate only a limited amount of bone growth as the unaffected osteoblasts fill in the holes that were dug by the now suppressed osteoclasts. They do not stimulate osteoblasts to make bone—they are *antiresorptives* not *bone anabolic* agents. (However, certain estrogen analogs and bisphosphates may stimulate bone growth to some extent by lengthening osteoblast working lives.) To grow new bone and restore bone strength lost in space and on Earth we must know what controls bone growth and destruction. Here we discuss the newest bone controllers and how they might operate. These include leptin from adipocytes and osteoblasts and the statins that are widely used to reduce blood cholesterol and cardiovascular damage. But the main focus of this article is necessarily the currently most promising of the anabolic agents, the potent parathyroid hormone (PTH) and certain of its 31- to 38-aminoacid fragments, which are either in or about to be in clinical trial or in the case of Lilly's Forteo [hPTH-(1–34)] tentatively approved by the Food and Drug Administration for treating osteoporosis and mending fractures. © 2002, Elsevier Science (USA).

I. INTRODUCTION

A small band of cosmonauts will soon be traveling to Mars and beyond. However, they are traveling with skeletons that were designed to bear terrestrial loads but not to withstand proloned exposure to microgravity for Earth-years in a space ship. The bones' strain-sensing mechanisms will respond to the lack of load by mobilizing teams of osteoclasts to get rid of the unneeded bone (Marie et al., 2000; Vico et al., 2001). This could result in breaking bones upon return to Earth, or even worse, stepping out onto an alien planet with weakened bones and only minimal medical backup to fix fractures.

Back on Earth, a "gray tidal wave" has already started to strain the medical facilities of the so-called "First World." Coming with it are failing body parts that are difficult and expensive to replace or repair. One of the greatest problems for aging men and women, but particularly for women during their first postmenopausal decade, is the severe loss of bone, especially cancellous (trabecular) bone, and the "spontaneous" fracturing of the remaining bone known as osteoporosis that can lead to extended hospitalization, crippling, and all too often despair and death because of broken hips (Baylink et al., 1999; Stevenson and Lindsay, 1999).

To develop a "chemical gravity" to treat bone loss in cosmonauts during their Earth-years-long exposure to microgravity and in aging or immobilized earthlings we must know what controls bone growth and resorption and how to manipulate these controls. In what follows we will focus on postmenopausal osteoporosis simply because it is the socially most important bone loss model from which most of our knowledge is being obtained.

II. WHAT IS OSTEOPOROSIS?

Osteoporotic postmenopausal women need not fall or hit something to break their fragile bones. Their wrists, hips, ribs, and vertebrae are apt to be broken or crushed by the muscular strains or minor blows of ordinary daily activities. Those women who were fortunate enough to have put enough bone in the bank when young to keep them above the "spontaneous" fracture threshold during the postmenopausal years of accelerated decline may become only "osteopenic" or pseudoosteoporotic. Their bones are more likely to be broken by falls and other bumps and blows due to poor eyesight and balance (Boxsein, 1999). Nevertheless, they still have enough mechanically strong bone that can resist being broken by their declining musculature during the commonly less strenuous activities of older people. However, weakening muscles are not unmixed blessings because there has to be some muscle-produced strain for optimal bone maintenance (Noble and Reeve, 2000; Skerry, 1999).

Osteoporosis in women is perhaps the most debilitating of the several consequences of the disappearance of estrogen at menopause. Perhaps surprisingly the osteoporosis that develops more slowly in men is also at least partly due to an estrogen decline, in this case to the smaller amounts of estrogen that can be made from the slowly dropping testosterone by the bone cells' aromatase (Klein, 1999). (Of course, the estrogen made by the fat cells' aromatase might moderate the effects of the lack of ovarian estrogen in the bones of those postmenopausal women whose adrenal cortices still make a significant amount of androgens.)

Clearly the aging populations need something to replace lost bone and restore bone strength.

It looks as if parathyroid hormone (PTH) and some of its fragments together with a potent physiological osteoclast inhibitor such as osteoprotegerin (OPG) are currently the most promising answers to this need, but as we shall see later there is conflicting evidence for the cholesterol-lowering statins being better (Whitfield, 2001a). To understand how these new tools work, we must first learn how bones are made and maintained.

III. BMUs: THE BONE REPAIR/REMODELING CREWS

A. THE CALL TO ACTION

Bone is the only organ that harbors cells whose job is to destroy it. Obviously, it must have a good reason for this. And it does! A load-bearing bone is like a busy interstate highway, which becomes cracked and worn with use and will eventually crumble unless the cracked and worn patches are detected and replaced by road maintenance crews. Bones too have maintenance crews— the *B*asic *M*ulticellular *U*nits (*BMUs*)—animated cartoons of which can be seen in Susan Ott's bone physiology website. Every 10 sec a crew is activated and at any time about 35,000,000 of them are at work removing and replacing about 500 mg of calcium from the skeleton each day. However, the remodeling rate varies widely throughout the adult skeleton. For example, the remodeling rate in adult cortical bone can be as low as 2% per year in the distal radius and as high as 50% per year in ilial trabecular bone (Noble and Reeve, 2000).

It appears that the continuous remodeling that is characteristic of big bones such as those in humans, horses, and elephants but not in the small bones of mice and rats has evolved to manage the continuous regional microcraking in the big bones caused by their owners' activities (R. B. Martin, 2000, 2002). The remodeling is restricted to microcracked patches by inhibitory signals maintained by normal bone straining from a network of osteoclasts locked inside the bone to lining cells on the bone surface (R. B. Martin, 2000). Microcracking in a patch of bone severs the local osteonetwork connections and cuts off the flow of inhibitory signals from the osteocytes each of which is locked in a tiny, fluid-filled cubicle (lacuna) and its processes are connected to the other osteocytes by extending through a network of canaliculi (little canals) inside the bone (Noble, 2000; Noble and Reeve, 2000; Skerry, 1999; Schaffler, 2000). Normal bone loading squeezes this "sponge," which causes the fluid in the cubicles and canaliculi to slosh around and stretch the osteocytes (Kufahl and Saha, 1990). This activates mechanosensitive Ca^{2+} channels (N. X. Chan *et al.*, 2000; Duncan and Misler, 1989) and pulls on signal-generating integrins that attach the cell and its processes to the osteopontin lining the walls of the cubicles and canaliculi. The result of this normal sloshing, shearing, and stretching is intracellular Ca^{2+} surges and pulses of signalers such as NO and glutamate (the well-known excitatory neurotransmitter until recently believed to be used only by central neurons) passing from cell to cell through gap junctions and ion channels and bouncing from receptor to receptor to prevent bone lining cells from signaling vascular endothelial cells and marrow stromal cells to start the repair process (N. X. Chan *et al.,* 2000; Daifotis *et al.,* 1992; Duncan and Misler, 1989; Edlich *et al.,* 2001; Kufahl and Saha, 1990; Manolagas, 2000; Martin, 2000; Noble, 2000; Noble and Reeve, 2000; Noda *et al.,* 1996; Nomura and Takano-Yamamoto, 2000; Pirola *et al.,* 1994; Ryder and Duncan, 2001; Schaffler, 2000; Skerry and Genever, 2001; Steers *et al.,* 1998). But if the stretching is great enough it might actually break the local connections of the osteointernet as well as trigger the suicidal process called apoptosis (from a Greek word meaning *to fall like autumn leaves*) (Noble, 2000; Noble and Reeve, 2000; Skerry, 1999; Schaffler, 2000). Although the death of the overstrained osteocytes would stop the inhibitory signaling to the lining cells and release the lining cells to activate the repair machinery, it would also cut off a source of the principal physiological osteoclast suppressor, osteoprotegerin (OPG), that the osteocytes had been making since its gene was turned on earlier in their development by Cbfa1, the master osteoblast-specifying transcription factor (Ducy, 2000; Karsenty, 2000a; Komori, 2000; Noble and Reeve, 2000; Thirunavukkarnasu and Halliday, 2000) This would remove a major obstacle to the generation of osteoclasts, the first members of the BMU to arrive on the job.

B. THE DIGGERS

As in a road repair crew, the first on the job are the diggers (Noble, 2000; Noble and Reeve, 2000; Skerry, 1999; Whitfield *et al.,* 1998a). And one of the first responses of the osteocytes to microcracking strain is an upsurge of osteopontin expression to glue the diggers to the signaling patch (Nomura and Takano-Yamamoto, 2000). But how might the osteocytes summon these diggers to the damaged patch?

Osteocytes are liable to die when exposed either to very low or very high strains, but function best when the straining is within the range produced by normal body movements (Noble and Reeve, 2000). Without being regularly stimulated by the strains of normal activity, osteocytes, and with them their products such as the osteoclast-suppressing OPG, may be lost because they are not stimulated enough to make antiapoptosis survival factors such as Bcl-2. This would cut off OPG production and the flow of inhibitory signals to the bone-lining cells, which then starts bone resorption by mobilizing osteoclasts. On the other hand, the stimulation of OPG-produciing osteocytes by unusually high strain would cause a massive production of osteogenic factors that reach the marrow stroma where they mobilize osteoblast precursors and start bone formation. But if the strain should exceed a certain level, the osteocytes would be killed, their osteoclast-suppressing OPG production would be lost, and the lining cells would activate osteoclast production to destroy bone (R. B. Martin, 2000).

There is growing evidence that glutamate is an "osteotransmitter" as well as a neurotransmitter that plays a major part in the mobilization of the responses to loading and microdamage (Skerry and Genever, 2001). But where are the glutamate secretors? Where are the "preosteosynaptic" cells? It might be released from strained osteocytes and/or glutamatergic nerves running along blood vessels and contacting marrow stromal cells and bone cells (Mason *et al.*, 1997; Skerry and Genever, 2001; Serre *et al.*, 1999).

Osteocytes are not glutamate targets. They do not express glutamate receptors, but they have GLAST (EAAT1) glutamate transporters, which they might use to reaccumulate glutamate they have released, for example, in response to Ca^{2+} surges through mechanosensitive Ca^{2+} channels (Skerry and Genever, 2001). Osteoblasts are glutamate targets because osteoblast precursor cells express the ionotropic 2-(aminomethyl)phenylacetic acid (AMPA)/kainate and *N*-methyl-d-aspartate (NMDA) receptor/channels and mature osteoblasts express these receptor/channels as well as the G-protein-coupled, metabotropic glutamate receptors that are related to the CaRs and can also be activated by the high Ca^{2+} concentrations in the excavation sites (Brown and MacLeod, 2001). Mature osteoblasts, like glutamatergic neurons, also express GLAST (EAAT1) transporters presumably, as in neurons, to recharge their glutamate stores and clear the pericellular area of released glutamate to avoid lethally prolonged signaling from their glutamate receptors. Inhibiting the AMPA/kainate and NMDA receptor signaling reduces the production and differentiation of osteoblasts and causes precursor cells to become adipocytes rather than osteoblasts (Dobson and Skerry, 2000; Skerry and Genever, 2001; Taylor *et al.*, 2000). Although a strain-induced glutamate surge stimulates the production of osteoblasts it can also stimulate osteoclast generation in two ways. It can stimulate osteoblastic stromal cells to express on their surfaces RANKL, the ligand for the osteoclast precursors' *R*eceptor *A*ctivator of *N*F-κB transactivator (RANK) receptors the signals from which stimulate osteoclast differentiation (Taylor *et al.*, 2000). And it can directly stimulate osteoclast generation and function via the NMDA receptor/channels on both osteoclast precursors and mature osteoclasts (Espinosa *et al.*, 1999; Itzstein *et al.*, 2000; Laketic-Ljubojevic *et al.*, 1999; Patton *et al.*, 1998; Peet *et al.*, 1999; Skerry and Genever, 2001). Osteoclasts seem to need the Ca^{2+} that surges through these activated receptor/channels to mature and to assemble the actin ring that will seal off the excavation pit (Itzstein *et al.*, 2000; Mason *et al.*, 1997).

According to a remodeling/repairing scenario for cortical bone proposed by Parfitt (1998, 2000a,b) preosteoclasts home onto a signaling patch using the same addressing system that other members of the monocyte/macrophage family of leukocytes use to home onto damaged tissue. According to this scenario, signals from the patch start the microcrack repair process by causing the endothelial cells of a local blood vessel to sprout a loop. The cells at the leading edge of the vascular loop switch on a set of genes the products of which make the cells' surfaces selectively sticky to grab appropriately addressed preosteoclasts from the passing blood (Ruoslathi and Rayotte, 2000; Springer, 1994), the flow of which into the patch has been increased by the local surge of the vasodilatory NO from the strained osteocytes (Whitfield *et al.*, 1998a). The preosteoclasts then squeeze between the loop cells to leave the blood and fuse with others to form large active multinuclear osteoclasts each of which dig for the next 2 weeks. These pathfinding osteoclasts, the first of several waves, start tunneling through the damaged patch at a speed of about 25 μm/day accompanied by the vascular loop extruding new preosteoclasts from its tip.

Things are different in cancellous and probabably endocortical bone directly facing the bone marrow. Something remarkable happens when the lining cells on a trabecula are uncoupled from the underlying osteocytes. A kind of blister or "bone-remodeling compartment (BRC)" forms on the trabecular surface (Hauge *et al.*, 2001; Parfitt, 2001)! It appears that the derepression of the lining cells causes them to start making collagenase to remove the matrix covering and lift off the surface to form a canopy over the future work site. This blister forms a pseudoblood vessel with a wall of osteoprogenitors and lining cells instead of true CD34-bearing endothelial cells (Hauge *et al.*, 2001). The blister vessel connects to the local marrow sinusoids to transport osteoclasts and their precursors to the worksite.

C. THE FILLERS

As the osteoclasts are digging out a large patch of bone around the microcrack with proteases such as cathepsin K, they release a pack of bone-growth-promoting cytokines [e.g., bone morphogenic protein (BMP)-2, fibroblast growth factor (FGF)-2, insulin-like growth factors (IGFs)-I and -II, IGF-binding protein (IGFBP)-5, transforming growth factor-βs (TGF-βs) that were deposited in the matrix 2 to 5 years earlier by the osteoblasts of a previous BMU. It is widely assumed that these liberated factors, particularly the IGFs, form an osteogenic factors "bank account" that can be drawn on by a subsequent BMU to start new bone growth for repairing a microcrack. However, they may merely be the remnants of the autocrine/paracrine factors that the osteoblasts were using to stimulate themselves and may not survive the proteases that liberated them.

A lot of Ca^{2+} is also released when the osteoclasts dissolve the bone mineral (hydroxyapatite) by directing a stream of H^+ ions into the hole with their vacuolar H^+-ATPase. This Ca^{2+} is a two-way switch—"off" for the osteoclasts and "on" for the bone-making osteoblasts coming behind them. When the Ca^{2+} concentration in a tunnel or trench rises above a certain level, it activates Ca^{2+} sensors on the osteoclast's apical H^+- and protease-pumping ruffles jutting into the hole (Kameda *et al.*,1998; Zaidi *et al.*, 1999). The resulting Ca^{2+} surges and signals cause the osteoclast to pull up the actin sealing ring it put around the hole and glide over to another part of the patch and start digging again, but if it cannot find another place to attach itself it will kill itself by apoptosing (Frisch and Screaton, 2001; Lorget *et al.*, 2000; Sakai *et al.*, 2000). By contrast the signals from the Ca^{2+}-sensing receptor expressed by the oncoming osteoblast precursors (Brown and MacLeod, 2001; Farzaneh-Far *et al.*, 2000; Hinson *et al.*, 1997; Pi *et al.*, 1997; T. Yamaguchi *et al.*, 1998a,b, 2000, 2001) stimulate the cells to express BMP-2 and -4, which, in turn, stimulate the expression of Cbfa1, the osteoblast-specifying transcription factor, which by triggering collagen I production starts the next stages of the bone-making process (Nakade *et al.*, 2001). At certain optimal levels the increased Ca^{2+} may also stimulate the expression of IGFs-I and -II, which would stimulate the proliferation of osteoprogenitors but not mature osteoblasts (Honda *et al.*, 1995; Sugimoto *et al.*, 1994). The signals from their Ca^{2+} sensors also cause osteoblast precursors to migrate to the excavation site (Brown and MacLeod, 2001; Yamaguchi *et al.*, 1998a).

When the leading endothelial cells are displaced from the tip of the advancing capillary loop, they turn off the genes for grabbing passing preosteoclasts and turn on the genes for products such as endothelin-1 that will stimulate osteoblast progenitors and preosteoblasts that came into the cortical or trabecular excavation site alongside the advancing capillary loop, or, in the case of trabecular or endocortical bone, through the lining cell "blister" from osteoblastic marrow stromal cells or from the bone lining cells themselves (Bianco *et al.*, 1993; Doherty and Canfield, 1999; Hauge *et al.*, 2002; Parfitt, 2000a,b; 2002; Schaffler, 2000). If any of it has escaped the osteoclasts' proteases, the BMP-2 liberated from the matrix by the osteoclasts would push these precursors along the osteoblast differentiation pathway (Chen *et al.*, 1998). They would then be stimulated to mature and start working by the signals from other receptors, which include the Ca^{2+} sensors (Brown and MacLeod, 2001) and type 1 PTH/PTHrP (PTHR1) receptors (Aubin 1998, 2000, 2001; McCauley *et al.*, 1996), the expression of which increases about 10-fold during the transition from mature osteoprogenitor to preosteoblast. Signaling by glutamate from invading or adjacent nerves is also needed. As we learned above, the osteoblasts, like the osteoclasts they are following, are equipped with glutamate receptors and they also have the two ionotropic receptor channels as well as the G-protein-coupled metabotropic mGluR1b receptors with their seven transmembrane α-helices, which belong to the same G-protein-coupled receptor (GPCR) family as the Ca^{2+} receptors (Gu and Publicover, 2000). Moreover, they have Na^+-dependent GLAST (EAA1) transporters to limit the duration of receptor activation by released glutamate by carrying the glutamate back into the cell (Mason *et al.*, 1997). The activated mGluRs, like activated Ca^{2+} sensors and PTHR1 receptors, trigger a burst of scaffold-bound phospholipase C (PLC) activity that triggers a prompt release of Ca^{2+} from internal stores followed by the flow of Ca^{2+} through opened membrane channels and a surge of protein kinase Cs (PKCs) activity (Gu and Publicover, 2000). The importance of glutamate as an osteogenic osteotransmitter is still uncertain, because although Dobson and Skerry (2000) and Taylor *et al.* (2000) have found that inhibiting the glutamate receptors reduce bone formation *in vivo* and *in vitro,* Gray *et al.* (2001) have reported that high (and possibly nonselective) doses of NMDA receptor/channel inhibitors did not affect bone formation by cultured rat osteoblasts and bone formation in GLAST transporter knockout mice was normal. However, osteoblasts also have AMPA/kainate receptors (Skerry and Genever, 2001), which might take over from blocked NMDA receptors, and we do not know whether GLAST activity and glutamate reuptake are needed for bone making.

The neural signaling similarities do not end with an array of glutamate receptors and transporters. Active osteoblasts also make brain-derived neurotrophic factor (BDNF) and its TrkB receptor (Yamashiro *et al.*, 2001). This BDNF might be used to stimulate the innervation of new bone. In addition, because they have TrkB receptors, osteoblasts may use their BDNF as an autocrine/paracrine driver of some function(s).

Before the osteoblasts arrive on the scene and start working, the lining cells have the janitorial task of sweeping up the litter left by the osteoclasts on the floor of the resorption cavity (Everts *et al.*, 2002). The osteoclasts have left collagen "bristles" sticking out of the floor of the cavity. The lining cells move onto the surface and shave off the "bristles" to make a smooth surface upon which to put a layer of osteopontin for cementing the new collagen from the osteoblasts to the old bone (Everts *et al.*, 2002; McKee and Nanci, 1996). As we shall see below, it is possible that residual phosphate in the resorption cavity is the stimulator of this first burst of osteopontin expression (Beck *et al.*, 2000).

The osteoblasts take about eight times longer to fill the tunnels and trenches than the osteoclasts took to dig them. Therefore, at any moment a mature bone is peppered with the holes and "blisters" from the microcrack-repairing BMUs—the *"remodeling space."* As the osteoblasts are slowly filling these holes with new factor-loaded matrix ("osteoid"), some of them will be trapped in it as it is gradually mineralized with apatite-like Ca-phosphate. Walled up in their apatite-lined cubicles they greatly reduce PTHR1 receptor expression (Aubin, 1998, 2000, 2001; Aubin and Triffitt, 2002), stop making bone, and become osteocytes that plug themselves into the osteointernet, messages from which can travel through gap junctions and via signaling receptors as far as lining cells and adjacent marrow stromal cells and blood vessels to summon BMUs to repair damage or mobilize Ca^{2+} if needed to restore the circulating level (Skerry, 1999; Whitfield *et al.*, 1998a; Yellowley *et al.*, 2000). When the new patch is finally in place 6 to 9 months later, the signaling stops and the approximately 3-month-old members of the last osteoblast crew are now out of work, redundant and dispensible. Those that cannot find a place to attach and activate their survival-promoting integrins commit apoptotic suicide (Frisch and Screaton, 2001; Whitfield *et al.*, 1998a, 2000a). This drastic downsizing process is known as *anoikis* from the Greek word for homelessness— homelessness-induced apoptosis (Frisch and Screaton, 2001). However, the apoptosing osteoblasts may make one last contribution to bone formation, specifically mineralization, by dumping alkaline phosphatase into the new matrix in vesicles released from their blebbing surfaces (Farley and Stilt-Coffing, 2001). Some of this alkaline phosphatase will get into the blood where its level would be an indicator of bone formation as well as osteoblast apoptosis.

Most of the dying (apoptosing) cells in bone are located in microcrack sites being repaired by BMUs. The large amount of P_i released along with Ca^{2+} from the apatite crystals dissolved by the osteoclasts may increase the chance of osteoblasts triggering apoptosis unless they are attached and protected by survival factors such as Bcl-2 (Meleti *et al.*, 2000). The mechanism of this is likely to be the Na^+ gradient-powered pumping of P_i into the osteoblasts by their type-III Pit 1 and 2 Na^+/P_i transporters, particularly the Pit 1 transporter that they selectively up-regulate during maturation for matrix mineralization (Adams *et al.*, 2001; Meleti *et al.*, 2000; Nielsen *et al.*, 2001; Takeda *et al.*, 1999). When the phosphate reaches a critical level, it would be carried into the mitochondria by the P_i^-/OH^- exchanger and trigger the assembly of the huge mitochondrial permeability transition pore and the leakage of cytochrome *c*, apoptosis-inducing factor (AIF), procaspase-9, and Diablo into the cytoplasm from the intermembrane space. Diablo neutralizes a group of caspase inhibitors and cytochrome *c* forms a complex with APAF-1, which causes caspase-9 to autoactivate and trigger a lethal cascade of so-called "executioner" caspase proteases (Crompton, 2000; Finkel, 2001). [This ability of P_i to greatly promote apoptogenesis was first noticed by one of us (Whitfield) nearly 40 years ago when searching for a way to enhance radiation-induced apoptosis of rat thymic lymphocytes for an ultrasensitive radiation biodosimeter (e g., Whitfield *et al.*, 1967).] Miletei *et al.* (2000) have shown that inhibiting the transporter with phosphonoformic acid (PFA) prevents P_i from killing primary human osteoblasts and Mansfield *et al.* (2001) have shown that PFA also prevents P_i from triggering apoptosis in chondrocytes. It follows from this that inhibiting the Na^+/P_i transporter might prolong the lifetime of osteoblasts by shielding them from the high P_i in their workplace. However, as we shall see later, this could be counterproductive because the transporter is also a key player in osteoid mineralization.

The apoptogenic action of P_i in cultured human osteoblasts from bone fragments and murine MC3T3-E1 is greatly enhanced by a small, by itself harmless, increase (0.1–1.0 m*M*) in the Ca^{2+} concentration in the culture medium (Adams *et al.*, 2001). Therefore, it is likely that the high Ca^{2+} concentrations in the excavation sites promotes the P_i apoptogenesis. How Ca^{2+} does this is unknown. It is not due to an increased flow through L-type Ca^{2+} channels, because blocking these channels does not prevent Ca^{2+} from enhancing P_i apoptogenesis (Adams *et al.*, 2001). It is also unlikely that Ca^{2+} and P_i would form an "endocytosable" apoptogenic particulate complex on the cell surface because P_i apoptogenesis is completely prevented in the presence of Ca^{2+} by inhibiting the P_i transporter with PFA, which means that P_i has to be delivered separately into the cell to trigger apoptosis. A more likely mechanism is the stimulation of CaRs, a burst of phospholipase C activity, a surge of inositol 1,4,5-trisphosphate (IP_3) from membrane phospholipid breakdown, and a highly localized release of Ca^{2+} from IP_3 receptor-bearing endoplasmic reticulum Ca^{2+} stores that are closely juxtaposed to clusters of mitochondria (Hajnoczky *et al.*, 2000; Mannella, 2000). This Ca^{2+} would enhance P_i action by targeting the mitochondrial Ca^{2+} uniporter and riding it into the adjacent mitochondria to join P_i in stimulating the formation of the permeability transition pore (Hajnoczky *et al.*, 2000).

Finally, although osteoblast recruitment and activity are tightly coupled to prior osteoclast activity in the BMUs of continuously remodeling mature human bone, they are not necessarily so (Karsenty, 1999). Clusters of osteoblasts operate independently from osteoclasts in the growing bones of rat pups and human children (Frost, 1997; Selye, 1932). As we shall see below, PTHs can stimulate a massive layering of osteoblasts on trabeculae and bone formation *without a prior activation of osteoclasts*. Indeed mutant mice, which cannot generate functional osteoclasts become osteopetrotic due to the unopposed bone building by osteoblasts; and mutant mice that cannot generate functional osteoblasts become osteoporotic due to the unopposed osteoclast activity (Karsenty, 1999; Whitfield *et al.*, 1998a).

D. THE DECLINE AND FALL OF BMU EFFICIENCY

Unfortunately, BMUs are not as good at patching as road repair teams, which is why once strong koric (from the Greek word *kore* meaning young woman) bone weakens with advancing age. First, the availability of osteoblasts for BMUs drops with advancing age as the number of osteoprogenitor cells in the bone marrow declines (Nishida *et al.*, 1999). Osteoblasts working on the inner surface of the cortical bone and the trabeculae in the cancellous (trabecular) compartment of bone do not completely refill the osteoclasts' excavations, but periosteal osteoblasts tend to overfill the holes (Eriksen, 1994a,b; 1994; Frost, 1997). Therefore, with advancing age the cortical shells thin as the endocortical and trabecular parts of the bones waste away. Fortunately the periosteal overfilling increases the diameters of load-bearing bones (such as the femur), which somewhat compensates for the overall thinning and loss by resisting an increase in the bone's vulnerability to bending and breaking (Einhorn, 1996). But the cortex of the femoral neck thins without the neck increasing in diameter because there are no periosteal BMUs (Einhorn, 1996). This combination of a thinning cortical shell with perforated trabecular plates and broken connecting rods in the cancellous compartment without an increase in the neck's diameter makes the aging hip especially vulnerable to bending and breaking which is the most serious of the consequences of osteoporosis (Einhorn, 1996). The amount of bone removed by BMU osteoclasts depends on the number of preosteoclasts that can be snatched from the circulation and the lifespans of the osteoclasts into which they fuse (Manolagas, 2000; Parfitt, 2000a). When the signaling from a repaired patch stops, osteoclast recruitment also stops and so should the digging. However, depending on the number of osteoclasts in the excavation, the digging may continue for some time after the signaling has stopped (Parfitt, 1998, 2000a,b). Therefore, any *antiresorptive* or *antiremodeling* agent that reduces the number of osteoclasts and/or their activity can reduce or stop resorption and remodeling. This would result in a limited amount of bone growth as the unaffected osteoblasts continue filling the existing holes, the resorption space without having to contend with osteoclasts digging more holes faster than they can be filled.

The amount of bone put into the osteoclast excavations is a function of the number of mature osteoblasts and how long they can work. Therefore, bone formation beyond the amount removed by osteoclasts can be stimulated by agents that increase the number and/or active lifespan of PTHR1 receptor-expressing osteoblasts (Dempster, 1997; Manolagas, 2000). As we shall see, the PTHs, some statins, and probably leptin, are just such *anabolic* bonebuilders.

IV. MENOPAUSE AND BONE LOSS

A. ESTROGEN, FAT, BRAINS, AND BONES

So far it has seemed that estrogen is the *primum inter pares* of an ever-growing number of agents that control bone growth and strength in both women and, perhaps surprisingly, men (Baylink *et al.,* 1999; Klein, 1999; Stevenson and Lindsay, 1999; Vanderschueren *et al.,* 2000). Now it appears that bones are also targets of the mechanism that controls the body's white fat energy reserves (Ducy *et al.,* 2000a,b; Fleet, 2000; Karsenty, 2000b) (Fig. 1). White fat cells (and, as we shall see below, osteoblasts) make a helical cytokine, the 167-amino acid, 16.7-kDa leptin (from the Greek word *leptos* meaning thin) in amounts depending on their fat load (Ahima and Flier, 2000; Friedman, 2000). Therefore, the amount of circulating leptin tells the neurons in the hypothalamic arcuate nucleus of the size of the fat stores. Leptin inhibits the secretion of the orexigenic (phagostimulatory) neuropeptide Y (NPY) by the arcuate neurons (Ahima and Flier, 2000; Friedman, 2000). If the fat load is at the body's optimal set point, the circulating leptin holds NPY secretion at the appropriate level. But if the fat load and thus leptin production drop, NPY secretion, and with it eating, increase to restore the fat load and the circulating leptin concentration to their optimal values.

Of particular importance for the bones are leptin's stimulation of gonadotropin-releasing hormone (GnRH) secretion, and the estrogen dependence of the *Lep* (originally called *Ob* for obese) gene expression (Ahima and Flier, 2000; Bann *et al.,* 1999; Chu *et al.,* 1999). Therefore, when either the *Lep* (*Ob*) gene or the LepRb receptor is disabled, the animal will be actually or functionally leptinless and consequently obese and hypoestrogenic, and therefore should be osteopenic.

Instead of being severely osteopenic because of having too little estrogen, $Ob(Lep)^{-/-}$ obese mice that cannot make leptin and $Db^{-/-}$ mice with disabeled LepR receptors have a *high* bone mass (HBM) (Ducy *et al.,* 2000a,b; Karsenty, 2000b). Because the lack of estrogen increases osteoclast production, they do have more osteoclasts than normal lean animals and only a normal number of osteoblasts. But these osteoblasts are hyperactive; they override the increased resorption by the osteoclasts by making twice as much bone matrix as the osteoblasts in lean mice (Ducy *et al.,* 2000a,b; Karsenty, 2000b).

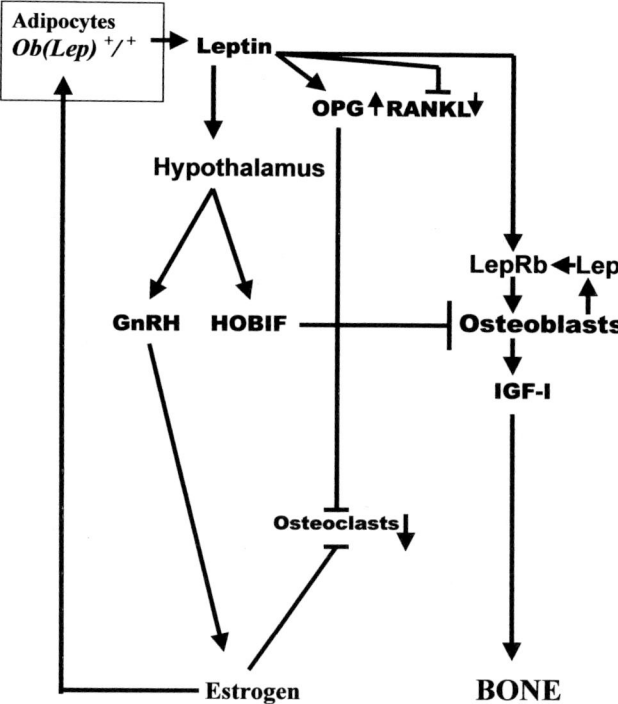

FIGURE 1 The two ways by which leptin controls bone growth. The size and strength of the bones are the net result of the amount of circulating that crosses the blood–brain barrier, reaches the hypothalamus, and maintains the secretion of the hypothalamic osteoblast inhibitory factor (HOBIF) that can reach the blood and restrain the ability of osteoblasts to make bone. However, signals from the osteoblasts' long leptin receptors (LepRb) activated by circulating leptin from fat cells and autocrine leptin from the osteoblasts themselves can directly trigger an IGF-I-mediated increase in bone growth and matrix mineralization. Leptin from both sources can also promote bone formation by inhibiting osteoclast generation by stimulating the expression of the osteoclast-suppressing osteoprotegerin (OPG) and inhiting the expression of the proosteoclast RANKL in osteoblastic stromal cells.

This supranormal bone production in the face of a soaring osteoclast population is due to a hypothalamic response to the failure of fat cells to make [e.g., $Ob(Lep)^{-/-}$ obese mice] or to respond to (e.g., leptin receptor-deficient $Db^{-/-}$ obese mice) leptin, and it can be stopped in $Ob(Lep)^{-/-}$ mice (which have functional leptin receptors) by intracerebroventricular injection of leptin (Ducy et al., 2000a,b; Karsenty, 2000b). The osteoblast hyperactivity might be due to the excessive amounts of NPY, but NPY causes bone loss (Ducy et al., 2000; Karsenty, 2000b). It might be due to increased endocannabinoid expression (DiMarzo et al., 2001), but we don't yet know what these agents do to bone. Thus, it seems that a setpoint level of leptin might induce hypothalamic neurons to produce a hypothalamic osteoblast inhibitory factor(s) (HOBIF) that somehow holds osteoblast matrix-producing activity at an optimal level (Ducy et al., 2000a,b; Karsenty, 2000b;

Whitfield et al., 2002) (Fig. 1). Without signals from leptin-activated receptors on hypothalamic neurons, there is no HOBIF, the brakes on osteoblast matrix production are released, and bone growth surges upward along with food consumption and body weight (Fig. 1). But the supranormal bone growth is not linked to the weight gain because it starts before the weight of the obese mouse starts rising and it also happens in leptinless A-ZIP/F-1 mice that do not have white-fat adipocytes and do not gain weight (Ducy et al., 2000a; Karsenty, 2000b).

On the basis of these observations, Karsenty (2000b) has called leptin the osteoblast-suppressing equivalent of the osteoclast-suppressing OPG. But it isn"t! Although the primary adult mouse osteoblasts in the experiments of Ducy et al. (2000a) did not express LepR receptors and thus could not respond directly to leptin, others have found that mouse bone cells can directly and positively respond to the cytokine both in culture and in the mouse. Liu et al. (1997) reported that leptin injection directly stimulates endocortical bone formation in obese mice. Steppan et al. (2000) reported the results of experiments showing that intraperitoneally injected leptin is a *potent stimulator* of bone growth in $Ob(Lep)^{-/-}$ mice. Moreover, Cornish et al. (2001) reported that leptin stimulates fetal rat osteoblasts as strongly as IGF-I and increases bone strength in mature male mice by 20% or more. And Maor et al. (2002) found that leptin causes IGF-I-mediated endochondral bone formation in cultured murine mandibular condyle and humeral growth plate.

Primary adult human osteoblasts, as well as primary rat osteoblasts and ROS 17/2.8 rat osteoblastic osteosarcoma cells express active LepRb receptors (Bassilana et al., 2000; Enjuanes et al., 2002; Evans et al., 2001; Lee et al., 2001; Reseland et al., 2002; Steppan et al., 1998, 2000). Signals from these receptors stimulate bone growth because Thomas et al. (2001) reported that intraperitoneally infusing human recombinant leptin prevents disuse-induced bone loss in tail-suspended rats. Burguerera et al. (2001) have shown that leptin promotes rat bone growth by suppressing the osteoclast-promoting RANKL expression and stimulating osteoclast-suppressing OPG expression in osteoblastic stromal cells (Fig. 1). Leptin also makes human marrow stromal cells differentiate into osteoblasts instead of adipocytes (Thomas et al., 1999), and it stimulates the proliferation of cultured human osteoblasts and causes human marrow stromal osteoprogenitor cells to express alkaline phosphatase, collagen I, and osteocalcin, and to mineralize matrix (Evans et al., 2001; Thomas et al., 1999). The ability of leptin to directly stimulate matrix mineralization by human osteoprogenitors fits in nicely with the fact that that human osteoblasts start making and secreting autocrine/paracrine leptin when they are either in the mineralization or early osteocyte stage (Reseland et al., 2001) (Fig. 1).

All of this means that human osteoblasts are under dual leptin-dependent control—indirect negative control by leptin-induced HOBIF from the brain and positive *direct* control by leptin itself from the white fat adipocytes, yellow marrow adipocytes, and the late-stage osteoblasts themselves (Laharrague et al., 1998; Reseland et al., 2001) (Fig. 1). Therefore in a person with common, as opposed to genetic (e.g., $Ob[Lep]^{-/-}$), obesity in whom large amounts of leptin are made in the fat but cannot be carried into the brain by their saturated transvascular transport mechanism, there would not be a correspondingly large HOBIF production but there would be a strong direct stimulation of osteoblast activity by the leptin entering the bone through marrow sinusoids.

It is likely that the postmenopausal estrogen drop turns off both of the leptin mechanisms (Brann *et al.*, 1999; Chu *et al.*, 1999) while causing an osteoclast population explosion. Thus, for example, ovariectomy silences *Ob(Lep)* gene expression in rats and leptin vanishes from their circulation (Chu *et al.*, 1999) while their weight increases and femoral trabecular bone is being destroyed (Whitfield *et al.*, 1995). The leptin drop and the resulting derepression of endocannabinoid and NPY secretion in the hypothalamus cause the hyperphagia and weight gain. At first this seems to conflict with the amount of circulating leptin expression being positively correlated to the body fat load—leptin should increase. But *Ob(Lep)* gene expression is estrogen dependent (Fig. 1) and giving estrogen to the Ovariectomized (OVXed) rats prevents the leptin drop, hyperphagia, and weight gain (Chu *et al.*, 1999). Therefore, without enough estrogen to maintain *Ob(Lep)* gene expression leptin virtually vanishes from the circulation by 7 weeks after ovariectomy (OVX), but the increasing load of white fat with its enlarging adipocytes eventually overrides the estrogen lack and leptin production rebounds and its level soars above the sham OVX level between 9 and 13 weeks after the operation (Chu *et al.*, 1999). Therefore, the bones of a hyperphagic OVXed rat do not at first have to face the double blow of leptin-induced HOBIF suppression of osteoblast activity and an osteoclast feeding frenzy, but the direct stimulation of osteoblasts by an eventual leptin surge might moderate the bone loss. It follows from this that it would be important to find out whether a leptin loss might also affect bone loss in postmenopausal women and whether leptin stimulates bone growth in humans as it does in rats and human bone cultures.

B. HOW DOES AN ESTROGEN LOSS CAUSE AN OSTEOCLAST POPULATION EXPLOSION?

Osteoclast progenitors respond to estrogen through their conventional "genomic" estrogen receptors that operate in the nucleus to activate target genes with so-called estrogen response elements (EREs) in their regulatory regions as well as through what appear to be "nongenomic" receptors that might fire signals into the cell from lipid rafts in plasma membrane caveolae (Chambliss *et al.*, 2000; Falkenstein *et al.*, 2000; Fionelli *et al.*, 1996; Kelly and Levine, 2001; Kim *et al.*, 1999; Krishnan *et al.*, 2001; Levin, 2000; Toran-Allerand, 2000). Signals from these receptors in a young woman might limit the size of her preosteoclast population by a TGF-β-mediated killing of the cells by apoptosis (Figs. 2 and 3; Manolagas, 2000; Whitfield *et al.*, 1998a; Zecchi-Orlandini *et al.*, 1999). Therefore, as the estrogen level falls with approaching menopause, less TGF-β is made and osteoclast precursors live longer to make more osteoclasts for more deeply excavating BMUs. However, the number of osteoclast progenitors in the

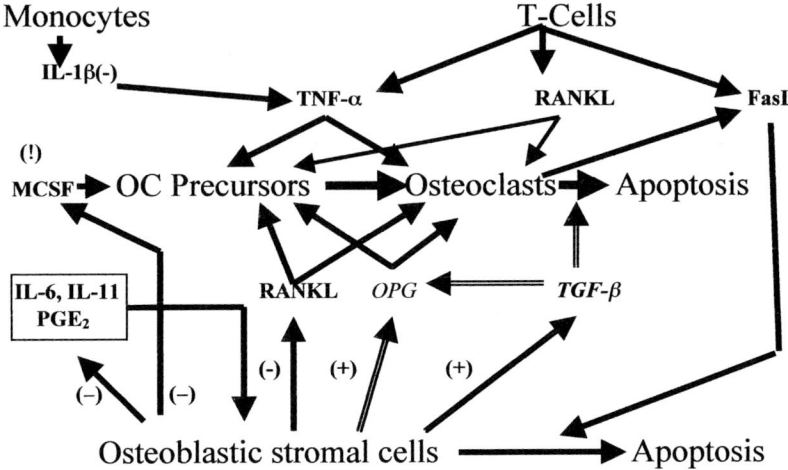

FIGURE 2 The positive (+) or negative (−) control by estrogen of the production by monocytes, osteoblastic stromal cells, and T cells of the several factors that control the osteoclast differentiation, activity, and lifespan. Solid lines (→) mean osteoclast stimulation, and open lines (⇒) mean osteoclast inhibition. MCSF, macrophage colony-stimulating factor; OC Precursors, osteoclast precursors.

microenvironment created by marrow stromal cells rises not only because of lifespan lengthening, but also because of a surge into the stromal microenvironment of factors such as interleukin (IL)-1β and tumor necrosis factor (TNF)-α from IL-1β-stimulated marrow monocytes and TNF-α from T-lymphocytes (Fig. 2; Cenci et al., 2000; Hofbauer et al., 2000; Jilka, 1998; Kobayashi et al., 2000;

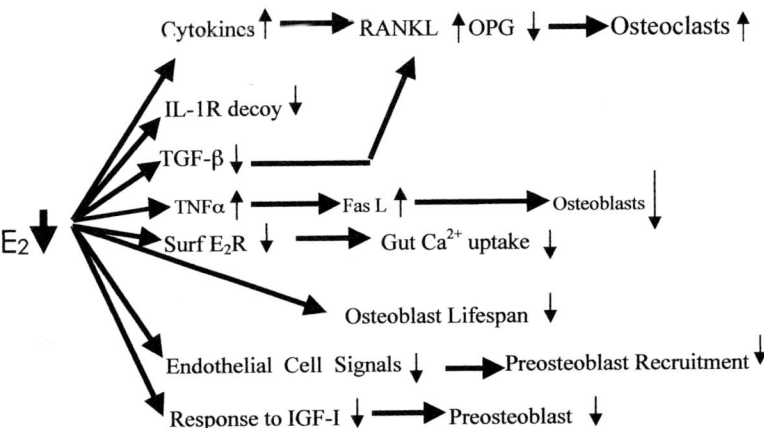

FIGURE 3 The known major effects of an estrogen (E_2, 17β-estradiol) loss on osteoclast and osteoblast production. A drop is designated by ↓ and a rise by ↑ These events are discussed in the text. " Surf E_2R" refers to "nongenomic" estrogen receptors on the cell surface and IL-1R refers to the nonsignaling IL-1 decoy receptor. FasL, refers to Fas receptor ligand.

Martin *et al.*, 1998; Pilbeam *et al.*, 1997). M-CSF steers uncommitted precursors onto the osteoclast pathway (Hofbauer *et al.*, 2000). In the estrogen-rich koric bones, the responsiveness of the osteoclast progenitors to the TNF-α-inducing IL-1β is blunted by the estrogen-promoted, nonsignaling type 2 IL-1 decoy receptors on their surfaces that vie with the signaling-competent receptors for IL-1β (Figs. 2 and 3; Hofbauer *et al.*, 2000; Manolagas, 2000; Pilbeam *et al.*, 1997; Suda *et al.*, 1999; Sunyer *et al.*, 1999). But as estrogen declines, so do the decoy recptors and IL-1β is no longer diverted from driving the development of the osteoclast progenitors. The osteoblastic stromal cells also start making the potent osteoclast differentiation stimulators IL-6 and IL-11 that have been suppressed by estrogen (Fig. 2; Jilka, 1998; Manolagas, 2000; Whitfield *et al.*, 1998a).

The various actions of these several interleukins, prostaglandins, and TNF-α culminate in osteoblastic marrow stromal cells making more RANKL and putting it out on their surfaces (Figs. 2 and 3; Aubin and Bonnelye, 2000; Greenfield *et al.*, 1999; Hofbauer *et al.*, 2000; Karsenty, 1999; Suda *et al.*, 1994). The RANKL emerging on the surfaces of osteoblastic cells binds to and activates the differentiation-stimulating RANK receptors on preosteoclasts when they dock on the marrow stromal cells (Aubin and Bonnelye, 2000; Hofbauer *et al.*, 2000; Suda *et al.*, 1999). [However, cell–cell contact is not necessary because free RANKL is also secreted by osteoblastic cells and other cells such as activated T-lymphocytes, which could cause of arthritic bone erosion (Fig. 2; Karsenty, 1999; Kong *et al.*, 1999).] The ability of the stromal osteoblastic precursor cells to make RANKL falls and the expression of the antiosteoclast OPG rises to protect bone from osteoclasts as their progeny differentiate into bone-making osteoblasts (Fig. 2; Gori *et al.*, 2000).

Estrogen controls the OPG/RANK expression ratio. Thus, a young woman's estrogen stimulates the production of TGF-βs, which, in addition to directly stimulating osteoclast apoptosis, increase the osteoblastic stromal cells' OPG/RANKL expression ratio to reduce osteoclast generation (Fig. 2; Aubin and Bonnelye, 2000; Hofbauer *et al.*, 1999, 2000). OPG is another decoy receptor, like the nonsignaling type 2 IL-1 receptor, that is designed to restrain osteoclast production. OPG is a floating piece of the RANK receptor that binds to RANKL on the osteoblastic stromal cells and prevents RANKL from binding to the differentiation-driving RANK receptors on prefusion osteoclasts (Aubin and Bonnelye, 2000; Hofbauer *et al.*, 1999, 2000; Makhluf *et al.*, 2000; Moonga *et al.*, 1998; Oyajobi *et al.*, 2001; Takai *et al.*, 1998).

The estrogen loss doesn't just increase the numbers of longer lived osteoclasts. The increased number of osteoclasts associated with stromal osteoblastic cells and cytokine-stimulated T-lymphocytes probably produces soluble and/or membrane-bound FasL (Fas ligand), which triggers apoptosis of the Fas receptor-bearing osteoblastic cells (Kawakami *et al.*, 1997, 1998). The TGF-β stores in the bone matrix are also depleted (Jilka, 1998; Whitfield *et al.*, 1998a). Therefore there is even less of it that can evade the proteases and reduce osteoclast generation and kill mature osteoclasts when they dig it out of the matrix (Jilka, 1998). And there is less of it to attract immature osteoblasts into the excavation sites, although TGF-β receptor signals may be less important than the migration-stimulating signals from the osteoblastic cells' Ca^{2+}-sensing receptors stimulated by the large amount of Ca^{2+} in the excavation sites (Brown amd MacLeod, 2001; Yamaguchi *et al.*, 1998). Because the endothelial cells in the BMU capillary loops express estrogen receptors and secrete osteoblast-stimulating factors such as endothelin-1 (Brandi *et al.*, 1993; Soerensen and Eriksen, 1998), a lack of estrogen receptor signaling in the endothelial cells of the capillary loops growing into the excavation sites could impair the ability of the capillary's endothelial cells to stimulate accompanying preosteoblastic pericytes and immature osteoblasts and thus reduce the size of the BMU osteoblast crews. Because estrogen also lengthens the working lives of osteoblasts and the microcrack-detecting osteocytes by suppressing apoptosis, the estrogen loss unleashes apoptosis and reduces the number of osteocytes (Manolagas, 2000; Manolagas *et al.*, 1999; Tomkinson *et al.*, 1998). This loss of osteoclasts would reduce the inhibitory signaling through the osteocyte-lining cell network, which would have the same effect as severing the signaling network by microcracks (R. B. Martin, 2000, 2002). This false microcrack signaling would increase BMU activation and thus bone turnover and the remodeling space.

Osteoblast working life and activities such as laying down bone matrix and secreting various factors may also be reduced, and the BMU replacement deficit thus increased, by the loss of signals from the nongenomic estrogen receptors such as those found on the surfaces of cultured female rat osteoblasts (Falkenstein *et al.*, 2000; Lieberherr *et al.*, 1993; Manolagas, 1999). These surface receptors appear either to be the products of the gene for the nucleus-seeking, gene-activating ERα genomic receptor that have been modified to stay at the cell surface and fire Ca^{2+} and cyclic AMP signals into the cell when activated by 17β-estradiol (Chambliss *et al.*, 2000; Falkenstein *et al.*, 2000; Fiorelli *et al.*, 1998; Kelly and Levin, 2001; Lieberherr *et al.*, 1993; Razandi *et al.*, 1999; Whitfield *et al.*, 1998a). Or they might be the entirely unrelated ER-X receptors (Toran-Allerand, 2000) or the "γ-adrenergic" receptors that are activated by dopamine, epinephrine, and norepinephrine as well as 17β-estradiol and various xenoestrogens (Benten *et al.*, 2001; Nadal *et al.*, 1998, 2000). The loss of Ca^{2+} influx-driving signals from these nongenomic receptors on the surfaces of intestinal epithelial cells may further reduce intestinal Ca^{2+} uptake and increase the bone loss due to the vitamin D deficiency that commonly develops in older persons (Chen and Kalu, 1998; Doolan *et al.*, 2000; Picotto *et al.*, 1999). Adding to this vitamin deficiency may be a drop in the estrogen-dependent expression of the 1α,25-dihydroxyvitamin D_3 receptor that drives Ca^{2+} uptake and transcellular transport to the blood by colon epithelial cells (Schwartz *et al.*, 2000).

In summary, the many different effects of the menopausal estrogen drop (Fig. 4) combine to produce BMUs with larger osteoclast and smaller osteoblast crews that cannot fill the larger number of deeper holes. This results in weakening bones and a dramatic upsurge of microcracks calling for repair (Bouxsein, 1999). But the BMU crews answering the calls are increasingly incompetent. The result is a vicious cycle

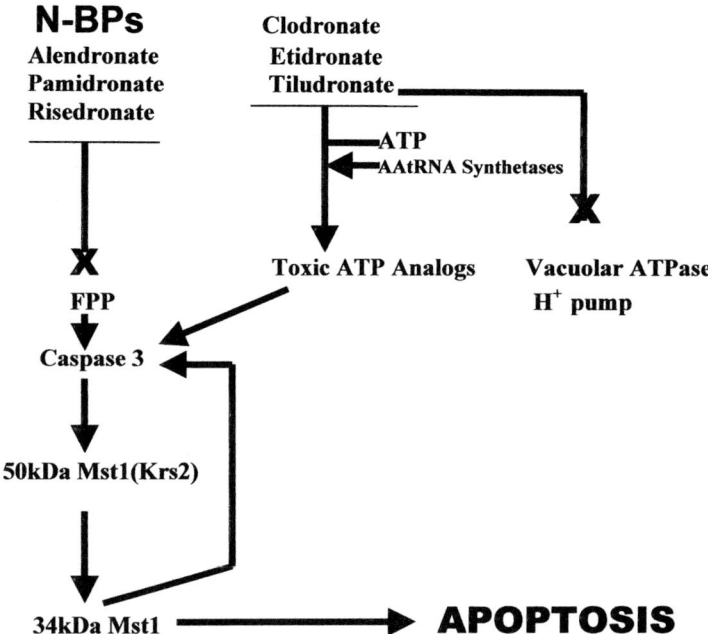

FIGURE 4 The mechanisms by which nitrogen-containing (N-BPs) and non-nitrogen-containing bisphosphonates trigger osteoclast apoptosis.

of a growing number of inadequately repaired cracks → increased turnover and remodeling space → weaker, more porous bone → increased microcracking → the summoning of more inept BMUs → more inadequately repaired cracks → etc.

C. HOW TO STOP MENOPAUSAL BONE LOSS

Obviously it can be stopped by giving the woman estrogens or one of the partially estrogen-mimicking *s*elective *e*strogen *r*eceptors *m*odulators (SERMs) such as raloxifene (Eli Lilly's Evista) (Roe et al., 2000). It can also be reduced or even stopped with an oral bisphosphonate such as alendronate (Merck & Co.'s Fosamax) (Fleisch, 1997, 1998), or oral or nasally sprayed calcitonin that stops osteoclast activity by activating receptors on the cells of the osteoclast lineage (Roodman, 1996; Suda et al., 1999). The various bisphosphonates kill osteoclasts in different ways (Amin et al., 1992; Benford et al., 2001a,b; Reszka et al., 1999; van Beek et al., 1999) (Fig. 4). The nitrogen-containing bisphosphonates (N-BPs) such as alendronate kill osteoclasts by inhibiting farnesyl pyrophosphate (FPP) synthase (van Beek et al., 1999) (Fig. 4). This inhibits the production of geranylgeranyl-pyrophosphate needed to isoprenylate and pin key G-proteins Cdc 42, Rac, the Rabs, Ras, and Rho to cell membranes. This disrupts the cytoskeleton and prevents the formation of the sealing ring, disables vesicular transport, impairs cell signaling, stimulates caspase-3, and, consequently, disables osteoclast function, inhibits osteoclast generation, and stimulates their apoptotic demise, all which can be prevented simply by giving the cells geranylgeraniol, a cell-permeating analogue of geranylgeranyl-pyrophosphate (Benford *et al.*, 2001a,b; Fisher *et al.*, 1999; Reszka *et al.*, 1999). Of all of these, it is the stimulation of the caspase that is lethal unless blocked in time by an inhibitor such as zVAD-fmk or SB-281277 (Benford *et al.*, 2001a,b; Reszka *et al.*, 1999). Caspase-3 cleaves the 50-kDa Mst 1 (Krs2) kinase into a regulatory C-terminal fragment and an active 34-kDa kinase fragment, which in turn activates more caspases (Fig. 4; Reszka *et al.*, 1999). The non-N-BPs (clodronate, etidronate, tiludronate) do not inhibit FPP, but they combine with ATP to form toxic ATP analogues [e.g., adenosine $5'$-$(\beta\gamma$-dichloromethylene)triphosphate] that kill the cell (Benford *et al.*, 2001a,b; Reszka *et al.*, 1999). Tiludronate has the additional ability to directly disable osteoclast digging by inhibiting the vacuolar ATPase proton pump (Fig. 4; David *et al.*, 1996).

Because these antiresorptives (or antiremodelers) suppress only osteoclasts they cause only a limited amount of bone growth in mature human bones as the unaffected osteoblasts continue filling in the remodeling holes, which account for 10% or more of the current total bone volume (Dempster, 1995, 1997; Fleisch, 1997, 1998). But bisphosphates such as alendronate do reduce microstructural damage and fracture rates by prolonging secondary mineralization without affecting trabecular thickness or number (Boivin *et al.*, 2000). They may also stimulate bone growth by lengthening osteoblast working lives (Manolagas, 2000) by preventing excessive, apoptogenic phosphate uptake by their P_i transporters (Mansfield *et al.*, 2001). Nevertheless, they can neither dramatically increase bone mass nor reduce the fracture risk in the mature human skeleton by more than 50% of the baseline risk (Rosen and Bilezekian, 2000; Seeman, 2001). However, a false impression of anabolic ability can be obtained by extrapolating the responses to antiresorptives of the still-growing bones of mice and rats to mature remodeling human bones, where, because of osteoblast generation's tight coupling to osteoclast activity, the reduction of osteoclast generation ultimately reduces osteoblast generation. Thus, rodents treated with bisphosphonates or estrogens are functionally equivalent to mutant mice, which cannot generate functional osteoclasts and become osteopetrotic as their independently generated and operating osteoblasts continue unopposed osteogenesis (Karsenty, 1999).

V. THE BONE ANABOLIC PTHs

A. ANABOLICS

Clearly the antiresorptives are far from being the "Holy Grails" of osteoporosis therapy. Because osteoporotic postmenopausal women have lost about 30% of their bone mass by the time of their first fracture, it is essential to find an agent that can increase bone mass more than the 6–10% that can be obtained from a 3-year treatment with antiresorptives. The ideal drug to treat osteoporosis and accelerate fracture healing in both women and men would be a true *anabolic* that goes beyond just filling the remodeling space. It would bypass the BMU remodeling cycle by directly and strongly stimulating bone formation. It would directly stimulate a buildup of longer-lived, less apoptosis-prone osteoblasts, which would restore the mechanical strength of bones such as vertebrae and femurs by thickening their remaining trabeculae and thus strengthening their connectivity (Dempster, 1997; Manolagas, 2000). The PTHs—the parathyroid hormone and certain of its fragments and at least one their analogues—are very close to being these urgently sought-after anabolics.

B. PTHs-INDUCED BONE GROWTH AND FRACTURE MENDING IN ANIMALS

Seventy-two years ago everyone knew that PTH's only job was to stimulate the resorption of bone in order to put Ca^{2+} into the blood when needed. But Bauer *et al.* (1929) got a surprise when they treated female rats with the only commercial source of PTH at that time, a bovine parathyroid gland extract from Eli Lilly and saw—*bone growth*! They did not follow this up, but Hans Selye did. Three years later, Selye (1932) reported that daily injections of the Eli Lilly extract into rat pups caused a massive accumulation of bone-building (osteogenic) osteoblasts on the surfaces of the trabeculae in cancellous bone. Moreover, the injections did it without first stimulating osteoclasts! But was it the PTHs [the native PTH and its fragments] in the extract that stimulated osteoblast accumulation and bone growth? Kalu *et al.* (1970) and Walker (1971) proved that it was. Since then the basic features of PTH's osteogenic action have been established by hundreds of experiments on cynomolgus monkeys, dogs, ferrets, rabbits, and sheep, but most of all on OVXed rats, using a supposedly fully bioactive piece of PTH, h(human)PTH-(1–34) [comprehensively reviewed by Dempster *et al.* (1993) and Whifield *et al.* (1998a, 2000d)]. But hPTH-(1–34) might not be fully bioactive because Divieti *et al.* (2001) found a specific apoptosis-triggering receptor for PTH fragments containing the hormone's C-terminal 19–84 region.

The native PTH is a string of 84 amino acids (Morley *et al.*, 1999; Whitfield *et al.*, 1998a). Injecting one small dose (e.g., 1–50 nmol/100 g of body weight) of the native hPTH-(1–84) or certain of its N-terminal fragments [[Leu^{27}]*cyclo*(Glu^{22}-Lys^{26})hPTH-(1–28)NH$_2$, hPTH-(1–30)NH$_2$, hPTH-(1–31)NH$_2$(Ostabolin), [Leu^{27}]*cyclo*(Glu^{22}-Lys^{26})hPTH-(1–31)NH$_2$ (Ostabolin C), b(bovine)PTH-(1–34), hPTH-(1–34)NH$_2$, [Leu^{27}]*cyclo*(Glu^{22}-Lys^{26})hPTH-(1–34)NH$_2$, [Asp^{35}]hPTH-(1–35), hPTH-(1–36), or hPTH-(1–38)] into OVXed rats each day stops the bone loss, stimulates cortical bone growth, acts synergistically with external loading to increase cortical bone strength, and increases the mean trabecular thickness in trabeculae-rich bones such as the distal femur and vertebrae as much as two or more times above the normal or baseline level (Andreassen *et al.*, 1999; Brommage *et al.*, 1999; Cosman and Lindsay, 1998; Dempster, 1997, 2000; Dempster *et al.*, 1993; Dobnig and Turner, 1995; Gasser, 1997; Hagino *et al.*, 1999; Hodsman *et al.*, 1999a,b; Jerome *et al.*, 1994, 1999, 2001; Jilka *et al.*, 1999; Kneissel *et al.*, 2001; Kostenuik *et al.*, 1999, 2000, 2001; Lane *et al.*, 1995; Leaffer *et al.*, 1995; Liang *et al.*, 1999; Ma *et al.*, 2000; Mohan *et al.*, 2000; Manolagas, 2000; Morley *et al.*, 1999, 2001; Mosekilde, 1997; Mosekilde *et al.*, 1997; Nakajima *et al.*, 2000; Nishida *et al.*, 1994; Opas *et al.*, 2000; Rixon *et al.*, 1994; Samnegard *et al.*, 2001; Sogaard *et al.*, 1997; Strein, 1994; Sung *et al.*, 2000; Tam *et al.*, 1982; Tormanoff *et al.*, 1997; Walker, 1971; Whitfield *et al.*, 1995, 1996b, 1997a–d, 1998a,b, 1999a,b, 2000a–d, 2001; Wronski and Li, 1997). But there was a good reason for the reluctance to believe that PTH could be osteogenic. Even the intermittent injections that strongly stimulate bone growth at the same time stimulate osteoclast production and start the resorption mechanism, but this can be overridden by the bone-building arm provided the injections are sufficiently far apart. However, if the PTH is injected at close intervals or continuously infused into rats the response is net bone resorption instead of net growth (Tam *et al.*, 1982).

According to the results of experiments on rats, humans, and cultured human and rat bone cells, PTHs owe their abilities to dramatically stimulate osteoblast accumulation and bone growth by stimulating the proliferation of marrow osteoprogenitor cells, by directly and reversibley stimulating the "reversion" of quiescent lining cells to active osteoblasts on quiescent bone surfaces without prompting by signaling from prior osteoclast activity, by increasing osteoblast activity, and/or by increasing the osteoblasts' lifespan by preventing apoptosis (Dobnig and Turner, 1995; Hock, 1999; Hodsman and Steer, 1993; Manolagas, 2000; Jilka *et al.*, 1999; Kostenuik, *et al.*, 1999; Leaffer *et al.*, 1995). The PTHs are most potently anabolic for the bones of the fast-growing fetal and newborn animals simply (Walker, 1971) because the marrow is red and thus loaded with osteoblast precursors and the growing bones are carpeted with clusters of mature osteoblasts loaded with PTHR1 receptors (Fermore and Skerry, 1995). Because the growing bones are in this superresponsive state, PTHs can so greatly stimulate trabecular bone growth that it fills the marrow spaces and stops blood formation in these young animals (and, as we shall soon see, in at least one historically very significant young French boy) (Walker, 1971). Of course, this is why Selye (1932) found that the PTH in the Lilly bovine parathyroid extract so greatly stimulated osteoblast accumulation and bone growth in his rat pups without first stimulating osteoclast activity. As yellow marrow displaces red marrow (Bianco and Riminucci, 1998) and bone growth drops with age so does the responsiveness to PTH in rats (Walker, 1971). In animals such as humans with microcracking remodeling bones (R. B. Martin, 2002), the supply of mature osteoblasts for an *immediate* or *first* response to PTH injection in the mature skeleton is a function of the number of stimulable lining cells as well as BMUs in the anabolic phase (Manolagas, 2000), which is in turn linked to osteoclast activation.

No PTH can restore the original bone structure! In rats, they cannot generate new trabeculae *de novo,* increase the trabecular number, or make broken trabeculae rejoin if they have drifted too far apart (Dempster, 2000; Jerome *et al.,* 2001; Lane *et al.,* 1995; Whitfield *et al.,* 1998b). Nor, it seems, can hPTH-(1–34) stop or reverse the drop in the number of central metaphyseal trabeculae in the distal femur when its injections are started 2 weeks after OVX, because it stimulates osteoclast generation as well as osteoblast bone building (Whitfield *et al.,* 1998b). In other words although the OVX- and PTH-(1–34)-stimulated osteoclasts are busily destroying centrally placed trabeculae in the femur, PTH-stimulated osteoblasts are thickening the more laterally placed, load-transmitting trabeculae (Fig. 5; Whitfield *et al.,* 1998a). The net result of this is an increased trabecular thickness (but not number) and increased total cancellous volume and a disproportionate strengthening of the OVX rat's load-bearing bones (Fig. 5; Dempster *et al.,* 1997; Whitfield *et al.,* 1998a,b).

But treating OVXed nonhuman primates such as *Macaca fascicularis* with PTH-(1–34) increases the number of trabeculae, though not by making them *de novo.* It does this by increasing the thickness of existing trabeculae, which are then cut in two by osteoclastic "tunneling" when they reach a certain size (Jerome *et al.,* 1994, 2001). Tunneling is thus a smart kind of homeostatic mechanism that both restores the original trabecular thickness and increases trabecular number.

Without estrogens, the new bone in a PTH-treated OVXed rat should be, and apparently is in some cases, as much a victim of overdigging BMUs as was the old bone, when the PTH injections are stopped (reviewed by Whitfield *et al.,* 1998a, 2000c,d). But according to the results of most of the relevant experiments on rats, the new "PTH bone," be it cortical or trabecular, can be effectively protected by an osteoclast suppressor such as a bisphosphonate, calcitonin, or an estrogen (Samnegard *et al.,* 2001; Whitfield *et al.,* 1998a, 2000c,d). However, the new bone's need for protection would depend on its rate of turnover (i.e., the size and activity of the osteoclast population). Thus, for example, Mosekilde *et al.* (1997) found that the new "PTH bone" in the slowly turning over bones of aged osteopenic OVXed rats did not need protection by the bisphosphonate, risedronate. And the positive effect of a 12-month treatment with PTH-(1–34) on the cancellous bone of femoral neck, iliac crest, and vertebrae of OVXed cynomolgus monkeys was maintained for 3–6 months after the injections were stopped (Jerome *et al.,* 2001). But new "PTH bone" can face another threat if it should exceed mechanical needs. It will be noticed by the mechanostatic controller as being superfluous and then be resorbed (Dempster *et al.,* 1997).

It would be reasonable to expect that cotreatment with an antiresorptive to block the resorption component would increase the size of the maximum response to the PTH. But the maximum increase in trabecular and cortical bone mass in rats treated with hPTH-(1–34) or hPTH-(1–38) alone is the same as in rats cotreated with PTH and a bisphosphonate, calcitonin, or an estrogen (Hodsman *et al.,* 1999a,b). However, cotreatment with the SERM raloxifiene reduces the dose of the PTH fragment needed to achieve this maximum effect. But Boyce *et al.* (1996) found

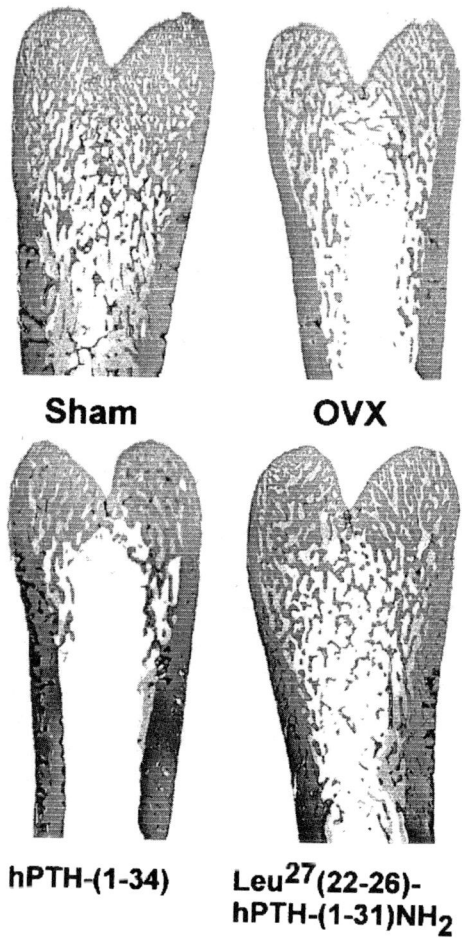

FIGURE 5 Typical specimens of demineralized distal femurs of OVXed rats showing that daily injections of 0.6 nmol of PTH-(1–34) (from the end of the second week to the end of the eight week after OVX) cannot stop the destruction and loss of metaphyseal trabeculae whereas [Leu^{27}]cyclo(Glu^{22}-Lys^{26})hPTH-(1–31)NH$_2$ (Ostabolin C) can reduce the loss. However, by 6 weeks of injections both fragments had nearly doubled (1.8–2.0) the mean thickness of the remaining trabeculae compared to the mean thickness of the trabeculae in the femurs of the vehicle-injected OVXed control rats. The specimens were prepared at the end of the 6 weeks of daily injections (8 weeks after OVX). These photographs are from Whitfield et al. (1998b) and are reproduced with permission of Springer-Verlag.

that cotreating beagles with hPTH-(1–34) and a bisphosphonate, risedronate, was osteogenically more effective than treatment with hPTH-(1–34) alone. On the other hand Delmas et al. (1995) found that a bisphosphonate, tiludronate, eliminated the otherwise strong osteogenic action of hPTH-(1–34) in old sheep with their slowly remodeling, hence sparsely BMUed, bones. Evidently in these old sheep, eliminating BMUs with the bisphosphonate reduced the pool of PTHR1 receptor-loaded osteoblasts below the critical size needed to trigger the osteogenic response.

OPG, the physiological inhibitor of osteoclast generation, should potently enhance PTH osteogenicity by selectively stopping the PTH from stimulating osteoclast generation without affecting the stimulation of osteoblast accumulation and activity (Capparelli *et al.*, 2000; Hofbauer *et al.*, 2000; Kostenuik and Shaloub, 2001; Kostenuik *et al.*, 2000a,b). Nor would it have any of the side effects of the other less selective artificial antiresorptives. This expectation has been confirmed by Kostenuik and his colleagues (Kostenuik and Shaloub, 2001; Kostenuik *et al.*, 2001). They OVXed 3-month-old rats and began treating them 15 months later. OVX significantly reduced the bone mineral density (BMD) in the distal femurs and lumbar vertebrae. Subcutaneously injecting hPTH-(1–34) (0.08 mg/kg) three times a week for $5\frac{1}{2}$ months into the aging rats *increased* the osteoclast surface by 50%, but it also significantly increased the BMD in the distal femurs and lumbar vertebrae. Subcutaneously injecting recombinant OPG (10 mg/kg of body weight) by itself reduced osteoclast activity by 99%. The femoral and vertebral mineral density then rose as the unaffected osteoblasts continued making bone. But the OPG was less effective than the PTH injections in increasing the BMD. Injecting the recombinant OPG along with the PTH stopped the PTH from increasing osteoclast activity without affecting its osteogenic activity, and consequently, caused a greater increase in BMD than did the injections of either peptide alone. These observations underscore the fact that in both young and aging rats a PTH does not need a prior activation of osteoclasts to stimulate bone growth (it stimulated bone growth despite an essentially total osteoclast shut down) and that OPG strongly enhances a PTH's bone-building action.

As expected, the osteogenic PTHs accelerate fracture mending in both normal and osteoporotic bones. For example, daily subcutaneous injections of 200 μg of hPTH-(1–34)/kg of body weight increased both the ultimate load that could be tolerated before breaking and the callus volume of rat tibial fractures by as much as 75% and 95%, respectively, over the control values by 20 days after fracturing and by 175% and 72%, respectively, over the controls during the next 20 days (Andreassen *et al.*, 1999). A similar acceleration of fracture mending by Ostabolin C [Leu^{27}]*cyclo*(Glu^{22}-Lys^{26})hPTH-(1–34)NH$_2$] has been reported by Andreassen *et al.* (2001). Kim and Jahng (1999) also reported that a daily injection of recombinant hPTH-(1–84) for 30 days significantly accelerated the mending of bilateral tibial shaft fractures in OVXed rats.

In a different and very elegant way to accelerate fracture mending, a degradable *g*ene-*a*ctivated collagen *m*atrix (GAM) sponge loaded with a plasmid containing DNA coding for hPTH-(1–34) was implanted into surgical breaks in rat femurs or beagle femurs and tibias that were so wide that they should not have healed or would have taken a very long time to do so (Bonadio, 2000; Bonadio *et al.*, 1999; Fang *et al.*, 1996; Goldstein and Bonadio, 1998). But the fibroblasts in the fractures' granulation tissue picked up the plasmid DNA and became transient minibioreactors making and secreting hPTH-(1–34) that dramatically stimulated fracture healing. Implanting a second GAM with DNA coding for BMP-4 (bone morphogenic protein-4) along with the PTH-GAM further accelerated mending.

So far we have been considering the effects of injected PTHs. What would happen if we silenced the parathyroid gland chief cells' CaRs? The cells would interpret this as being due to a drop in the blood Ca^{2+} concentration that must be corrected by releasing PTH. NPS 2143 is a CaR silencer (a "calcilytic"), which when given orally in a dose of 100 μmol/kg causes the PTH concentration in the blood of OVXed Sprague–Dawley rats to rise from 25 pg/ml to about 110 pg/ml in 30 min and then stay there for at least 4 h (Gowan *et al.*, 2000). This prolonged surge of endogenous PTH dramatically increases both bone turnover and bone formation in the proximal tibia without a net bone loss or gain or significantly affecting the OVX-induced drop in trabecular number. However, bone mineral density as well as trabecular thickness and area, but not trabecular number, can be increased by giving 17β-estradiol along with the calcilytic to reduce resorption. But the NPS 2143/estrogen combination is not nearly as osteogenic as the subcutaneous injections of 10 nmol/kg of hPTH-(1–34)NH$_2$ or [Leu^{27}]*cyclo*(Glu^{22}-Lys^{26})hPTH-(1–31)NH$_2$ (Ostabolin C) into OVXed rats (Whitfield *et al.*, 1998b). However, the osteogenicity of the calcilytics can undoubtedly be improved by shortening their circulating half-life and therefore the duration of the PTH surge, which determines the balance between bone building and resorption.

C. PTHs-INDUCED BONE GROWTH IN HUMANS

It was a paper describing the case of a 8-year-old boy who died from osteopetrosis-induced aplastic anemia (Péhu *et al.*, 1931) that prompted Selye's demonstration of the dramatic osteogenic action of the Eli Lilly bovine parathyroid extract in rat pups (Selye, 1932). The observation that particularly caught Selye's eye was that the osteopetrotic boy had a parathyroid adenoma, which the authors believed caused the lethal bone growth by secreting abnormal amounts of PTH.

Forty-four years later a reasonably cheap synthetic tool, hPTH-(1–34), became available for determining whether PTH is as osteogenic in humans as it is in animals. In 1976 Reeve *et al.* (1976a,b) reported that single daily injections of 100 μg/kg of hPTH-(1–34) stimulated iliac trabecular bone growth, but they did not affect femoral BMD. In a subsequent uncontrolled, multicenter study, they injected 16 osteoporotic women and 5 osteoporotic men once each day with 50–100 μg/kg of hPTH-(1–34) for 6 months to 2 years (Reeve *et al.*, 1980). The treatment produced a significant amount of new, normally mineralized iliac lamellar trabecular bone.

Since 1980, the results of 20 or more studies have been published (Cosman and Lindsay, 1998b; Dempster *et al.*, 1993; Hodsman *et al.*, 1997a; Meunier, 1999; Netelenbos, 1998; Whitfield *et al.*, 1998a, 2000b,d), but we will mention only the more recent, better controlled examples. The most commonly used measure of a PTH's osteogenic effectiveness in these human studies has been the size of the increase it causes in the overall BMD as determined by dual-energy X-radiation absorptiometry (DEXA) or more recently quantitative computed tomography (QCT) with which the cortical and cancellous (trabecular) compartments can be separately measured (Genant et al., 1982). A low BMD means a higher fracture risk (Faulkner, 2000).

In a 3-year study by Lindsay et al., single daily subcutaneous injections of 25 μg of hPTH-(1–34) increased the vertebral BMD by 12.8% in 27 postmenopausal osteoporotic women who also received estrogen, which by itself did not affect the BMD in the 25 women of the control group (Lindsay et al., 1997). As expected, the vertebral BMD rise was accompanied by a drop in vertebral fracturing as indicated by the PTH-treated women not losing as much radiographic vertebral height as the untreated women. As in the animal models, the increases were less in bones with smaller cancellous compartments than vertebrae such as hip (4.4%) and forearm (1.0%). The whole body BMD increased by 8%.

In a follow-up to this study, Cosman et al. (2001a) reported that the BMD did not change during the 3-year treatment and for 1 year after the end of the treatment in the control group receiving hormone (estrogen) replacement therapy (HRT). In the PTH–HRT group, the PTH-stimulated biochemical markers—bone formation (bone-specific alkaline phosphatase) and resorption (urinary cross-linked collagen N-telopeptides)—peaked at 6 months and then fell to baseline levels by 30 months. During this period, spinal bone mass increased 13.4%, hip bone mass increased 4.4%, and total body bone mass increased 3.7%. The bone masses did not change significantly during the first year after stopping the PTH injections but continuing HRT. The PTH–HRT treatment dramatically and significantly reduced the number of incident ("spontaneous" or normal body activity-inflicted) vertebral fractures during the follow-up year. The frequency of incident fractures in the PTH–HRT group was reduced to as little as 0% or 25% of the frequency in the HRT-only group depending on the height reduction (from vertebral crushing) "cut point" used.

A larger increase in vertebral BMD was seen by Hesch et al. (1989) in a smaller group of 13 osteoporotic women. After 14 months of single daily subcutaneous injections of 54 μg (about 700–750 U) of hPTH-(1–38) the mean vertebral BMD had increased by 20%. This large response might have been due to the post-PTH prevention of a resumption of resorption by the nasally sprayed calcitonin that was also given to the women.

In a randomized controlled 2-year trial with 30 osteoporotic women receiving six cycles, each of which consisted of 1 month (28 days) of single daily subcutaneous injections of 800 U (or about 60 μg)/day of hPTH-(1–34) followed by a 3-month rest period (to restore full responsiveness to the PTH for the next round of injections), there was a 10.1% increase in BMD, of lumbar vertebrae, a nonsignificant 2.4% increase in the femoral neck BMD and a substantial 80% drop in vertebral fracture incidence (Hodsman et al., 1997). However, in this study treating the osteoporotic women with calcitonin during the rest periods did not enhance PTH's osteogenicity.

Rittmaster *et al.* (2000) reported the results of a study in which 75 postmenopausal women were treated for 1 year with daily injections of 50, 70, or 100 μg of recombinant hPTH-(1–84) and then with a daily dose of 10 mg of alendronate for the following year. This treatment stopped cortical bone loss and increased the spinal bone mass by 14% as compared to a 6.9–9.2% increase in patients without subsequent alendronate treatment.

In a much larger study Fujita *et al.* (1999) treated 220 osteoporotic patients with weekly subcutaneous doses of 50 U (3.7 μg; 0.8 nmol), 100 U (7.4 μg; 1.6 nmol), or 200 U (14.8 μg; 3.2 nmol) of hPTH-(1–34) for 48 weeks. The highest dose caused a significant 8.1% increase in the lumbar vertebral BMD, without affecting cortical thickness or metacarpal BMD. The increase in vertebral BMD was accompanied by a 30–40% reduction in "backache."

In another study (Rose *et al.*, 1999) osteoporotic women self-injected a placebo or 400 U (about 30 μg) of hPTH-(1–34) subcutaneously once each day for 2 years and also took an oral estrogen (0.625 mg Premarin/day) with or without methoxyprogesterone. They all received 800 U of vitamin D and 1500 mg of calcium/day. By the end of 2 years the overall (cortical plus trabecular) lumbar vertebral BMD in the PTH/estrogen-treated group was a dramatic 28.3% higher than in the placebo/estrogen-treated control group. The femoral neck BMD in the PTH/estrogen-treated group was 10.8% higher than the femoral neck BMD in the placebo/estrogen control group. The density of the trabecular (cancellous) compartments of the L1 and L2 vertebrae in the PTH/estrogen-treated women (as measured selectively by QCT) had increased even more dramatically—by 74% during the 2 years, whereas the mean density in the vertebrae of the placebo/estrogen-treated controls had *dropped* 2.1%.

At the end of 1998, Lindsay *et al.* (1998) summarized the results of a 1-year, Phase II, multicenter (18 centers in Canada and the United States), placebo controlled, double-blind study of the effects of recombinant rhPTH-(1–84) (NPS Pharmaceuticals Inc.'s ALX-1-11) on BMD and serum indicators of bone resorption and formation in 217 postmenopausal osteoporotic women between 50 and 75 (average 64.5) years of age. The results of the prior Phase I study had indicated that one subcutaneous injection of 0.02–5.0 μg of ALX-1-11/kg into healthy postmenopausal women did not produce a frank hypercalcemia even at the highest dose, which indicated that the hormone was safe and well-tolerated in this dose range (Schweitert *et al.*, 1997). The Phase II patients were given single daily injections of 50, 75, or 100 μg of hormone ($\approx$0.8–1.6 μg/kg). The largest dose increased the vertebral BMD by 6.9%, whereas the densities of the placebo-treated women's vertebrae did not change. However, the BMD *dropped* in the arms and legs by 0.3% (75 μg) and 0.9% (100 μg), respectively. The serum levels of formation indicators (bone-specific alkaline phosphatase, osteocalcin) and resorption (deoxypyridinolines, N-terminal cross-linked collagen peptides) increased 100–200%. There were no serious side effects nor did the patients make antibodies to the recombinant human hormone. The authors believed that the drops in arm and leg BMDs were only transient and that PTH appears to be a promising novel treatment for osteoporosis.

Transient though they might have been, the drops in the BMDs of the arms and legs of the patients of Lindsay et al. are reminiscent of the results of a study by Neer et al. (1991) that nearly eliminated the PTHs as credible therapeutics for osteoporosis. As expected, giving hPTH-(1–34) and calcitriol [$1\alpha,25$-$(OH)_2$ vitamin D_3] to 15 osteoporotic women for 1 to 2 years dramatically increased the lumbar vertebral BMD by 32%. But at the same time there was a net 4% *drop* in the radial BMD, which was what set off alarm bells. Although PTH can reduce the unpleasant though not crippling fracturing of vertebrae, it might increase the far deadlier hip fracturing. To explain this alarming result they suggested that PTH somehow steals cortical bone to make trabecular bone. Luckily for the PTHs the "cortical steal" hypothesis is simply wrong, although as we shall now see there is a reason why some people see a transient drop in cortical BMD.

Neer et al. (2001) reported the results of a 2-year study of the effects on fracturing and BMD of daily subcutaneous injections of a placebo, or 20 or 40 μg of Eli Lilly's recominant hPTH-(1–34) ("Forteo") into 1637 postmenopausal women who had already suffered vertebral fractures. These were the results of the fragment's Phase III trial. The 20 μg injections of the PTH reduced the relative risk of vertebral fracturing to 0.35 [95% confidence interval (CI) 0.22–0.55] and the 40 μg injections reduced the relative fracturing risk to 0.31 (95% CI 0.19–0.50). The relative risk of nonvertebral fracturing was 0.5 in women injected with 20 μg of the PTH (95% CI 0.25–0.88) and 0.5 in women injected with 40 μg of the PTH (95% CI 0.25–0.86). Both doses of the PTH increased the BMD of hip and spine by 2–4%, but again there was a significant drop in the BMD of the radial shaft (i.e., cortical bone) during the first year. The authors concluded that hPTH-(1–34) decreases the risk of vertebral and nonvertebral fracturing, increases femoral and vertebral BMDs, and is well tolerated although there was some (29%) hypercalcemia at the higher dose.

The answer to the crucial question of what the PTH treatment does to cortical bone has been provided by Cann et al. (1999). They used 3D-QCT to separately track the hPTH-(1–34)-induced changes in the density and mass of the cortical and trabecular (cancellous) parts of the proximal femurs of 34 osteoporotic women (on estrogen therapy). By the end of the second year the PTH injections had increased the spinal BMD by 74% and the BMD of the trabecular (cancellous) compartments of the proximal femur from 10.5% in the trochanter to 12.4% in the whole hip. On the other hand, the cortical BMD *dropped* by less than 3.5% during the first year, but, as in Lindsay et al.'s 1998 study, the drop was only transient—the density did not drop further during the next year. So were Neer et al. (1991) right? Yes, but PTH treatment is certainly not hazardous for femoral necks. Despite the small drop in cortical bone *density,* the use of 3D-QCT enabled Cann et al. to see that the cortical bone *mass* in the proximal femurs had risen by 10.9% in the trochanter and by a dramatic 20.7% in the femoral neck by the end of the second year. These large increases in cortical mass took place on the endocortical surfaces [just as Andreassen and Oxlund (2000) found in the femurs of hPTH-(1–34)-treated OVXed rats] because they, like the trabecular surfaces, are adjacent to the sources of various PTH/PTHrP-induced stromally generated factors and osteoblastic progenitors in the hemopoietic red marrow. Thus, as expected in a mature skeleton, PTH undoubtedly rapidly increased BMU activation, which would reduce the density by increasing the cortical porosity and remodeling space, but this was eventually more than balanced by a large production of endocortical bone by increased numbers of the more slowly working, longer lived osteoblasts.

The same changes were observed in OVXed female cynomolgus monkeys (*Macaca fascicularis*) by Burr *et al.* (2001) that received one daily subcutaneous injection of 1 or 5 μg of hPTH-(1–34)/kg for 12 or 18 months. The injections increased the cortical porosity, especially in the inner third of the middiaphysis of the left humerus where it was increased 5 to 16 times above the porosity in the sham-operated monkeys. This should have reduced the bone's mechanical strength, but it did not. The strength actually increased because although the osteoclasts dug more holes in the endocortical zone the PTH-stimulated osteoblasts more than compensated for this by increasing the mineralizing surface 11-fold and the bone formation rate 4-fold on this inner cortex.

A similar conclusion was reached by Hirano *et al.* (2000) from a study of the effects of an osteogenic dose of hPTH-(1–34) on cortical bone porosity and bending rigidity of rabbit tibias. They found that the PTH greatly increased the porosity mostly on the endocortical surface, but this was more than offset by the formation of new endocortical bone as well as new periosteal bone, which *increased* the tibial bending rigidity. The same coincidence of bone loss and large bone building has been seen by Whitfield *et al.* (1998b) in OVXed rats where hPTH-(1–34) injections did not stop the loss of femoral central metaphyseal trabeculae but they increased thickness of the surviving laterally sited trabeculae 1.65 times.

Accumulating osteoblasts should release into the blood indicators of their presence such as osteocalcin. Indeed this is what happens in PTH-treated osteoporotic humans. The hPTH-(1–34) injections in the study of Lindsay *et al.* (1997) caused serum osteocalcin levels to rise for the first 6 months, after which they slowly returned to normal levels. Cosman *et al.* (1998, 2000) found that the PTH injections in these experiments caused the serum osteocalcin and procollagen IC-terminal propeptide (a byproduct of collagen fiber processing and assembly) to rise during the first 4–6 weeks, which indicated a rapid build-up of osteoblasts. There should also have been evidence of PTH-stimulated osteoclasts. However, the osteoclast response, as indicated by the release of cross-linked *N*-telopeptides from osteoclast-cleaved collagen, was not detectable until 6 months and then returned to baseline values by 24 months.

Reeve *et al.* (2001) reported the effects of treating severely osteoporotic women with hPTH-(1–34) or hPTH-(1–38). They found that the PTHs improve the body calcium balance and "impressively" increase the spinal BMD and stimulate smaller increases in proximal femur and radius. They concluded that "hPTH or comparable

PTH receptor activators remain the most promising anabolic treatment for osteoporosis currently under clinical evaluation."

Glucocorticoid therapy can also cause osteoporosis. It does this by inhibiting osteoblastogenesis [by suppressing Cbfa1 and Cbfa1-stimulated TGF-β type I receptor expression (Chang *et al.*, 1998; Ducy, 2000)], promoting osteoblast and osteocyte apoptosis, and inhibiting intestinal Ca^{2+} uptake (Lane *et al.* 2000; Weinstein *et al.*, 1998). Injecting a PTH can overcome the glucocorticooid action and stimulate bone growth. This is shown in a recent experiment involving 51 women (mean age of 63 years) receiving glucocorticoids and estrogens (Lane *et al.*, 2000). Twenty-eight women injected themselves once each day for 12 months with 40 μg (400 U) of hPTH-(1–34) whereas 23 simply continued receiving estrogen. Here is another case of a PTH preferentially stimulating the growth of cancellous (trabeculae-rich), as opposed to cortical, bone. By 12 months the PTH injections had increased the trabecular *mass* (as measured by QCT) in the lumbar vertebrae by about 35%, which continued rising to 45% during the next 12 months as the PTH-enhanced remodeling space (remember that PTH also increases BMU activity and thus remodeling space) continued to be filled up after the injections were stopped. By 24 months the overall BMD (as measured by dual energy X-radiation absorptiometry rather than QCT), which is a function of the extent of mineralization and the amount of remodeling space in the increasingly massive vertebral bone, had increased by only 12.6%. The hip mass rose by 4.7% over the 24-month period. However, the treatment did not affect the cortical bone mass in the forearm.

PTH can also treat idiopathic osteoporosis in middle-aged men (Kurland *et al.*, 2000). Unlike postmenopausal osteoporosis in women, which is associated with high bone turnover due to osteoclastic overactivity, male idiopathic osteoporosis is associated with low bone turnover and thus is not amenable to the drugs designed to suppress osteoclasts. Kurland *et al.*(2000) reported the results of an experiment on 23 men (50 ± 1.9 years of age) with idiopathic osteoporosis. They gave 10 of the men a daily subcutaneous injection of 400 IU of hPTH-(1–34) for 18 months. Although the lumbar spine bone mass did not change in the 13 placebo-treated control patients, it increased by a significant 13.5% in the PTH-treated patients. The PTH treatment also significantly increased the femoral neck BMD by 2.9%. The PTH fragment increased the indicators of bone turnover such as serum osteocalcin (growth) and urinary *N*-telopeptides (resorption) by 300–400% without causing hypercalcemia.

D. THE MECHANISM OF PTHs' ANABOLIC ACTION

1. The PTHR1 Receptor and Its Signaling

To understand how the PTHs stimulate bone growth in humans, rodents, and other animals we must know what signals they send into their target cells via the PTHR1 receptors to trigger osteogenesis. But how do we know that the PTHR1 receptor emits the osteogenic signal? To show this, Calvi et al. (2001) coupled a mutant human PTHR1 gene (HKrk-H223R) encoding a constitutively active receptor to the mouse α(I) collagen promoter to drive the continuous expression of the gene specifically in osteoblasts of transgenic mice. The signaling PTHR1, like PTH, increased endosteal bone formation and caused an accumulation of intertrabecular preosteoblastic marrow stromal cells and a build-up of hyperactive trabecular osteoblasts in both proximal tibia and cranial bone. As must happen with PTH infusion or, as in this case, continuous PTHR1 signaling, there was also a large increase in cortical and trabecular osteoclasts, a lack of periosteal growth, increased cortical porosity, and a resulting decrease in cortical bone mass.

The PTHR1 receptor (Fig. 6) is a type II G-protein-coupled receptor (GPCR) that has seven transmembrane α-helices and, like all type II GPCRs, has a medium-size [about 160 amino acids as compared to less than 40 for type I receptors and 360 for a type II receptor such as the Ca^{2+} receptor (Brown and MacLeod, 2001)] extracelluar N-terminal domain (Bockaert et al., 2002; Hoare and Usdin, 2001; Hoare et al., 2001; Whitfield et al., 1998a). PTH-(1–34) dispenses most of its binding energy when attaching first to the receptor's N-terminus (the "N-interaction") with its C-terminal portion and then binding its N-terminus to the receptor's juxtamembrane domain (the "J interaction") (Fig. 6; Hoare and Usdin, 2000; Hoare et al., 2001). The "J interaction" of hPTH-(1–34)'s N-terminus generates active GTP · Gsα that stimulates adenylyl cyclase (AC) to make cyclic AMP while at the same time the "N-interaction" generates active GTP · Gqα that stimulates phospholipase-Cβ1 (PLC-β1), which in turn cleaves a minor membrane phospholipid, phosphatidylinositol 4,5-bisphosphate (PIP_2), into diacylglycerols and inositol 1,4,5-trisphosphate (IP_3) (Hoare and Usdin, 2001; Hoare et al., 2001; Morley et al., 1999; Whitfield et al., 1998a) (Fig. 6). However, although N-terminally truncated fragments such as hPTH-(3–34) or hPTH-(13–34) with intact C-tails cannot have an AC-activating "J-interaction," they can still activate PLC-β1. It is a ligand-induced switching of the receptor's C-tail that activates PLC-β1.

As shown in Fig. 6, PTHR1, like the β_2-AR(adrenergic receptor), has a so-called PDZ-binding domain in its C-tail with which it attaches to NHERF-2, the Na^+–H^+ exchanger's regulatory factor-2 (Bockaert et al., 2002; Harris and Lim, 2001; Mahon et al., 2002). PLC-β1 is also attached to the PDZ-2 domain of NHERF-2 which serves as a membrane-associated scaffold upon which enzymes such as this PLC and receptors such as PTHR1 can be assembled into functional signaling complexes (Harris and Lim, 2001). It appears that the NHERF-2 scaffold holds PLC-β1 on its PDZ1 domain (Mahon et al., 2002). It appears that when the target cell has NHERF-2 and PTHR1 is attached to it, AC stimulation is reduced and PLC-β1 stimulation is greatly amplified. The mechanism seems to be due to the receptor activating $G_{i/o}$ the α subunit of which inhibits adenylyl cyclase while the $\beta\gamma$ subunits stimulate PLC-β1 on NERF-2's PDZ1 domain (Mahon et al., 2002). This means that cells well supplied with NHERF-2 will respond very differently from those without. Thus, ROS 17/2.8 rat osteosarcoma cells without

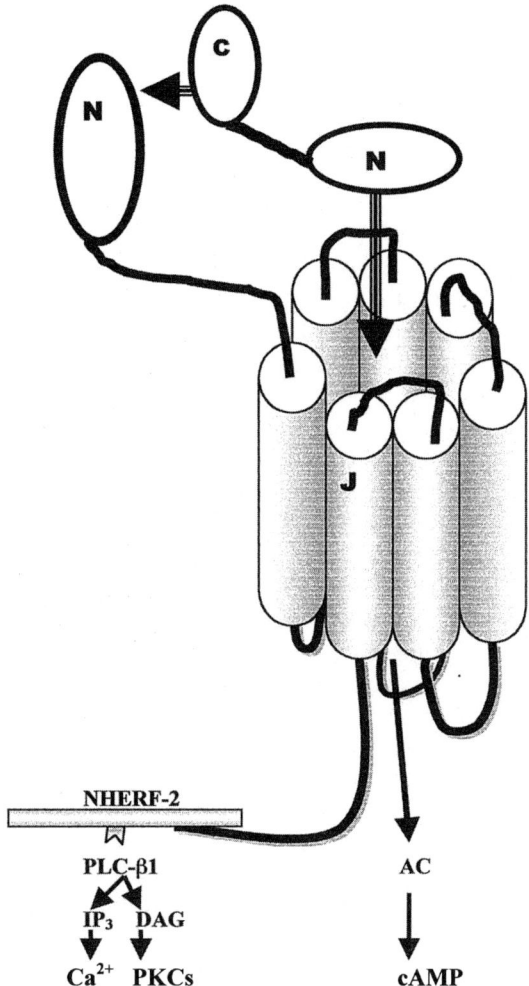

FIGURE 6 The PTHR1 receptor has seven transmembrane α-helices arranged in a "barrel" the staves of which are connected by extracellular and cytoplasmic loops. An adenylyl cyclase-stimulating PTH such as hPTH-(1–34) first attaches by its C-terminal tail (C) to the receptor's extracellular N-terminus (the "N-interaction") and then by its N-terminal nose (N) to the receptor's juxtamembrane region (the "J-interaction"). The result of this two-step binding is a configuration shift in the cytoplasmic loops that causes the conversion of inactive GDP · Gsα · $\beta\gamma$ to the active, adenylyl cyclase-activating GTP · Gsα and the release of $\beta\gamma$. The receptor's C-tail is associated with the NHERF-2 scaffold to which is bound PLC-β1. The shift caused by the "N-interaction" also causes the formation of GTP · Gqα, which stimulates the NHERF-2-bound PLC-β1. The activated PLC-β1 chops Ptd Ins(1,4,5)P$_2$ into inositol 1,4,5-trisphosphate (IP$_3$), which releases Ca^{2+} from internal stores, and diacylglycerols (DAGs), which stimulate PKC isoforms. The one-step "N-interaction" of an N-terminally truncated fragment such as hPTH-(3–34) or hPTH-(13–34) can stimulate formation of PLC-β1-activating GTP · Gqα and PKC activity, but not adenylyl cyclase.

NHERF-2 respond to PTH-(1-34) with a powerful burst of AC activity but no PKC-β1 activity while lymphocytes with large complements of NHERF-2 respond to the PTH with no burst of AC activity but a burst of PLC-β1 and membrane-associated PKC activity (mPKC) (Mahon et al., 2002; Whitfield et al., 1998a, 1999a). In other words, one cannot predict the signals a PTH sends to a target cell—the signal depends on the way its PTHR1 receptors are connected to other components of the cell's membrane.

It is also possible that the receptors stimulate phospholipase D (PLD), which breaks the major membrane phospholipid phosphatidylcholine (PtdCh) into choline and phosphatidic acid, which is eventually converted into diacylglycerol (Friedman et al., 1999; Singh et al., 1999). If the AC response is not muted by PTHR1 being tethered to NHERF-2, the cyclic AMP it makes activates the available cyclic AMP-dependent protein kinase isoforms (PKAs), the diacylglycerols from PIP_2 or PtdCh breakdown activate the available and responsive protein kinase-C isoforms (PKCs), and the IP_3 from PIP_2 stimulates the release of Ca^{2+} from internal endoplasmic reticulum stores (Fig. 6). These PTH-induced signalers trigger a veritable tsunami of events. For example, some of the endoplasmic reticulum's IP_3-activated Ca^{2+} release sites are closely apposed to mitochondrial Ca^{2+} uniporters, which carry the released Ca^{2+} into the mitochondria to increase ATP production by stimulating key enzymes such as pyruvate dehydrogenase, NAD^+-isocitrate dehydrogenase, and 2-oxoglutarate dehydrogenase (Ashby and Tepikin, 2001; Hajnoczky et al., 2000). Perhaps the most important of these PTH-triggered events is the expression of FGF-2 and IGF-I (Fig. 7).

If it is cyclic AMP and bursts of PKA activity that trigger bone growth, and if a PTH fragment could be produced that needs only to activate AC to stimulate bone growth, it could have potentially fewer side effects resulting from PKC activity in the many different cells expressing PTHR1 receptors. Such a fragment could be produced if the multisignaling hPTH-(1–34) could be cut into smaller pieces each of which stimulates one or the other mechanism. And this seems to have been done under certain circumstances. Thus, the potently osteogenic (in OVXed rats) Ostabolin [hPTH-(1-31)NH_2] and Ostabolin C [[Leu^{27}]$cyclo$(Glu^{22}-Lys^{26})hPTH-(1–31)NH_2] are as effective AC stimulators as hPTH-(1–34)NH_2, but unlike hPTH-(1–34)NH_2 they do not stimulate membrane-associated PKC (mPKC) activity in cultured nontransformed murine MC3T3-E1 osteoblasts, rat ROS 17/2, osteoblast-like osteosarcoma cells, UMR 106-01 rat osteoblast-like osteosarcoma cells, as well as primary rat osteoblasts (Barbier et al., 1997; Jouishomme et al., 1992, 1994; Ryder and Duncan, 2001; Singh et al., 1999; Swarth et al., 2000; Whitfield et al., 1996b, 1999a). Cultured human fetal osteoblasts (hFOB cells) appear to respond to hPTH-(1–31)NH_2 in the same way. Thus, hPTH-(1–31)NH_2 does not stimulate the PKC-dependent expression of TGF-β1 (and presumably not PKC activity) in these human fetal osteoblasts, although it stimulates the cyclic AMP-dependent expression of TGF-β2 as strongly as hPTH-(1–34) (Wu and Kumar, 2000).

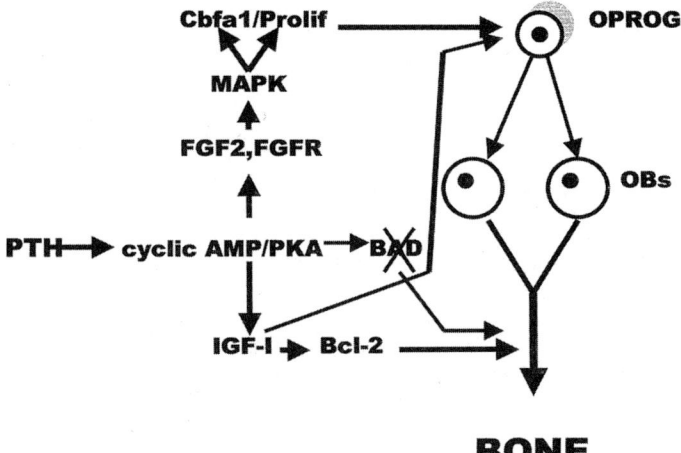

FIGURE 7 How intermittent injections or infusion for no loner than 1 h of an adenylyl cyclase-stimulating PTH can induce a large accumulation of long-lived, postproliferative osteoblasts. Each injection produces a cyclic AMP transient and burst of cyclic AMP-dependent protein kinase (PKA) activity in mature osteoblasts (OBs) with the maximum number of PTHR1 receptors. The burst of PKA activity stimulates the expression of the FGF-2, FGF receptor (FGFR), and IGF-I genes. FGF-2 stimulates the MAP kinase cascade, which in turn stimulates the proliferation and the expression of Cbfa1, the master osteoblast differentiation driver in marrow osteoblast precursors (PreOb), which either have no or very few PTHR1 receptors. In this case, the PKA and FGF-2 pulses are sufficiently spaced to stimulate the proliferation and differentiation of osteoprogenitors without significantly affecting the expression of collagenase-3 in the mature osteoblasts. The PTH-induced burst of PKA (cyclic AMP-dependent protein kinase) activity can prolong the working lives of mature osteoprogenitors, preosteoblasts, and osteoblasts by directly phosphorylating and inactivating the proapoptotic BAD protein and by stimulating the expression of IGF-I, which stimulates the expression of the antiapoptosis Bcl-2. The IGF-I also reaches into the bone marrow to stimulate the proliferation of immature osteoprogenitors, which cannot make it for themselves (Kostenuik *et al.*, 1999).

Whether such C-terminally truncated fragments can also stimulate mPKC activity depends on which cells the PTHR1 receptors find themselves. For example, hPTH-(1–31)NH$_2$ can stimulate *m*PKC activity in pig kidney cells engineered to express a large number (950,000) of human PTHR1 receptors on their surfaces and in normal human newborn foreskin fibroblasts that have only a very few (in fact not measurable by competitive binding assay) human PTHR1 receptors (Whitfield *et al.*, 2001). The failure of hPTH-(1–31)NH$_2$ and [Leu^{27}]*cyclo*(Glu^{22}-Lys^{26})hPTH-(1–31)NH$_2$(Ostabolin C) to stimulate *m*PKC activity in ROS 17/2 osteoblast-like cells and fetal human osteoblasts means that although PTHR1 receptors on cultured osteoblasts cannot activate PLC or PLD, they can activate one or both of these enzymes in densely human-receptored pig kidney cells and very sparsely receptored cultured human foreskin fibroblasts that are differently equipped with the necessary messengers and kinase isoforms.

PTH signaling may be more complex than hitherto believed. GPCRs such as PTHR1 do not directly activate receptor tyrosine kinase (RTK) receptors, but they can transactivate RTK receptors such as the EGF receptor (Cole, 1999; Daub et al., 1997; Iwamoto and Mekada, 2000; Prenzel et al., 1999). At first, it seemed that in these cases the EGF receptor was not activated by its EGF ligand, but it now appears that the activated GPCR stimulates a transmembrane metalloproteinase that releases sHB-EGF, a heparin-binding protein with EGF-like motifs, from membrane-anchored pro-HB-EGF (pro-heparin-binding EGF) (Iwamoto and Mekada, 2000; Prenzel et al., 1999). The liberated sHB-EGF then activates EGF receptors on the liberator cell or adjacent cells. When the PTH/PTHR1-transactivated EGF receptors autophosphorylate, they become scaffolds for assembling the components of the Ras/Raf-1 mechanism that triggers the multipurpose MAP kinase cascade (Cole, 1999; Daub et al., 1997). However, we don't know what the Ras/Raf/MAP kinase cascades triggered by PTHR1-transactivated RTKs might contribute to osteogenesis. But the important message is that the signals triggered by a PTH can come from both PTHR1 receptors and any transactivatable RTKs the cell happens to be expressing.

PTH signaling may be still more complex!! After having sent its first signals, the PTH · PTHR1 complex is rapidly silenced and endocytosed with a half-time of 3–5 min but then slowly returns to the surface, ready to operate again, with a half-time of about 2–4 h (Chauvin et al., 2001). The endocytosis mechanism operates independently of the G-protein activation—blocking GDP · Gsα and GDP · Gq11α → GTP · Gsα and GTP · Gq11α activation does not affect internalization (Vilardaga et al., 2002). The PTH · PTHR1 complex's endocytosis starts when the receptor's ligand-triggered configurational shift causes β-arrestin-2 to translocate to the cell membrane. The receptor's phosphorylated (by a kinase for the activated receptor) C-tail then binds to the β-arrestin-2 scaffolding protein and stabilizes the association of arrestin's amphipathic α-helix 1 with the cell membrane (Chauvin et al., 2001; Pouysségur, 2000; Vilardaga et al., 2002). The PTH · PTHR1 · β-arrestin-2 complex is packaged in a clathrin-coated pit and carried into the cytoplasm (Chauvin et al., 2001; Pouysségur, 2000). By analogy with the type IA angiotensin II receptor, although the endocytosis of the PTHR1 receptor silences the AC and PLC-β1 signaling, it might at the same time trigger another signal(s) because the β-arrestin-2 scaffold has an RRSLHL (Arg-Arg-Ser-Leu-His-Leu) motif that brings Raf-1, ASK-1, MAP Kinase Kinase-4, and a JNK MAP kinase together with the PTH · PTHR1 complex (Keenan and Baldassare, 2001; Luttrell et al., 2001; Miller et al., 2001; Pouysségur, 2000). When β-arrestin-2 attaches to the PTHR1 of the PTH · PTHR1 complex, AK1 could activate MAP Kinase Kinase-4, which in turn activates the JNK, which is released and surges into the nucleus where it phosphorylates and activates gene transactivators such as c-Jun (Keenan and Baldassare, 2001).

The complexity of PTHR1 signaling is increased even further by PTHR1's nuclear localization signal. This means that either newly made PTHR1 might be transported into the nucleus without reaching the cell membrane or after being released from the β-arrestin-2 complex, the endocytosed receptor might go into the nucleus along with the JNK to stimulate some as yet unknown nuclear function. The possibility of a receptor directly stimulating genes after being transported from the cell surface into the nucleus after binding and activation by its ligand has recently been demonstrated with EGF and ErbB-1 and ErbB-4 receptors (Lin et al., 2001; Ni et al., 2001). The EGF receptor, with a strong gene transactivation domain in its C-tail, gets into the nuclei of a variety of cells where it can bind to AT-rich sequences such as the consensus site in the cyclin D1 gene promoter (Lin et al., 2001). When the ErbB-4 receptor on a mammary carcinoma cell binds its ligand, Heregulin, its cytoplasmic tail is clipped off by γ-secretase and travels to the nucleus where it activates a gene(s), the product(s) of which inhibits the growth and stimulates the differentiation of the cell (Ni et al., 2001).

To find out whether the osteogenic signal from PTHR1 receptors requires the stimulation of AC, mPKCs, or both we must also have fragments that stimulate mPKC but not AC. Unlike the C-terminally truncated fragments, no N-terminally truncated PTH fragment can stimulate AC. However, as we learned above the ability of an N-terminally truncated PTH to stimulate mPKC activity also depends on the kind of cell expressing the PTHR1 receptors. For example, b(bovine)PTH-(3–34)NH$_2$, hPTH-(3–34)NH$_2$, hPTH-(13–34)OH, hPTH-(28–48), hPTH-(28–34), hPTH-(29–32), and hPTH-(8–84) can stimulate mPKC activity in cultured ROS17/2 osteoblastic cells, UMR-106 rat osteoblastic cells, human fetal osteoblasts, and human dermal fibroblasts, and hPTH-(8–84) can strongly stimulate mPKC activity in rats, but they are at best only very weak stimulators of mPKC activity in the pig cells with nearly a million human PTHR1 receptors (Fujimori et al., 1992; Jouishomme et al., 1992, 1994; Neugebauer et al., 1995; Swarth et al., 2000; Whitfield et al., 2001; Wu and Kumar, 2000).

hPTH-(1–31)NH$_2$ and its lactam derivative, Ostabolin C, which stimulate AC, but not mPKC activity, in cultured rat and human fetal osteoblasts, strongly stimulate bone growth in OVXed rats (Rixon et al., 1994; Whitfield et al., 1996b, 1997b–d, 1998b, 1999c, 2000a–c). On the other hand, the N-terminally disabled 1-desamino-hPTH-(1–34) and the N-terminally truncated hPTH-(28–48) and hPTH-(8–84) which strongly stimulate mPKC activity, but not AC, in cultured osteoblasts and in rats, do not stimulate bone growth in the OVXed rats (Rixon et al., 1994; Strein, 1994). Therefore, as things stand at the moment, activating the AC/cyclic AMP/PKA mechanism seems to be all a PTH needs to do to start, but not subsequently drive, bone formation (at least in rats). This is supported by the ability of selective cyclic AMP-elevating, cyclic nucleotide phosphodiesterase inhibitors (pentoxyfylline and rolipram) to stimulate bone growth in normal male mice (Kinoshita et al., 2000).

There are reasons to believe that hPTH-(1–31)NH$_2$ may also stimulate AC only in adult human bones as it seems to do in cultured fetal human osteoblasts (Wu and Kumar, 2000). If hPTH-(1–31)NH$_2$ were a dual-signaler like hPTH-(1–34) in adult human bones, it should affect adult human bone metabolism such as hPTH-(1–34). But it doesn't! Although the two fragments equally stimulate AC in human volunteers, hPTH-(1–31)NH$_2$ is a much weaker stimulator of bone resorption than hPTH-(1–34) as indicated by a failure to cause hypercalcemia and increase urinary collagen breakdown products (Fraher et al., 1999). hPTH-(1–31)NH$_2$ is also as strong a stimulator of adenylyl cyclase and bone growth as hPTH-(1–34) in OVXed mice, but it is only half as strong a stimulator of osteoclast activity as hPTH-(1–34) (Mohan et al., 2000). There seems to be some support for this in the observations of Whitfield et al. (1998b) (Fig. 5). They found that daily injections of a low dose (0.6 nmol/100 g of body weight) of hPTH-(1–34) into 3-month-old rats, starting 2 weeks after OVX, did *not* reduce the loss of femoral trabeculae (56% loss compared to a 52% loss in the untreated OVXed rats), particularly the central metaphyseal trabeculae, although they raised the mean thickness of the surviving lateral trabeculae by about 1.7 times above the level in the sham-operated control rats. On the other hand, although injections of the same dose of Ostabolin C ([Leu^{27}]*cyclo*(Glu^{22}-Lys^{26})hPTH-(1–31)NH$_2$) raised the trabecular thickness 1.5 times above the sham-operated value, they also significantly reduced the OVX-induced loss of central trabeculae from 56% to 29% (Fig. 5). However, when the trabecular number was allowed to drop for 9 weeks before starting the 6 weeks of daily injections, neither fragment could appreciably stop any further drop, although both still strongly increased the mean thickness of the remaining lateral trabeculae to 1.6–1.7 times the mean thickness of the trabeculae in the femurs of the sham-operated control animals. The responses to a higher dose of Ostabolin C or hPTH-(1–34) (2 nmmol/100 g of body weight) were identical.

This difference between the effects of hPTH-(1–34) and the two C-terminally truncated fragments in humans and rodents might be due to lower doses of hPTH-(1–31)NH$_2$ and Ostabolin C not having hPTH-(1–34)'s ability to stimulate osteoblastic stromal cells to express the osteoclast differentiation stimulator RANKL and reduce the expression of the osteoclast-inhibiting OPG (Hofbauer et al., 2000; Suda et al., 1999). Indeed maybe Ostabolin C raises the OPG/RANKL expression ratio to a level that actually inhibits OVX-induced osteoclast activation.

2. The Opposite Responses of Bone to PTHR1 Signaling

When an adenylyl cyclase or adenylyl cyclase/*m*PKC-stimulating PTH is injected subcutaneously it hits target cells in both the bone and the hematopoietic compartment of the bone marrow. First, like the estrogen loss, it stimulates the production of osteoclasts and hence increases the remodeling space. It stimulates the proliferation of the pluripotent colony-forming unit stem cells (CFU-S) and other members of the hematopoietic cell populations, which need the microenvironment provided by docking on osteoblastic stromal cells (Gallien-Lartigue and Carrez, 1974; Perris et al., 1971). This expands, via a cyclic AMP-driven mechanism, the population of PTHR1 receptor-expressing, GM- and M-CSF-responsive early osteoclast progenitors, which will continue making PTHR1 mRNA but will later stop making PTHR1 receptors as differentiation progresses (Kanatami et al., 1998; Langub et al., 2001). The burst of cyclic AMP formation triggered by PTH also stimulates osteoclast formation by causing marrow osteoblastic stromal cells to stop making the osteoclastogenesis inhibiting OPG and stimulating the expression of the osteoclastogenesis-stimulating RANKL (Fu *et al.*, 2001; Galvin *et al.*, 2001; Kanazawa *et al.*, 2000; Stilgren *et al.*, 2001). The PTH also induces cultured osteoblasts and possibly osteoblastic stromal cells to make autocrine/paracrine IL-6, which causes osteoblastic stromal cells to make RANKL (Fig. 2; Greenfield *et al.*, 1999; Tamura *et al.*, 1993).

But what about the PTHs' potent anabolic capability? Unlike estrogen loss, the PTHs also stimulate osteoblast production and activity. However, they do not directly stimulate the proliferation of osteoblast lineage cells. In fact, the switching on of the *pthrp* gene and the *pthr1* receptor gene (which in murine and human osteoblasts is run from its P2 promoter and maybe its P3 promoter) requires collagen expression and is therefore probably driven by signals from collagen-bound/activated integrin receptors (Goltzman and White, 2000; McCauley *et al.*, 1996). These expressions start in mature osteoprogenitors and peak in preosteoblasts, which coincides with the initiation of the production and secretion of autocrine/paracrine PTHrP polyprotein, the PTH-(1–34)-like N-terminal portion of which stimulates the emerging PTHR1 receptors, shuts down the proliferative machinery, and starts the procession of mature osteoblast-specific genes (Aubin, 1998, 2000, 2001; Aubin and Triffitt, 2002; Whitfield *et al.*, 1998a). The PTH injections and signals from the emerging PTHR1 receptors stimulate the expression and secretion of factors such as FGF-2, FGF receptors, and IGF-I, and further stimulate the expression PTHrP and one or more of its cleavage-produced components (Fig. 7; Amizuka *et al.*, 1996; Goltzman and White, 2000; Kartsogiannis *et al.*, 1997, Walsh *et al.*, 1997; Zhang *et al.*, 1995), and TGF-βs [TGF-β1 and β2 if the PTH is hPTH-(1–34) that stimulates adenylyl cyclase and mPKC or only TGF-β2 if the PTH is hPTH-(1–31)NH$_2$ that only stimulates adenylyl cyclase (Wu and Kumar, 2000)].

The production of mRNAs for PTHrP-(1–139), -(1–141), and -(1–173) in primary human osteoblasts increases 38-fold by 45 to 90 min after the addition of hPTH-(1–34) to the culture medium (Walsh *et al.*, 1997). Although we know virtually nothing about the contributions of PTHrP and its autocrine/paracrine components in osteoblast activity and the osteogenic response, there are reasons to believe that they are important. PTHrP's PTH-like 1–34 N-terminal region would bind to and activate the PTHR1 receptors as efficiently as a PTH [see Goltzman and White (2000), Lam *et al.* (1999), and Whitfield *et al.* (1998a) for reviews]. But very much unlike PTH, PTHrP or a PTHrP fragment having the 87–107 nuclear-targeting sequence (NTS) that is either endogenous or rides into the cell from the surface on its receptor can function as an 'intrakine' or intracrine (Aarts *et al.*, 1999; Goltzman and White, 2000; Lam *et al.*, 1999, 2000; Re, 2002; Watson *et al.*, 2000 a,b). This means that the peptide or the peptide–receptor complex is transported into the nucleus where it can *directly* affect nuclear functions. However, this nuclear intrakine stimulation would happen only in the mature, postmitotic osteoblast because the cyclin-dependent protein kinases that drive the cell cycle would disable the NTS by phosphorylating Thr^{85} (Goltzman and White, 2000; Lam et al., 1999, 2000). Once in the nucleus and nucleolus the peptide targets RNA production and certain genes, one of which is the antiapoptosis Bcl-2 protein (Amling et al., 1997; Lam et al., 2000; Re, 2002). That would extend the cell's matrix-making life by preventing it from killing itself by apoptosis (Antonsson and Martinou, 2000; Fadeel et al., 1999). But in murine MC3T3-E1 osteoblasts the PTHR1 receptor is loaded into the nucleus during the late G_1 phase of the cell cycle (when the cyclin-dependent protein kinases are active and would stop NTS-bearing PTHrP from entering the nucleus) and then stays there throughout the rest of the cycle (Watson et al., 2000b). Therefore, the PTHR1 receptor either alone or with an N-terminal fragment lacking the Thr^{85}-containing region seems to be able to get into the preosteoblast's nucleus to carry out some job related to DNA replication.

Adding to the PTHrP polyprotein's many activities is the specific ability of one of its components, hPTHrP-(107–139) (osteostatin) with its highly conserved active 107–111 region, to stimulate the proliferation of cultured fetal rat osteoblasts (Cornish et al., 1999) and possibly bone formation in 7- to 8-month-old OVXed rats (Roufflet et al., 1994). Moreover, concentrations of PTHrP-(107–139) and PTHrP-(107–111) as low as 10^{-15} and 10^{-13} M inhibit osteoclast generation and activity in vivo and in vitro (Fenton et al., 1991; Cornish et al., 1997; Zheng et al., 1994). And PTHrP-(107–139) is angiogenic: it stimulates human osteoblasts to make the angiogenic vascular endothelial growth factor (VEGFs), which would provide the blood vessels needed by the new bone (Esbrit et al., 2000). And it might be the PTHrP-induced modulator of vascular invasion into cartilage that does not operate through the PTHR1 receptor (Lanske et al., 1999). These fragments operate by stimulating *m*PKCs not by activating PTHR1 but by activating an apparently very high-affinity "TRSAW" [Thr$(T)^{107}$-Arg$(R)^{108}$-Ser$(S)^{109}$-ALA$(A)^{110}$-Trp$(W)^{111}$] receptor (Cuthbertson et al., 1999; Moonga and Dempster, 1995; Whitfield et al., 1996a, 1998a).

Stromal cells are put on the osteoblast pathway when sonic hedgehog (Shh) stimulates them via FGF-4 to express BMP-2, which stimulates them (via BMP receptor-induced Smad second messengers) to express the type 2 isoform of Cbfa1, the master osteoblast-specifying gene (Fig. 8; Banerjee et al., 2001; Ducy, 2000; Karsenty, 2000a; Komori, 2000). Another hedgehog, Indian hedgehog (Ihh), produced by prehypertrophic and hypertrophic chondrocytes together with BMP-2, BMP-6, and Cbfa1, stimulates differentiation of a distinct population of perichondrial cells into osteoblasts in the collar of the developing bone (Chung et al., 2001; Karsenty, 2001). The osteoblast-determining potency of the Cbfa1 gene product is illustrated by the fact that forcing nonosteoblastic, primary fibroblasts to express their Cbfa1 gene switches on the set of osteoblastic genes (Ducy, 2000). One of the most important things an injected adenylyl cyclase-stimulating osteogenic PTH does is push progenitor cells along the osteogenic path by stimulating them to express the Cbfa1 gene and then load their nuclei with its protein product (Figs. 7 and 8; Fujita et al., 2000; Moore et al., 2000; Selvamurugan et al., 2000). Cbfa1 has a nuclear matrix-binding domain and a DNA-binding "Runt" domain that

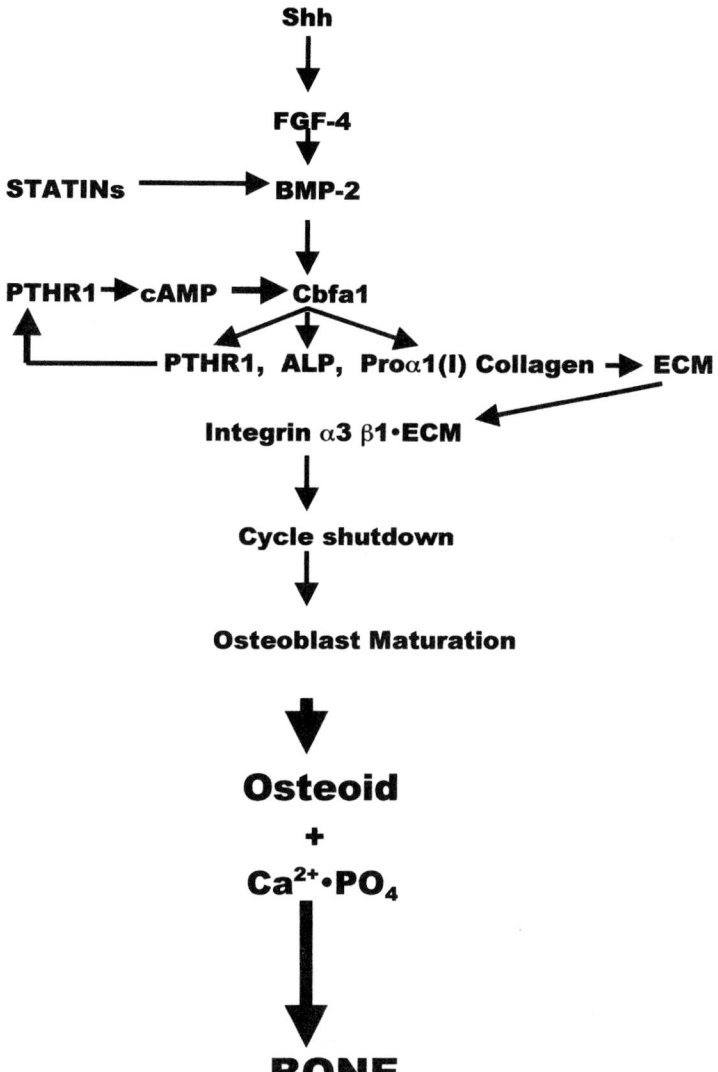

FIGURE 8 The mechanisms by which signals from PTHR1 receptors activated by injected PTH or autocrine PTHrP and statins stimulate bone growth. ALP, bone-specific alkaline phosphatase; BMP-2, bone morphogenic protein; ECM, extracellular matrix; Shh, sonic hedgehog.

are homologous to the *Drosophila runt* gene, the product of which is needed for segmentation without which the larva is a "runt" (Ducy, 2000; Ito, 1999). Cbfa1 binds to so-called OCE2 upstream sites of still silent osteoblast-specific genes such as the genes for alkaline phosphatase (ALP), bone sialoprotein (BSP), osteocalcin (OC), osteopontin (OP), OPG, proα1(I) and proα2(I) procollagen, PTHR1, and type I TGF-β receptor (Franceschi, 1999; Karperien *et al.*, 2000; Kern *et al.*, 2000; Komori, 2000). Cbfa1 then pins these genes to sites on the nuclear matrix containing the RNA polymerase II-containing transcription devices. The transcriptional activity of the master Cbfa1/Runx2 is prevented from working and the osteoprogenitors kept proliferating by the homeobox-targeting protein Msx2 (Dodig *et al.*, 1999; Ryoo *et al.*, 1997; Shirakabe *et al.*, 2001). But eventually the rising signals from collagen-stimulated integrins release Cbfa1/Runx2 from Msx2's grip followed by the downstream expression of the osteoblastic-specific zinc finger protein transcription factor Osterix without which there can be no osteoblast maturation or bone formation (Nakashima *et al.*, 2002). The now mature osteoprogenitors start expressing PTHrP and the PTHR1 receptor (Aubin, 2000, 2001; Aubin and Triffitt, 2002).

The appearance of PTHrP and PTHR1 receptors coincides with the secretion of autocrine/paracrine PTHrP polyprotein, the PTH-(1–34)-like N-terminal portion of which both activates and stimulates the further expression of PTHR1 receptors, shuts down of the proliferative machinery, and starts the procession of mature osteoblast-specific genes (Aubin, 1998, 2000, 2001; Aubin and Triffitt, 2002). Schiller *et al.* (1999) have provided a good example in growing MC3T3-E1 murine calvarial prosteoblast cultures of the coincidence of PTHR1 receptor upregulation with decreased proliferation followed about 5 hours later by the initiation of bone nodule formation. The cyclicAMP signals from the PTHR1 receptors activated by the emerging autocrine/paracrine PTHrP shut down the cell cycle machinery, further increase the expression of PTHR1 (Lu *et al.*, 2002). They also push Cbfa1/Runx2 into the cell nucleus (Fujita *et al.*, 1999, 2001a) and the embrace of pRb, in this case a co-transactivator, to promote the effectiveness of the cycle-blocking $p21^{CIP}$ and $p27^{KIP\ 1}$ proteins and start the expression of the package of mature osteoblast gene expressions (Thomas *et al.*, 2001). The role of pRb of as a Cbfa1/Runx2 co-transactivator that shuts down of proliferation and promotes the expression of the mature post-proliferative osteoblast gene package could explain the 500-times higher incidence of osteosarcoma associated with the disabled pRb in patients with retinoblastoma patients (Gurney *et al.*, 1995; Thomas *et al.*, 2001). This critical step in the progression of an osteoprogenitor cell such as MC3T3-E1 toward bone-making osteoblast is associated with a transient surge of the expression of the pRB-dependent gene TIS21/BTG2/PC2 gene (Raouf and Seth, 2002), the product of which blocks the G1 build-up to chromosome replication (Guardavaccaro *et al.*, 2000). This cycle-stopping expression of TIS21 seems to be part of a common mechanism for shifting a proliferating precursor cell's progeny into a post-proliferative stage of development. Thus, a mouse or rat neuroectodermal cell in the ventricular zone of the neural tube that has reached the point of generating a post-proliferative neuron starts expressing its gene for the TIS21 protein (Iacopetti *et al.*, 1994, 1999). One of its daughters will shut its TIS21 gene off, but will inherit the mother's TIS21 protein and keep it during migration and the initial stage of neuronal differentiation—the appearance of TIS21 marks a neuron's birthday which over the subsequent years the cell would celebrate as its "TIS21 day" (Iacopetti et al., 1994). It appears that TIS21 causes the dephosphorylation of pRB and because of this down-regulates cyclin D1gene expression, the initiator of the cell cycle-driving parade of cyclin-dependent protein kinases (Whitfield and Chakravarthy, 2001). Knocking out pRB prevents neuronal differentiation, keeps the neuroblasts proliferating which ultimately results in a massive apoptosis of accumulating neuroblasts (Guardavaccaro et al., 2000); Lee et al., 1994). By the time the TIS21 protein has disappeared, the cell, be it a neuron or osteoblast, is permanently locked into becoming a terminally differentiated neuron or osteoblast. TIS21 is not the only participant in the locking of MC3T3-E1 cells into terminal maturation. The quiescence-induced Q(quiescin)6 protein peaks and drops in step with TIS21 as does HAX-1 which is part of the EDC (epithelial differentiation gene) complex (e.g., involucrin, loricrin, S100s, trichohyalin) on human chromosome 1q21-22 (Coppock et al., 2000; Marenholz et al., 2001; Raouf and Seth, 2002; Whitfield and Chakravarthy, 2001) (Fig. 10). The Ca^{2+}-triggered expression of this gene complex is part of the terminal differentiation of keratinocytes (Whitfield and Chakravarthy, 2002).

When the cells have arrived at their workplace and deposited enough collagen, the signals from emerging PTHrP-activated PTHR1 receptors and collagen-bound/activated integrins shut down the cycle-driving machinery and up-regulate the differentiation-related genes (Fig. 8). But although they may look alike, the resulting postmitotic mature osteoblasts do not have the same gene expression profile. The cells' gene expression profiles are specifically tailored to the different needs for making bone in different regions of the bone on which they are operating. Thus, for example, the relative expressions of BSP, OP, OC, and PTHrP at the mRNA and protein levels depend on which part of a developing bone (e.g., endocranial or ectocranial surface of mouse calvaria) the cell is located as well as its immediate neighbors and consequently the microenvironment in which it finds itself. But all osteoblasts, no matter where they are or who their neighbors are, express the same three genes at about the same levels, one of which is the PTHR1 receptor (the other two are the type I collagen and alkaline phosphatase genes) (Candeliere et al., 2001).

The PTHR1-expressing mature osteoblasts in modeling (as opposed to microcrack repairing/remodeling) clusters making new bone for growing skeletons or in BMUs filling osteoclasts' remodeling holes in mature skeletons are the prime targets of injected PTH. By the time they have dismantled their proliferation machinery and matured into "postmitotic" working osteoblasts they have greatly (10-fold) increased the expression of PTHR1 receptors, which probably enables the autocrine/paracrine N-terminal PTHrPs that they have made and secreted to give themselves and their neighbors the cyclic AMP signals needed to drive their further maturation and bone building (Aubin, 1998, 2000, 2001; Aubin and Triffitt, 2002).

PTH, and the pulses of PKA activity it triggers, only *starts* the osteogenesis but it is its products that take over and drive it (Fig. 7; Whitfield et al., 1998a). As we previously said, the single daily cyclic AMP surges triggered by PTH injections do not stimulate either the marrow osteoprogenitors or mature osteoblasts to proliferate (Kostenuik *et al.,* 1999; reviewed by Whitfield *et al.,* 1998a). But the PKA pulses do increase the productivity of mature osteoblasts by lengthening their working lives by phosphorylating and inactivating the proapoptosis BAD protein that releases the antiapoptosis Bcl-X_L from inactive complexes with BAD (Bellido *et al.,* 2001; Chao and Korsmeyer, 1998), by stimulating sphingosine kinase that produces the antiapoptosis sphingosine phosphate, and by stimulating the expression of IGF-I, which in turn stimulates the expression of the apoptosis-preventing Bcl-2 family proteins (Calvi *et al.,* 2001; Hill *et al.,* 1997; Johanson and Rosen, 1997; Machwate *et al.,* 1998; Pfeilschrifter *et al.,* 1995; Pugazhenthi *et al.,* 1999; Tovar-Sepulveda *et al.,* 2002; Tumber *et al.,* 2000; Virdee *et al.,* 2000; Watson *et al.,* 1995, 1999) (Fig. 7). The IGF-I secreted by the mature PTH-stimulated osteoblasts also increases the osteoblast generation by traveling to the bone marrow where it stimulates the proliferation of osteoprogenitors, which, although they have PTH receptors, are not yet mature enough to make their own IGF-I (Calvi *et al.,* 2001; Kostenuik *et al.,* 1999). However, neither PTH nor the IGF-I it stimulates can stimulate growth if the bones are unloaded (Kostenuik *et al.,* 1999) because the lack of loading somehow disables the responsiveness of the osteoprogenitor cells to IGF-I (Bikle *et al.,* 1994; Kostenuik *et al.,* 1999).

But cyclic AMP-stimulated IGF-I might not be as important for the stimulating bone growth in humans as it seems to be in rodents (Johanson and Rosen, 1997). Human osteoblasts make much more IGF-II than IGF-I (Mohan and Baylink, 1999). Moreover, PTHs do not stimulate human osteoblasts to make IGF-II. Indeed cyclic AMP actually seems to inhibit IGF-II expression. However, the PTHs and cyclic AMP do trigger some production of IGF-I in these human cells (Mohan and Baylink, 1999).

The cyclic AMP surges also induce osteoblasts to make another important factor in addition to IGF-I—the large proliferogenic, NLS-bearing isoform of FGF-2 [also known as bFGF (Okada-Ban *et al.,* 2000; Ornitz and Itoh, 2001)] (Hurley *et al.,* 1999; Liang *et al.,* 1999; Montero *et al.,* 2000) (Fig. 7). FGF-2, like holoPTHrP and its NLS-bearing fragments, is an intrakine that by escaping from the cell and activating its receptor can trigger a proliferogenic MAP kinase cascade and when endocytosed can also get into the nucleus either alone or bound to its receptor and stimulate the expression of a set of genes that includes Cbfa1 (Fig. 7). Therefore, PTH-induced surges of FGF-2 and IGF-I contribute to the PTH-induced accumulation of osteoblasts by both stimulating the proliferation of immature mature osteoprogenitors and accelerating their ultimate differentiation into osteoblasts with IGF-I-lengthened working lives (Fig. 7).

The striking PTH-induced layering of rat bones with mature osteoblasts may not be due only to the stimulation of preosteoblast proliferation and maturation into longer-lived osteoblasts. In rats, it seems to be at least partly due to a substantial, reversible, reversion of the thin, flat, postmitotic, osteogenically inactive bonelining cells into plump, bone-making, but still proliferatively disabled, osteoblasts (Dobnig and Turner, 1995; Leaffer *et al.,* 1995). This transformation could be due to a direct stimulation by PTH because although the lining cells reduce PTHR1 receptor expression and other vestiges of their past osteoblast lives they still express some receptors (at least in humans) (Langub *et al.*, 2001). And the reverting cells probably raise their PTHR1 expression to the level in mature osteoblasts. This reversion might also be triggered by the FGF-2, IGF-I, and TGF-β(s) secreted by the PTHR1 receptor-loaded osteoblasts in microcrack-repairing BMUs or modeling clusters (Whitfield *et al.*, 1998a). When the PTH injections stop, the ex-lining cells stop making bone and return to being lining cells (Dobnig and Turner, 1995; Leaffer *et al.*, 1995). The resulting drop in IGF-I production and secretion by the shrinking lining cells and osteoblasts would restore their vulnerability to apoptosis.

The osteoblast's Na$^+$ gradient-driven, type III Pit1 phosphate transporter is also involved in bone formation (Caverzasio and Bonjour, 1996; Nielsen *et al.*, 2001; Selz *et al.*, 1989; Takeda *et al.*, 1999). When the preosteoblasts stop proliferating and start maturing they up-regulate the expression of the Pit1 transporter instead of the "housekeeping" Pit2 transporter (Nielsen *et al.*, 2001). They also start expressing their alkaline phosphatase (ALP) gene and inserting ALP into their cell membranes with its catalytic subunit protruding into the extracellular space (Beck *et al.*, 2000). ALP then catalyzes the release of phosphate from external substrates such as the β-glycerophosphate added to culture media and matrix breakdown products. Pit1 then dumps this phosphate into the cell where it *directly* activates the osteopontin gene's phosphate-responsive promoter (Beck *et al.*, 2000). The cells use the resulting multipurpose osteopontin to cement the new matrix to the old bone and with signals from their osteopontin-attached integrins prevent the cells from committing apoptotic suicide (Denhardt and Noda, 1998; McKee and Nanci, 1996). [If they are lucky enough to survive to become osteocytes, they will trigger a second burst of osteopontin synthesis to attach themselves and their processes to their cubicles and canaliculi (Denhardt and Noda, 1998; McKee and Nanci, 1996).] The transporter is also used to mineralize the new osteoid. Osteoid is mineralized when transporter-delivered phosphate and Ca^{2+} flowing through channels formed by annexin V (Anx V) combine to form apatite-like (Ca-P) crystals in matrix-bound vesicles (Bandorowicz-Pikula *et al.*, 2001; Caverzasio and Bonjour, 1996; Selz *et al.*, 1989). These vesicles have budded from the surfaces of the osteoblasts and when they are loaded, they rupture and dump the accumulated apatite-like crystals onto the osteoid (Caverzasio and Bonjour, 1996; Nielsen *et al.*, 2001). Because PTH-induced cyclic AMP transients stimulate type III transporters, well-separated boluses of AC-stimulating PTHs promote mineralization by selectively increasing the V_{max} of the transporter, which persists even after the vesicles bud from the osteoblast's surface (Caverzasio and Bonjour, 1996).

PTH might also sensitize the strain-responsive osteocytes [which still express PTHR1 receptors, though at much lower levels than when they were osteoblasts (Aubin, 2000; Langub *et al.*, 2001)] to strain (Duncan *et al.*, 1992). Thus, adenylyl cyclase-stimulating PTHs phosphorylate the mechanosensitive Ca^{2+} channels (MSCCs) of MC3T3-E1 mouse preosteoblasts and UMR 106.01 rat osteoblasts by cyclic AMP-stimulated PKA (Duncan and Misler, 1989; Duncan *et al.*, 1992;

Ryder and Duncan, 2001). This causes the MSCCs to open wider (increase their conductance) and stay open longer when they are stimulated by shear stress. However, at high levels of shearing both the adenylyl cyclase and PLC pathways may be required to increase the MSCCs' responsiveness (Ryder and Duncan, 2001). A need for PTH to maintain MSCC responsiveness to strain could be why thyroparathyroidectomy eliminates the responsiveness of rat bones to mechanical stress (Chow et al., 1998). This shearing also stimulates the cells to pump out adenosine triphosphate (ATP) and uridine triphosphate (UTP), which by stimulating their P2 nucleotide receptors sensitize the cells to signals from PTHR1 receptors (Bowler et al., 2001). Because these nucleotides are very short-lived, they accumulate at high levels only around the producing cells (Bowler et al., 2001), which focuses and magnifies PTH's action on the strained cells and their MSCCs. A practical use of this sensitization of MSCCs by PTH would be to increase bone formation upon return to macrogravity or remobilization after prolonged inactivity (Chow et al., 1998; Ma et al., 1999).

While the osteogenic action of single daily injections of an adenylyl cyclase-stimulating PTH predominates over their osteoclast-stimulating, bone-demolishing action, the balance shifts to resorption if the PTH is continuously infused or injected at close intervals. Why?

3. The Delicate Balance Between PTH-Induced Bone Formation and Bone Loss

Whereas a daily injection or no more than a 1-h infusion of an adenylyl cyclase-stimulating PTH optimally stimulates net bone formation, prolonging the exposure to the peptide beyond 1 h causes net bone loss (Dobnig and Turner, 1997). Watson et al. (1999) found part of the reason for this in rats. Intermittent injections of hPTH-(1–84) caused a massive accumulation of IGF-I-making and bone-making osteoblasts (Watson et al., 1995, 1999). Continuously infusing the hormone still caused osteoblasts to accumulate, but now they make less autocrine/paracrine IGF-I and start making IGF-binding proteins, IGFBP-3, IGFBP-4, and IGFBP-5 (Fig. 9). The IGFBPs, particularly IGFBP-4 (Mohan et al., 1999), would have prevented the smaller amounts of IGF-I from stimulating bone formation. Added to this blockade of IGF-I would have been the switching on of the osteoblasts' matrix-destroying collagenase-3 by an excessive production of FGF-2 (Fig. 9; Hurley et al., 1995, 1999). PTH itself would also have stimulated collagenase-3 expression by stimulating Cbfa1 expression (Selvamurugan et al., 2000), but as noted in the following paragraph this might not be the case with excessive exposure to PTH. This combination of IGF-I blockage and a surging, matrix-destroying collagenase would have been enough to shift the response to bone demolition.

Single daily boluses of adenylyl cyclase-stimulating PTHs can stimulate the expression of about 158 genes in rat femoral metaphyseal bone cells and the loading of nuclei with the master osteoblast differentiator Cbfa1 drive osteogenesis (Fujita et al., 2000, 2001a; Moore et al., 2000; Onyia et al., 2001a). But continuously

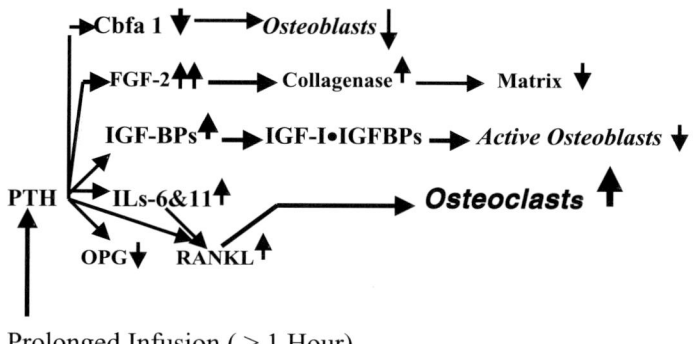

Prolonged Infusion (> 1 Hour)

FIGURE 9 Several bone resorption-stimulating events triggered and driven by PTH infused for more than 1 h. A drop is designated by ↓ and a rise by ↑ These events are discussed in the text.

infusing these PTHs into rats selectively switches on a different set of 759 genes in the metaphyseal cells, which include proteases, which can destroy bone matrix (Onyia et al., 2001a,b). This prolonged exposure can cause the shutdown of Cbfa1-dependent osteoblast-specific genes such as OPG and a PKA-stimulated ubiquitination of Cbfa1 and its destruction in the proteasome with a consequent shutdown Cbfa-1-dependent genes such as OPG (Fig. 9; Tintut et al., 1999).

How could continuous PTH infusion turn on so many more genes than single daily boluses of the peptide? Logically, continuous infusion must somehow activate additional transcription factors. According to Chauvin et al. (2001), spacing the boluses 24 h apart would enable an osteoblast's PTHR1 receptors to fully recycle once and enable the cell to present a fully receptored surface to the next PTH bolus. Each bolus would cause a brief burst of cyclic AMP/PKA signaling from the full complement of PTHR1s on the various target cells. But a continuously infused PTH might cause a continuous, β-arrestin-2-mediated endocytic cycling of PTHR1s with a reduced steady-state level of receptors and adenylyl cyclase signaling but with a steady, β-arrestin-2-mediated flow of MKK4-activated JNK kinase (and maybe PTHR1 receptor itself) into the nucleus. Such a sustained flow of the transcription factor(s)-activating JNK into the nucleus could massively stimulate the expression of a larger set of genes and their resorption-driving products.

But this is not all. A prolonged stimulation of osteoblasts by a continuously infused AC-stimulating PTH or by closely spaced injections of the peptide also causes them to make preosteoblast-inhibiting and/or osteoclast-stimulating IL-6 and IL-11 cytokines (Fig. 9; Kroll, 2000). Moreover, as noted above, a prolonged PTH exposure might prevent osteoblastic cells from making the osteoclast-blocking OPG (Fig. 9; Fu et al., 2001; Galvin et al., 2001; Hofbauer et al., 1999, 2000; Kanazawa et al., 2000; Oniya et al., 2000; Stilgren et al., 2001). It might also cause an atypical Cbfa1-mediated stimulation of RANKL expression (O'Brien et al., 2000). Thus, one injection of the AC-stimulating hPTH-(1–38) into rats

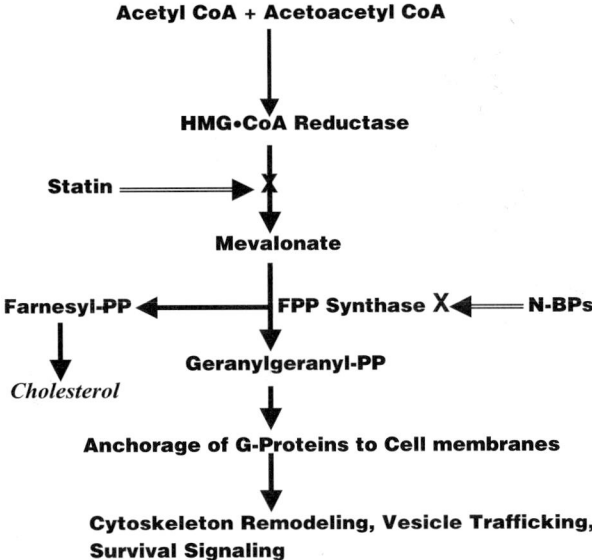

FIGURE 10 Statins and nitrogen-containing bisphosphonates (N-BPs) both inhibit geranyl-geranyl-pyrophosphate generation and the isoprenylation and cell membrane attachment that key proteins need to function. This inhibition results in the disabling and killing of osteoclasts (Fig. 4).

causes a prompt (within 1 h) drop in OPG mRNA production in the distal femoral metaphysis and diaphysis, which rebounds in 3 h (Oniya et al., 2000). Therefore, closely spaced injections of, or a continuous exposure to, a PTH would likely prevent OPG production from rebounding and thus increase osteoclast production (Fig. 10).

Yet another factor will come into play if the PTH boluses are too closely spaced. Adenylyl cyclase-stimulating PTHs increase the velocity (V_{max}) of the mature, working osteoblasts' Pit1 phosphate transporter (Caverzasio and Bonjour, 1996; Nielsen et al., 2001; Selz et al., 1989). Normally this transporter provides the inflow of phosphate needed to drive the expression of osteopontin and the accumulation of mineral in the matrix vesicles budded off the osteoblasts (Beck et al., 2000; Caverzasio and Bonjour, 1996; McKee and Nanci, 1996; Selz et al., 1989). Well-spaced cyclic AMP/PKA pulses would promote osteoid mineralization by causing phosphate to surge into the matrix vesicles. However, continuous infusion or more closely spaced boluses of a PTH might boost the flow of phosphate into the cell to an apoptosis-triggering level (Mansfield et al., 2001; Meleti et al., 2000).

E. THE CLINICAL PROSPECTS OF THE PTHs

One of the adenylyl cyclase-stimulating PTHs, Lilly's recombinant hPTH-(1–34) (Forteo), will soon be available for treating established osteoporosis. It is also likely that the PTHs will be used to accelerate fracture mending in people of all ages, but especially children with their osteoblast-loaded bones (Walker, 1971), and they should serve as a side effect-free "chemical gravity" to strengthen the depleted bones of astronauts returning from long space missions. [The PTHs may not be effective *during* long space missions because of a progressive loss of the ability of the osteoprogenitor cells in unloaded bone to respond to PTH-induced IGF-I (Bikle *et al.*, 1994; Kostenuik *et al.*, 1999).] The PTHs should also be able to strengthen bone eroded by arthritis.

The first-generation PTHs, Eli Lilly's "Forteo" and recombinant hPTH-(1–84) (NPS Pharmaceuticals Inc.'s ALX 1-11, which has finished its Phase II trial), will be followed into the clinic by the smaller second-generation fragments such as ParaTecH Therapeutic's synthetic Ostabolin-C ([Leu^{27}]*cyclo*(Glu^{22}-Lys^{26})hPTH-(1–31)NH$_2$], which, if it is like Ostabolin, its hPTH-(1–31)NH$_2$ parent, may have the considerable advantage of being as potent a bone builder as the larger PTH family members, but a poorer stimulator of resorption (Fraher *et al.*, 1999; Mohan *et al.*, 2000; Whitfield *et al.*, 1998b). In this case there would not be the worrisome possibility of the PTH causing a damaging hypercalcemia in some patients. Then may come the third generation of oral, nasal, or topical analogues or mimetics. In fact, a recent report by Mehta *et al.* (2001) suggests that a "PTH pill" may be on the horizon. They have found that the osteogenic second generation hPTH-(1–31)NH$_2$ [Whitfield *et al.*'s Ostabolin (Whitfield *et al.*, 1996b)] incorporated into an orally administered solid capsule formulation is much more effective in getting into the blood of dogs than Lilly's hPTH-(1–34) (Forteo) in the same type of capsule. The maximum blood concentration of hPTH-(1–31)NH$_2$ was an impressive 2155 ± 456 (SEM) pg/ml compared to a modest 359 ± 152 (SEM) pg/ml for equimolar hPTH-(1–34) (Forteo).

Other agents such as BMPs, FGF-2, and IGF-I are being considered as anabolic therpeutics. However, the remarkable fact is that the PTHs actually operate through the same Cbfa1 as BMP-2 and stimulate osteoblastic cells to make FGF-2 and IGF-I to mediate their osteogenic action. In other words the patient gets the full value of three anabolic agents in each PTH injection.

A drawback of the PTHs is that a lot of cells, if not most cells, express PTHR1 receptors at various stages of differentiation mainly for responding to locally produced PTHrP. Therefore, it would be worthwhile to try to develop a PTH analogue that localizes to bone. Such an analog would have minmal side effects.

It is also possible that a "calcilytic" with a short cirulating half-life will be invented that can trigger short, osteogenic bursts of endogenous PTH secretion by blocking the CaRs on the patient's own parathyroid cells (Gowan *et al.*, 2000). If supplemented with OPG, an estrogen, or a SERM to prevent the stimulation of osteoclasts, the calcilytic-induced pulses of endogenous PTH might be strongly osteogenic. However, blocking CaRs may be too risky because a host of other cells use these sensors to control major body functions ranging from brain to gut and kidneys (Brown and MacLeod, 2001).

Thus, we foresee a time in the near future when an osteoporotic woman will be given (or like a diabetic, will give herself) a single daily subcutaneous injection of a PTH, perhaps along with an occasional subcutaneous injection of the potent, long-acting, osteoclast-inhibiting OPG (Bekker *et al.*, 2001) for 1 or 2 years to bring her bones back to optimal strength and then either continue with occasional OPG injections or switch to one of the oral antiresorptive drugs to keep her new "PTH bone."

VI. THE STATINS

Although the PTHs are currently the most promising anabolic agents for treating osteoporosis, another family of drugs is now challenging their position (Mundy, 2000; Whitfield, 2001a). These are old drugs. They are the statins, which have been used for decades to lower blood cholesterol and prevent heart attacks and stroke.

Although the PTHs are very close to the ideal anabolic drugs, they must still be injected subcutaneously and they affect the various cells that share PTHR1 receptors with bone cells. The ideal drug would be a small molecule that stimulates bone growth as strongly as a PTH by selectively stimulating bone cells when given orally or topically. Such a drug should stimulate osteoblast progenitors to make BMP-2, which, in turn, would stimulate them to express Cbfa1 and differentiate into mature osteoblasts (Ji *et al.*, 2000; A. Yamaguchi *et al.*, 2000). Therefore, Mundy and his colleagues tested more than 30,000 small "natural" compounds for their ability to activate the BMP-2 gene (Garrett *et al.*, 2001b; Mundy, 2000; Mundy *et al.*, 1999). They used 2T3-BMP-2-LUC mouse osteoblasts that had been immortalized by carrying the SV-40 tumor virus's large T-antigen attached to a BMP-2 gene promoter and a firefly luciferase gene that was also coupled to a BMP-2 gene promoter (Ghosh-Choudhury *et al.*, 1996; Mundy *et al.*, 1999). Their effort was rewarded by one of the compounds stimulating the luciferse gene's BMP-2 promoter at 0.1–5 μM (Garrett *et al.*, 2001c; Mundy *et al.*, 1999). This compound was lovastatin, a member of a family of very widely used, serum cholesterol-lowering drugs. Afterward they found that other lipophilic statins, atorvastatin, mevaststin, and simvastatin, were also effective. Currently the synthetic cerivastatin is the most potent, but the hydrophilic pravastatin does not stimulate the BMP-2 promoter because it was designed to only get into hepatocytes, the main cholesterol makers (Garrett *et al.*, 2001c).

Sugiyama *et al.* (2000), using human osteosarcoma cells with a BMP-2 or BMP-4 promoter-driven luciferase reporter gene, reported that simvastatin stimulated the BMP-2 promoter but not the BMP-4 promoter. Significantly, the statin did not affect BMP-2 expression in Chinese hamster ovary cells. Evidently osteoblasts contain something needed specifically for statins to stimulate a BMP-2 promoter. More recently, Maeda *et al.* (2001) reported that simvastatin, at concentrations between 0.01 and $0.1 \mu M$, stimulated MC3T3-E1 mouse osteoblastic cells to express BMP-2, make alkaline phosphatase activity, and mineralize the extracellular matrix.

This discovery was, and is, very exciting because the statins were found at the end of the 1970s and have since become very familiar drugs that have been given to thousands of men and women throughout the world for long periods of time to reduce cholesterol levels and the risk of heart attacks and other cardiovascular problems (Herrington and Klein, 2001; Moghadasian and Frohlich, 2001; Whitfield, 2001a). After being converted in the body from lactone prodrugs to the β-hydroxy acid drugs, they stop cholestrol synthesis by inhibiting the synthesis of its mevalonic acid precursor from HMG-CoA by HMG-CoA reductase (Fisher *et al.*, 1999; Garrett *et al.*, 2001b; Whitfield, 2001a) (Fig. 10).

Even more exciting is the consistently reduced incidence of Alzheimer's dementia (odds ratio of 0.6 to 0.73) in people treated with the statins, but not other lipid-lowering drugs (i.e., benzafibrate, ciprofibrate, clofibrate, fenofibrate, gemfibrozil, colestipol, cholestyramine, acipimox, niacin/nicotinic acid) probably by reducing the cleavage of the amyloid precursor protein into plaque-forming $\beta 42$ fragments (Fassbender *et al.*, 2001; Jick *et al.*, 2000; Panegyres, 2001; Wolozin *et al.*, 2000).

A. ANIMAL STUDIES

Mundy and his colleagues (Garrett *et al.*, 2001c) first confirmed that the lipophilic statins, atorvastatin, cerivastatin, fluvaststain, mevaststatin, and simvastatin, but not the hydrophilic pravastatin, stimulated bone cells to make BMP-2. They then showed that 0.25 and 0.5 μM lovastatin and simvaststain stimulated bone formation in mouse calvarial organ cultures by 2.8- and 4.3-fold, respectively (reviewed by Garrett *et al.*, 2001c; Mundy *et al.*, 1999). Remarkably, only one 24-h exposure to cerivastatin was enough to kick-start the osteogenic mechanism into operating for the following 2 weeks! They also gave simvastatin to ovary-intact and OVXed rats by oral gavage for 35 days. Giving the animals 5 or 10 mg of simvastatin/kg/day nearly doubled trabecular volume and increased the bone formation rate by about 50% (Garrett *et al.*, 2001c; Mundy *et al.*, 1999).

Oral delivery of statins is not the best way to stimulate bone growth in mice and rats because the drugs can be converted into inactive metabolites by the cytochrome P-450 system during their first passage through the liver (Gutierrez *et al.*, 2000; Shepherd, 2000). In fact, the pharmaceutical industry has designed these drugs to selectively target the liver—the main site of cholesterol synthesis (Garrett *et al.*, 2001c). The ideal would be to deliver them by topical application. And this can be done. Gutierrez *et al.* (2000) found that lovastatin was, on a per milligram basis, a 50 times more potent osteogen when applied topically to rat skin than when given orally. Whang *et al.* (2000) reported that lovastatin's potency as a mouse calvarial osteogen was increased as much as 80 times when delivered continuously from a poly[lactide-co-glycolide] scaffold implanted into the skin over the mouse skulls.

B. OSTEOGENIC MECHANISM

The first known action of the statins is to inhibit HMG-CoA reductase (Fig. 10; Fisher *et al.*, 1999; reviewed by Garrett *et al.*, 2001c; Whitfield, 2001a). In addition to preventing cholesterol synthesis, the inhibition of HMG-CoA reductase by a statin such as lovastatin will inhibit osteoclast generation and trigger osteoclast apoptosis as does the inhibition of FPP by an N-containing bisphosphonate (N-BP) such as alendronate (Figs. 4 and Fig. 10; Benford *et al.,* 2001a,b; Fisher *et al.,* 1999; Rezka *et al.,* 1999; van Beek *et al.,* 1999). However, the statins may do something else, because they also selectively stimulate the expression of BMP-2, which in turn mediates the osteogenic response probably by stimulating the expression of Cbfa1. In other words in rodents the statins are "double-barreled" bone anabolics that kill osteoclasts and stimulate osteoblasts. To get the same effect with a PTH it would have to be given along with OPG or an antiresorptive such as alendronate to prevent PTH from stimulating osteoclasts.

However, a problem has arisen with the responses of rodents to statins. Some groups very recently reported that they could not get statins to stimulate bone growth or prevent OVX-induced bone loss in mice and rats (Crawford *et al.,* 2001; Gasser, 2001; Sato *et al.,* 2001; Yao *et al.,* 2001). We will offer a possible explanation for this conflicting evidence further on.

At first sight it would seem that the stimulation of BMP-2 and osteogenesis when it happens should be independent from HMG-CoA-reductase inhibition. But Garrett *et al.* (2001b) reported that mevalonate or geranylgeranyl-pyrophosphate prevents lovastatin from stimulating BMP-2 expression and bone formation in rat bone cultures. An exciting abstract by Garrett *et al.* (2000) suggested that a separate osteoblast-stimulating mechanism had been found. But as we shall see all that came from that was the discovery of different kinds of stimulator!

The clue was the resemblance of the statin lactone ring to that of lactacystin, an inhibitor of the 26S proteasome. They then found that lactacystin and other unrelated proteasome inhibitors (the peptide aldehydes PS-1 and MG132; the α,β-epoxyketone epoxomicin), *which* (according to them) *are unlikely to be HMG-CoA inhibitors,* were bone anabolics. They all stimulated BMP-2 expression and the growth of cultured neonatal mouse bone and the growth of trabecular bone in rats and mice. Moreover, preosteoblasts and osteoblasts could also be induced to express alkaline phosphatase and osteocalcin and make bone by disabling the ubiquitin-proteasome system (UPS) either by having the cells overexpress a dominant negative mutant of the E3 ubiquitin ligase gene or by inhibiting the E3 ligase with N-Leu-Ala. Significantly, the chymotrypsin component of the osteoblasts' proteasome was more responsive to the inhibitors than the chymotrypsin component of other cells. This disabling of osteoblast UPS would be expected to stop the cytoplasmic turnover of the β-catenin transcription enhancer, which would then accumulate in the nucleus and stimulate genes for BMP-4, cyclin D1, and Myc (Hershko *et al.,* 2000). And Chen *et al.* (2001) have shown that disabling the osteoblast UPS with epoxomicin or proteasome inhibitor-1 also stops the Slimb ubiquitin ligase-induced cleavage of active big Gli3 into inactive little (the N-terminal short form) Gli3, which results in big Gli3 building up and getting into the nucleus to stimulate the *bmp*-2 gene and the production of autocrine/paracrine BMP-2. Operating through Smad second messengers, BMP-2, like the osteogenic

PTHs operating through cyclic AMP (Fujita et al., 2000; Moore et al., 2000; Selvamurugan et al., 2000), would stimulate the cells to express Cbfa1 and drive the differentiation of the cells into mature, bone-making osteoblasts (Fig. 8 and Fig. 11). As expected, inhibiting BMP-2 gene expression with Noggin prevented both the UPS inhibitors and statins from stimulating bone formation (Garrett et al., 2001a).

So far, Garrett and his group have not reported whether statins did in fact inhibit the UPS and cause a Gli3 buildup. And if the statins do stimulate bone formation by inhibiting the UPS it is hard to understand why their anabolic actions, like the antiresorptive actions of N-bisphosphonates such as alendronate, are prevented by mevalonate or geranylgeranyl-pyrophosphate (Garrett et al., 2001a). It would be expected that alendronate, for example, would be a potent stimulator of BMP-2 expression and bone growth, which it is not. And finally it has been reported at the Phoenix meeting of the American Association for Bone and Mineral Research that one of the statins, lovastatin, does not inhibit the UPS. In fact, lovastatin actually *stimulates* proteasome activity in MC3T3-E1 mouse calvarial preosteoblasts (Murray and Murray, 2001). As so often happens, a beautiful idea has been murdered by an ugly fact. Obviously we must look elsewhere for the source of the statins' anabolic action.

Because statins stimulate endothelial nitric oxide synthase, the constitutive Ca^{2+}–calomodulin-activatable eNOS (Feron et al., 2001); because knocking out the eNOS gene reduces bone formation in mice and the osteogenic response of calvarial osteoblasts from these eNOS$^{-/-}$ mice (Aguirre et al., 2001; Armour et al., 2001; Garrett et al., 2001b; van't Hof and Ralston, 2001); and because inhibiting eNOS activity prevents statins from stimulating BMP-2 expression and bone formation in murine calvarial cultures (Garrett et al., 2001b), an obvious place to look for the statin targets are the eNOS-sequestering caveolae, which are small (60–100 μm in diameter) clathrin-free, flask-shaped invaginations of the membranes of most cells including human and murine osteoblastic cells (Solomon et al., 2000).

Caveolae contain a lot of free cholesterol, eNOS, and various other signalers (e.g., calmodulin, G-proteins such as H-Ras and Rho, nonreceptor tyrosine kinases, tyrosine kinase adaptor proteins, platelet-derived growth factor, protein kinase A) bound by their caveolin-binding motifs to the scaffolding domain of caveolin, a 18- to 24-kDa integral membrane protein plugged into the caveolar membrane (Anderson, 1998; Couet et al., 2001; Solomon et al., 2000) (Fig. 11). It is a general rule that binding to caveolin and caveolar sequestration silence eNOS and other signaling kinases (Anderson, 1998; Couet et al., 2001; Feron et al., 2001; Fielding and Fielding, 2001; Xu et al., 2001) (Fig. 11). Because caveolin expression is controlled by the free cholesterol level (Fielding and Fielding, 2001), reducing cholesterol production with a statin should, and does, stop caveoli formation, liberate eNOS, and thus raise the basal and Ca^{2+}-stimulable eNOS activity and NO production that stimulates BMP-2 expression and bone growth (Feron et al., 2001) (Fig. 11).

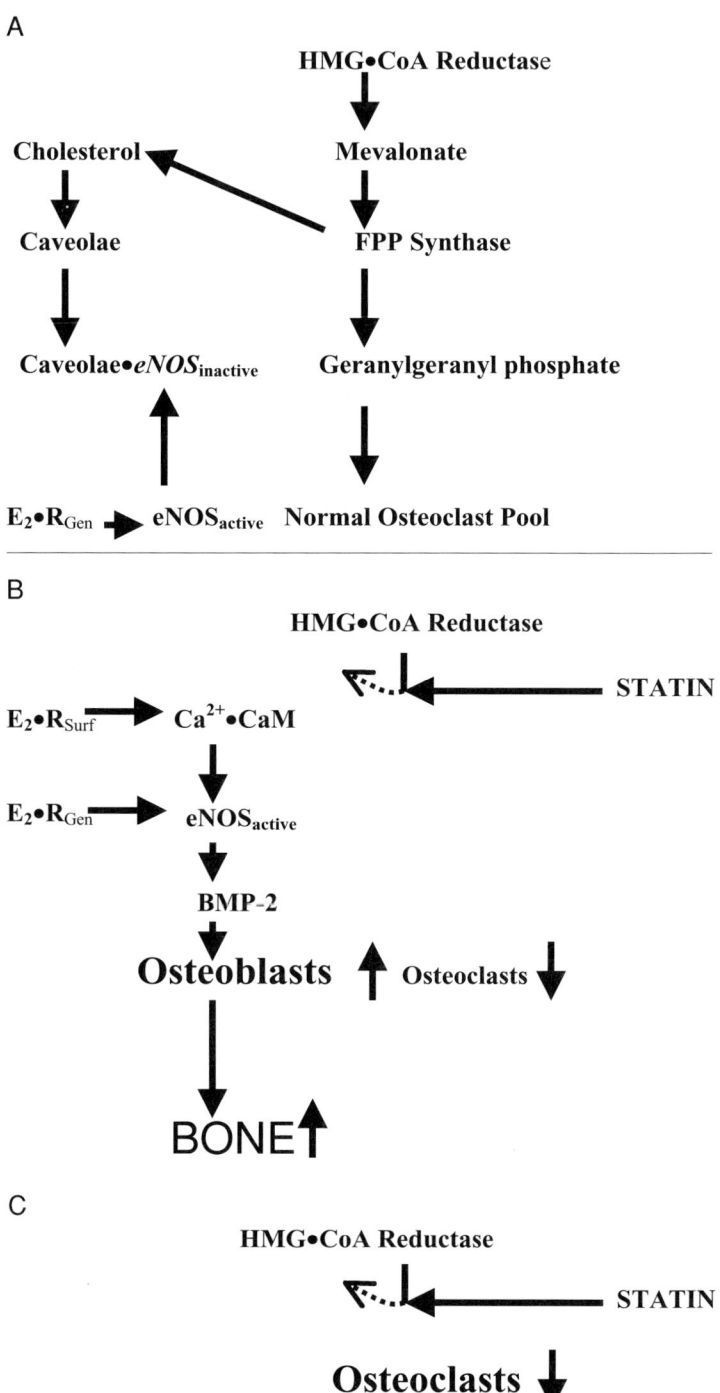

But there is a problem with eNOS being responsible for statin-induced BMP-2 expression and osteogenesis. While inhibiting eNOs in mouse calvarial cells with L-NAME prevents the osteogenic response to statins, the same does not seem to apply to human cells. Ohnaka *et al.* (2001) have reported that inhibiting rho-associated protein kinase, probably by preventing rho-GTPase's isoprenylation and tethering to the caveolar lipid raft mediated the pitavastatin-stimulated productions of BMP-2 and especially osteocalcin mRNAs in both normal human osteoblasts from surgically discarded bone chips and MG-63 human osteosarcoma cells. As would be expected if the stimulated gene expressions were the consequences of a failure of prenylation, it was prevented by giving the cells geranylgeranyl pyrophosphate. But the statin's actions were unaffected by inhibiting eNOS with L-NAME or, for that matter, by inhibitng various protein kinases such as PKCs, protein tyrosine kinases and MAP kinases. Mediation by rho-kinase was supported by its specific inhibitor, hydroxyfasudil (Nakamura *et al.*, 2001; Shimokawa *et al.*, 19990, also increasing BMP-2 and osteocalcin gene expressions.

The identity of the rho-kinase is unknown. But it seems that geranylgeranylated rho-protein kinase membrane plugged into a lipid-raft phosphorylates and inhibits a protein that otherwise would stimulate BMP-2 expression and osteoblastic maturation. A model for this mechanism has been found in the rabbit basilar artery (Nakamura *et al.*, 2001). Rho-kinase inhibits MLC(myosin light chain)phosphatase and thus promotes myosin light chain phosphorylation and the artery's contraction. But inhibiting the rho-kinase with hydroxyfasudil derepresses MLCphosphatase which dephosphorylates myosin light chain and causes the artery to relax.

C. HUMAN STUDIES

It would be reasonable to expect from some of the results of the mouse and rat studies that the lipophilic statins (except for the hydrophilic pravastatin) would be potent bone anabolic agents in osteoporotic humans. But as we shall see, anabolic effects of statins in postmenopausal women, like the effects in OVXed rodents, are not certain.

FIG. 11. A possible mechanism by which statins stimulate bone growth and kill osteoclasts. (A, B). By inhibiting HMG-CoA reductase, statins inhibit the production of the large amounts of cholesterol needed for caveolin synthesis and the formation of caveolae that sequester several signaling enzymes such as in an inactive state. Among these enzymes is constitutive or endothelial nitric oxide synthase (eNOS$_{inactive}$). The lack of sufficient cholesterol inhibits *caveolin*-I gene expression, which causes the dissolution of caveolae and the release of their Ca^{2+} · CaM (calmoduin)-activatable eNOS$_{active}$. The resulting NO surge produced by activated eNOS$_{active}$ stimulates BMP-2 expression and bone formation. The downstream lack of geranylgeranyl-phosphate also disables and kills osteoclasts. (C). Because eNOS expression is estrogen (E$_2$) dependent (via the genomic E$_2$R$_{gen}$), eNOS expression and the bone-anabolic action, but not the osteoclast-killing antiresorptive action, of statins are reduced in OVXed rats and presumably in postmenopausal women. E$_2$ · R$_{gen}$, gene-stimulating estrogen–genomic estrogen receptor complex; E$_2$ · R$_{Surf}$, surface-signaling estrogen–receptor complex.

1. Statins Reduce Fracturing

Bauer *et al.* (1999) reported the results of an analysis of the data from a study of osteoporotic fractures (8412 women 65 years and older) and a fracture intervention trial (6459 women aged 55 to 80 years). Most of the 599 statin users among the 14,871 women in the two studies had been given lovastatin. The relative risk of hip fracture was only 0.30 (95% CI 0.08–1.18) and the relative risk of nonspine fractures was 0.83 (95% CI 0.61–1.15).

Chan *et al.* (2000) reported that according to the records from six U.S. HMOs (health maintenance organizations) the odds ratio of fracturing for patients who had had more than 13 dispensings of statins was 0.48 (95% CI 0.27–0.83). Meier *et al.* (2000) found the same from 300 practices in the U.K.-based General Practice Research Database. Their base population was 91,611 patients who were at least 50 years old. Of these 28,340 were taking lipid-lowering drugs. The adjusted odds ratio for fracturing in statin users relative to controls was 0.55 (95% CI 0.44–0.69). Other lipid-lowering drugs did not affect the odds ratio. Wang *et al.* (2000) got similar results from the records of 6110 New Jersey residents aged 65 years or more. Statin users had a hip fracture risk of 0.54 (95% CI 0.36–0.82). Chung *et al.* (2000) examined the records of 69 type 2 diabetics, 33 of whom had not been given statins and 36 of whom had been given lovastatin, pravastatin (which would not have affected bone cells), and simvastatin. The spinal BMD rose 2.9% in the statin-treated group during 14 months. During the same time the BMDs in the different regions of the femur and in the total hip increased by a significant 0.88–2.32% in the statin-treated men but by only 1.8% in the statin-treated women. Finally, Chan *et al.* (2001) reported that giving simvastatin (20 mg/day for 4 weeks) to 17 hypercholesterolemic, nonosteoporotic patients increased one indicator of bone formation, the serum osteocalcin concentration. However, it did not increase another formation indicator, the bone-specific alkaline phosphatase activity.

It seems from these reports that the statins are indeed safe, effective bone anabolics that can reduce the risk of hip and spine fractures by as much as 71% depending on the extent of use. However, there are other reports that just as convincingly indicate that the statins do not stimulate bone growth or affect the fracture risk.

2. Statins Don't Reduce Fracturing

One of the problems with retrospectively assessed effects of giving statins on fracturing is that one statin, the popular hydrophilic pravastatin, does not affect bone cells (Garrett *et al.,* 2001). This explains the findings of Reid *et al.* (2001) on the effects of long-term pravastatin treatment on fracturing in 9014 patients with ischemic heart disease. There were 107 pravastatin-treated (40 mg/day) patients admitted to hospital for fractures compared with 101 in the placebo-treated group [fracture risk 1.05 (95% CI 0.80–1.16)]. When fractures not requiring hospital admission were also included, the fracture risk was 0.94 (95% CI 0.77–1.16).

LaCroix et al. (2000) searched the very large database in the Women's Health Initiative Observational Study (WHI-OS) and found no reducton of fracturing in statin (atorvastatin, lovastatin, pravastatin, fluvastatin)-treated women. In a study funded by Proctor and Gamble, van Staa et al. (2000) also found that cholesterol-lowering doses of atorvastatin, pravastatin, and simvastatin did not significantly lower the fracture risk. This was a much more extensive analysis of the U.K. General Practice Research database than that of Meier et al. which indicated that statins nearly halved the fracture risk. The conclusion of van Staa et al. that were was no reduction was based on a population of 216,062 fracture patients from 686 general practices, whereas the conclusion of Meier et al. was based on 91,611 fracture patients from only 300 practices. Sirola et al. (2002) and Solomon et al. (2001) also found no evidence in the clinical databases that statins significantly lower fracture risk. Finally, Cauley et al. (2000) used the WHI-OS database to find out whether statins could enhance the BMDs of postmenopausal women. They found that although statin treatment did not affect fracture risk, treatment with a high-potency (based on cholesterol-lowering ability) statin such as atorvastatin or simvastatin for more than 3 years modestly protected hip and lumber vertebral BMD. Edwards et al. (2000) also found that treatment of 41 women with various statins (27 received the high-potency simvastatin or atorvastatin) did not reduce fracturing, but there were 8.5–12% increases in the BMDs of hip and spine.

Cosman et al. (2001), like Chan et al. (2001), carried out a small, short-term experiment on 14 postmenopausal women (mean age of 58 years) designed to find out what a 12-week treatment with 0.4 mg of cerivastatin/day might do to bone instead of serum cholesterol levels. Bone formation markers (type I pro-collagen propeptide and osteocalcin) did not change, but the resorption markers (urinary N- and C-terminal telopeptides) did drop slightly ($<20\%$) within 6 weeks in the cerivastatin-treated group. They concluded that this statin did not detectably stimulate bone formation, but it might have had a modest bisphosphonate-like antiresorptive action.

D. STATINS' CLINICAL PROSPECTS

Some, but not all, groups have found statins to be potent bone anabolics for rats and mice and their anabolic potency for human bones is also uncertain. There could be two reasons for this uncertainty. First, the statins may simply be potent antiresorptives like the N-BP alendronate and owe their osteogenic effectiveness in young rodents to their potent osteoclast-killing abilities, which would leave the still-growing, nonremodeling bones to the osteoblasts. The situation could be different in mature humans with their tighly coupled osteoclast–osteoblast re-modeling teams: killing osteoclasts could reduce the recruitment of osteoblasts. Alternatively, data such as those in the large WHI-OS and the U.K.'s General Practice Research Database might be misleading because they come from patients treated for various lengths of time with different doses and with statins meant to reduce cholesterol rather than fracturing.

However there may be another explanation. As we previously learned, the anabolic actions of statins may be mediated through a burst of eNOS activity. If true, (but one must take note of Ohnaka *et al.* [2001]) the anabolic effectiveness of statins would be affected by the lack of estrogen in OVXed rodents and postmenopausal women simply because eNOS expression is estrogen dependent (Farhat *et al.,* 1996; Pavo *et al.,* 2000; Stefano and Peter, 2001; Tan *et al.,* 1999; Whitfield *et al.,* 2002). In other words, in an estrogen-depleted OVXed rat or postmenopausal woman, there may not be enough eNOS for an optimal anabolic, fracture-lowering response to statins, but they could still kill osteoclasts and produce an N-BP-like antiresorptive response (Fig. 11C).

Clearly there must be prospective placebo-controlled trials designed specifically to assess the osteogenicity rather than the lipid-lowering abilites of various doses of statins with reduced hepatoselectivity in postmenopausal patients. Moreover, the estrogen and eNOS statuses of the patients should be assessed along with BMD and fracturing. If topical delivery of a potent bone-anabolic statin turns out to be as feasible in humans as it has been in some rat experiments we will have a double-barreled osteoclast-killing/osteoblast-stimulating drug with the invaluable bonus of protecing the patient from a wide variety of other things ranging from Alzheimer's dementia to cardiovascular diseases. However, it may turn out that the statins require help from estrogens to be reliably anabolic for postmenopausal women, which would make them less attractive than the potently anabolic PTHs, which do not need estrogen to stimulate bone growth and may soon be orally deliverable.

VII. CONCLUSION

Our aging and now space-faring world needs drugs that can directly stimulate bone growth instead of merely stopping bone loss. As we have seen, there are now some promisingly potent anabolic drugs with the PTHs leading the way. One of the PTHs, the venerable hPTH-(1–34), has reached the end of its clinical trials and others such as Ostabolin C will soon be starting their trials. They are safe peptides that use two other candidate anabolic agents, FGF-2 and IGF-I, to mediate their potent osteogenicities. But they suffer from having to be injected subcutaneously for a couple of years interspersed with rest periods to grow new bone before the injections can be stopped and the patient switched over to an oral antiresorptive to protect the new bone. (However, the experience from the several trials indicates that autoinjection by osteoporotics is not a significant deterrent and it is now likely that hPTH-(1–31) can be delivered orally [Mehta *et al.,* 2001].) Close behind the PTHs may be topically applicable statins full of bone anabolic promise from experiments on mice and rats and with known beneficial cardiovascular and neurological side-effects in humans. Then just on the horizon is the hormone/cytokine leptin, which, unlike the PTHs is directly osteogenic with the twin abilities to stimulate osteoblasts and inhibit osteoclasts. Clearly we are at the cutting edge of bone anabolic drug discovery.

ACKNOWLEDGMENT

We thank Diane Candler of the NRC Press for her help in preparing the manuscript.

REFERENCES

Aarts, M. M., Rix, A., Guo, J., et al. (1999). The nucleolar targeting signal (NTS) of parathyroid hormone-related protein mediates endocytosis and nucleolar translocation. *J. Bone Miner. Res.* **14**, 1493–1503.
Adams, C., Mansfield, K., Perlot, R. L., et al. (2001). Matrix regulation of skeletal cell apoptosis: Role of calcium and phosphate atoms. *J. Biol. Chem.* **276**, 20316–20322.
Aguirre, J., Buttery, L., O'Shaughnessy, M., et al. (2001). Endothelial nitric oxide synthase gene-deficient mice demonstrate marked retardation in postnatal bone formation, reduced bone volume, and defects in osteoblast maturation and activity. *Am. J. Pathol.* **158**, 247–257.
Ahima, R. S., and Flier, J. S. (2000). Leptin. *Annu. Rev. Physiol.* **62**, 413–437.
Amalric, F., Baldin, V., Bosc-Bierne, I., et al. (1991). Nuclear translocation of basic fibroblastic growth factor. *Ann. N.Y. Acad. Sci.* **638**, 127–138.
Amin, D., Cornell, S. A., Gustafson, S. K., et al. (1992). Bisphosphonates used for the treatment of bone disorders inhibit squalene synthase and cholesterol biosynthesis. *J. Lipid Res.* **33**, 1657–1663.
Amizuka, N., Karaplis, A. C., Henderson, J. E., et al. (1996). Haploinsufficiency of parathyroid hormone related peptide (PTHrP) results in abnormal postnatal development. *Dev. Biol.* **175**, 166–176.
Amling, M., Neff, L., Tanaka, S., et al. (1997). Bcl-2 lies downstream of parathyroid hormone-related peptide in a signaling pathway that regulates chondrocyte maturation during skeletal development. *J. Cell Biol.* **136**, 205–213.
Anderson, R. G. W. (1998). The caveolar membrane system. *Annu. Rev. Biochem.* **67**, 199–225.
Andreassen, T. T., and Oxlund, H. (2000). The influence of combined parathyroid hormone and growth hormone treatment on cortical bone in aged ovariectomized rats. *J. Bone Miner. Res.* **15**, 2266–2275.
Andreassen, T. T., Ejersted, C., and Oxlund, H. (1999). Intermittent parathyroid hormone (1-34) treatment increases callus formation and mechanical strength of healing rat fractures. *J. Bone Miner. Res.* **14**, 960–968.
Andeassen, T. T., Willick, G. E., Morley, P., et al. (2001). Treatment with parathyroid hormone fragments increases fracture strength and enhances callus amount—Strength and callus amount normalize after treatment withdrawal (abstract). *J. Bone Miner. Res.* **16**, S425.
Antonsson, B., and Martinou, J. C. (2000). The Bcl-2 protein family. *Exp. Cell Res.* **256**, 50–57.
Armour, K. E., Armour, K. J., Gallagher, M. E., et al. (2001). Defective bone formation and anabolic response to exogenous estrogen in mice with targeted disruption of endothelial nitric oxide synthase. *Endocrinology* **142**, 760–766.
Ashby, M. C., and Tepikin, A. V. (2001). ER calcium and the functions of intracellular organelles. *Cell Dev. Biol.* **12**, 11–17.
Aubin, J. E. (1998). Advances in the osteoblast lineage. *Biochem. Cell Biol.* **76**, 899–910.
Aubin, J. E. (2000). The role of osteoblasts. In "The Osteoporosis Primer" (J. E. Henderson and D. Goltzman, Eds.), pp. 18–35. Cambridge University Press, Cambridge.
Aubin, J. E. (2001). Regulation of osteoblast formation and function. *Rev. Endocrine Metab. Dis.* **2**, 81–94.
Aubin, J. E., and Bonnelye, E. (2000). Osteoprotegerin and its ligand: A new paradigm for regulation of osteoclastogenesis and bone resorption. *Medscape Women's Health* **5**, http://www.medscape.com.
Aubin, J. E., and Triffitt, J. E. (2002). Mesenchymal stem cells and osteoblast differentiation. In "Principles of Bone Biology (2ed)", Vol. 1, pp. 59–91. San Diego, Academic Press.
Bandorowicz-Pikula, J., Buchet, R., and Pikula, S. (2001). Annexins as nucleotide-binding proteins: Facts and speculations. *BioEssays* **23**, 170–178.

Banerjee, C., Javed, A., Choi, J. Y., et al. (2001). Differential regulation of the two principal Runx2/Cbfa1 n-terminal isoforms in response to protein morphogenic protein-2 during development of the osteoblast phenotype. *Endocrinology* **142,** 4026–4039.

Barbier, J.-R., Neugebauer, W., Morley, P., et al. (1997). Bioactivities and secondary structures of constrained analogues of human parathyroid hormone: Cyclic lactams of the receptor-binding region. *J. Med. Chem.* **40,** 1373–1380.

Bassilana, F., Susa, M., Keller, H. J., et al. (2000). Human mesenchymal cells undergoing osteogenic differentiation express leptin and and functional leptin receptors (abstract). *J. Bone Miner. Res.* **15,** S378.

Bauer, D. C., Mundy, G. R., Jamal, S. A., et al. (1999). Satatin use, bone mass and fracture: An analysis of two prospective studies (abstract). *J. Bone Miner. Res.* **14,** S179.

Bauer, E., Aub, J. C., and Albright, F. (1929). Studies of calcium and phosphorus metabolism: V. Study of the bone trabeculae as a readily available reserve supply of calcium. *J. Exp. Med.* **49,** 145–162.

Baylink, D. J., Strong, D. D., and Mohan, S. (1999). The diagnosis and treatment of osteoporosis: Future prospects. *Mol. Med. Today* **5,** 133–140.

Beck, G. R., Jr., Zerler, B., and Moran, E. (2000). Phosphate is a specific signal for induction of osteopontin gene expression. *Proc. Natl. Acad. Sci. USA* **97,** 8352–8357.

Bekker, P. J., Holloway, D., Nakanishi, A., et al. (2001). The effect of a single dose of osteoprotegerin in postmenopausal women. *J. Bone Miner. Res.* **16,** 348–360.

Bellido, T., Plotkin, L. I., Davis, J., et al. (2001). Protein kinase A-dependent phosphorylation and inactivation of the pro-apoptotic protein Bad mediates the anti-apoptotic effect of PTH on osteoblastic cells (abstract). *J. Bone Miner Res.* **16,** S203.

Benford, H. L., Frith, J. C., Auriola, S., et al. (2001a). Farnesol and geranylgeraniol prevent activation of caspases by aminobisphosphonates: Biochemical evidence for two distinct pharmacological classes of bisphosphonate drugs. *Mol. Pharmacol.* **56,** 131–140.

Benford, H. L., McGowan, N. W., Helfrich, M. H., et al. (2001b). Visualization of bisphosphonate-induced caspase-3 activity in apoptotic osteoclasts *in vitro*. *Bone* **28,** 465–473.

Benten, W. P., Stephan, C., Lieberherr, M., et al. (2001). Estradiol signaling via sequestrable surface receptors. *Endocrinology* **142,** 1669–1677.

Bianco, P., and Riminucci, M. (1998). The bone marrow stroma *in vivo:* Ontogeny, structure, cellular composition and changes in disease. *In* "Marrow Stromal Cell Culture" (J. N. Beresford and M. E. Owen Eds.), pp.10–25. Cambridge University Press, Cambridge.

Bianco, P., Bradbeer, J. N., Riminucci, M., et al. (1993). Marrow stromal cells: Identification, morphometry, confocal imaging and changes in disease. *Bone* **14,** 315–320.

Bianco, P., Riminucci, M., Kuznetsov, S., et al. (1999). Multipotential cells in the bone marrow stroma: Regulation in the context of organ physiology. *Crit. Rev. Euk. Gene Express.* **10,** 159–173.

Bikle, D. D., Harris, J., Halloran, B. P., et al. (1994). Skeletal unloading induces resistance to insulin-like growth factor. *J. Bone Miner. Res.* **9,** 1789–1796.

Bockaert, J., Claeysen, S., Bôcamel, C., et al. (2002). G protein-coupled receptors: dominant players in cell-cell communication. *Int. Rev. Cytol.* **212,** 63–132.

Boivin, G. Y., Chavassieux, P. M., Santors, A. C., et al. (2000). Alendronate increases bone strength by increasing the mean degree of mineralization of bone tissue in osteoporotic women. *Bone* **27,** 687–694.

Bonadio, J. (2000). Tissue engineering via local gene delivery: Update and future prospects for enhancing the technology. *Adv. Drug Del. Rev.* **44,** 185–194.

Bonadio, J., Smiley, E., Path, P., et al. (1999). Localized, direct plasmid gene delivery *in vivo*: Prolonged therapy results in reproducible tissue regeneration. *Nat. Med.* **5,** 753–759.

Bouxsein, M. L. (1999). Etiology and biomechanics of hip and vertebral fractures. *In* "Atlas of Clinical Endocrinology, Vol. 3, Osteoporosis" (R. Marcus Ed.), pp. 139–148. Current Medicine, Philadelphia.

Bowler, W. H., Buckley, K. A., Gartland, A., et al. (2001). Extracellular nucleotide signaling: A mechanism for integrating local and systemic responses in the activation of bone remodeling. *Bone* **28,** 507–512.

Boyce, R. W., Paddock, C. L., Franks, A. F., *et al.* (1996). Effects of intermittent hPTH-(1–34) alone and in combination with 1,25(OH)$_2$D$_3$ or risedronate on endosteal bone remodeling in canine cancellous and cortical bone. *J. Bone Miner. Res.* **11,** 600–613.

Brandi, M. L., Crescioli, C., Tanini, A., *et al.* (1993). Bone endothelial cells as estrogen targets. *Calcif. Tissue Int.* **53,** 312–317.

Brann, D. W., De Sevilla, L., Zamorano, P. L., *et al.* (1999). Regulation of leptin gene expression and secretion by steroid hormones. *Steroids* **64,** 659–663.

Brommage, R., Hotchkiss, C. E., Lees, C. J., *et al.* (1999). Daily treatment with human recombinant parathyroid hormone-(1–34), LY33333, for 1 year increases bone mass in ovariectomized monkeys. *J. Clin. Endocrinol. Metab.* **84,** 3757–3763.

Brown, E. M., and MacLeod, R. J. (2001). Extracellular calcium sensing and extracellular calcium signaling. *Physiol. Rev.* **81,** 240–297.

Burguera, B., Hofbauer, L. C., Thomas, T., *et al.* (2001). Leptin reduces ovariectomy-induced bone loss in rats. *Endocrinology* **142,** 3546–3553.

Burr, D. B., Hirano, T., Turner, C. H., *et al.* (2001). Intermittently administered human parathyroid hormone (1-34) treatment increases intracortical bone turnover and porosity without reducing bone strength in the humerus of ovariectomize cynomolgus monkeys. *J. Bone Miner. Res.* **16,** 157–165.

Calvi, L. M., Sims, N. A., Hunzelman, J. L., *et al.* (2001). Activated parathyroid hormone/parathyroid hormone-related protein receptor in osteoblastic cells differentially affects cortical and trabecular bone. *J. Clin. Invest.* **107,** 277–286.

Candelicre, G. A., Liu, F., and Aubin, J. E. (2001). Individual osteoblasts in the developing calvaria express different gene repertoires. *Bone* **28,** 351–361.

Cann, C. E., Roe, E. B., Sanchez, S. D., *et al.* (1999). PTH effects in the femur: Envelope-specific responses by 3DQCT in postmenopausal women (abstract). *J. Bone. Miner. Res.* **14,** S137.

Capparelli, C., Morony, S., Warmington, K. S., *et al.* (2000). Pharmacological effects of a single treatment with osteoprotegerin (OPG) on bone remodeling and bone density in normal rats (abstract). *J. Bone Miner. Res.* **15,** S171.

Caverzasio, J., and Bonjour, J. R. (1996). Characteristics and regulation of Pi transport in osteogenic cells for bone metabolism. *Kidney Int.* **49,** 975–980.

Cenci, S., Weitzmann, M. N., Roggia, C. R., *et al.* (2000). Estrogen deficiency induces bone loss by enhancing T-cell production of TNF-α. *J. Clin. Invest.* **106,** 1229–1237.

Chambliss, K. L., Yuhanna, I. S., Mineo, C., *et al.* (2000). Estrogen receptor alpha and endothelial nitric oxide synthase are organized into functional signaling module in caveolae. *Circ. Res.* **87,** E44–E52.

Chan, K., Andrade, S. E., Boles, M., *et al.* (2000). Inhibitors of hydroxymethylglutaryl-coenzyme A reductase and risk of fracture among older women. *Lancet* **355,** 2185–2188.

Chan, M. H., Mak, T. W., Chiu, R. W., *et al.* (2001). Simvastatin increases serum osteocalcin concentration in patients treated for hypercholesterolaemia. *J. Clin. Endocrinol. Metab.* **86,** 4556–4559.

Chan, N. X., Ryder, K. D., Pavalko, F. M., *et al.* (2000). Ca^{2+} regulates fluid shear-induced cyto-skeletal reorganization and gene expression in osteoblasts. *Am. J. Physiol. Cell Physiol.* **278,** C989–997.

Chang, D. J., Ji, C., Kim, K. K., *et al.* (1998). Reduction in transforming growth factor beta receptor I expression and transcription factor CBFa1 on bone cells by glucocorticoid. *J. Biol. Chem.* **273,** 4892–4896.

Chang, W., Tu, C., Chen, T. H., *et al.* (1999). Expression and signal transduction of calcium-sensing receptors in cartilage and bone. *Endocrinology* **140,** 5883–5893.

Chao, D. T., and Korsmeyer, S. J. (1998). BCL-2 family: Regulators of cell death. *Annu. Rev. Immunol.* **16,** 395–419.

Chauvin, S., Vilardaga, J., Benecsik, M., *et al.* (2001). Parathyroid hormone receptor recycling: Regulation by specific structural features of the receptor (abstract). *J. Bone Miner. Res.* **16,** S228.

Chen, C., and Kalu, D. N. (1998). Modulation of intestinal estrogen receptor by ovariectomy, estrogen and growth hormone. *J. Pharmacol. Exp. Ther.* **286,** 328–333.

Chen, D., Ji, X., Harris, M. A., *et al.* (1998). Differential roles for bone morphogenic protein (BMP) receptor type IB and IA in differentiation and specification of mesenchymal precursor cells to osteoblast and adipocyte lineages. *J. Cell Biol.* **142,** 295–305.

Chen, D., Garrett, I. R., Qiao, M., *et al.* (2001). Proteasome inhibitors stimulate osteoblast differentiation and bone formation by inhibiting GLI3 degradation and enhancing BMP-2 expression (abstract). *Bone* **28,** S74.

Chow, J. W., Fox, S., Jagger, C. J., *et al.* (1998). Role of parathyroid hormone in mechanical responsiveness of bone. *Am. J. Physiol.* **274,** E146–E154.

Chu, S.-C., Chou, Y.-C., Liu, J.-Y., *et al.* (1999). Fluctuation of serum leptin level in rats after ovariectomy and the influence of estrogen supplement. *Life Sci.* **64,** 2299–2306.

Chung, U., Schipani, E., McMahon, A. P., *et al.* (2001). Indian hedgehog couples chondrogenesis to osteogenesis in endochondral bone development. *J. Clin. Invest.* **107,** 295–304.

Chung, Y. S., Lee, M. D., Lee, S. K., *et al.* (2000). HMG-CoA reductase inhibitors increase BMD in type 2 diabetes mellitus patients. *J. Clin. Endocrinol. Metab.* **85,** 1137–1142.

Cole, J. A. (1999). Parathyroid hormone activates mitogen-activated protein kinase in opossum kidney cells. *Endocrinology* **140,** 5771–5779.

Coppock, D., Kopman, C., Gudas, J., *et al.* (2000). Regulation of the quiescence-induced genes: quiescin Q6, decorin, and ribosomal protein S29. *Biochem. Biophys. Res. Commun.* **269,** 604–610.

Cornish, J., Callon, K. E., Nicholson, G. C., *et al.* (1997). Parathyroid hormone-related protein-(107-139) inhibits bone resorption *in vivo. Endocrinology* **138,** 1299–1304.

Cornish, J., Callon, K. E., Lin, C., *et al.* (1999). Stimulation of osteoblast proliferation by C-terminal fragments of parathyroid hormone-related protein. *J. Bone Miner. Res.* **14,** 915–922.

Cornish, J., Callon, K. E., Bava, U., *et al.* (2001). The direct actions of leptin on bone cells increase bone strength *in vivo*: An explanation of low fracture rates in obesity (abstract). *Bone* **28,** S88.

Cosman, F., and Lindsay, R. (1998a). Is parathyroid hormone a therapeutic option for osteoporosis? A review of the clinical evidence. *Calcif. Tissue Int.* **62,** 475–480.

Cosman, F., and Lindsay, R. (1998b). Parathyroid hormone as an anabolic treatment. *In* "Osteoporosis" (J. C. Stevenson and R. Lindsay, Eds.), pp. 293–307. Chapman & Hall, London.

Cosman, F., Nieves, J., Woelfert, L., *et al.* (1998). Alendronate does not block the anabolic effect of PTH in postmenopausal osteoporotic women. *J. Bone. Miner. Res.* **13,** 1051–1055.

Cosman, F., Nieves, J., Formica, L., *et al.* (2000). Parathyroid hormone in combination with estrogen dramatically reduces vertebral fracture risk (abstract). *Osteoporosis Int.* **11**(Suppl. 2) S176.

Cosman, F., Nieves, J., Woelfert, L., *et al.* (2001a). Parathyroid hormone added to established hormone therapy: Effects on vertebral fracture and maintenance of bone mass after parathyroid hormone withdrawal. *J. Bone Miner. Res.* **16,** 925–931.

Cosman, F., Nieves, J., Zion, M., *et al.* (2001b). Effects of short-term cerivastatin on bone turnover (abstract). *J. Bone Miner. Res.* **16,** S296.

Crawford, D. T., Qi, H., Chisey-Frink, K. L., *et al.* (2001). Statin increases cortical bone in young male rats by single, local administration but fails to restore bone to ovariectomized (OVX) rats by daily systemic administration (abstract). *J. Bone Miner. Res.* **16,** S295.

Crompton, M. (2000). The mitochondrial permeability transition pore and its role in cell death. *Biochem. J.* **341,** 233–249.

Cuthbertson, R. M., Kemp, B. E., and Barden, J. A. (1999). Structure of osteostatin PTHrP[Thr 107] (107-139). *Biochim. Biophys. Acta* **1432,** 64–72.

Daifotis, A. G., Weir, E. C., Dreyer, B. E., *et al.* (1992). Stretch-induced parathyroid hormone-related peptide gene expression in the rat uterus. *J. Biol. Chem.* **267,** 23455–23458.

Daub, H., Wallasch, C., Lankenau, A., *et al.* (1997). Signal characteristics of G protein-transactivated EGF receptor. *EMBO J.* **16,** 7032–7044.

David, P., Nguyen, H., Barbier, A., *et al.* (1996). The bisphosphonate tiludronate is a potent inhibitor of the osteoclast vacuolar H^+-ATPase. *J. Bone Miner. Res.* **11,** 1498–1507.

Delmas, P. D., Vergnaud, P., Arlot, M. E., *et al.* (1995). The anabolic effect of human PTH (1-34) on bone formation is blunted when bone resorption is inhibited by the bisphosphonate tiludronate—Is activated resorption a prerequisite for the *in vivo* effect of PTH on formation in a remodeling system? *Bone* **16,** 603–610.

Dempster, D. W. (1995). Bone remodeling. *In* "Osteoporosis: Etiology, Diagnosis, and Management," 2nd ed. (B. L. Riggs and L. J. Melton III Eds.), pp. 67–91. Lippincott-Raven, Philadelphia.

Dempster, D. W. (1997). Exploiting and bypassing the bone remodeling cycle to optimize the treatment of osteoporosis. *J. Bone Miner. Res.* **12,** 1152–1154.

Dempster, D. W. (2000). The contribution of trabecular architecture to cancellous bone quality. *J. Bone Miner. Res.* **15,** 20–23.

Dempster, D. W., Cosman, F., Parisien, M., *et al.* (1993). Anabolic action of parathyroid hormone. *Endocrine Rev.* **14,** 690–709.

Denhardt, D. T., and Noda, M. (1998). Osteopontin expression and function: Role in bone remodeling. *J. Cell. Biochem. Suppl.* **30–31,** 92–102.

DiMarzo, V., Goparaju, S. K., Wang, L., *et al.* (2001). Leptin-regulated endocannabinoids are involved in maintaining food intake. *Nature* **12,** 822–825.

Divieti, P., Inomata, N., Chapin, K., *et al.* (2001). Receptors for the carboxyl-terminal region of PTH(1–84) are highly expressed in ostocytic cells. *Endocrinology* **142,** 916–925.

Dobnig, H., and Turner, R. T. (1995). Evidence that intermittent treatment with parathyroid hormone increases bone formation in adult rats by activation of bone lining cells. *Endocrinology* **136,** 3632–3638.

Dobnig, H., and Turner, R. T. (1997). The effects of programmed administration of human parathyroid hormone fragment (1–34) on bone histomorphometry and serum chemistry in rats. *Endocrinology* **138,** 4607–4612.

Dobson, K. B., and Skerry, T. M. (2000). The NMDA-type glutamate receptor antagonist MK801 regulates the differentiation of rat bone marrow osteoprogenitors and influence adipogenesis (abstract). *J. Bone Miner. Res.* **15,** S272.

Dodig, M., Tadic, T., Kronenberg, M. S., *et al.* (1999). Ectopic Msx2 overexpression inhibits and Msx2 antisense stimulates calvarial osteoblast differentiation. *Dev. Biol.* **209,** 298–307.

Doherty, M. J., and Canfield, A. E. (1999). Gene expression during vascular pericyte differentiation. *Crit. Rev. Euk. Gene Express.* **9,** 1–17.

Doolan, C. M., and Condliffe, S. B., and Harvey, B. J. (2000). Rapid non-genomic activation of cytosolic cyclic AMP-dependent protein kinase activity and $[Ca^{2+}]_I$ by 17β-oestradiol in female rat distal colon. *Br. J. Pharmacol.* **129,** 1375–1386.

Ducy, P. (2000). Cbfa1: A molecular switch in osteoblast biology. *Dev. Dynam.* **219,** 461–471.

Ducy, P., Amling, M., Takeda, S., *et al.* (2000a). Leptin inhibits bone formation through a hypothalamic relay: A central control of bone mass. *Cell* **100,** 197–207.

Ducy, P., Schinke, T., and Karsenty, G. (2000b). The osteoblast: A sophisticated fibroblast under central surveillance. *Science* **289,** 1501–1504.

Duncan, R., and Misler, S. (1989). Voltage-activated and stretch-activated Ba^{2+}-conducting channels in an osteoblast–like cell line (UMR 106). *FEBS Lett.* **251,** 17–21.

Duncan, R. L., Hruska, K. A., and Misler, S. (1992). Parathyroid hormone activation of stretch-activated cation channels in osteosarcoma cells (UMR 106.01). *FEBS Lett.* **307,** 219–223.

Edlich, M., Yellowley, C. E., Jacobs, C. R., *et al.* (2001). Oscillating fluid flow regulates cytosolic calcium concentration in bovine articular chondrocytes. *J. Biomech.* **34,** 59–65.

Edwards, C. J., and Hart, D. J., and Spector, T. D. (2000). Oral status and increased bone mineral density in postmenopausal women. *Lancet* **355,** 2218–2219.

Einhorn, T. A. (1996). Biomechanics of bone. *In* "Principles of Bone Biology" (J. P. Bilzekian, L. G. Raisz, and G. A. Rodan Eds.), pp. 25–37. Academic Press, San Diego.

Enjuanes, A., Supervia, A., Noguѕ̈s, X., *et al.* (2002). Leptin receptor (OB-R) gene expression in human primary osteoblasts : confirmation. *J. Bone Miner. Res.* **17,** 1135.

Eriksen, E. F., Axelrod, D. W., and Melsen, F. (1994). "Bone Histomorphometry." Raven Press, New York.

Eriksen, E. F., Langdahl, B., and Klassen, M. (1994b). The cellular basis of osteoporosis. *Spine: State Arts Rev.* **8,** 23–62.

Esbrit, P., Alvarez-Arroyo, M. V., De Miguel, F., *et al.* (2000). C-terminal parathyroid hormone-related protein increases vascular endothelial growth factor in human osteoblastic cells. *J. Am. Soc. Nephrol.* **11,** 1085–1092.

Espinosa, L., Itzstein, C., Cheynel, H., *et al.* (1999). Active NMDA glutamate receptors are expressed by mammalian osteoclasts. *J. Physiol.* **518,** 47–53.

Evans, B. A. J., Elford, C., and Gregaory, J. W. (2001). Leptin control of bone metabolism (abstract). *Bone* **28,** S149.

Everts, V., Delaissϑ, J. M., Korper, W., Jansen, D., Tigchelaar-Gutter, W., Saftig, P., and Beertsen, W. (2002). The bone lining cell: Its role in cleaning Howship's lacunae and initiating bone formation, *J. Bone Miner. Res.* **17,** 77–90.

Fadeel, B., Zhivotovsky, B., and Orrenius, S. (1999). All along the watchtower: On the regulation of apoptosis regulators. *FASEB J.* **13,** 1647–1657.

Falkenstein, E., Tillmann, H.-C., Christ, M., *et al.* (2000). Multiple actions of steroid hormones—a focus on rapid, nongenomic effects. *Pharmacol Rev.* **52,** 513–555.

Fang, J., Zhu, Y. Y., Smiley, E., *et al.* (1996). Stimulation of new bone formation by direct transfer of osteogenic plasmid genes. *Proc. Natl. Acad. Sci. USA* **93,** 5753–5758.

Farhat, M. Y., Lavigne, M. C., and Ramwell, P. W. (1996). The vascular protective effects of estrogen. *FASEB J.* **10,** 615–624.

Farley, J. R., and Stilt-Coffing, B. (2001). Apoptosis may determine the release of skeletal alkaline phosphatase activity from human osteoblast-line *cells. Calcif. Tissue Int.,* published online 23 February.

Farzaneh-Far, A., Proudfoot, D., Weisberg, P. L., *et al.* (2000). Matrix gla protein is regulated by a mechanism functionally related to the calcium-sensing receptor. *Biochem. Biophys. Res. Commun.* **277,** 736–740.

Fassbender, K., Simons, M., Bergmann, C., *et al.* (2001). Simvaststin strongly reduces levels of Alzheimer's disease β amyloid peptides Aβ42 and Aβ40 *in vivo* and *in vitro. Proc. Natl. Acad. Sci. USA* **98,** 5856–5861.

Faulkner, K. (2000). Bone matters: Are density increases necessary to reduce fracture risk? *J. Bone Miner Res.* **15,** 183–187.

Fenton, A. J., Kemp, B. E., Hammonds, R. G., *et al.* (1991). A potent inhibitor of osteoclastic bone resorption within a highly conserved pentapeptide region of parathyroid hormone-related protein; PTHrP[107–111]. *Endocrinology* **129,** 3424–3426.

Fermore, B., and Skerry, T. M. (1995). PTH/PTHrP receptor expression in osteoblasts and osteocytes but not resorbing bone surfaces in growing rats. *J. Bone Miner. Res.* **10,** 1935–1943.

Feron, O., Dessy, C., Desager, J.-P., *et al.* (2001). Hydroxy-methylglutaryl-coenzyme A reductase inhibition promotes endothelial nitric oxide synthase activation through a decrease in caveolin abundance. *Circulation* **103,** 113–118.

Fielding, C. J., and Fielding, P. E. (2001). Caveolae and intracellular trafficking of cholesterol. *Adv. Drug Del. Rev.* **49,** 251–264.

Finkel, E. (2001). The mitochondrion: Is it central to apoptosis? *Science* **292,** 624–626.

Fiorelli, G., Gori, F., Frediani, U., *et al.* (1996). Membrane-binding sites and non-genomic effects of estrogen in cultured human preosteoclastic cells. *J. Steroid Biochem. Mol. Biol.* **59,** 233–240.

Fisher, J. E., Rogers, M. J., Halasay, J. M., *et al.* (1999). Alendronate mechanism of action: Geranylgeraniol, an intermediate in the mevalonate pathway, prevents inhibition of osteoclast formation, bone resorption, and kinase activation *in vitro. Proc. Natl. Acad. Sci. USA* **96,** 133–138.

Fleet, J. C. (2000). Leptin and bone: Does the brain control bone biology? *Nutr. Rev.* **58,** 209–211.

Fleisch, H. (1997). "Bisphosphonates in Bone Disease." The Parthenon Publishing Group, London.

Fleisch, H. (1998). Bisphosphonates: Mechanisms of action. *Endocrine Rev.* **19,** 80–100.

Fraher, L. J., Avram, R., Watson, P. H., *et al.* (1999). A comparison of the biochemical responses to 1-31 hPTH and 1-34 hPTH given to healthy humans by slow infusion. *J. Clin. Endocrinol. Metab.* **84,** 2739–2743.

Franceschi, R. T. (1999). The developmental control of osteoblast-specific gene expression: Role of specific transcription factors and the extracellular matrix environment. *Crit. Rev. Oral Biol. Med.* **10**, 40–57.

Friedman, J. M. (2000). Obesity in the new millennium. *Nature* **404**, 632–634.

Friedman, P. A., Gesek, F. A., Morley, P., *et al.* (1999). Cell-specific signaling and structure-activity relations of parathyroid analogs in mouse kidney cells. *Endocrinology* **140**, 301–309.

Frisch, S. M., and Screaton, R. A. (2001). Anoikis mechanisms. *Curr. Opin. Cell Biol.* **13**, 555–562.

Frost, H. M. (1997). Osteoporoses. *In* "Anabolic Treatments for Osteoporosis" (J. F. Whitfield and P. Morley Eds.), pp. 1–27. CRC Press, Boca Raton, FL.

Fu, Q., Jilka, R. L., Manolagas, S. C., *et al.* (2001). Stimulation of RANKL gene transcription and mRNA stability in stroma/osteoblastic cells by PTH: A direct effect mediated by the protein kinase pathway (abstract). *J. Bone Miner. Res.* **16**, S485.

Fujimori, A., Cheng, S. L., Avioli, L. V., *et al.* (1992). Structure-function relationship of parathyroid hormone: Activation of phospholipase-C, protein kinase-A and -C in osteosarcoma cells. *Endocrinology* **130**, 29–36.

Fujita, T., Inoue, T., Morii, H., *et al.* (1999). Effect of an intermittent weekly dose of human parathyroid hormone (1-34) on osteoporosis: A randomized double-masked prospective study using three dose levels. *Osteoporosis Int.* **9**, 296–306.

Fujita, T., Fukuyama, R., Izumo, N., *et al.* (2000). Enhanced Cbfa1 transactivation as a basic mechanism to trigger bone anabolic action of PTH(1-34) (abstract). *J. Bone Miner. Res.* **15**, S373.

Fujita, T., Fukuyama, R., Izumo, N., *et al.* (2001a). Transactivation of of core binding factor $\alpha 1$ as a basic mechanism to trigger parathyroid hormone-induced osteogenesis. *Jpn. J. Pharmacol* **86**, 405–416.

Fujita, T., Izumo, N., Fukuyama, R., *et al.* (2001b). Phosphate provides an extracelluar signal that drives nuclear export of Runx2/Cbfa1 in bone cells. *Biochem. Biophys. Res. Commun.* **280**, 348–352.

Fukumoto, S. (1998). Localization and function of calcium-sensing receptors in bone cells. *Nippon Rinsho* **56**, 1419–1424.

Gallien-Lartigue, O., and Carrez, D. (1974). Induction *in vitro* de la phase S dans les cellules souches multipotentes de la moelle osseuse par l'hormone parathroidienne. *C. R. Acad. Sci. Paris [D]* **278**, 1765–1768.

Galvin, R. J. S., Fuson, T. R., Yang, X., *et al.* (2001). hPTH (1-38) stimulation of bone resorption in murine calvariae is mediated in part by its ability to decrease OPG (abstract). *J. Bone Miner. Res.* **16**, S449.

Garrett, I. R., Gutierrez, G., Chen, D., *et al.* (2000). Specific inhibitors of the chymotryptic component of the proteasome are potent bone anabolic agents (abstract). *J. Bone Miner. Res.* **15**, S197.

Garrett, I. R., Chen, D., Zhao, M., *et al.* (2001a). Statins mediate bone formation by enhancing BMP-2 expression (abstract). *Bone* **28**, S75.

Garrett, I, R., Gutierrez, G., Chen, D., *et al.* (2001b). Statins stimulate bone formation by enhancing eNOS expression (abstract). *J. Bone Miner. Res.* **16**, S75.

Garrrett, I. R., Gutierrez, G., and Mundy, G. R. (2001c). Statins and bone formation. *Curr. Pharm. Design* **7**, 715–736.

Gasser, J. A. (1997). Quantitative assessment of bone mass and geometry by pQCT in rats *in vivo* and site specificity of changes at different skeletal sites. *J. Jpn. Soc. Bone Morphom.* **7**, 107–114.

Gasser, J. A. (2001). Fluvastatin and cerivastatin are notanabolic for bone after local or systemic administration of non-toxic doses in mice and rats (abstract). *J. Bone. Miner. Res.* **16**, S295.

Genant, H. K., Cann, C. E., Etting, B., *et al.* (1982). Quantitative computed tomography of vertebral spongiosa: A sensitive method for detecting early bone loss after oophorectomy. *Ann. Intern. Med.* **97**, 699–705.

Ghosh-Choudhury, N., Windle, J. J., Koop, B. A., *et al.* (1996). Immortalized murine osteoblasts derived from BMP2-T-antigen expressing transgenic mice. *Endocrinology* **137**, 331–339.

Gohel, A., and Gronowicz, G. (1997). Glucocorticoids induce apoptosis in osteoblasts in mice by the regulation of BCL-2 and other cell cycle factors (abstract). *J. Bone Miner. Res.* **12**, S284.

Goldstein, S. A., and Bonadio, J. (1998). Potential role for direct gene transfer in the enhancement of fracture healing. *Clin. Orthop.* **355**(Suppl.), S154–162.

Goltzman, D., and White, J. H. (2000). Developmental and tissue-specific regulation of parathyroid hormone (PTH)/PTH-related peptide receptor gene expression. *Crit. Rev. Euk. Gene Express.* **10**, 135–149.

Gori, F., Hofbauer, L. C., Dunstan, C. R., *et al.* (2000). The expression of osteoprotegerin and RANKL ligand and the support of osteoclast formation by stromal-osteoblast lineage cells is developmentally regulated. *Endocrinology* **141**, 4768–4776.

Gowen, M., Stroup, G. B., Dodds, R. A., *et al.* (2000). Antagonizing the parathyroid calcium receptor stimulates parathyroid hormone secretion and bone formation in osteopenic rats. *J. Clin. Invest.* **105**, 1595–1604.

Gray, C., Marie, H., Arora, M., *et al.* (2001). Glutamate does not play a major role in controlling bone growth. *J. Bone Miner. Res.* **16**, 742–749.

Greenfield, E. M. (1999). ODF/OPGL is increased by IL-1 and IL-6 (abstract). *J. Bone Miner Res.* **14**, S361.

Gu, Y., and Publicover, S. J. (2000). Expression of functional metabotropic glutamate receptors in primary cultured rat osteoblasts. *J. Biol. Chem.* **275**, 34252–34259.

Guaradvaccaro, D., Corrente, G., Covone, F., *et al.* (2000). Arrest of G1-S progression is Rb dependent and relies on the inhibition of of cyclin D1 transcription. *Mol. Cell. Biol.* **20**, 1797–1815.

Gurney, J. G., Severson, R. K., Davis, S., *et al.* (1995). Incidence of cancer in children in the United States. Sex-, race- and 1-year age-specific rates by histologic type. *Cancer* **75**, 2186–2195.

Gutierrez, G., Garrett, J. R., Rossini, G., *et al.* (2000). Dermal application of lovaststin to rats causes greater increases in bone formation and plasma concentration than when administered by oral gavage (abstract). *J. Bone Miner. Res.* **15**, S427.

Hagino, H., Okano, T., Enokida, M., *et al.* (2000). The effect of parathyroid hormone on cortical bone response to *in vivo* external loading (abstract). *J. Bone Miner. Res.* **15**, 806.

Harris, B. Z., and Lim, W. A. (2001). Mechanism and role of PDZ domains in signaling complex assembly. *J. Cell Sci.* **114**, 3219–3231.

Hauge, E. M., Qvesel, D., Eriksen, E. F., *et al.* (2001). Cancellous bone remodeling occurs in specialized compartments lined by cells expressing osteoblast markers. *J. Bone Miner. Res.* **16**, 1575–1582.

Hajoczky, G., Csordas, G., and Krishnamurthy, R. (2000). Mitochondrial calcium signaling driven by the IP_3 receptor. *J. Bioenerg. Biomembr.* **32**, 15–25.

Hall, R. A., Ostedgaard, L. S., Premont, R. T., *et al.* (1998). A C-terminal motif found in the β_2-adrenergic receptor, P2Y1 receptor and cyctic fibrosis transmembrane conductance regulator determines binding to the Na^+/H^+ exchanger regulatory factor family of PDZ proteins. *Proc. Natl. Acad. Sci. USA* **95**, 8496–8501.

Herrington, D. M., and Klein, K. P. (2001). Statins, hormones, and women: Benefits and drawbacks for athersclerosis and osteoporosis. *Curr. Atherosclerosis Rep.* **3**, 35–42.

Hershko, A., Ciechanover, A., and Varshavsky, A. (2000). The ubiquitin system. *Nat. Med.* **6**, 1073–1081.

Hesch, R. D., Busch, U., Prokop, M., *et al.* (1989). Increase of vertebral density by combination therapy with pulsatile 1-38 PTH and sequential addition of calcitonin nasal spray in osteoporotic patients. *Calcif. Tissue Int.* **44**, 176–180.

Hill, P. A., Tumber, A., and Meikle, M. C. (1997). Multiple extracellular signals promote osteoblast survival and apoptosis. *Endocrinology* **138**, 3849–3858.

Hinson, T. K., Damodaran, T. V., Chen, J., *et al.* (1997). Identification of putative trans-membrane receptor sequences homologous to the calcium-sensing G-protein-coupled receptor. *Genomics* **45**, 279–289.

Hirano, T., Burr, D. B., Cain, R. L., *et al.* (2000). Changes in geometry and cortical porosity in adult, ovary-intact rabbits after 5 months treatment with LY333334 (hPTH 1-34). *Calcif. Tissue Int.* **66**, 456–460.

Hoare, S. R. J., and Usdin, T. B. (2001). Molecular mechanisms of ligand recognition by parathyroid hormone 1 (PTH1) and PTH2 receptors. *Curr. Pharm. Design.* **7**, 689–713.

Hoare, R. J., Gardella, T. J., and Usdin, T. B. (2000). Evaluating the signal transduction mechanism of the parathyroid hormone 1 receptor: Effect of receptor G-protein interaction on the ligand binding mechanism and receptor conformation. *J. Biol. Chem.* Online, December 6.

Hock, J. M. (1999). Stemming bone loss by suppressing apoptosis. *J. Clin. Invest.* **104**, 371–373.

Hodsman, A. B., and Steer, B. M. (1993). Early histomorphometric changes in response to parathyroid hormone therapy in osteoporosis: Evidence for de novo bone formation on quiescent cancellous surfaces. *Bone* **14**, 523–527.

Hodsman, A. B., Fraher, L. J., and Watson, P. H. (1997a). Parathyroid hormone: The clinical experience and prospects. *In* "Anabolic Treatments for Osteoporosis" (J. F. Whitfield and P. Morley Eds.), pp. 83–108. CRC Press, Boca Raton, FL.

Hodsman, A. B., Fraher, L. J., Watson, P. H., *et al.* (1997b). A randomized controlled trial to compare the efficacy of cyclical parathyroid hormone versus cyclical parathyroid hormone and sequential calcitonin to improve bone mass in post-menopausal women with osteoporsis. *J. Clin. Endocrinol. Metab.* **82**, 620–628.

Hodsman, A. B., Drost, D., Fraher, L. J., *et al.* (1999a). The addition of raloxifene analog (LY117018) allows for reduced PTH(1-34) dosing during reversal of osteopenia in ovariectomized rats. *J. Bone Miner. Res.* **14**, 675–679.

Hodsman, A. B., Watson, P. H., Drost, D., *et al.* (1999b). Assessment of maintenance therapy with reduced doses of PTH(1-34) in combination with a raloxifene analogue (LY117018) following anabolic therapy in the ovariectomized rat. *Bone* **24**, 451–455.

Hodsman, A. B., Watson, P. H., Fraher, L. J., *et al.* (2000). Increased bone turnover without change in cortical thickness or porosity after 2 years of cyclical PTH therapy for postmenopausal osteoporosis (abstract). *J. Bone Miner. Res.* **15**, 799.

Hofbauer, L. C., Khosla, S., Dunstan, C. R., *et al.* (1999). Estrogen stimulates gene expression and protein production of osteoprotegerin in human osteoblastic cells. *Endocrinology* **140**, 4367–4370.

Hofbauer, L. C., Khoslab, S., Dunstan, C. R., *et al.* (2000). The roles of osteoprotegerin and osteoprotegerin ligand in the paracrine regulation of bone resorption. *J. Bone Miner. Res.* **15**, 2–12.

Honda, Y., Fitzsimmons, R., Baylink, D. J., *et al.* (1995). Effects of extracellular calcium on insulin-like growth factor II in human bone cells. *J. Bone Miner. Res.* **10**, 1660–1665.

Hurley, M. M., Marcello, K., Abreu, C., *et al.* (1995). Transcriptional regulation of the collagenase gene by basic fibroblast growth factor in osteoblastic MC3T3-E1 cells. *Biochem. Biophys. Res. Commun.* **214**, 331–339.

Hurley, M. M., Tetradis, S., Huang, Y.-F., *et al.* (1999). Parathyroid hormone regulates the expression of fibroblast growth factor-2 mRNA and fibroblast growth factor receptor mRNA in osteoblastic cells. *J. Bone Miner. Res.* **14**, 776–783.

Iacopetti, P., Barsacchi, G., Tirone, F., *et al.* (1994). Developmental expression of PC3 gene is correlated with neuronal cell birthday. *Mech. Devel.* **47**, 127–137.

Iacopetti, P., Michelini, M., Stuckmann, I., *et al.* (1999). Expression of the anti[rpliferative gene TIS21 at the onset of neurogenesis identifies single neuroepithelial cells that switch from proliferative to neuron-generating division. *Proc. Natl. Acad. Sci. USA* **96**, 4639–4644.

Ito, Y. (1999). Molecular basis of tissue-specific gene expression mediated by the Runt domain transcription factor PEBP2/CBF. *Genes Cells* **4**, 685–696.

Itzstein, C., Espinosa, L., Delmas, P. D., *et al.* (2000). Specific agonists of NMDA receptors prevent osteoclasts sealing zone formation required for bone resorption. *Biochem. Biophys. Res. Commun.* **268**, 201–209.

Iwamoto, R., and Mekada, E. (2000). Heparin-binding EGF-like growth factor: A juxtacrine growth factor. *Cytokine Growth Factor Rev.* **11**, 335–344.

Jans, D. A., and Hassan, G. (1998). Nuclear targeting by growth factors, cytokines, and their receptor: A role in signaling? *BioEssays.* **29**, 400–411.

Jerome, C. P., Carlson, C., S., Register, T. C., *et al.* (1994). Bone functional changes in intact, ovariectomized, and ovariectomized, hormone-supplemented adult cynomolgus monkeys (*Macaca fascicularis*) evaluated by serum markers and dynamic histomorphometry. *J. Bone Miner. Res.* **9**, 527–540.

Jerome, C. P., Johnson, C. S., Vafai, H. T., *et al.* (1999). Effect of treatment for 6 months with human parathyroid hormone (1-34) peptide in ovariectomized cynomolgus monkeys (*Macaca fascicularis*). *Bone* **25**, 301–309.

Jerome, C. P., Burr, D. B., Van Bibber T., *et al.* (2001). Treatment with human parathyroid hormone (1-34) for 18 months increases cancellous bone volume and improves trabecular architecture in ovariectomized cynomolgus monkeys (*Macaca fascicularis*). *Bone* **28**, 150–159.

Ji, X., Chen, D., Xu, C., *et al.* (2000). Patterns of gene expression associated with BMP-2-induced osteoblasts and adipocyte differentiation of mesenchymal progenitor cell 3T3-F442A. *J. Bone Miner. Res.* **15**, 132–139.

Jick, H., Zornberg, G. L., Jick, S. S., *et al.* (2000). Stains and the risk of dementia. *Lancet* **356**, 1627–1631.

Jilka, R. L. (1998). Cytokines, bone remodeling, and estrogen deficiency. *Bone* **23**, 75–78.

Jilka, R. L., Weinstein, R. S., Bellido, T., *et al.* (1999). Increase bone formation by prevention of osteoblast apoptosis with parathyroid hormone. *J. Clin. Invest.* **104**, 439–446.

Johansson, A., Rosen, C. J., and *et al.* (1997). The insulin-like growth factors: Potential anabolic agents for the skeleton. *In* "Anabolic Treatments for Osteoporosis" (J. F. Whitfield and P. Morley Eds.) pp.185–205. CRC Press, Boca Raton, FL.

Jouishomme, H., Whitfield, J. F., Chakravarthy, B., *et al.* (1992). The protein kinase-C activation domain of the parathyroid hormone. *Endocrinology* **130**, 53–60.

Jouishomme, H., Whitfield, J. F., Gagnon, L., *et al.* (1994). Further definition of the protein kinase C activation domain of the parathyroid hormone. *J. Bone Miner. Res.* **9**, 943–949.

Kalu, D. N., Doyle, F. H., Pennock, J., *et al.* (1970). Parathyroid hormone and experimental osteosclerosis. *Lancet* **1**, 1363–1366.

Kameda, T., Mano, H., Yameda, Y., *et al.* (1998). Calcium-sensing receptors in mature osteoclasts which are bone-resorbing cells. *Biochem. Biophys. Res. Commun.* **245**, 419–422.

Kanatami, M., Sugimoto, T., Takahashi, Y., *et al.* (1998). Estrogen via the estrogen receptor blocks cAMP-mediated parathyroid hormone (PTH)-stimulated osteoclast function. *J. Bone Miner Res.* **13**, 854–862.

Kanazawa, M., Sugimoto, T., Kanatami, M., *et al.* (2000). Involvement of osteoprotegerin/ osteoclastogenesis inhibitory factor in the stimulation of osteoclast formation by parathyroid hormone in mouse bone cells. *Eur. J. Endocrinol.* **142**, 661–664.

Karperien, M., Farih-Sips, H., Papapoulos, S. E., *et al.* (2000). Involvement of Cbfa1 in transcritional regulation of the type I PTH/PTHrP-receptor in KS483 osteoblasts (abstract). *J. Bone Miner. Res.* **15**, S175.

Karsenty, G. (1999). The genetic transformation of bone biology. *Genes Dev.* **13**, 3037–3051.

Karsenty, G. (2000a). Role of Cbfa1 in osteoblast differentiation and function. *Semin. Cell. Dev. Biol.* **11**, 343–346.

Karsenty, G. (2000b). The central regulation of bone remodeling. *Trends Endocrinol. Metab.* **11**, 437–439.

Karsenty, G. (2001). Chondrogenesis just ain't what it used to be. *J. Clin. Invest.* **107**, 405–407.

Kartsogiannis, V., Moseley, J., McKelvie, B., *et al.* (1997). Temporal expression of PTHrP during endochondral bone formation in mouse and intramembranous bone formation in an *in vivo* rabbit model. *Bone* **21**, 385–392.

Kawakami, A., Eguchi, K., Matsuoka, N., *et al.* (1997). Fas and Fas ligand interaction is necessary for human osteoblast apoptosis. *J. Bone Miner. Res.* **12**, 1637–1646.

Kawakami, A., Nakashima, T., Tsuboi, M., *et al.* (1998). Insulin-like growth factor I stimulates proliferation and Fas-mediated apoptosis of human osteoblasts. *Biochem. Biophys. Res. Commun.* **247**, 46–51.

Keenan, S. M., and Baldassare, J. J. (2001). Molecular scaffold protein and cellular responses. *Trends Endocrinol. Metab.* **12**, 184–186.

Kelly, M. J., and Levin, E. R. (2001). Rapid actions of plasma membrane estrogen receptors. *Trends Endocrinol. Metab.* **12**, 152–156.

Kern, B., Shen, J., Starbuck, M., *et al.* (2000). Cbfa1 contributes to the osteoblast-specific expression of type I collagen genes. *J. Biol. Chem.* Published online December 5.

Kim, H. P., Lee, J. Y., Jeong, J. K., *et al.* (1999). Nongenomic stimulation of nitric oxide release by estrogen is mediated by estrogen receptor alpha localized in caveolae. *Biochem. Biophys. Res. Commun.* **263,** 257–262.

Kim, H. W., and Jahng, J. S. (1999). Effect of intermittent administration of parathyroid hormone on fracture healing in ovariectomized rats. *Iowa Orthop J.* **19,** 71–77.

Kinoshita, T., Kobayashi, S., Ebara, S., *et al.* (2000). Phosphodiesterase inhibitors, pentoxyfylline and rolipram, increase bone mass mainly by promoting bone formation in normal mice. *Bone* **27,** 811–817.

Klein, R. F. (1999). Osteoporosis in men. In "Atlas of Clincal Endocrinology," Vol. 3, Osteoporosis. (R. Marcus Ed.), pp. 85–99. Current Medicine, Philadelphia.

Kneissel, M., Boyde, A., and Gasser, J. A. (2001). Bone tissue and its mineralization in aged estrogen-depleted rats after long-term intermittent treatment with parathyroid hormone (PTH) analog SDZ PTS 893 or human PTH-(1-34). *Bone* **28,** 237–250.

Kobayashi, K., Takahashi, N., Jimi, E., *et al.* (2000). Tumor necrosis factor alpha stimulates osteoclast differentiation by a mechanism independent of the ODF/RANKL-RANK interaction. *J. Exp. Med.* **191,** 275–286.

Komori, T. (2000). A fundamental transcription factor for bone and cartilage. *Biochem. Biophys. Res. Commun.* **276,** 813–816.

Kong, Y.-Y., Felge, U., Sarosi, I., *et al.* (1999). Activated T cells regulate bone loss and joint destruction in adjuvant arthritis through osteoprotegerin ligand. *Nature* **402,** 304–309.

Kostenuik, P. J., and Shalhoub, V. (2001). Osteoprotegerin: A physiological and pharmacological inhibitor of bone resorption. *Curr. Pharm. Design.* **7,** 613–635.

Kostenuik, P. J., Harris, J., Halloran, B. P., *et al.* (1999). Skeletal unloading causes resistance of osteoprogenitor cells to parathyroid hormone and to insulin-like growth factor-I. *J. Bone Miner. Res.* **14,** 21–31.

Kostenuik, P. J., Capparelli, C., Morony, S., *et al.* (2001). OPG and PTH-(1-34) have additive effects on bone density and mechanical strength in osteopenic ovariectomized rats. *Endocrinology* **142,** 4295–4304.

Krishnan, V., Heat, H., and Bryant, H. U. (2001). Mechanism of action of estrogens and selective estrogen receptor modulators. *Vitamins Hormones* **60,** 123–147.

Kroll, M. H. (2000). Parathyroid hormone temporal effects on bone formation and resorption. *Bull. Math. Biol.* **62,** 163–187.

Kufahl, R. H., and Saha, S. (1990). A theoretical model for stress-generated fluid flow in the canaliculi-lacunae network in bone tissue. *J. Biomech.* **23,** 171–180.

Kurland, E. H., Cosman, F., McMahon, D. J., *et al.* (2000). Parathyroid hormone as a therapy for idiopathic osteoporosis in men: Effects on bone mineral density and bone markers. *J. Clin. Endocrinol. Metab.* **85,** 3069–3076.

LaCroix, A. Z., Cauley, J. A., Jackson, R., *et al.* (2000). Does statin use reduce the risk of fracture in postmenopausal women? Results from the Women's Health Initiative Observational Study (WHI-OS) (abstract). *J. Bone Miner. Res.* **15,** S155.

Laharrague, P., Larrouy, D., Fontanilles, A. M., *et al.* (1998). High expression of leptin by human bone marrow adipocytes in primary culture. *FASEB J.* **12,** 747–752.

Laketic-Ljubojevic, I., Suva, L. J., *et al.* (1999). Functional characterization of the N-methyl-D-aspartic acid-gated channels in bone. *Bone* **25,** 631–637.

Lam, M. H., House, C. M., Tiganis, T., *et al.* (1999). Phosphorylation at the cyclin-dependent kinase site (Thr 85) of parathyroid hormone-related protein negatively regulates its nuclear localization. *J. Biol. Chem.* **274,** 18559–18566.

Lam, M. H. C., Thomas, R. J., Martin, T. J., *et al.* (2000). Nuclear and nucleolar localization of parathyroid hormone-related protein. *Immunol. Cell Biol.* **78,** 395–402.

Lane, N. E., Thompson, J. M., Strewler, G. J., *et al.* (1995). Intermittent treatment with parathyroid hormone (hPTH 1-34) increased trabecular bone volume but not connectivity in osteopenic rats. *J. Bone Miner. Res.* **10,** 1470–1477.

Lane, N. E., Sanchez, S., Modin, G. W., *et al.* (2000). Bone mass continues to increase at the hip after parathyroid hormone treatment is discontinued in glucocorticoid-induced osteoporosis: Results of a randomized controlled clinical trial. *J. Bone Miner. Res.* **15,** 944–951.

Langub, M. C., Monier-Faugere, M.-C., Qi, Q., *et al.* (2001). Parathyroid hormone/parathyroid hormone-related peptide type 1 receptor in human bone. *J. Bone Miner. Res.* **16,** 448–456.

Lanske, B., Amling, M., Neff, L., *et al.* (1999). Ablation of the PTHrP gene or the PTH/PTHrP receptor gene leads to distinct abnormalities in bone development. *J. Clin. Invest.* **104,** 399–407.

Leaffer, D., Sweeny, M., Kellerman, L. A., *et al.* (1995). Modulation of osteogenic cell ultrastructure by RS-23581, an analog of human parathyroid hormone (PTH)-related peptide-(1-34), and ovine PTH-(1-34). *Endocrinology* **136,** 3624–3631.

Levin, E. R. (2000). Nuclear receptor versus plasma membrane oestrogen receptor. *Novartis Found. Symp.* **230,** 41–55.

Liang, H., Pun, S., and Wronski, T. J. (1999). Bone anabolic effects of basic fibroblast growth factor in ovariectomized rats. *Endocrinology* **140,** 5780–5788.

Liang, J. D., Hock, J. M., Sandusky, G. E., *et al.* (1999). Immunohistochemical localization of selected early response genes expressed in trabecular bone of young rats given PTH 1-34. *Calcif. Tissue Int.* **65,** 369–373.

Lieberherr, M., Grosse, B., Kachkache, M., *et al.* (1993). Cell signaling and estrogens in female rat osteoblasts: A possible involvement of unconventional non-nuclear receptors. *J. Bone Miner. Res.* **8,** 1365–1376.

Lin, S.-Y., Makino, K., Xia, W., *et al.* (2001). Nuclear localization of EGF receptor and its potential new role as a transcription factor. *Nat. Cell Biol.* **3,** 802–808.

Lindsay, R., Nieves, J., Formica, C., *et al.* (1997). Randomised controlled study of effect of parathyroid hormone on vertebral bone mass and fracture incidence among postmenopausal women on oestrogen with osteoporosis. *Lancet* **550,** 555.

Lindsay, R., Hodsman, A., Genant, H., *et al.* (1998). A randomized controlled multicenter study of 1–84hPTH for treatment of postmenopausal osteoporosis (abstract). *Bone* **23,** S175.

Lorget, F., Kamel, S., Mentaverri, R., *et al.* (2000). High extracellular calcium concentrations directly stimulate osteoclast apoptosis. *Biochem. Biophys. Res. Commun.* **268,** 899–903.

Lu, S. S., Ducayen-Knowles, M., Dempster, D. W., *et al.* (2001). Effects of parathyroid hormone on gene expression of RANK ligand (RANKL), osteoprotegerin (OPG) and the cognate receptor for PTH in mice. *J. Bone Miner. Res.* **16,** S426.

Luttrell, L. M., Roudabush, F. L., Choy, E. W., *et al.* (2001). Activation and targeting of extracellular signal-regulated kinases by beta-arrestin scaffolds. *Proc. Natl. Acad. Sci. USA.* **98,** 2449–2454.

Ma, Y., Jee, W., Yuan, Z., *et al.* (1999). Parathyroid hormone and mechanical usage have a synergistic effect in rat tibial diaphyseal cortical bone. *J. Bone Miner. Res.* **14,** 439–448.

Ma, Y. L., Zeng, Q. Q., Cain, R. L., *et al.* (2000). PTH induces similar anabolic effects in lumbar vertebrae of adult ovary-intact and osteopenic ovariectomized rats (abstract). *J. Bone Miner. Res.* **15,** 813.

Machwate, M., Rodan, S. B., Rodan, G. A., *et al.* (1998). Sphingosine kinase mediates cyclic AMP suppression of apoptosis in rat periosteal cells. *Mol. Pharmacol.* **54,** 70–77.

Maeda, T., Matsunuma, A., Kawane, T., *et al.* (2001). Simvastatin promotes osteoblast differentiation and mineraliuzation in MC3T3 cells. *Biochem. Biophys. Res. Commun.* **280,** 874–877.

Mahon, M. J., Donowitz, M., Yun, C. C., *et al.* (2002). Na+/H+ exchanger regulatory factor 2 directs parathyroid hormone 1 receptor signaling. *Nature.* **417,** 858–861.

Makhluf, H. A., Mueller, S. M., Mizuno, S., *et al.* (2000). Age-related decline in osteoprotegerin expression expression by human bone marrow cells cultured in three-dimensional collagen sponges. *Biochem. Biophys. Res. Commun.* **268,** 669–672.

Mannella, C. A. (2000). Our changing views of *mitochondria*. *J. Bioener. Biomembr.* **32,** 1–4.
Manolagas, S. C. (1999). Advances in the treatment of osteoporosis. *Medscape Endocrinol. J.* **1,** http://www.medscape.com.
Manolagas, S. C. (2000). Birth and death of bone cells: Basic regulatory mechanisms and implications for the pathogenesis and treatment of osteoporosis. *Endocrine Rev.* **21,** 115–137.
Manolagas, S. C., Weinstein, R. S., Bellido, T., *et al.* (1999). Activators of non-genomic estrogen-like signaling (ANGELS): A novel class of small molecules with bone anabolic properties (abstract). *J. Bone Miner. Res.* **14,** S180.
Mansfield, K., Teixeira, C. C., Adams, C. S., *et al.* (2001). Phosphate ions mediate chondrocyte apoptosis through a plasma membrane transporter mechanism. *Bone* **28,** 1–8.
Maor, G. A., Rochwerger, M., Segev, Y., *et al.* (2002). Leptin acts as a skeletal growth factor on chondrocytes of skeletal growth centers. *J. Bone Miner. Res.* **17,** 1034–1043.
Marenholz, I., Zirra, M., Fischer, D. F., *et al.* (2001). Identification of human epidermal differentiation complex (EDC)-encoded genes by subtractive hybridization of entire YACs to a gridded keratinocyte cDNA library. *Genome Res.* **11,** 341–355.
Marie, P. J., Jones, D., Vico, L., *et al.* (2000). Osteobiology, strain, and microgravity: Part 1. Studies at the cellular level. *Calcif. Tissue Int.* **67,** 2–9.
Martin, R. B. (2000). Toward a unifying theory of bone remodeling. *Bone* **26,** 1–6.
Martin, R. B. (2002). Is all cortical bone remodeling initiated by microdamage? *Bone* **30,** 8–13.
Martin, T. J., Romas, E., and Gillespie, M. T. (1998). Interleukins in the control of osteoclast differentiation. *Crit. Rev. Euk. Gene Express.* **8,** 107–123.
Mason, D. J., Suva, L. J., Genever, P. G., *et al.* (1997). Mechanically regulated expression of a neural glutamate transporter in bone: A role for excitatory amino acids as osteotropic agents? *Bone* **20,** 199–205.
McCauley, L. K., Koh, A. J., Beecher, C. A., *et al.* (1996). PTH/PTHrP receptor is temporally regulated during osteoblasts differentiation and is associated with collagen synthesis. *J. Cell. Biochem.* **61,** 638–647.
McKee, M. D., and Nanci, A. (1996). Osteopontin: An interfacial extracellular matrix protein in mineralized tissues. *Connect. Tissue Res.* **35,** 197–205.
Mehta, N., Stern, W., Sturmer, A., *et al.* (2001). Oral delivery of PTH analogs by a solid dosage formulation (abstract). *J. Bone Miner. Res.* **16,** S540.
Meir, C. R., Schlienger, R. C., Kraenzlin, M. E., *et al.* (2000). HMG-CoA reductase inhibitors and the risk of fractures. *JAMA* **283,** 3205–3210.
Meleti, Z., Shapiro, I. M., and Adams, C. S. (2000). Inorganic phosphate induces apoptosis of osteoblast-like cells in culture. *Bone* **27,** 359–366.
Meunier, P. J. (1999). Evidence-based medicine and osteoporosis: A comparsion of fracture risk reduction data from osteoporosis randomized clinical trials. *Int. J. Clin. Pract.* **53,** 122–129.
Miller, W. E., McDonald, P. H., Cai, S. F., *et al.* (2001). Identification of a motif in the carboxyl terminus of β-arrestin 2 responsible for activation of JNK3. *J. Biol. Chem.* **276,** 27770–27777.
Moghadasian, M. H., and Frohlich, J. J. (2001). Statins and bones. *Can. Med. Assoc. J.* **164,** 803–805.
Mohan, S., and Baylink, D. J. (1999). IGF system components and their role in bone metabolism. *In* "The IGF System. Molecular Biology, Physiology, and Clincal Applications" (R. G. Rosenfeld and C. T. Roberts, Jr., Eds.), pp. 457–496. Humana Press, Totowa, NJ.
Mohan, S., Kutilek, S., Zhang, C., *et al.* (2000). Comparison of bone formation responses to parathyroid hormone(1-34), (1-31), and 2-34 in mice. *Bone* **27,** 471–478.
Montero, A., Okada, Y., Tomita, M., *et al.* (2000). Disruption of the fibroblast growth factor-2 gene results in decreased bone mass and bone formation. *J. Clin. Invest.* **105,** 1085–1093.
Moonga, B. S., and Dempster, D. W. (1998). Effects of peptide fragments of protein kinase C on isolated rat osteoclasts. *Exp. Physiol.* **83,** 717–725.
Moonga, B. S., Sun, L., Corisdeo, S., *et al.* (1998). Novel mechanistic insights into the regulation of mature osteoclasts by osteoprotegerin (OPG) and osteoprotegerin-ligand (OPGL). Evidence for

resorption stimulation through inhibition of extracellular Ca^{2+} sensing (abstract). *J. Bone Miner. Res.* **14,** S363.

Moore, T. L., Krishnan, V. G., Onyia, J. E., *et al.* (2000). A mechanism for bone anabolic activity of PTH through Cbfa1/Osf-2 (abstract). *J. Bone Miner. Res.* **15,** S158.

Morley, P., Whitfield, J. F., and Willick, G. (1999). Design and application of parathyroid hormone analogues. *Curr. Med. Chem.* **11,** 1095–1106.

Morley, P., Whitfield, J. F., Willick, G. E., *et al.* (2001). The effect of monocyclic and bicyclic analogs of human parathyroid hormone hPTH-(1-31)NH_2 on bone formation and mechanical strength in ovariectomized rats. *Calcif. Tissue. Int.* **68,** 95–101.

Mosekilde, L. (1997). Osteoporosis: Mechanisms and models. *In* "Anabolic Treatments for Osteoporosis" (J. F. Whitfield and P. Morley, Eds.), pp. 31–58. CRC Press, Boca Raton, FL.

Mosekilde, L., Thomsen, J. S., and McOsker, J. E. (1997). No loss of biomechanical effects after withdrawal of short-term PTH treatment in an aged, osteopenic, ovariectomized rat model. *Bone* **20,** 429–437.

Mundy, G. (2000). Pathogenesis of osteoporosis and challenges for drug delivery. *Adv. Drug Del. Res.* **42,** 165–173.

Mundy, G., Garrett, R., Harris, S., *et al.* (1999). Stimulation of bone formation *in vitro* and in rodents by statins. *Science* **286,** 1946–1949.

Murray, E. J. B., and Murray, S. S. (2001). Lovastatin stimulates rather than inhibits proteasome *in vitro* and in osteoblasts (abstract). *J. Bone Miner. Res.* **16,** S206.

Nadal, A., Rovira, J. M., Laribi, O., *et al.* (1998). Rapid insulinotropic effect of 17β-estradiol via a plasma membrane receptor. *FASEB J.* **12,** 1341–1348.

Nadal, A., Ropero, A. B., Laribi, O., *et al.* (2000). Nongenomic actions of estrogens and xenoestrogens by binding at a plasma membrane recepor unrelated to estrogen receptor α and estrogen receptor β. *Proc. Natl. Acad. Sci. USA* **97,** 11603–11608.

Nakade, O., Takahashi, K., Takuma, T., *et al.* (2001). Effect of extracellular calcium on the gene expression of bone morphogenic protein-2 and -4 of normal human bone cells. *J. Bone Miner. Metab.* **19,** 13–19.

Nakajima, M., Ejiri, S., Tanaka, M., *et al.* (2000). Effects of intermittent administration of human parathyroid hormone (1-34) on mandibular condyle of ovariectomized rats. *J. Bone Miner. Metab.* **18,** 9–17.

Nakamura, K., Nishimura, J., Hirano, K., *et al.* (2001). Hydroxyfasudil, an active metabolite of fasudil hydrochloride, relaxes the rabbit basilar artery by disinhibition of myosin light chain phosphatase. *J. Cereb. Blood Flow Metab.* **21,** 876–885.

Nakashima, K., Zhou, X., Kunkel, G., *et al.* (2002). The novel zinc finger-containing transcription factor osterix is required for osteoblast differentiation and bone formation *Cell* **108,** 17–29.

Neer, R., Slovik, D., Daly, M., *et al.* (1991). Treatment of postmenopausal osteoporosis with daily parathyroid hormone plus calcitriol. *In* "Osteoporosis" (C. Christiansen and K. Overgaard, Eds.), pp. 1314–1317. Osteopress, Copenhagen.

Neer, R. M., Arnaud, C. D., Zanchetta, J. R., *et al.* (2001). Effect of parathyroid hormone (1-34) on fractures and bone mineral density in postmenopausal women with osteoporosis. *N. Engl. J. Med.* **344,** 1434–1441.

Netelenbos, C. (1998). Osteoporosis: Intervention options. *Maturitas* **30,** 235–239.

Neugebauer, W., Barbier, J.-R., Sung, W. L., *et al.* (1995). Solution structure and adenylyl cyclase stimulating activities of C-terminal truncated human parathyroid hormone analogues. *Biochemistry* **34,** 8835–8842.

Ni, C.-Y., Murphy, M. P., Golde, T. E., *et al.* (2002). γ-Secretase cleavage and nuclear localization of ErbB-4 receptor tyrosine kinase. *Science* **294,** 2179–2181.

Nielsen, L. B., Pedersen, F. S., and Pedersen, L. (2001). Expression of type III sodium-dependent phosphate transporters/retroviral receptors mRNAs during osteoblast differentiation. *Bone* **28,** 160–166.

Nishida, S., Yamaguchi, A., Tanizawa, T., *et al.* (1994). Increased bone formation by intermittent parathyroid hormone administration is due to the stimulation of proliferation and differentiation of osteoprogenitor cells in bone marrow. *Bone* **15,** 717–723.

Nishida, S., Endo, N., Yamagiwa, H., *et al.* (1999). Number of osteoprogenitor cells in human bone marrow markedly decreases after skeletal maturation. *J. Bone Miner. Metab.* **17,** 171–177.

Noble, B. S. (2000). Osteocyte death: Its biological significance (abstract). *J. Bone Miner. Res.* **15,** 823.

Noble, B. S., and Reeve, J. (2000). Osteocyte function, osteocyte death and bone fracture resistance. *Mol. Cell. Endocrinol.* **159,** 7–13.

Noda, M., Takuwa, Y., Katoh, T., *et al.* (1996). Mechanical force regulation of vascular parathyroid hormone-related peptide expression. *Kidney Int. Suppl.* **55,** S154–S1555.

Nomura, S., and Takano-Yamamoto, T. (2000). Molecular events caused by mechanical stress in bone. *Matrix Biol.* **19,** 91–96.

O'Brien, C. A., Kern, B., Gubrij, I., *et al.* (2000). Analysis of the role of Cbfa1 in RANKL expression. *J. Bone Miner. Res.* **15,** S270.

Okada-Ban, M., Thiery, J. P., and Jouanneau, J. (2000). Fibroblast growth factor-2. *Int. J. Biochem. Cell Biol.* **32,** 263–267.

Ohnaka, K., Shimoda, S., Nawata, H., *et al.* (2001). Pitavastatin enhanced BMP-2 and osteocalcin expression by inhibition of rho-associated kinase in human osteoblasts. *Biochem. Biophys. Res. Commun.* **287,** 337–342.

Oniya, J. E., Miles, R. R., Halladay, D. L., *et al.* (2000). In vivo demonstration that parathyroid hormone (hPTH 1-38) inhibits the expression of osteoprotegerin (OPG) in bone with the kinetics of an immediate early gene. *J. Bone Miner. Res.* **15,** 863–871.

Onyia, J. E., Gelbert, I., Zhang, M., *et al.* (2001a). Analysis of gene expression by DNA microarray reveals novel clues to the mechanism of the catabolic and anabolic actions of PTH in bone (abstract). *J. Bone Miner. Res.* **16,** S227.

Onyia, J. E., Ma, Y. L., Galbreath, E., *et al.* (2001b). ADAMTS-1: A cellular disinctgrin and metalloprotease with thrombospondin motifs is essential for normal bone growth and PTH regulated bone metabolism (abstract). *J. Bone Miner. Res.* **16,** S158.

Opas, E. E., Gentile, M. A., Rossert, J. A., *et al.* (2000). Parathyroid hormone and prostaglandin E_2 preferentially increase luciferase levels in bone of mice harboring a luciferase transgene controlled by the elements of the pro-$\alpha 1$(I) collagen promoter. *Bone* **27,** 27–32.

Ornitz, D. M., and Itoh, N. (2001). Fibroblast growth factors. *Genome Biol.* **2,** http://genome-biology.com/2001/2/3/reviews/3995.1.

Ott, S. Osteoporosis. http://courses.washington.edu/bonephys/.

Oyajobi, B. O., Anderson, D. M., Traianedes, K., *et al.* (2001). Therapeutic efficacy of a soluble receptor activator of nuclear factor κB-IgG Fc fusion protein in suppressing bone resorption and hypercalcemia in a model of humoral hypercalcemia of malignancy. *Cancer Res.* **61,** 2572–2578.

Panegyres, P. K. (2001). The functions of the amyloid precursor gene. *Rev. Neurosci.* **12,** 1–39.

Parfitt, A. M. (1998). Osteoclast precursors as leukocytes: Importance of the area code. *Bone* **23,** 491–494.

Parfitt, A. M. (2000a). BMU origination and progression: Relationship to targeted and nontargeted remodeling (abstract). *J. Bone Miner. Res.* **15,** 823.

Parfitt, A. M. (2000b). The mechanism of coupling: A role for the vasculature. *Bone* **26,** 319–323.

Parfitt, A. M. (2001). The bone remodeling compartment: acirculatory function for bone lining cells. *J. Bone Miner. Res.* **16,** 1583–1585.

Patton, A. J., Genever, P. G., Birch, M. A., *et al.* (1998). Expression of an N-methyl-D-aspartate receptor by human and rat osteoblasts and osteoclasts suggests a novel glutamate signaling pathway in bone. *Bone* **22,** 645–649.

Pavo, I., Laszlo, F., Morschl, E., Newmcsik, J., *et al.* (2000). Raloxifene, an estrogen-receptor modulator, prevents decreased constitutive nitric oxide and vasoconstriction in ovariectomized rats. *Eur. J. Pharmacol.* **410,** 101–104.

Peet, N. M., Grabowski, P. S., Laketic-Ljubojevic, I., *et al.* (1999). The glutamate receptor antagonist MK801 modulates bone resorption *in vitro* by a mechanism predominantly involving osteoclast differentiation. *FASEB J.* **13,** 2179–2185.

Péhu, M., Policard, A., and Dufort, A. (1931). L'Ostéopetrose ou maladie des os marmoréens. *La Presse Méd.* **53,** 999–1003.

Perris, A. D., MacManus, J. P., Whitfield, J. F., *et al.* (1971). Parathyroid glands and mitotic stimulation in rat bone marrow after hemorrhage. *Am. J. Physiol.* **220**, 773–778.

Pfeilschrifter, J., Laukhuf, F., Mηller-Beckmann, B., *et al.* (1995). Parathyroid hormone increases the concentration of insulin-like growth factor-1 and transforming growth factor beta 1 in rat bone. *J. Clin. Invest.* **96**, 767–774.

Pi, M., and Hinson, T. K., and Quarles, L. D. (1999). Failure to detect the extracellular calcium-sensing receptor (CasR) in human osteoblast cell lines. *J. Bone Miner. Res.* **14**, 1310–1319.

Pi, M., Garner, S. C., Flannery, P., *et al.* (2000). Sensing of extracellular cations in CasR-deficient osteoblasts. Evidence for a novel cation-sensing mechanism. *J. Biol. Chem.* **275**, 3256–3263.

Picotto, G., Vazquez, G., and Boland, R. (1999). 17β-Oestradiol increases intracellular Ca^{2+} concentration in rat enterocytes. Potential role of phospholipase C-dependent store-operated Ca^{2+} influx. *Biochem. J.* **339**, 71–77.

Pilbeam, C., Rao, Y., Alander, C., *et al.* (1997). Downregulation of mRNA expression for the 'decoy' interleukin-1 receptor 2 by ovariectomy (abstract). *J. Bone Miner. Res.* **12**(Suppl. 1) S433.

Pirola, C. J., Wang, H. M., Strgacich, M. I., *et al.* (1994). Mechanical stimuli induce vascular parathyroid hormone-related protein gene expression *in vivo* and *in vitro*. *Endocrinology* **134**, 2230–2236.

Pouysségur, J. (2000). Signal transduction: An arresting start for MAPK. *Science* **290**, 1574–1577.

Prenzel, N., Zwick, E., Daub, H., *et al.* (1999). EGF receptor transactivation by G-protein-coupled receptors requires metalloproteinase cleavage of pro-HB-EGF. *Nature* **402**, 884–888.

Pugazhenthi, S., Miller, E., Sable, C., *et al.* (1999). Insulin-like growth factor-I induces bcl-2 promoter through the transcription factor cAMP-response element-binding protein. *J. Biol. Chem.* **274**, 27529–27535.

Raouf, A., Seth, A., *et al.* (2002). Discovery of osteoblast-associated genes using cDNA microarrays. *Bone* **30**, 463–471.

Razandi, M., Pedram, A., Greene, G. L., *et al.* (1999). Cell membrane and nuclear estrogen receptors (ERs) originate from a single transcript: Studies of ERα and ERβ expressed in Chinese hamster ovary cells. *Mol. Endocrinol* **13**, 307–319.

Re, R. N. (2002). Toward a theory of intracrine hormone action. *Regul. Pept.* **106**, 1–6.

Reeve, J., Hesp, R., Williams, D., *et al.* (1976a). Anabolic effect of low doses of a fragment of human parathyroid hormone fragment on the skeleton in postmenopausal osteoporosis. *Lancet.* **I**, 1035–1036.

Reeve, J., Tregear, G. W., and Parsons, J. A. (1976b). Preliminary trial of low doses of human parathyroid hormone fragment 1-34 in treatment of osteoporosis. *Clin. Endocrinol.* **21**, 469–477.

Reeve, J., Meunier, P. J., Parsons, J A., *et al.* (1980). Anabolic effect of human parathyroid hormone fragment on trabecular bone in involutional osteoporosis: A multicentre trial. *Br. Med. J.* **280**, 1340–1344.

Reeve, J., Mitchell, A., Tellez, M., *et al.* (2001). Treatment with parathyroid peptides and estrogen replacement for severe postmenopausal vertebral osteoporosis; prediction of long-term responses in spine and femur. *J. Bone Miner. Metab.* **19**, 102–114.

Reid, I. R., Hague, W., Emberson, J., *et al.* (2001). Effect of pravastatin on frequency of fracture in the LIPID study: Secondary analysis of a randomized controlled trial. Long-term intervention with pravastatin in ischaemic disease. *Lancet* **357**, 509–512.

Reseland, J. E., Gordeladze, J. O., and Drevon, C. A. (2002). Leptin receptor (OB-R) gene expression in human primary osteoblasts:reaffirmation. *J. Bone Miner. Res.* **17**, 1136.

Reseland, J., Syversen, U., Bakke, I., *et al.* (2001). Leptin is expressed in and secreted from primary cultures of human osteoblasts and promotes bone mineralization. *J. Bone Miner. Res.* **16**, 1426–1433.

Reszka, A. A., Halasy-Nagy, J. M., Masarachia, P. J., *et al.* (1999). Bisphosphoinates act directly on the osteoclast to induce caspase cleavage of Mst1 kinase during apoptosis. *J. Biol. Chem.* **274**, 34967–34973.

Riggs, B. L. (2000). The mechanism of estrogen regulation of bone resorption. *J. Clin. Invest.* **106**, 1203–1204.

Rittmaster, R. S., Bolognese, M., Ettinger, M. P., et al. (2000). Enhancement of bone mass in osteoporotic women with parathyroid hormone followed by alendronate. *J. Clin. Endocrinol. Metab.* **85,** 2129–2134.

Rixon, R. H., Whitfield, J. F., Gagnon, L., et al. (1994). Parathyroid hormone fragments may stimulate bone growth in ovariectomized rats by activating adenylyl cyclase. *J. Bone Miner. Res.* **9,** 1179–1189.

Roe, E. B., Chiu, K. M., and Arnaud, C. D. (2000). Selective estrogen receptor modulators and postmenopausal health. *Adv. Intern. Med.* **45,** 259–278.

Roodman, G. D. (1996). Advances in bone biology: The osteoclast. *Endocrine Rev.* **17,** 308–332.

Rose, E. B., Sanchez, S. D., Del Puerto, G. A., et al. (1999). Parathyroid hormone 1–34 (hPTH 1–34) and estrogen produce dramatic bone density increases in postmenopausal osteoporosis (abstract). *J. Bone Miner. Res.* **14,** S137.

Rosen, C. J., and Bilezikian, J. P. (2001). Anabolic therapy for osteoporosis. *J. Clin. Endocrinol. Metab.* **86,** 957–964.

Roufflet, J., Coxam, V., Gaumet, N., et al. (1994). Preserved bone mass in ovarectomized rats treated with parathyroid-hormone-related peptide (1–34) and (107–111) fragments. *Reprod. Nutr. Dev.* **34,** 473–481.

Ruoslahti, E., and Rajotte, D. (2000). The address system in the vasculature of normal tissues and tumors. *Annu. Rev. Immunol.* **18,** 813–827.

Ryder, K. D., and Duncan, R. L. (2001). Parathyroid hormone enhances fluid shear-induced $[Ca^{2+}]_I$ signaling in osteoblastic cells through activation of mechanosensitive and voltage-sensitive Ca^{2+} channels. *J. Bone Miner. Res.* **16,** 240–248.

Ryoo, H. M., Hoffmann, H. M., Beumer, T., et al. (1997). Stage-specific expression of Dlx-5 during osteoblast differentiation: involvement in regulation of osteocalcin gene expression. *Mol. Endocrinol.* **11,** 1681–1694.

Sakai, H., Kobayashi, Y., Sakai, E., et al. (2000). Cell adhesion is a prerequisite for osteoclast survival. *Biochem. Biophys. Res. Commun.* **270,** 550–556.

Samnegard, E., Iwaniec, U. T., Cullen, D. M., et al. (2001). Maintenance of cortical bone in human parathyroid hormone(1–84)-treated ovariectomized rats. *Bone* **28,** 251–260.

Sato, M., Schmidt, A., Cole, H., et al. (2001). The skeletal efficacy of statins do not compare with low-dose parathyroid hormone (abstract). *Bone* **28,** S80.

Schaffler, M. B. (2000). The role of osteocytes in targeting microdamage-induced remodeling (abstract). *J. Bone Miner. Res.* **15,** 823.

Schiller, P. C., D'Ippolito, G., Roos, B. A., et al. (1999). Anabolic or catabolic responses of MC3T3-E1 osteoblastic cells to parathyroid hormone depends on time and duration of treatment. *J. Bone Miner. Res.* **14,** 1504–1512.

Schwartz, B., Smirnoff, P., Shany, S., et al. (2000). Estrogen controls expression and bioresponse of 1,25-dihydroxyvitamin D receptors in the rat colon. *Mol. Cell. Biochem.* **203,** 87–93.

Schweitert, H. R., Groen, E. W. J., Sollie, F. A. E., et al. (1997). Single dose subcutaneous administration of recombinant human parathyroid hormone [rhPTH(1–84)] in healthy postmenopausal volunteers. *Clin. Pharmacol. Ther.* **61,** 360–376.

Seeman, E. (2001). How do antiresorptive agents reduce fracture? IBMS BoneKEy 2001 January 23 10.1138/ibmske; 20011012.

Selvamurugan, N., Pulumati, M. R., Tyson, D. R., et al. (2000). Parathyroid hormone regulation of the rat collagenase-3 promoter by protein linase A-dependent transactivation of core binding factor $\alpha 1$. *J. Biol. Chem.* **275,** 5037–5042.

Selye, H. (1932). On the stimulation of new bone formation with parathyroid extract and irradiated ergosterol. *Endocrinology* **16,** 547–558.

Selz, T., and Caverzasio, J., and Bonjour, J.-P. (1989). Regulation of Na-dependent P_i transport by parathyroid hormone in osteoblast-like cells. *Am. J. Physiol.* **256,** E93–E100.

Serre, C. M., Farlay, D., Delmas, P. D., et al. (1999). Evidence for a dense and intimate innervation of the bone tissue, including glutamate-containing fibers. *Bone* **25,** 623–629.

Shepherd, J. (2000). Lipid lowering: Statins and the future. *Heart* **84,** 46–47.
Shimokawa, H., Seto, M., Katsumata, N., *et al.* (1999). Rho-kinase-mediated pathway induces enhanced myosin light chain phosphorylation in a swine model of coronary artery spasm. *Cardiovasc. Res.* **43,** 1029–1039.
Shirakabe, K., terasawa, K., Miyama, K., *et al.* (2001). Regulation of the activity of the transcription factor Runx2 by two homeobox proteins, Msx2 and Dlx5. *Genes Cells* **6,** 851–856.
Singh, A. T., Kunnel, J. G., Strieleman, P. J., *et al.* (1999). Parathyroid hormone (PTH)-(1–34), [Nle(8,18),Tyr34]PTH-(3–34) amide, PTH-(1–31)amide, and PTH-related peptide-(1–34) stimulate phosphatidylcholine hydrolysis in UMR-106 osteoblastic cells: Comparison with effects of phorbol 12,13-dibutyrate. *Endocrinology* **140,** 131–137.
Sirola, J., Honkanen, R., Kröger, H., *et al.* (2002). Relation of statin use and bone loss: a prospective population-based cohort study in early postmenopausal women. *Osteoporosis Int.* **13,** 537–541.
Skerry, T. M. (1999). Identification of novel signaling pathways during functional adaptation of the skeleton to mechanical loading: The role of glutamate as a paracrine signaling agent in the skeleton. *J. Bone Miner. Metab.* **17,** 66–70.
Skerry, T. M., and Genever, P. G. (2001). Glutamate signaling in non-neuronal tissues. *Trends Pharmacol. Sci.* **22,** 174–181.
Soerensen, L., and Eriksen, E. F. (1998). Endothelial cells mediate osteoblastic responses to sex steroids and selective estrogen receptor modulators (SERMS) via NO independent pathways (abstract). *Bone* **23,** S369.
Sogaard, C. H., Mosekilde, L., Thomsen, J. S., *et al.* (1997). A comparison of the effects of two anabolic agents (fluoride and PTH) on ash density and bone strength assessed in an osteopenic rat model. *Bone* **20,** 439–449.
Solomon, D. H., Finkelstein, J. S., Wang, P. S., *et al.* (2001). Statin lipid-lowering drugs and bone density (abstract). *J. Bone Miner. Res.* **16,** S293.
Springer, T. A. (1994). Traffic signals for lymphocyte recirculation and leukocyte emigration: The multistep paradigm. *Cell* **76,** 301–314.
Steers, W. D., Broder, S. R., Persson, K., *et al.* (1998). Mechanical stretch increases secretion of parathyroid hormone-related protein by cultured bladder smooth muscle cells. *J. Urol.* **160,** 908–912.
Stefano, G. B., and Peter, D. (2001). Cell surface estrogen receptors coupled to eNOS mediate immune and vascular tissue regulation: Therapeutic implications. *Med. Sci. Monit.* **7,** 1066–1074.
Steppan, C. M., Crawford, D. T., Chidsey-Frink, K. L., *et al.* (2000). Leptin is a potent stimulator of bone growth in *ob/ob* mice. *Regul. Peptides* **92,** 73–78.
Stevenson, J. C., and Lindsay, R. (1999). "Osteoporosis." Chapman & Hall Medical, London; Current Medicine, Philadelphia.
Stilgren, L. S., Reppe, S., Abrahamsen, B., *et al.* (2001). Differential effects of PTH peptides on OPG and RANK-L mRNA expression in human OHS osteosarcoma cells: A possible pathway of osteoblast dependent bone resorption (abstract). *J. Bone Miner. Res.* **16,** S545.
Strein, K. (1994). Are animal studies with bisphosphonates and PTH fragments predictive for the clinical studies? *In* "Bone Diseases and Osteoporosis" (R. G. Russell, Ed.), article 12. IBC Technical Services Limited, London.
Suda, T., Takahashi, N., Udagawa, N., *et al.* (1999). Modulation of osteoclast differentiation and function by the new members of the tumor necrosis factor receptor and ligand families. *Endocrine Rev.* **20,** 345–357.
Sugimoto, T., Kanatani, M., Kano, J., *et al.* (1994). IGF-I mediates the stimulatory effect of high calcium concentration on osteoblastic cell proliferation. *Am. J. Physiol.* **266,** E709–E716.
Sugiyama, M., Kodama, T., Konishi, K., *et al.* (2000). Compactin and simvastatin, but not pravastatin, induce bone morphogenic protein-2 in human osteosarcoma cells. *Biochem. Biophys. Res. Commun.* **271,** 688–692.

Using GFP–Ligand Fusions to Measure Receptor-Mediated Endocytosis in Living Cells

Lali K. Medina-Kauwe and Xinhua Chen

Department of Biochemistry, Institute for Genetic Medicine, University of Southern California Keck School of Medicine, Los Angeles, California 90033

I. Introduction
II. Background
   A. GFP
   B. Ligands
   C. Protein Production
III. Applications
   A. Detecting Receptor Binding
   B. Detecting Receptor-Mediated Endocytosis
   C. Visual Monitoring of Cell Binding and Internalization
IV. Summary and Conclusions
   References

Recombinant DNA technology has enabled the production of many types of chimeric proteins containing heterologous functional domains that have served a variety of useful capacities for cell biology research. Among proteins gaining wide use as a fusion partner is *Aequorea victoria* green fluorescent protein (GFP). GFP has been employed by numerous groups as a reporter gene for cell transfection and as an autofluorescent tag by recombinant fusion to foreign sequences. Here we describe the use of GFP as a tag for ligands, and provide examples of how purified recombinant GFP–ligand fusion proteins may be used to detect ligand-receptor interactions, including receptor-mediated endocytosis. Both its utility and limitations are discussed. © 2002, Elsevier Science (USA).

I. INTRODUCTION

Current methods of detecting endocytosis, or the internalization of molecules from the cell surface to endosomes, rely on the binding of labeled ligands to their cognate receptors. The major route of ligand internalization is by receptor-mediated trafficking from the plasma membrane into clathrin-coated vesicles (Mellman, 1996; Mukherjee et al., 1997). These vesicles are formed by the pinching off of invaginations produced in the plasma membrane at sites of ligand–receptor interaction. Receptor mediated endocytosis is temperature dependent and is inhibited at 4°C. To detect receptor-binding or internalization activity, ligands may be labeled with radioisotopes (i.e., ^{125}I) or fluorophores through biochemical modification of amino acid side chains, which in turn may or may not alter the binding activity of the ligand, depending on the location of the receptor-binding site (Opas, 1999; Schumacher and Tsomides, 1995).

Internalization of radiolabeled ligands can be detected by the determination of cell-associated radioactivity after cells have been washed in a mild acid solution. As acidic conditions interfere with ligand–receptor interations, molecules remaining on the cell surface are removed during acid treatment whereas internalized molecules, including radiolabeled molecules, are protected. This methodology has been established using radiolabeled transferrin bound to the transferrin receptor, which has been a well-studied system for analysis of receptor-mediated endocytosis (Dautry-Varsat et al., 1983; Klausner et al., 1983).

Alternatively, the receptor binding and endocytosis of fluorescently labeled ligands can be observed in living cells using fluorescence microscopy. Fluorescence-activated cell sorting (FACS) analysis permits the counting of fluorescent cells, or events, from among a population of cells and thus allows the quantification of receptor-bound or internal fluorescence. The choice of fluorophore is an important consideration as the acidic conditions of the endosome may alter the fluorescence of certain pH-sensitive labels (Carraway and Cerione, 1993).

A more recent alternative approach to radioactive or biochemical labeling relies on recombinant DNA technology for the production of chimeric fusion proteins containing the jellyfish *Aequorea victoria* green fluorescent protein (GFP) as one of the fusion partners. Many soluble proteins fold independently of one another when produced as a fused chimera, thus retaining many if not all of the natural characteristics of each partner of the fusion (LaVallie et al., 1993; Maina et al., 1988; Smith and Johnson, 1988). This feature is useful for the production of recombinant ligand proteins fused to GFP, which, after protein isolation, require no further modifications for use in receptor-binding and internalization assays. The autofluorescence of GFP makes it an ideal natural label for tagging ligands. As with other types of fluorescently labeled molecules, GFP-tagged ligands can be used to detect receptor binding and internalization in live cells. Moreover, this activity can be observed by fluorescence microscopy and quantified by FACS analysis.

This discussion will (1) describe the use of recombinant techniques to produce and utilize GFP-tagged ligands for measuring receptor activities, (2) provide examples of how this technology has been used to measure receptor binding and endocytosis, and (3) critically discuss its utility and limitations for the analysis of ligand–receptor interactions.

II. BACKGROUND

A. GFP

GFP is an inherently fluorescent 27-kDa protein that transduces chemiluminescence from the calcium-ion-activated protein, aequorin, into green fluorescent light in the jellyfish *Aequorea victoria* (Chalfie *et al.,* 1994; Chalfie, 1995; Yang *et al.,* 1996). When aequorin is in an excited state, GFP undergoes autocatalytic oxidation at a serine–tyrosine–glycine motif found in the middle of the protein amino acid sequence, transforming this peptide sequence into a chromophore. This biochemical reaction requires no prosthetic groups or enzyme activities, thus expanding the potential of GFP as an autofluorescent tag. Isolation and cloning of the GFP gene have allowed the production of recombinant protein for biochemical analyses (Prasher *et al.,* 1992; Stearns, 1995) and have opened up a wide range of applications for cellular studies. For example, a GFP-myosin chimeric protein retained the normal ATPase activities and *in vitro* motility of wild-type myosin, and could be used to visually detect intracellular localization of myosin during certain cellular stages (Moores *et al.,* 1996). GFP fusion to a leucine zipper domain has allowed the direct observation of cytoskeletal reorganization by modular dimerization with recombinant proteins localized to specific sites of the cytoskeleton (Katz *et al.,* 1998). GFP has also been fused to viral proteins for detecting viral spread in infected cells (Elliott and O'Hare, 1997; Epel *et al.,* 1996). Fusion to the Herpesvirus structural protein, VP22, permits the detection of VP22 protein translocation and spread from infected cells to neighboring cells, and suggests that GFP does not hinder the translocation across the plasma membrane (Elliott and O'Hare, 1997). More recently, cell membrane to nuclear translocation of a STAT1–GFP fusion could be observed microscopically, thus permitting the observation of intracellular trafficking events following receptor activation (Lillemeier *et al.,* 2001). In such studies using recombinant GFP, the protein appeared to form a stable folded structure that is relatively heat resistant and tolerates fusion to heterologous protein sequences. Although wild-type GFP forms insoluble, nonfluorescent inclusions when produced as a recombinant protein in bacteria, several mutant variations of wild-type GFP have been developed that have an improved folding ability and brightness (Crameri *et al.,* 1996). One commonly used mutant encodes a red-shifted protein with an excitation maxima at 488 nm and an emission maxima at 507 nm (Delagrave *et al.,* 1995).

B. LIGANDS

The methodology of using GFP–ligand fusion proteins for receptor-binding and endocytosis assays has been validated using two types of ligands: heregulin and the adenovirus knob protein.

Heregulin-α is a 45-kDa protein isolated from the conditioned media of MDA-MB-231 human breast cancer cells (Holmes *et al.,* 1992). To date, four heregulin isomers have been identified, all of which have structural similarity to the epidermal growth factor (EGF) family members. Recombinant heregulin binds with high affinity ($K_d = 105 \pm 15$ pM) to heterodimers of HER2/3 or HER2/4 receptor subunits, which are overexpressed on certain mammary tumor cell lines (Holmes *et al.,* 1992; Hung *et al.,* 1995). The heregulin receptors, encoded by the erbB gene family, are heterodimers of growth factor receptor subunits containing tyrosine kinase activity (Bacus *et al.,* 1994). ErbB2, or HER2, is amplified in many human cancers, including breast and ovarian cancer (Carraway and Cantley, 1994; Carraway *et al.,* 1994; Press *et al.,* 1990; Slamon and Clark, 1988; Slamon *et al.,* 1987), and on the surface of several human breast cancer cell lines such as MDA-MB-453 cells (Kraus *et al.,* 1987; Plowman *et al.,* 1993). Despite the similarity to EGF receptors, HER2 is not bound or activated by EGF or transforming growth factor-α (TGF-α)(Goldman *et al.,* 1990; Yarden and Weinberg, 1989).

The knob protein is actually the receptor-binding domain of the adenovirus capsid fiber protein, which initiates cell binding and viral infection (Xia *et al.,* 1994). The adenovirus serotype 5 (Ad5) fiber is comprised of a long shaft at the amino (N)-terminal two-thirds of the protein ending in a globular knob domain at the carboxy (C)-terminal one-third (Chroboczek and Jacrot, 1987; Devaux *et al.,* 1990; Novelli and Boulanger, 1991). The knob binds with high affinity ($K_d = $ 1.7 nM) to the coxsackievirus adenovirus receptor (CAR), which is expressed on the cell surfaces of a broad range of tissues and cell lines, including HeLa and 293 cells, which are infectable by Ad5 (Bergelson *et al.,* 1997). The knob domain can be produced as a soluble protein by itself, in the absence of the shaft domain, and retains its natural properties of trimerization and receptor binding (Henry *et al.,* 1994; Medina-Kauwe *et al.,* 2000).

C. PROTEIN PRODUCTION

Both the receptor-binding domain of heregulin-α, designated eHRG (Medina-Kauwe *et al.,* 2000) or Her (Medina-Kauwe *et al.,* 2001b), and the knob protein have similar molecular weights, are soluble, and have been produced as chimeric fusion proteins to heterologous sequences. When fused to GFP, both GFP–Her (GFP–eHRG) and GFP–Knob retain solubility and functional activity. These proteins have been expressed in bacteria by recombinant methods from DNA constructs containing the genes encoding GFP and each ligand as a contiguous transcript. The translation of this message produces a peptide comprising two functional domains that appear to fold independently of one another. Both GFP–Her and GFP–Knob have been expressed in *Escherichia coli* from plasmid constructs encoding the proteins as C-terminal fusions to a polyhistidine sequence. This sequence enables the affinity purification of recombinant proteins by metal chelation to a resin containing immobilized nickel (Hochuli, 1988; Hoffmann and Roeder, 1991; Petty, 1996).

The expression of GFP and GFP-tagged proteins in bacteria is evident by the detectable green fluorescence of the cells during protein production. This fluorescence persists throughout cell lysis and column chromatography when performed under nondenaturing conditions. Purified GFP-tagged proteins can be stored frozen with little to no loss of fluorescence. Using this method, up to hundreds of milligrams of GFP–ligand have been produced per liter of bacterial culture.

The powerful methodology of recombinant DNA techniques allows the fusion of GFP to nearly any given protein, thus permitting a wide array of applications for cell biology research. Typically, proteins that are soluble in their wild-type form or normally located in the cytosol remain soluble when produced as recombinant proteins, including GFP-fusion proteins. Thus, it is highly likely that a given ligand will accommodate chimeric fusion to GFP and undergo recombinant protein production with little difficulty.

III. APPLICATIONS

The examples provided describe the use of GFP for receptor binding and endocytosis assays. The recombinant ligands, GFP–Her and GFP–Knob, have been used extensively to both validate and measure ligand–receptor interactions on cultured cells. The human breast cancer cell line, MDA-MB-453, which overexpresses the heregulin receptor, has proven to be a useful cell line for studying GFP–Her binding and endocytosis. Likewise, the cell lines HeLa and 293 (Graham *et al.,* 1977), which are infectable by Ad5, have been useful lines for studying GFP–Knob binding.

A. DETECTING RECEPTOR BINDING

1. Dose Curve

Increasing concentrations of GFP–ligands bound to their respective receptors on living cells will increase the cell-associated green fluorescence, which can be quantified by FACS. A simple FACS histogram can show the mean fluorescence

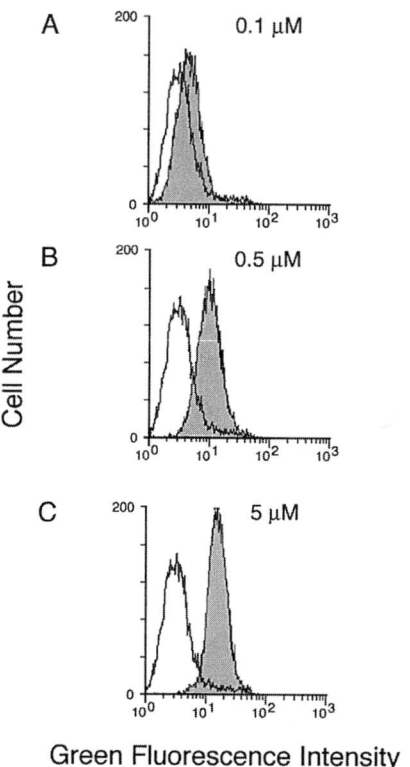

FIGURE 1 FACS scan of dosed GFP–Knob binding to HeLa cells. The concentration of GFP–Knob is shown with each FACS scan (A–C). Untreated cells and cells bound by GFP–Knob are shown by open and filled histograms, respectively.

intensity of a population of cells. When compared to a population of untreated cells, the mean fluorescence intensity value of green fluorescent cells is relatively higher. Thus, cell populations containing progressively increasing numbers of bound fluorescent molecules will result in a progressive increase in green fluorescence intensity per cell. The example provided shows the sequentially enhanced green fluorescence of HeLa cells when bound by increasing concentrations of GFP–Knob ligand (Fig. 1). This assay has been used to show dosed binding of GFP–Her on MDA-MB-453 human breast cancer cells (Medina-Kauwe et al., 2000), and to compare the binding of increasing concentrations of recombinant GFP-tagged adenoviral capsid penton proteins on 293 cells (Medina-Kauwe et al., 2001a).

Cellular fluorescence intensity is saturatable as well, reflecting a dose-dependent increase in binding up to maximum receptor occupation. A saturation curve may be plotted using the mean fluorescence intensity values obtained from each cell population quantified by FACS. This has been used to compare levels of saturating

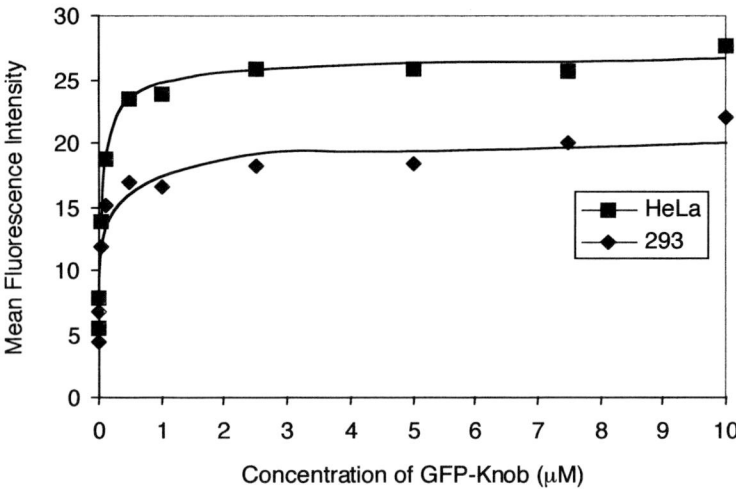

FIGURE 2 Saturation curve of GFP–Knob binding to HeLa and 293 cells. Mean fluorescence intensity values were obtained by FACS scan of cell populations treated at the indicated GFP–Knob concentrations.

GFP–ligand binding to different cell lines, thus reflecting differences in relative receptor numbers. Cell lines such as HeLa and 293 that are infectable by Ad5 express the CAR protein on their cell surfaces, to which the Ad5 knob domain binds (Bergelson et al., 1997). Differing levels of receptor saturation are evident when the two lines are bound by equivalent concentrations of GFP–Knob, reflecting differing CAR levels between the two cell lines (Fig. 2). In the example given, the maximum fluorescence intensity of 293 cells at receptor saturation is approximately 30% lower than that of HeLa, inferring that 293 cells express approximately 30% fewer CAR proteins than HeLa cells. As FACS analysis counts a predetermined number of total cells, or events, differences due to cell number are eliminated. Furthermore, the fluorescence intensity of each population of treated cells is compared to that of untreated cells, thus eliminating differences that may be due to cell morphology. GFP–ligands are added to cells in the presence of a nonspecific blocking agent (i.e., 3% milk powder) to reduce nonspecific binding, and washed thoroughly to remove free and nonspecifically bound protein prior to FACS analysis. Thus, the results obtained are generally reflective of specific binding activity.

The examples just described demonstrate that comparative assessments can be made about relative receptor levels between cell lines by dosed binding of specific GFP–ligands. Although this method has not been used to determine absolute receptor numbers, comparing the results obtained with published absolute values determined by conventional methods on the same control cell line may allow the receptor number of a test line of cells to be deduced.

2. Competitive Inhibition

As with classic receptor binding assays using radiolabeled and "cold" ligands, specific binding can be determined by using non-GFP-tagged proteins as "cold" competitors. This was demonstrated by the sequential reduction of cellular fluorescence intensity, as measured by FACS, resulting from GFP–Her binding to MDA-MB-453 cells in the presence of successively increasing concentrations of the untagged competitor, Her (Medina-Kauwe *et al.*, 2000). The amount of bound GFP–Her inversely correlated with the concentration of untagged Her competitor, indicating specific binding of GFP–Her to its receptor target. Saturating levels of Her reduced cellular green fluorescence by 80%. The inability of a nonspecific competitor (Knob) to reduce GFP–Her binding to its receptor is consistent with the binding specificity. Similarly, a 100-fold molar excess of Knob reduced HeLa fluorescence due to bound GFP–Knob by 85% (Fig. 3). In both cases, equivalent concentrations of GFP-tagged and -untagged proteins reduced cellular green fluorescence by nearly 50%, suggesting that the binding activity of the GFP-tagged ligands were unchanged from that of their untagged cognates.

This method can be used to confirm the binding specificity of a GFP–ligand by competition with its untagged competitor. Steric effects or ligand domain misfolding due to the presence of the GFP moiety may alter binding activity, thus confounding results. Therefore, the ability of a GFP-tagged ligand to retain the binding activity of its untagged cognate must be determined empirically for each protein. As GFP usually retains its fluorescence whether expressed as an N- or C-terminal fusion, it is conceivable to test all of the possible orientations of protein fusion to obtain the best candidate for binding analysis.

3. Analysis of Receptor-Binding Mutants

GFP-tagged ligands may be useful for determining the effect of mutations on receptor binding. As competitive inhibition of GFP–Knob by untagged Knob has already been established, testing the competitive binding of mutated versions of the knob can determine whether certain amino acid replacements interfere with receptor binding.

In the example given, several soluble knob proteins containing point mutations were produced and designated S576A, K513I, N515L, S572A, T574V, and Y577F. These mutations were introduced at specific sites to determine whether interference with oligomerization inhibited receptor binding. The ability of each mutant to competitively inhibit the receptor binding of GFP–Knob on HeLa cells was measured by FACS. Using equivalent molar concentrations of each untagged protein, the competitive receptor binding of each mutant could be assessed in comparison to wild-type soluble knob. Of the six mutants tested, two exhibited an inability to reduce cellular green fluorescence resulting from bound GFP–Knob, indicating that these mutants were binding defective. Equivalent concentrations of the remaining four mutants reduced cellular green fluorescence similarly to

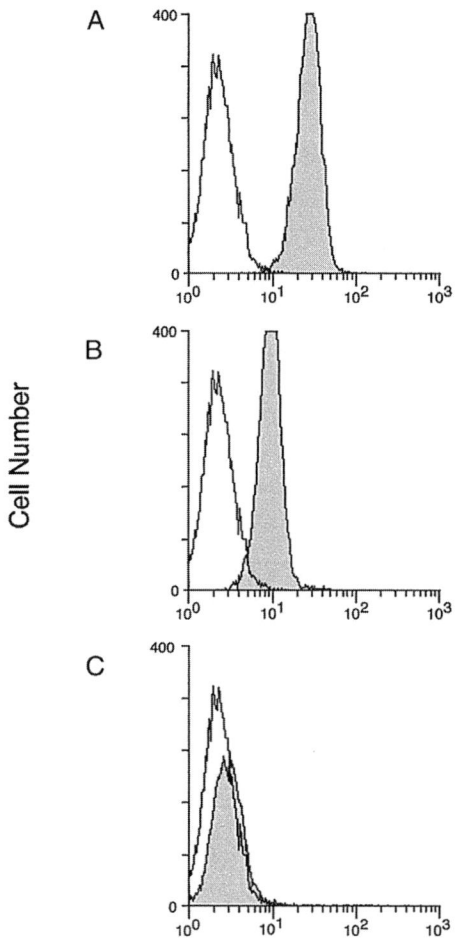

Green Fluorescence Intensity

FIGURE 3 FACS scans of HeLa cells showing competitive inhibition of GFP–Knob binding by untagged Knob. Cells were treated with equivalent concentrations of GFP–Knob in each FACS scan in the presence of 0 (A), 3.2 (B), or 32 (C) μM Knob. Untreated cells and cells bound by GFP–Knob are shown by open and filled histograms, respectively.

wild-type knob and, thus, exhibited binding abilities similar to wild-type soluble knob (Fig. 4).

This assay provided an assessment of the effect of mutations on receptor binding. Dissociation constants of either wild-type or mutant ligands have not been determined here. Thus, the results obtained can be used to make relative comparisons to a control ligand, such as the wild-type knob, but may not necessarily determine absolute rate constant values.

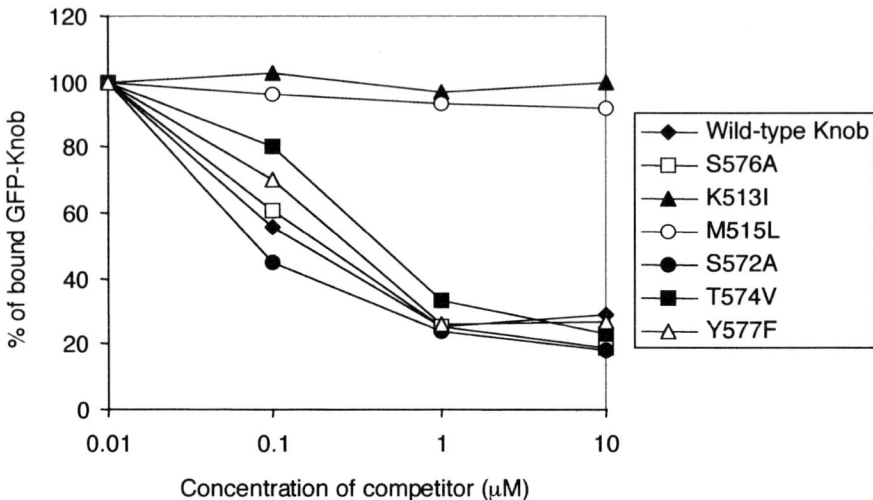

FIGURE 4 Analysis of receptor-binding mutants. HeLa cell populations were incubated with GFP–Knob and the indicated competitor concentrations, then measured by FACS. Mean fluorescence values obtained from FACS scan of cell populations were converted to a percentage of the control (HeLa bound by GFP–Knob alone).

B. DETECTING RECEPTOR-MEDIATED ENDOCYTOSIS

A modification of the cell-binding assay described earlier may be used to detect and measure the rate of endocytosis of GFP-tagged proteins after receptor binding. This has been demonstrated using GFP–Her on MDA-MB-453 human breast cancer cells (Medina-Kauwe et al., 2000). Incubation of GFP–Her with the cells on ice or at 4°C promoted receptor binding, after which cell washing removed free and nonspecifically bound proteins. Warming at 37°C to promote endocytosis was followed by cell harvesting at sequential time points. Each cell harvest was immediately followed by treatment with trypsin protease to remove residual cell surface-bound proteins, then analyzed by FACS. In this assay, internalized GFP protected from trypsinization can produce a fluorescence that is detectable by FACS. At each time point, a progressively enhanced fluorescence shift in the treated population of cells reflects an increased accumulation of intracellular GFP over time (Fig. 5). In contrast, GFP–ligands continuously retained on the cell surface and thus removed by trypsin treatment produce no fluorescence signal at any time point, as with GFP–Knob, which binds to the CAR cell surface protein but does not undergo receptor-mediated endocytosis after CAR binding (Wickham et al., 1993). The cell fluorescence at each time point in comparison to the total receptor-bound fluorescence prior to endocytosis permits the determination of the relative endocytic rate. The assay just described has demonstrated that of the total GFP–Her bound to the surface of MDA-MB-453 cells at 0 min, 80% internalized by 30 minutes at 37°C, whereas GFP–Knob remained unetectable at all time points, despite its high level of cell surface binding (Medina-Kauwe et al., 2000).

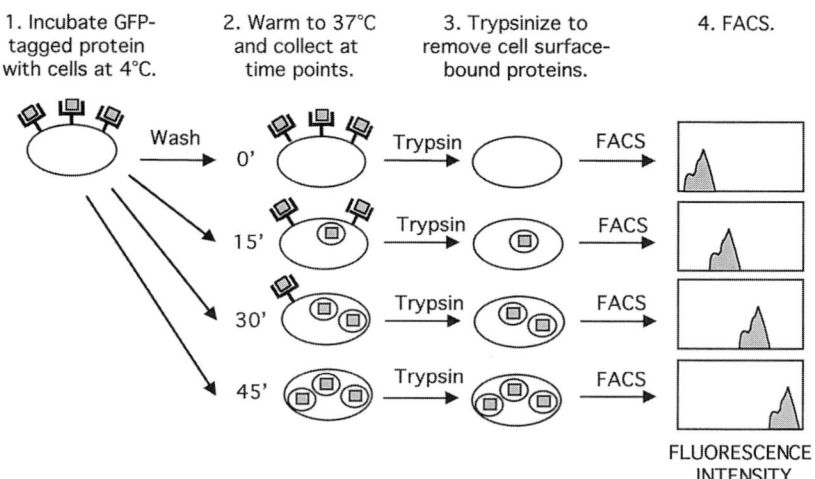

FIGURE 5 Schematic of internalization assay.

This method can be useful to determine the rate of endocytosis of certain ligands on certain cells. It has been shown that GFP–Her, which is approximately 50 kDa, is capable of cellular internalization, thus GFP-fusion proteins of comparable sizes or structures may not hinder the process of receptor-mediated endocytosis. The upper size limit for endocytosis of GFP-fusion proteins, however, is not known.

C. VISUAL MONITORING OF CELL BINDING AND INTERNALIZATION

GFP-tagged ligands can be detected microscopically without additional cell treatment, thus allowing the visual monitoring of receptor-bound or internalized fluorescence on live cells. MDA-MB-453 human breast cancer cells, which overexpress the heregulin receptor, exhibit bright punctate foci on their cell surfaces when bound by GFP–Her. Competitive inhibition by soluble, free Her ligand reduces the numbers of foci that can be seen by ultraviolet (UV) microscopy, thus relative quantification may be accomplished by visual scoring (Fig. 6). Increasing the concentration of specific competitor from 0 (Fig. 6A) to 10-fold molar excess reduces the numbers of bright green foci by approximately 81% (Fig. 6C). An 82-fold molar excess further reduces the foci count by 91% (Fig. 6E).

Internalized GFP–ligands can also be microscopically detectable, as seen with internalized GFP–Her (Medina-Kauwe *et al.*, 2000). Whereas breast cancer cells containing bound but not internalized GFP–Her exhibited little to no fluorescence after trypsin treatment, cells containing internalized GFP–Her retained punctate foci after trypsinization, suggesting transport into cellular endosomes.

Procedures to colocalize other cellular proteins, such as endocytic markers, with GFP require careful consideration. Histological staining requiring cellular

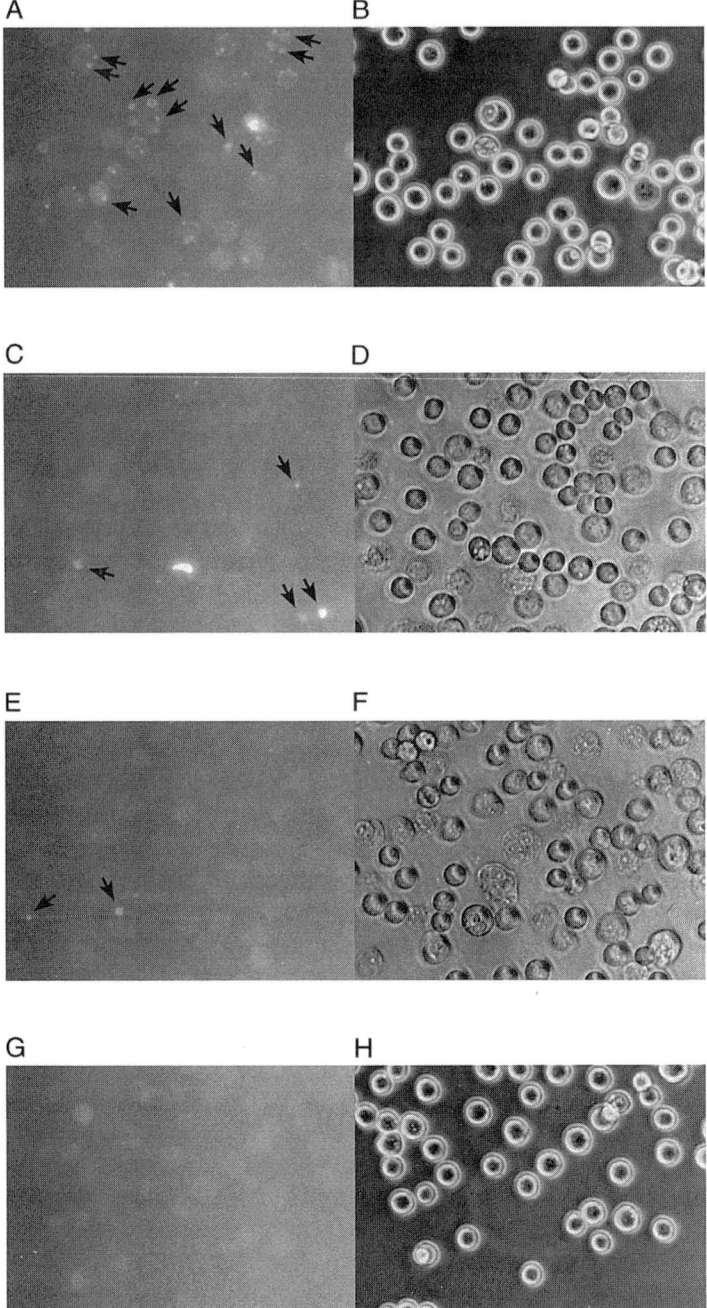

FIGURE 6 Fluorescence microscopic images of GFP–Her bound to MDA-MB-453 cells. Arrows point to representative foci indicating sites of localized GFP–Her binding. Micrographs were captured at 10× magnification under UV light (A, C, E, and G) or regular light (B, D, F, and H). MDA-MB-453 cells were incubated with GFP–Her in the absence (A and B), or presence of a Her competitor at a 10-fold (C and D) or 82-fold (E and F) molar excess. (G and H), untreated cells.

permeabilization may alter the intracellular levels of GFP, reducing detectable fluorescence. The commercial availability, however, of mutagenized forms of GFP that luminesce at alternative wavelengths, and thus emit different spectral colors, may enable the coexpression and detection of two different fluorophores in live cells.

IV. SUMMARY AND CONCLUSIONS

The use of GFP as a protein tag greatly facilitates trafficking studies of recombinant proteins. Current methods for studying intracellular localization of molecules involve time-consuming fixation methods that may distort both cells and proteins, whereas GFP-labeled molecules can be observed in live cells under "real-time" conditions. In addition, GFP serves as a safer and simpler alternative to handling radioactively labeled proteins that are usually required for studying receptor binding and translocation. GFP and GFP-tagged ligands can be produced and purified by recombinant methods as soluble proteins. Additionally, recombinant protein production can provide nearly limitless quantities of tagged ligand that can be stored long term with little to no loss in fluorescence as long as GFP retains proper folding. Recombinant GFP retains its autofluorescent activity and solubility when fused to most soluble proteins. GFP–ligand binding and cellular internalization can be observed microscopically, requiring no treatment prior to detection, and can be quantified by FACS.

Although the use of GFP–ligands has many advantages, its application is limited here to relative comparative analysis. GFP–ligand binding has not been used to determine dissociation constants or absolute receptor numbers, nor are the protein concentrations used here intended for determining these values. In addition, GFP may interfere with the oligomer formation of certain proteins, thus potentially confounding receptor-binding results. Thus, the ability of GFP–ligands to retain proper oligomerization and ligand binding must be determined experimentally for each protein.

Despite certain limitations, GFP has proven to be a useful tool to researchers and continues to be developed for diverse applications. Methodology such as cell sorting that require the use of fluorescently tagged molecules may be easily replaced by GFP-tagged proteins. Ongoing rapid progress in recombinant DNA technology may very well produce many more improvements in both GFP and recombinant protein production that will continue to lend themselves well to cell biology research.

REFERENCES

Bacus, S. S., Zelnick, C. R., Plowman, G., and Yarden, Y. (1994). Expression of the erbB-2 family of growth factor receptors and their ligands in breast cancers. Implication for tumor biology and clinical behavior. *Am. J. Clin. Pathol.* **102,** S13–24.

Bergelson, J. M., Cunningham, J. A., Droguett, G., Kurt-Jones, E. A., Krithivas, A., Hong, J. S., Horwitz, M. S., Crowell, R. L., and Finberg, R. W. (1997). Isolation of a common receptor for Coxsackie B viruses and adenoviruses 2 and 5. *Science* **275**, 1320–1323.

Carraway, K. L., 3rd, and Cerione, R. A. (1993). Fluorescent-labeled growth factor molecules serve as probes for receptor binding and endocytosis. *Biochemistry* **32**, 12039–12045.

Carraway, K. L., 3rd, and Cantley, L. C. (1994). A neu acquaintance for erbB3 and erbB4: A role for receptor heterodimerization in growth signaling. *Cell* **78**, 5–8.

Carraway, K. L., 3rd, Sliwkowski, M. X., Akita, R., Platko, J. V., Guy, P. M., Nuijens, A., Diamonti, A. J., Vandlen, R. L., Cantley, L. C., and Cerione, R. A. (1994). The erbB3 gene product is a receptor for heregulin. *J. Biol. Chem.* **269**, 14303.

Chalfie, M. (1995). Green fluorescent protein. *Photochem. Photobiol.* **62**, 651–656.

Chalfie, M., Tu, Y., Euskirchen, G., Ward, W. W., and Prasher, D. C. (1994). Green fluorescent protein as a marker for gene expression. *Science* **263**, 802–805.

Chroboczek, J., and Jacrot, B. (1987). The sequence of adenovirus fiber: Similarities and differences between serotypes 2 and 5. *Virology* **161**, 549–554.

Crameri, A., Whitehorn, E. A., Tate, E., and Stemmer, W. P. (1996). Improved green fluorescent protein by molecular evolution using DNA shuffling. *Nat. Biotechnol.* **14**, 315–319.

Dautry-Varsat, A., Ciechanover, A., and Lodish, H. F. (1983). pH and the recycling of transferrin during receptor-mediated endocytosis. *Proc. Natl. Acad. Sci. USA* **80**, 2258–2262.

Delagrave, S., Hawtin, R. E., Silva, C. M., Yang, M. M., and Youvan, D. C. (1995). Red-shifted excitation mutants of the green fluorescent protein [see comments]. *Bio/Technology* **13**, 151–154.

Devaux, C., Adrian, M., Berthet-Colominas, C., Cusack, S., and Jacrot, B. (1990). Structure of adenovirus fibre. I. Analysis of crystals of fibre from adenovirus serotypes 2 and 5 by electron microscopy and X-ray crystallography. *J. Mol. Biol.* **215**, 567–588.

Elliott, G., and O'Hare, P. (1997). Intercellular trafficking and protein delivery by a herpesvirus structural protein. *Cell* **88**, 223–233.

Epel, B. L., Padgett, H. S., Heinlein, M., and Beachy, R. N. (1996). Plant virus movement protein dynamics probed with a GFP-protein fusion. *Gene* **173**, 75–79.

Goldman, R., Levy, R. B., Peles, E., and Yarden, Y. (1990). Heterodimerization of the erbB-1 and erbB-2 receptors in human breast carcinoma cells: A mechanism for receptor transregulation. *Biochemistry* **29**, 11024–11028.

Graham, F. L., Smiley, J., Russell, W. C., and Nairn, R. (1977). Characteristics of a human cell line transformed by DNA from human adenovirus type 5. *J. Gen. Virol.* **36**, 59–74.

Henry, L. J., Xia, D., Wilke, M. E., Deisenhofer, J., and Gerard, R. D. (1994). Characterization of the knob domain of the adenovirus type 5 fiber protein expressed in Escherichia coli. *J. Virol.* **68**, 5239–5246.

Hochuli, E. (1988). Large-scale chromatography of recombinant proteins. *J. Chromatogr.* **444**, 293–302.

Hoffmann, A., and Roeder, R. G. (1991). Purification of his-tagged proteins in non-denaturing conditions suggests a convenient method for protein interaction studies. *Nucleic Acids Res.* **19**, 6337–6338.

Holmes, W. E., Sliwkowski, M. X., Akita, R. W., Henzel, W. J., Lee, J., Park, J. W., Yansura, D., Abadi, N., Raab, H., Lewis, G. D., *et al.* (1992). Identification of heregulin, a specific activator of p185erbB2. *Science* **256**, 1205–1210.

Hung, M. C., Matin, A., Zhang, Y., Xing, X., Sorgi, F., Huang, L., and Yu, D. (1995). HER-2/neu-targeting gene therapy—a review. *Gene* **159**, 65–71.

Katz, B. Z., Krylov, D., Aota, S., Olive, M., Vinson, C., and Yamada, K. M. (1998). Green fluorescent protein labeling of cytoskeletal structures—novel targeting approach based on leucine zippers. *Biotechniques* **25**, 298–302, 304.

Klausner, R. D., Ashwell, G., van Renswoude, J., Harford, J. B., and Bridges, K. R. (1983). Binding of apotransferrin to K562 cells: Explanation of the transferrin cycle. *Proc. Natl. Acad. Sci. USA* **80**, 2263–2266.

Kraus, M. H., Popescu, N. C., Amsbaugh, S. C., and King, C. R. (1987). Overexpression of the EGF receptor-related proto-oncogene erbB-2 in human mammary tumor cell lines by different molecular mechanisms. *EMBO J.* **6,** 605–610.

LaVallie, E. R., DiBlasio, E. A., Kovacic, S., Grant, K. L., Schendel, P. F., and McCoy, J. M. (1993). A thioredoxin gene fusion expression system that circumvents inclusion body formation in the E. coli cytoplasm. *Bio/Technology* **11,** 187–193.

Lillemeier, B. F., Koster, M., and Kerr, I. M. (2001). STAT1 from the cell membrane to the DNA. *EMBO J.* **20,** 2508–2517.

Maina, C. V., Riggs, P. D., Grandea, A. G., 3rd, Slatko, B. E., Moran, L. S., Tagliamonte, J. A., McReynolds, L. A., and Guan, C. D. (1988). An Escherichia coli vector to express and purify foreign proteins by fusion to and separation from maltose-binding protein. *Gene* **74,** 365–373.

Medina-Kauwe, L. K., Leung, V., Wu, L., and Kedes, L. (2000). Assessing the binding and endocytosis activity of cellular receptors using GFP-ligand fusions. *Biotechniques* **29,** 602–609.

Medina-Kauwe, L. K., Kasahara, N., and Kedes, L. (2001a). 3PO, a novel non-viral gene delivery system using engineered Ad5 penton proteins. *Gene Ther.* **8,** 795–803.

Medina-Kauwe, L. K., Maguire, M., Kasahara, N., and Kedes, L. (2001b). Nonviral gene delivery to human breast cancer cells by targeted Ad5 penton proteins. *Gene Ther.* **8,** 1753–1761.

Mellman, I. (1996). Endocytosis and molecular sorting. *Annu. Rev. Cell Dev. Biol.* **12,** 575–625.

Moores, S. L., Sabry, J. H., and Spudich, J. A. (1996). Myosin dynamics in live Dictyostelium cells. *Proc. Natl. Acad. Sci. USA* **93,** 443–446.

Mukherjee, S., Ghosh, R. N., and Maxfield, F. R. (1997). Endocytosis. *Physiol. Rev.* **77,** 759–803.

Novelli, A., and Boulanger, P. A. (1991). Deletion analysis of functional domains in baculovirus-expressed adenovirus type 2 fiber. *Virology* **185,** 365–376.

Opas, M. (1999). Fluorescence tracing of intracellular proteins. *Biotech. Histochem.* **74,** 294–310.

Petty, K. J. (1996). Metal-chelate affinity chromatography. *In* "Current Protocols in Molecular Biology" (F. M. Ausubel, R. Brent, R. E. Kingston, D. D. Moore, J. G. Seidman, J. A. Smith, and K. Struhl, Eds.), pp. 10.11.10–10.11.24. John Wiley & Sons, Inc., New York.

Plowman, G. D., Culouscou, J. M., Whitney, G. S., Green, J. M., Carlton, G. W., Foy, L., Neubauer, M. G., and Shoyab, M. (1993). Ligand-specific activation of HER4/p180erbB4, a fourth member of the epidermal growth factor receptor family. *Proc. Natl. Acad. Sci. USA* **90,** 1746–1750.

Prasher, D. C., Eckenrode, V. K., Ward, W. W., Prendergast, F. G., and Cormier, M. J. (1992). Primary structure of the Aequorea victoria green-fluorescent protein. *Gene* **111,** 229–233.

Press, M. F., Jones, L. A., Godolphin, W., Edwards, C. L., and Slamon, D. J. (1990). HER-2/neu oncogene amplification and expression in breast and ovarian cancers. *Prog. Clin. Biol. Res.* **354A,** 209–221.

Schumacher, T. N. M., and Tsomides, T. J. (1995). In vitro radiolabeling of peptides and proteins. *In* "Current Protocols in Protein Science" (J. E. Coligan, B. M. Dunn, H. L. Ploegh, D. W. Speicher, and P. T. Wingfield, Eds.), pp. 3.3.1–3.3.19. John Wiley & Sons, Inc., New York.

Slamon, D. J., and Clark, G. M. (1988). Amplification of c-erbB-2 and aggressive human breast tumors? *Science* **240,** 1795–1798.

Slamon, D. J., Clark, G. M., Wong, S. G., Levin, W. J., Ullrich, A., and McGuire, W. L. (1987). Human breast cancer: Correlation of relapse and survival with amplification of the HER-2/neu oncogene. *Science* **235,** 177–182.

Smith, D. B., and Johnson, K. S. (1988). Single-step purification of polypeptides expressed in Escherichia coli as fusions with glutathione S-transferase. *Gene* **67,** 31–40.

Stearns, T. (1995). Green fluorescent protein. The green revolution. *Curr. Biol.* **5,** 262–264.

Wickham, T. J., Mathias, P., Cheresh, D. A., and Nemerow, G. R. (1993). Integrins alpha v beta 3 and alpha v beta 5 promote adenovirus internalization but not virus attachment. *Cell* **73,** 309–319.

Xia, D., Henry, L. J., Gerard, R. D., and Deisenhofer, J. (1994). Crystal structure of the receptor-binding domain of adenovirus type 5 fiber protein at 1. 7 A resolution. *Structure* **2,** 1259–1270.

Yang, F., Moss, L. G., and Phillips, G. N., Jr. (1996). The molecular structure of green fluorescent protein. *Nat. Biotechnol.* **14,** 1246–1251.

Yarden, Y., and Weinberg, R. A. (1989). Experimental approaches to hypothetical hormones: detection of a candidate ligand of the neu protooncogene. *Proc. Natl. Acad. Sci. USA* **86,** 3179–3183.

Mitochondrial Energy Dissipation by Fatty Acids
Mechanisms and Implications for Cell Death

Paolo Bernardi,[*] Daniele Penzo,[*] and Lech Wojtczak[†]

[*]Department of Biomedical Sciences and Venetian Institute of Molecular Medicine, University of Padova, I-35131 Padova, Italy, and
[†]Department of Cellular Biochemistry, Nencki Institute of Experimental Biology, 02-093 Warsaw, Poland

I. Introduction
II. Mitochondrial Effects of Fatty Acids and Related Molecules
  A. Fatty Acids as Protonophores
  B. Fatty Acids and the Mitochondrial Permeability Transition Pore
  C. Acyl-CoA
  D. N-Acylethanolamines
  E. Quinones
III. Cellular Effects of Fatty Acids
  A. Effects of Fatty Acids on Mitochondrial Membrane Potential in Situ
  B. Saturated Fatty Acids and Cell Death
  C. Unsaturated Fatty Acids and Cell Death
IV. Conclusions and Perspectives
References

For most cell types, fatty acids are excellent respiratory substrates. After being transported across the outer and inner mitochondrial membranes they undergo β-oxidation in the matrix and feed electrons into the mitochondrial energy-conserving respiratory chain. On the other hand, fatty acids also physically interact with mitochondrial membranes, and possess the potential to alter their permeability. This occurs according to two mechanisms: an increase in proton conductance of the inner mitochondrial membrane and the opening of the permeability transition pore, an inner membrane high-conductance channel that may be involved in the release of apoptogenic proteins into the cytosol. This article addresses in some detail the mechanisms through which fatty acids exert their protonophoric action and how they modulate the permeability transition pore and discusses the cellular effects of fatty acids, with specific emphasis on their role as potential mitochondrial mediators of apoptotic signaling. © 2002, Elsevier Science (USA).

I. INTRODUCTION

A major fraction of long-chain fatty acids is present in eukaryotic cells in the esterified form, mostly as components of phospholipids and, in some types of cells, as triglyceride deposits. Nonesterified fatty acids are present in much lower amounts, mostly bound to specific or unspecific fatty acid-binding proteins. A small proportion of nonesterified fatty acids is associated with cellular membranes and only a minor part remains virtually free. In mammals, including humans, the main reservoir of nonesterified fatty acids is circulating blood where they are present as serum albumin complexes. Their amount increases significantly under particular physiological (e.g., fasting, exhausting exercise, high-fat diet) and pathological (e.g., diabetes) conditions (Seitz *et al.*, 1977; Barakat *et al.*, 1982; Bode *et al.*, 1990).

To become metabolically active, fatty acids must form thioesters with coenzyme A (CoA), a reaction requiring the hydrolysis of two pyrophosphate ("high-energy") bonds of adenosine triphosphate (ATP) per molecule of acyl-CoA formed. Only polyunsaturated fatty acids can enter metabolic pathways, e.g., prostaglandin synthesis, without such "activation." After forming thioesters with CoA, fatty acids are excellent respiratory substrates in most types of cells, feeding electrons to the mitochondrial energy-conserving respiratory chain. On the other hand, fatty acids also interact with mitochondrial membranes and alter their permeability. In particular, they (1) increase proton conductance of the inner mitochondrial membrane, causing dissipation of the electrochemical proton gradient that is the driving force for ATP synthesis (Wojtczak and Więckowski, 1999), and (2) can promote opening of the permeability transition pore (PTP), a high-conductance channel (Bernardi, 1999) that may cause release of mitochondrial apoptogenic proteins into the cytosol (Kroemer and Reed, 2000; Bernardi *et al.*, 2001). Thus, nonesterified fatty acids may play an important role in the regulation of mitochondrial energy metabolism as dual effectors with the potential to improve energy production (as substrates) or to cause energy dissipation (as uncoupling agents and modulators of ion channels).

Evidence has also accumulated that fatty acids can modulate the pathways to cell death. In the case of polyunsaturated fatty acids (such as arachidonic acid) the pro- and antiapoptotic effects have generally been ascribed to downstream products of the cyclooxygenase (COX) and lipooxygenase (LOX) pathways, which vary dramatically depending on the cell type; yet a possible role for nonesterified free arachidonic acid is emerging (Chan *et al.*, 1998; Cao *et al.*, 2000; Brash, 2001), and our recent results suggest that the PTP may be one of the proapoptotic targets of arachidonic acid (Scorrano *et al.*, 2001; Petronilli *et al.*, 2001). In the case of saturated fatty acids (such as palmitic acid), the effects vary among different cell types, with prominent mitochondrial effects in cardiac cells (Wenzel and Hale, 1978a; de Vries *et al.*, 1997; Sparagna and Hickson-Bick, 1999; Hickson-Bick *et al.*, 2000; Kong and Rabkin, 2000; Sparagna *et al.*, 2000). The issue is extremely complex, however, because the proapoptotic effect may also be due to increased synthesis of ceramide and sphingolipids, which are key molecules in the activation of the cell death cascade (Paumen *et al.*, 1997; Blazquez *et al.*, 2000; Perry *et al.*, 2000) and to a direct effect of palmitoylcarnitine on caspases (Mutomba *et al.*, 2000). In this respect, regulation of the activity of mitochondrial carnitine palmitoyltransferase I (Kerner and Hoppel, 2000) may be critical, because it controls the rate of fatty acid oxidation by mitochondria and, hence, indirectly dictates the subcellular distribution of fatty acids between mitochondria and other subcellular compartments.

In this article we will focus on the mechanisms through which fatty acids can cause mitochondrial energy dissipation, and discuss how these mechanisms can affect the course of cell death in the context of lipid signaling.

II. MITOCHONDRIAL EFFECTS OF FATTY ACIDS AND RELATED MOLECULES

A. FATTY ACIDS AS PROTONOPHORES

Fatty acids have long been known as uncouplers of oxidative phosphorylation. This is manifested by dissipation of the mitochondrial protonmotive force, increase of latent mitochondrial ATPase and resting state respiration, decrease of the respiratory control ratio, and lowering of the efficiency of oxidative phosphorylation expressed by the ADP/O ratio (Wojtczak and Schönfeld, 1993). For a long time, however, the mechanism by which these effects occurred remained unclear, and often the subject of controversy. In the late 1980s Skulachev and co-workers found that the uncoupling effect of fatty acids was alleviated by both inhibitors (Andreyev *et al.*, 1989) and substrates (Dedukhova *et al.*, 1991) of the adenine nucleotide translocase (ANT). These observations enabled Skulachev to formulate the hypothesis that the ANT was directly involved in the protonophoric mechanism of fatty acids (Skulachev, 1991).

1. Role of ANT

Undissociated long-chain fatty acids can undergo a spontaneous transbilayer movement ("flip-flop") in biological membranes due to their relatively high hydrophobicity. In contrast, the transbilayer passage of dissociated, anionic, forms of fatty acids is strongly hindered by the presence of the negatively charged carboxylic group (Kamp and Hamilton, 1992). Although the hydrocarbon chain is embedded in the lipid core of the membrane, the ionized carboxylic group protrudes into the aqueous phase and is associated with water dipoles. The fatty acid cycling model (Skulachev, 1991) assumes that fatty acid anions are transported across the inner mitochondrial membrane by the ANT. As a result, undissociated fatty acid molecules undergo a spontaneous flip-flop from the outer to the inner leaflet of the inner mitochondrial membrane where they release protons because of the alkaline milieu of the matrix. Then, in the form of anions, they are transported back to the external leaflet by ANT. The outward direction of this transport is promoted by the transmembrane electric potential difference ($\Delta\Psi$), positive outside. At the external surface of the inner membrane, fatty acid anions become protonated and can flip-flop to the inside, thus closing the cycle (Fig. 1). As a result, one proton is transferred from the external space to the matrix compartment per molecule of the fatty acid per cycle.

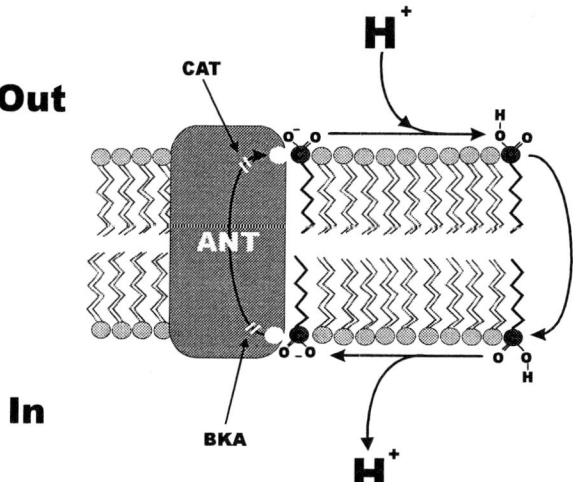

FIGURE 1 "Fatty acid cycling" model of proton conductance in the inner mitochondrial membrane. A fatty acid molecule in the protonated (undissociated) from performs a spontaneous transbilayer movement (flip-flop) within the membrane and dissociates a proton in the matrix compartment, and the fatty acid anion is electrophoretically transported to the external leaflet of the membrane by the adenine nucleotide translocase (ANT) or another mitochondrial carrier protein (not shown). As result, a net transfer of proton from the external compartment to the mitochondrial matrix compartment occurs. The ANT-mediated transport can be inhibited by carboxyatractyloside (CAT) and bongkrekic acid (BKA). Reproduced from Wojtczak and Więckowski (1999) with permission of Plenum Publishing Corporation, New York.

The crucial role of ANT in the fatty acid-mediated proton transfer across the inner mitochondrial membrane is further supported by the following observations: (1) the activity of ANT in mitochondria from various tissues (Schönfeld, 1990) and from animals of different thyroid hormone status (Schönfeld et al., 1997) correlates with the ability of carboxyatractyloside, a specific inhibitor of ANT, to abolish fatty acid-induced uncoupling; (2) photomodification of ANT by azido derivatives of fatty acids partly protects mitochondria against fatty acid-induced uncoupling (Schönfeld et al., 1996); (3) fatty acids increase proton conductance of proteoliposomes reconstituted with ANT (Brustovetsky and Klingenberg, 1994); and (4) mitochondria from ANT-deficient yeast mutants are insensitive to the uncoupling by fatty acids (Polčic et al., 1997). The fatty acid-cycling model requires that ANT is not absolutely specific for adenosine diphosphate (ADP) and ATP but can also transfer fatty acid anions. In fact, it has been observed that ANT can transport some other anionic compounds, such as pyrophosphate, phosphoenolpyruvate, and creatine phosphate, though with a much lower rate and/or affinity (Wojtczak and Schönfeld, 1993; Wojtczak and Więckowski, 1999). Further evidence that fatty acids, in their anionic form, are substrates for ANT is provided by the fact that they inhibit the transport of ATP and ADP (Wojtczak and Załuska, 1967; Pfaff and Klingenberg, 1970; Schönfeld et al., 1996).

2. Role of UCPs

Mitochondria possess a class of proteins structurally related to the ANT and to the P_i carrier, the *uncoupling proteins* (UCP) (Aquila et al., 1987; Klingenberg, 1990). The first and best characterized (UCP1) is specifically designed to increase proton conductance of the inner mitochondrial membrane. UCP1 is present exclusively in mitochondria of the brown adipose tissue, the unique thermogenic organ of most mammals, where it participates in fast dissipation of the mitochondrial proton electrochemical gradient, a process that requires fatty acids (Nicholls and Locke, 1984). When UCP1 is active, most of the energy produced by the mitochondrial respiratory chain is not utilized for ATP synthesis but is dissipated in the form of heat (Nicholls and Locke, 1984; Klingenberg, 1990). According to the generalized fatty acid cycling concept (Skulachev, 1991, 1999), UCP functions in a manner similar to the other mitochondrial carriers in transporting fatty acid anions across the inner mitochondrial membrane (Garlid et al., 1996; Ježek et al., 1998; Ježek, 1999). More recently, new members of the UCP family (designated as UCP2, UCP3, and UCP4) have been found in liver, skeletal muscle, brain, white adipose tissue, and some other mammalian organs (Boss et al., 1997; Fleury and Sanchis, 1999; Mao et al., 1999) as well as in mitochondria of plants (Vercesi et al., 1995; Laloi et al., 1997; Ježek et al., 1997a), protozoa (Jarmuszkiewicz et al., 1999; Uyemura et al., 2000), and fungi (Jarmuszkiewicz et al., 2000). The physiology of these new UCPs and their role as bona fide uncoupling proteins are the subject of active investigation and some controversy (Ježek et al., 1998; Ricquier and Bouillaud, 2000; Rial and Gonzalez-Barroso, 2001; Jarmuszkiewicz, 2001).

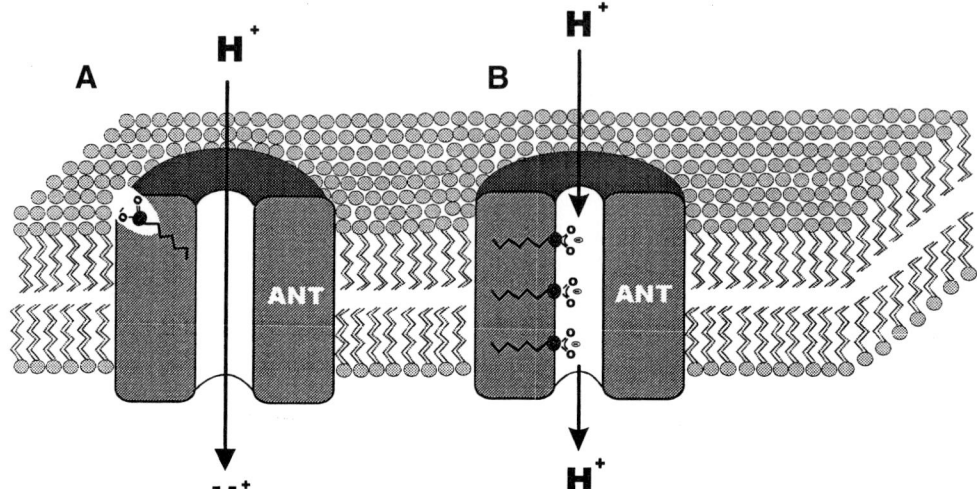

FIGURE 2 "Static" models of fatty acid-facilitated proton conductance. (A) "Conformational" model: Binding of fatty acid molecule(s) to ANT induces a conformational change of the translocase, resulting in the formation of a proton channel within the carrier molecule. (B) "Channel" model: Fatty acid molecule(s) anchored to ANT form a channel lined with negatively charged carboxylic groups that enable protons to pass through the carrier molecule. According to Winkler and Klingenberg (1994) and Klingenberg and Huang (1999), modified. Reproduced from Wojtczak and Więckowski (1999) with permission of Plenum Publishing Corporation, New York.

In contrast to the fatty acid cycling model of Skulachev (1991), Klingenberg and co-workers (Winkler and Klingenberg, 1994; Klingenberg and Huang, 1999; Klingenberg, 1999) proposed a "stationary" mechanism for the protonophoric activity of fatty acids for UCP, and possibly also for ANT. According to this concept, mainly based on studies of UCP1, fatty acids mediate proton transport across the inner mitochondrial membrane by inducing a proton channel within the carrier molecule. Thus, fatty acid molecules may bind to the carrier protein, be it UCP or ANT, either changing its conformation so as to open a proton channel within the protein molecule (Fig. 2A) or forming such a channel "lined" with negatively charged carboxylic groups (Fig. 2B). Quite recently, Klingenberg and co-workers reported an obligatory role of coenzyme Q (ubiquinone, CoQ), along with a fatty acid, in enabling the protonophoric function of UCP1, UCP2, and UCP3 expressed in *Escherichia coli* and reconstituted into phospholipid vesicles (Echtay *et al.*, 2000, 2001). These authors propose that fatty acids may cooperate with CoQ in interacting with UCP, presumably by forming a hydrogen bond between its undissociated carboxylic group and the quinone oxo group of CoQ. Only CoQs with a long isoprenoid chain, e.g., CoQ_{10} and to a lesser extent CoQ_6, and only their oxidized forms are active. It remains to be elucidated whether CoQ is also obligatory in the protonophoric function of fatty acids mediated by ANT and other mitochondrial carriers.

The "channel" concept is based on Klingenberg's studies on the conformation of UCP (Klingenberg and Huang, 1999; Klingenberg, 1999). However, the cycling model is favored by experiments with azido derivatives of fatty acids. Photoactivation and covalent immobilization of these derivatives irreversibly inhibit proton transport by UCP (Ježek et al., 1996) and ANT (Schönfeld et al., 1996). Another argument for the cycling model is provided by studies on the protonophoric effect of various fatty acid derivatives. As discussed above, only those derivatives that are able to flip-flop within the membrane exhibit protonophoric properties (Ježek et al., 1997b,c; Wojtczak et al., 1998). An exception seems to be ω-glucopyranoside-palmitate, found to be a good protonophore (Klingenberg and Huang, 1999) despite the fact that its bipolar molecule is unlikely to flip-flop within the membrane. Strong arguments in favor of the fatty acid cycling model are provided by the finding that nonesterified fatty acids strongly inhibit adenine nucleotide transport (Wojtczak and Załuska, 1967; Pfaff and Klingenberg, 1970; Schönfeld et al., 1996) and by the recent observation of Garlid and co-workers (Jabůrek et al., 2001) that undecanesulfonate, a strongly ionized fatty acid analog, is able to transport protons in UCP1 proteoliposomes by forming an ion pair with an amphiphilic organic base.

3. Other Carriers

Recent investigations have revealed that fatty acid cycling, and thus the protonophoric effect, can also be mediated, though to a lesser extent, by other mitochondrial anion carriers such as the glutamate/aspartate (Samartsev et al., 1997; Więckowski and Wojtczak, 1997), dicarboxylate (Więckowski and Wojtczak, 1997), monocarboxylate (Schönfeld et al., 2000), and phosphate (Žačková et al., 2000; Engstová et al., 2001) carriers (Wojtczak and Więckowski, 1999; Skulachev, 1999). Moreover, inhibition by fatty acids of the phosphate carrier (Wojtczak, 1976) and of the dicarboxylate carrier (Więckowski and Wojtczak, 1997) was also reported. The inhibition could be reversed when the fatty acid was removed by serum albumin. However, and similar to the ANT (Schönfeld et al., 1996), the dicarboxylate carrier became irreversibly inhibited by azido derivatives of long-chain fatty acids upon their photoactivation and covalent binding to the carrier (Więckowski and Wojtczak, 1997).

4. Structural Requirements

The structural requirements for the protonophoric and uncoupling effects of fatty acids and their derivatives have been carefully examined. The key features are the presence of an unsubstituted carboxylic group and a hydrocarbon chain of a proper length. The strongest activity was found for saturated fatty acids of chain lengths varying between 12 and 16 carbon atoms, and with cis-unsaturated C_{18} acids (Rottenberg and Hashimoto, 1986; Schönfeld et al., 1989; Schönfeld, 1992; Winkler and Klingenberg, 1994). Because the cis configuration produces a bend within the hydrocarbon chain, it reduces the distance between the two ends of the molecule and makes it comparable to that of C_{12} to C_{16} saturated fatty acids. It can therefore be speculated that the transbilayer mobility of the undissociated molecule of the fatty acid and/or of its anionic form is highest when the hydrodynamic length of its molecule is close to half the thickness of the hydrophobic core of the membrane (Wojtczak and Schönfeld, 1993).

A further requirement for the protonophoric capability of fatty acids or their analogs is the undisturbed hydrophobicity of the hydrocarbon chain. The presence of an OH group in close proximity to the carboxylic group presents no obstacle. However, a second carboxylic group or hydroxylic group(s) at or around the ω position changes the conformation of the molecule within the membrane from linear to U-shaped, and prevents the spontaneous flip-flopping. This has been shown for both artificial phospholipid membranes (Ježek *et al.*, 1997c), intact mitochondria (Wojtczak *et al.*, 1998), and liposomes reconstituted with UCP1 (Ježek *et al.*, 1997b). Interestingly, hydrophilic moieties that are shielded by vicinal hydrophobic structures, such as in 12-hydroxystearic acid (Ježek *et al.*, 1997a,c) or bis-methyl-substituted hexadecane α,ω-dioic acid (Hermesh *et al.*, 1998; Wojtczak *et al.*, 1998) do not prevent flip-flopping. Other structural requirements are described elsewhere (Ježek *et al.*, 1997b,c; Wojtczak *et al.*, 1998; Klingenberg and Huang, 1999).

Fatty acid analogs whose acidic groups are strongly ionized, such as long-chain sulfonates and sulfates, displayed little activity as protonophores in mitochondrial membranes (Wojtczak *et al.*, 1998). This is understandable, as at neutral pH these compounds are present almost completely in the anionic form and are therefore unable to flip-flop.

5. Summary

In summary, fatty acids can lead to energy dissipation through a *selective protonophoric action* mediated by coupling of transmembrane movement of the fatty acyl anion (via the ANT, the UCPs, and/or other inner membrane carriers) with the flip-flop of the undissociated fatty acid. These events result in dissipative proton cycling that decreases the transmembrane proton electrochemical gradient thereby affecting respiration, ATP synthesis, and ion homeostasis.

B. FATTY ACIDS AND THE MITOCHONDRIAL PERMEABILITY TRANSITION PORE

A potentially important effect of fatty acids that is relevant to cell death is exerted at the PTP. The PTP is an unselective, large conductance channel that enables proton and solute equilibration between the matrix and intermembrane spaces causing mitochondrial swelling and outer membrane rupture *in vitro* (Bernardi *et al.*, 1999). Considerable evidence has accumulated indicating that fatty acids modulate the PTP (Petronilli *et al.*, 1993b; Broekemeier and Pfeiffer, 1995; Więckowski and Wojtczak, 1998; Więckowski *et al.*, 2000).

Opening of the PTP was presumably the underlying cause of the large amplitude mitochondrial swelling already described by Albert L. Lehninger in the late

1950s, which among other factors was indeed stimulated by long-chain fatty acids (Lehninger and Remmert, 1959; Wojtczak and Lehninger, 1961). The swelling-inducing ability of fatty acids displayed a sharp dependence on the chain length (Zborowski and Wojtczak, 1963), similar to that of the protonophoric activity of fatty acids (see Section II.A.4). Large-amplitude swelling was widely studied in the 1960s, but the nature of the permeability pathway remained elusive (Azzone and Azzi, 1965).

The idea that large-amplitude swelling was mediated by opening of a *regulated* channel was first put forward by Haworth and Hunter, who carried out a thorough characterization of this phenomenon and defined its all-or-nothing, i.e., *transitional* nature (Hunter *et al.*, 1976; Haworth and Hunter, 1979; Hunter and Haworth, 1979a,b). The PTP has been reevaluated and studied by several authors (Zoratti and Szabó, 1995; Bernardi, 1999; and references therein) and is today considered an important player in cell death (Crompton, 1999; Kroemer and Reed, 2000; Bernardi *et al.*, 2001). The PTP is inhibited by nanomolar concentrations of cyclosporin A (Crompton *et al.*, 1988; Broekemeier *et al.*, 1989; Davidson and Halestrap, 1990; Szabó and Zoratti, 1991), a fungal product with strong immunosuppressive properties, that helps differentiate the PTP from other mitochondrial channels. The PTP is regulated by a variety of factors that have pathophysiological relevance such as matrix Ca^{2+}, P_i, and pH, the redox status of pyridine nucleotides and glutathione, and adenine nucleotides (Bernardi, 1999). The feature that best explains the effects of fatty acids, however, is the PTP voltage dependence (Bernardi, 1992; Szabó and Zoratti, 1993).

1. Mechanism of Induction by Fatty Acids

The PTP open–closed transitions are affected by the transmembrane voltage in the sense that inside more negative (physiological) potentials tend to stabilize the pore in the closed conformation, whereas depolarization tends to favor PTP opening (Bernardi, 1992). The voltage dependence can be observed by modulating the membrane potential with proton (Bernardi, 1992; Petronilli *et al.*, 1993a) or potassium currents (Scorrano *et al.*, 1997), as well as by modulating the applied voltage in patch-clamp experiments (Szabó and Zoratti, 1993). Many pore effectors, however, do not modify the membrane potential as such but rather affect the threshold voltage (the "gating potential") at which opening occurs (Petronilli *et al.*, 1993b). Pore inducers shift the apparent gating potential to more negative values, thereby favoring pore opening, whereas pore inhibitors have the opposite effect and favor its closure (Petronilli *et al.*, 1994a). Through this mechanism the PTP may also sense changes of the surface potential induced by amphiphilic molecules (Bernardi *et al.*, 1994), including fatty acids (Broekemeier and Pfeiffer, 1995). Fatty acids could indeed favor PTP opening according to two non-mutually exclusive mechanisms: (1) an *indirect* mechanism, which would depend on the protonophoric activity causing mitochondrial depolarization (Schönfeld and Bohnensack, 1997) (see Section II.A), and (2) a *direct* mechanism. The latter could be due to a decrease of the surface potential (Broekemeier and Pfeiffer,

1995), possibly through an effect on the hypothetical voltage sensor (Petronilli *et al.*, 1993b), or to interaction of the fatty acid with ANT, a potential regulator of PTP (Schönfeld and Bohnensack, 1997; Więckowski and Wojtczak, 1998). A direct effect of fatty acids on the PTP is strongly supported by the observation that cyclosporin A-sensitive swelling of mitochondria followed the addition of low concentrations of fatty acids but not of the chemical protonophore carbonyl cyanide *m*-chlorophenylhydrazone (CCCP), even though conditions were carefully chosen so that the initial depolarization caused by both compounds was identical (Więckowski and Wojtczak, 1998). This direct interaction of fatty acids with PTP components has recently been confirmed using a partially purified preparation enriched with rat brain ANT, porin, and hexokinase reconstituted into phospholipid vesicles that maintain several properties of the PTP (Beutner *et al.*, 1996). Indeed, fatty acids enabled cyclosporin A-sensitive vesicle permeabilization in the presence of low concentrations of Ca^{2+} that by themselves were not sufficient to open the PTP (Więckowski *et al.*, 2000).

2. Consequences of PTP Opening

The consequences of long-lasting PTP openings *in vitro* are relatively well understood. Pore opening causes membrane depolarization, due to proton backflow through the permeability defects, and this is followed by loss of ion homeostasis. Due to the PTP cutoff of about 1500 Da solutes may be lost according to their concentration gradient, and this leads to depletion of substrates and pyridine nucleotides (Vinogradov *et al.*, 1972). The initial uncoupling is thus followed by respiratory inhibition, which is also favored by the depletion of cytochrome *c* that follows matrix swelling and outer membrane rupture (Jacobs and Sanadi, 1960; Petronilli *et al.*, 1994b). It must be stressed that PTP opening could also favor release of cytochrome *c* without outer membrane rupture through an indirect effect on intercristal subcompartments (Frey and Mannella, 2000) that may normally restrict cytochrome *c* diffusion to the intermembrane space (Bernardi *et al.*, 2001). Release of cytochrome *c* interrupts electron flow at cytochromes *b*, and this creates a situation that is similar to that seen after mitochondrial poisoning with antimycin A, which indeed causes a very high rate of production of reactive oxygen species (Kowaltowski *et al.*, 1996). The increased rate of generation of the superoxide anion that follows cytochrome *c* release (Cai and Jones, 1998) may cause a surge in the production of hydrogen peroxide, leading in turn to glutathione oxidation. Because the PTP can be stabilized in the open conformation by oxidation of critical dithiols and of pyridine nucleotides (Costantini *et al.*, 1996), PTP opening could be both the cause and the consequence of cytochrome *c* release. In other words, PTP opening could follow the release of cytochrome *c* even when the initiating event is outer membrane insertion of tBid following its caspase-8-dependent cleavage, with Bax/Bak clustering in the outer mitochondrial membrane (Li *et al.*, 1998; Luo *et al.*, 1998; Gross *et al.*, 1999; Eskes *et al.*, 2000).

The very occurrence and the consequences of PTP opening *in vivo* are much more controversial. There is little doubt that long-lasting PTP openings *in vivo* could cause a bioenergetic catastrophe. Mitochondria would cease to make ATP, and would rather start hydrolyzing it at the maximal attainable rates, thus contributing to ATP depletion (Di Lisa and Bernardi, 1998). Ca^{2+} dysregulation would ensue both as a direct consequence of mitochondrial dysfunction and as an indirect consequence of ATP depletion. Respiration would slow down following release of mitochondrial NAD^+, and its hydrolysis by outer membrane glycohydrolase (Boyer et al., 1993; Di Lisa et al., 2001) could provide ADP-ribose with production of cyclic ADP-ribose and further release of Ca^{2+} from intracellular stores (Di Lisa and Ziegler, 2001). In addition to ATP depletion and Ca^{2+} dysregulation a potential role for PTP opening is emerging in the context of apoptosis. Intermembrane cytochrome c (Liu et al., 1996), Smac-Diablo (Du et al., 2000; Ekert et al., 2000), apoptosis-inducing factor (Susin et al., 1999), and endonuclease G (Li et al., 2001) exert their proapoptotic effects only after they have been released in the cytosol, and PTP opening could be instrumental in this process (Bernardi et al., 2001).

The controversies about the potential role of the PTP in cell death arise in part at least from methodological issues, in the sense that mitochondrial function *in situ* is monitored with indirect tools whose limits are not always adequately appreciated (Bernardi et al., 1999). For example, it is widely assumed that the occurrence of a permeability transition *in situ* can be documented as a mitochondrial depolarization, yet a decrease of the mitochondrial transmembrane potential may rather be the physiological consequence of an increased cellular demand for ATP. On the other hand *transient* PTP openings are likely to be missed by potentiometric probes because PTP flickering occurs in a time scale below the range of milliseconds (Szabó and Zoratti, 1991), which may not be detected by the slowly equilibrating potentiometric probes. To circumvent the problem of probe redistribution a method has been developed in which mitochondria are loaded with the acetomethoxyl ester of calcein within intact cells, which generates free, fluorescent intracellular calcein upon cleavage of the ester bond. After quenching of cytosolic calcein fluorescence with cobalt, intramitochondrial calcein allows visualization of mitochondria as bright fluorescent bodies over a dark background (Petronilli et al., 1999). When the PTP opens mitochondrial calcein fluorescence is quenched by cobalt, and this technique allowed detection of PTP opening even when potentiometric probes reported a constant $\Delta\psi$ (Petronilli et al., 1999). By using the calcein-loading cobalt-quenching technique we have documented that a permeability transition is a causative event in cell death induced by GD3 ganglioside (Scorrano et al., 1999); and with a very sensitive *in situ* method to detect cytochrome c release we have shown that PTP opening is causally related to cell death induced by arachidonic acid and tumor necrosis factor-α (TNF-α) (Petronilli et al., 2001; Scorrano et al., 2001). Furthermore, Di Lisa and co-workers have recently been able to detect PTP openings *in situ* in whole hearts subjected to ischemia-reperfusion, and to prove its causative role in reperfusion damage (Di Lisa et al., 2001). On balance, we think that the PTP remains a very good candidate for induction of cell death, particularly when lipid mediators such as polyunsaturated fatty acids are released as part of the intracellular signaling process or by neighboring inflammatory cells under pathophysiological conditions.

3. Structural Requirements

The structural requirements for induction of the PTP by fatty acids have not been addressed as thoroughly as the uncoupling effects. Analysis of earlier literature is complicated by the fact that PTP opening could have contributed to the uncoupling effect and vice versa. Furthermore, the incubation conditions (in particular the ionic composition of the media) can affect the fatty acid "concentration," which in turn makes comparisons problematic. We have therefore carried out a specific study of the relative potency of lauric ($C_{12:0}$), myristic ($C_{14:0}$), palmitic ($C_{16:0}$), stearic ($C_{18:0}$), linoleic ($C_{18:2}$), linolenic ($C_{18:3}$), and arachidonic ($C_{20:4}$) acids at (1) induction of the PTP and (2) uncoupling of mitochondrial respiration under

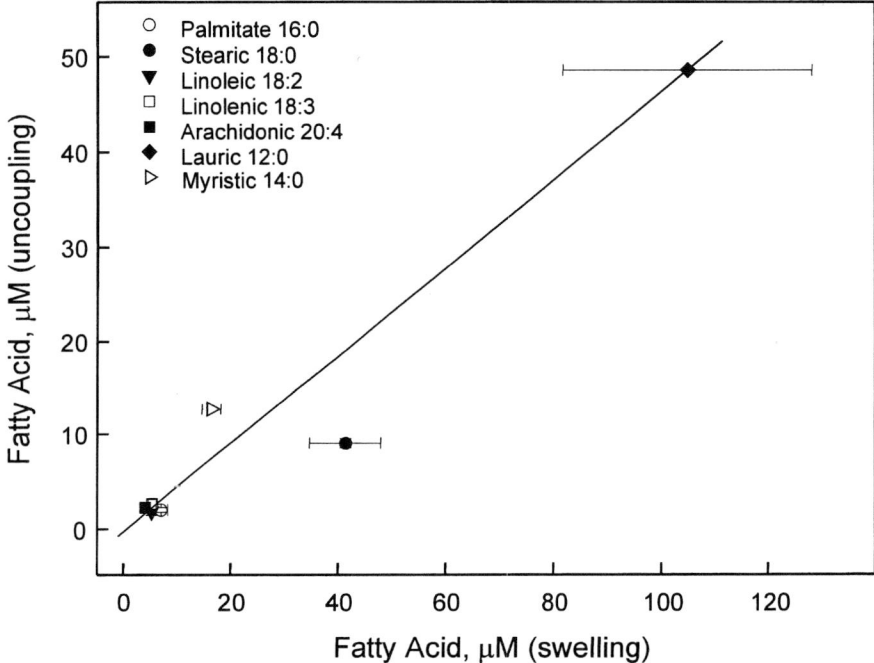

FIGURE 3 Effect of selected fatty acids on mitochondrial respiration and cyclosporin A-sensitive swelling. Rat liver mitochondria were prepared by standard differential centrifugation (Costantini *et al.*, 1995). Respiration was determined with a Clark oxygen electrode and swelling was determined as the decrease of light scattering at 540 nm exactly as described (Petronilli *et al.*, 1993b). In both cases mitochondria (0.5 mg protein ml^{-1} in a final volume of 2.0 ml) were incubated in 0.25 M sucrose, 10 mM Tris-Mops, 5 mM glutamate-Tris, 2.5 mM malate-Tris, 1 mM P$_i$-Tris, 20 μM EGTA (25°C at pH 7.4). For measurements of respiration the incubation medium was supplemented with 1 μM cyclosporin A. After 2 min 30 μM Ca^{2+} was added, followed 1 min later by the addition of increasing concentrations of lauric, myristic, palmitic, stearic, linoleic, linolenic, or arachidonic acids. Values on the ordinate refer to the concentration of fatty acid required to double the resting rate of respiration; values on the abscissa refer to the concentration of fatty acid able to induce swelling of 50% of the mitochondria. Data are from three separate experiments ± SD.

identical conditions, except for the presence of cyclosporin A in the respiration assays. The results depicted in Fig. 3 demonstrate that a higher concentration of fatty acid was always required to induce a half-maximal effect on PTP opening relative to uncoupling, yet for these fatty acids the relative potency was the same. The ranking order, from the most potent, was arachidonic = linolenic = linoleic = palmitic > stearic > myristic > lauric. These experiments suggest that to interact with the PTP these fatty acids must possess the same physicochemical properties as those conferring the uncoupling effect.

4. Summary

In summary, fatty acids can also lead to energy dissipation through *opening of the PTP*. This can be achieved partly because of the protonophoric action described in Section II.A, and partly through a direct effect on the PTP, which in turn may cause cell death through ATP depletion, Ca^{2+} dysregulation, and release of apoptogenic proteins.

C. ACYL-CoA

Potent effects on mitochondrial membranes and, in particular, on their transporting functions are exerted not only by nonesterified fatty acids but also by their thioesters with coenzyme A. Being highly amphiphilic, acyl-CoAs bind to biological membranes with high affinity, their hydrophobic hydrocarbon chain strongly anchored within the lipid core of the membrane and the large polar group of the CoA moiety protruding to the water phase. Long-chain acyl-CoAs are long known inhibitors of ANT (Pande and Blanchaer, 1971; Shug *et al.*, 1971; Lerner *et al.*, 1972; Wojtczak, 1976). The inhibition is of a competitive character (Duszyński and Wojtczak, 1975) and strongly depends on the carbon chain length of the fatty acyl moiety (Morel *et al.*, 1974). Acyl-CoAs also inhibit mitochondrial dicarboxylate, citrate, and phosphate carriers (Morel *et al.*, 1974) with a similar dependence on carbon chain length. Interestingly, the strongest inhibition is exerted by CoA esters of fatty acids with 12 to 16 carbon atoms, a similar dependence as that for the protonophoric (Rottenberg and Hashimoto, 1986; Schönfeld, 1992; Winkler and Klingenberg, 1994) and swelling-inducing (Zborowski and Wojtczak, 1963) abilities of nonesterified fatty acids. We speculated that this carbon chain length presumably ensures the "best fit" of the molecule to the mitochondrial membrane, as it corresponds to half of the lipid core thickness (Wojtczak, 1976).

Because long-chain acyl-CoAs apparently interact with ANT, it could be expected that they affect the PTP. In fact, Siliprandi and co-workers (Di Lisa *et al.*, 1989; Siliprandi *et al.*, 1992) reported a potentiation by palmitoyl-CoA of $\Delta\psi$ dissipation in the presence of Ca^{2+} and P_i. This effect could be prevented by l-carnitine, which competes with acyl-CoA for the acyl moiety, and by cyclosporin A, the well known blocker of PTP (see Section II.B). Moreover, Farber and co-workers observed that palmitoyl-CoA was able to induce PTP opening under conditions in which oxidation of long-chain acyl-CoA esters was prevented by anoxia or rotenone (Pastorino et al., 1993,1995). Again, this effect was prevented by cyclosporin A or l-carnitine, which also prevented PTP opening and cell death otherwise induced by anoxia or rotenone in intact hepatocytes (Pastorino et al., 1993, 1995).

It can thus be concluded that long-chain acyl-CoAs may also induce PTP opening and trigger apoptotic cell death when their level within the cell increases. However, it has to be stressed that acyl-CoAs do not exert protonophoric activity and do not uncouple oxidative phosphorylation (Wojtczak, 1976), as they are unable to cross the inner mitochondrial membrane. Although they insert within the membrane by their fatty acid hydrocarbon chains, they do not flip-flop because of the large hydrophilic CoA "head."

D. N-ACYLETHANOLAMINES

These are derivatives of fatty acids in which the carboxylic group of the fatty acid is linked to the amino group of ethanolamine (their proper chemical name is *fatty acylethanolamides*). N-Acylethanolamines (NAE) are natural compounds and are formed from phospholipids in a two-step enzymatic reaction—transacylation between phosphatidylcholine (I) and phosphatidylethanolamine (II) followed by hydrolysis of the resulting N-acyl-phosphatidylethanolamine (III), as depicted in the following scheme (Hansen et al., 2000b). N-Arachidonoylethanolamine (20:4-NAE), also called *anandamide*, has attracted particular attention as it appeared to be an endogenous ligand of the brain cannabinoid receptor (Devane et al., 1992; Di Marzo et al., 1994).

NAEs are present in various tissues at amounts ranging from about 0.1 to over 20 nmol/g (H. S. Hansen et al., 2000). Their content increases up to 500 nmol/g

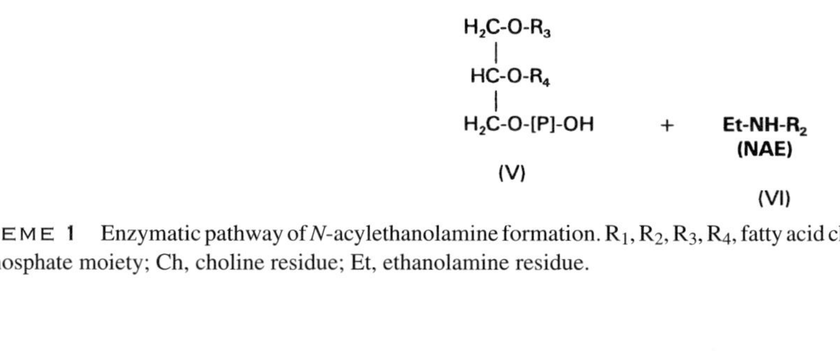

$$
\begin{array}{cccc}
\text{H}_2\text{C-O-R}_1 & \text{H}_2\text{C-O-R}_3 & \text{H}_2\text{C-O-R}_3 & \text{H}_2\text{C-O-R}_1 \\
| & | & | & | \\
\text{HC-O-R}_2 \quad + & \text{HC-O-R}_4 \quad \rightarrow & \text{HC-O-R}_4 \quad + & \text{HC-OH} \\
| & | & | & | \\
\text{H}_2\text{C-O-[P]-Ch} & \text{H}_2\text{C-O-[P]-Et-NH}_2 & \text{H}_2\text{C-O-[P]-Et-NH-R}_2 & \text{H}_2\text{C-[P]-Ch} \\
(\text{I}) & (\text{II}) & (\text{III}) & (\text{IV})
\end{array}
$$

$$
\downarrow
$$

$$
\begin{array}{cc}
\text{H}_2\text{C-O-R}_3 & \\
| & \\
\text{HC-O-R}_4 & \\
| & \\
\text{H}_2\text{C-O-[P]-OH} \quad + & \text{Et-NH-R}_2 \\
& (\text{NAE}) \\
(\text{V}) & (\text{VI})
\end{array}
$$

SCHEME 1 Enzymatic pathway of N-acylethanolamine formation. R_1, R_2, R_3, R_4, fatty acid chains; [P] phosphate moiety; Ch, choline residue; Et, ethanolamine residue.

tissue in canine heart during ischemia (Epps *et al.*, 1979). NAEs and their phospholipid precursors (compound III in Scheme 1) also accumulate in cortical neurons as result of glutamate-induced neurotoxicity (Hansen *et al.*, 1999a) and of damage produced by hydrogen peroxide (Hansen *et al.*, 1999b) and sodium azide (H. H. Hansen *et al.*, 2000). Because NAEs (in contrast to their phospholipid precursors) can easily diffuse from injured cells, in which they are presumably formed, and can be accumulated in other cells or tissues (Hillard and Jarrahian, 2000), it has been speculated that they could have signaling or cytoprotective effects. Support for such an assumption comes from the observations that NAEs can prevent increased permeability of the inner mitochondrial membrane in mitochondria (Epps *et al.*, 1982) as well as in the partially reconstituted PTP (Więckowski and Wojtczak, unpublished observations).

Anandamide and other NAEs have been found to promote apoptosis and/or inhibit cell proliferation in various types of cells (Schwarz *et al.*, 1994; De Petrocellis *et al.*, 1998; Bannerman *et al.*, 2000; Maccarrone *et al.*, 2000; Sarker *et al.*, 2000). One of the underlying mechanisms for this effect may be the inhibition of ceramidase activity by NAE (Spinedi *et al.*, 1999). For more information on this subject the reader is referred to a recent review by H. S. Hansen *et al.* (2000).

E. QUINONES

As mentioned in Section II.A.2, CoQ has striking effects on the activity of the reconstituted UCPs (Echtay *et al.*, 2000, 2001). A striking modulation of the PTP by quinones has also been discovered (Fontaine *et al.*, 1998a,b). Although the structural requirements needed for UCP and PTP regulation differ, we briefly summarize here the major findings on the structure–function correlation for PTP regulation by quinones for the sake of completeness. Quinones could be assigned to three functional classes: (1) PTP inhibitors [ubiquinone 0; decylubiquinone; ubiquinone 10; 2,3-dimethyl-6-decyl-1,4-benzoquinone; 2,3,5-trimethyl-6-geranyl-1,4-benzoquinone]; (2) PTP inducers [2,3-dimethoxy-5-methyl-6-(10-hydroxydecyl)-1,4-benzoquinone; 2,5-dihydroxy-6-undecyl-1,4-benzoquinone]; and (3) PTP-inactive quinones that counteract the effects of both inhibitors and inducers [ubiquinone 5; 2,3,5-trimethyl-6-(3-hydroxyisoamyl)-1,4-benzoquinone] (Walter *et al.*, 2000). The structure–function correlation indicates that minor modifications in the isoprenoid side chain can turn an inhibitor into an activator, and that the methoxy groups are not essential for the effects of quinones on the PTP. The above quinone analogs have a similar midpoint potential and decrease mitochondrial production of reactive oxygen species to the same extent, supporting the hypothesis that quinones may modulate the PTP through a common binding site rather than through oxidation–reduction reactions (Walter *et al.*, 2000). The possibility that endogenous quinones may modulate the PTP and that fatty acids may interfere with this interaction is currently under investigation. For further information on the effects of quinones we refer the interested reader to reviews by Fontaine and Bernardi (1999) and Walter *et al.* (2001).

III. CELLULAR EFFECTS OF FATTY ACIDS

As briefly mentioned in the Introduction, the effects of fatty acids *in situ* are predictably complex because of the multiplicity of metabolic pathways that can be involved. As specifically discussed for arachidonic acid (Brash, 2001), key issues are the physical properties of the fatty acid, the mechanisms of cellular uptake and intracellular transfer, the competition of added fatty acids with intracellular stores, and the bioactivities of the fatty acids as such and of their metabolic products. Given that all the above factors may vary in different cellular systems, a legitimate question is whether predictions can be made about the *in situ* acute effects of fatty acids on mitochondria based on their well characterized *in vitro* effects discussed above.

A. EFFECTS OF FATTY ACIDS ON MITOCHONDRIAL MEMBRANE POTENTIAL *IN SITU*

The answer to the above question is complicated by the fact that the first target of fatty acids applied to isolated cells or to perfused organs is the plasma membrane. Interaction of fatty acid with this membrane may modify its structure and permeability, and thus indirectly affect composition of the intracellular milieu. To partly overcome this problem, fatty acids have often been applied in the form of serum albumin complexes (Scholz *et al.*, 1984; Schönfeld *et al.*, 1988). For example, maintaining the oleate:serum albumin ratio below 7 (the concentration of free oleate in such a fatty acid–serum albumin "buffer" does not exceed a few micromolar) enabled us to preserve hepatocyte energetics and gluconeogenesis (Schönfeld *et al.*, 1988), whereas increasing this ratio above that value resulted in a progressive permeabilization of the plasma membrane, partial loss of cytosolic proteins (Wojtczak *et al.*, 1988), and, presumably, drastic alterations in concentrations of low-molecular-weight solutes in the cytosol.

Because the concentration of free fatty acid in equilibrium with the serum albumin complex is not easy to determine, experiments have often been performed with free, uncomplexed, fatty acids, keeping their concentration in the micromolar range. The experiment depicted in Fig. 4 was meant to study the effect of various saturated and unsaturated fatty acids on the *in situ* mitochondrial membrane potential in Morris hepatoma 1C1 cells, a minimal deviation hepatoma of the rat. The mitochondrial membrane potential was monitored in intact cells by cationic fluorescent probe tetramethylrhodamine methyl ester that enters the cells and accumulates inside the mitochondrial matrix in response to the inside-negative membrane potential (Bernardi *et al.*, 1999). After the mitochondrial fluorescence reading had stabilized, 75 μM of each fatty acid was added to the cells, and the fluorescence signal was monitored at 1 min intervals. It can be seen that in keeping with previous results (Scorrano *et al.*, 2001), arachidonic acid caused an almost complete mitochondrial depolarization within about 20 min that was matched only

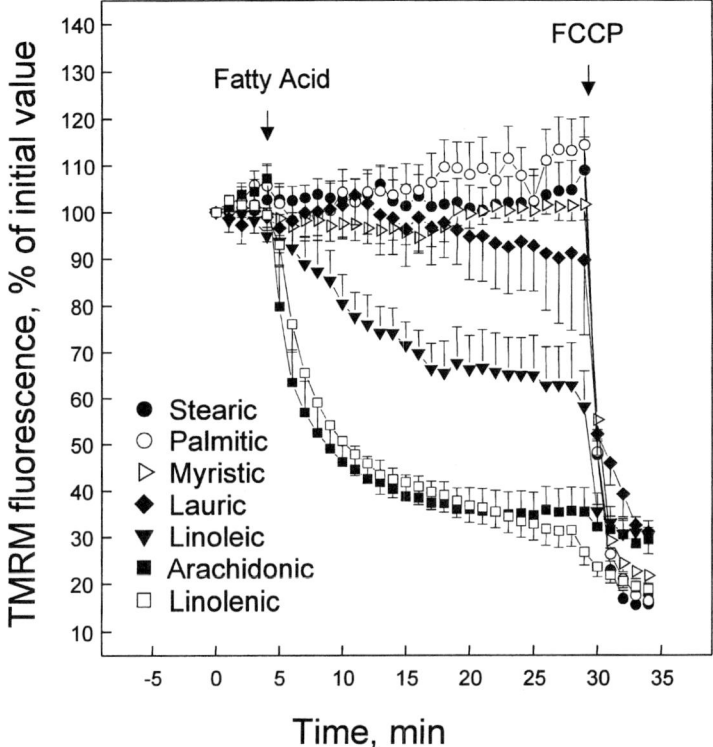

FIGURE 4 Effect of selected fatty acids on mitochondrial TMRM fluorescence in MH1C1 cells. Morris hepatoma (MH) 1C1 cells were grown in Ham's F-10 nutrient mixture containing 40 mM NaHCO$_3$ and supplemented with 10% fetal calf serum in a humidified atmosphere of 95% air–5% CO$_2$ at 37°C. Cells were rinsed with Hanks' balanced salt solution supplemented with 10 mM HEPES-NaOH pH 7.4 and 1.6 μM cyclosporin H, which inhibits the multidrug resistance pump (Bernardi *et al.*, 1999) without affecting the PTP (Nicolli *et al.*, 1996), and then loaded with 10 mM TMRM. Cellular fluorescence images were acquired exactly as described (Petronilli *et al.*, 2001). Clusters of several mitochondria (10–30) were identified as regions of interest, and fields not containing cells were taken as the background. Sequential digital images were acquired every 60 sec. The initial mitochondrial fluorescence intensities minus background are reported after normalization for comparative purposes, and they represent the mean of 10 regions of interest. Values on the ordinate report the mean ± SD of four independent experiments. Where indicated (arrows) 75 μM of the indicated fatty acids and 2 μM FCCP were added.

by the addition of linolenic acid, whereas linoleic acid was less effective. Remarkably, saturated fatty acids, lauric, myristic, palmitic, and stearic, were practically without effect. Thus, there was no obvious correlation between the effects of these fatty acids on mitochondrial energy coupling in isolated rat liver mitochondria and their acute effects on the mitochondrial membrane potential of MH1C1 cells *in situ*.

The observed effect could be, in principle, due either to a direct interaction of the fatty acids with mitochondria or a result of a changing intracellular milieu because of a defect to the plasma membrane. However, the fact (not presented in Fig. 4) that mitochondrial depolarization caused by arachidonic and linolenic acids was fully inhibited by cyclosporin A argues for the former possibility and indicates that PTP opening was the underlying cause. As expected, mitochondrial depolarization, and hence a disruption of cell energy supply, resulted in cell death, which could be detected within 90 min by propidium iodide or annexin-V staining (D. Penzo *et al.*, unpublished results).

B. SATURATED FATTY ACIDS AND CELL DEATH

Despite the above findings, it must be stressed that the situation may be different in other cell types, in particular depending on the metabolic fate of the added fatty acid. For example, rat neonatal ventricular myocytes are particularly sensitive to saturated but not monounsaturated C_{16} and C_{18} fatty acids, a finding that has been ascribed to membrane alterations (de Vries *et al.*, 1997). Neonatal cardiomyocytes appear to be very prone to PTP opening and cell death via the mitochondrial pathway following the addition of palmitate (Hickson-Bick *et al.*, 2000; Kong and Rabkin, 2000), like T cell hybridomas (de Pablo *et al.*, 1999), and these results are entirely consistent with a pioneering study of cytotoxicity of saturated fatty acids to cardiomyocytes and endothelioid cells (Wenzel and Hale, 1978a). Also the ceramide-independent stimulation of apoptosis recently reported in Chinese hamster ovary cells upon addition of palmitate, which was accompanied by increased production of reactive oxygen species, could have been due to opening of the PTP (see Section II.B), although this possibility was not formally addressed (Listenberger *et al.*, 2001). Palmitate also stimulated apoptosis of the murine hematopoietic, IL-3-dependent LyD9 cell line (Paumen *et al.*, 1997), yet in this case cell death was enhanced rather than decreased by triacsin C, an inhibitor of acyl-CoA synthesis that should have blocked mitochondrial transport of palmitoyl-CoA (Paumen *et al.*, 1997). Furthermore, and consistent with a nonmitochondrial mechanism, in the same model stearate was as effective as palmitate at inducing apoptosis, whereas unsaturated fatty acids with chain lengths between C_{16} and C_{20} were inactive (Paumen *et al.*, 1997). This is exactly the opposite of what we found for the mitochondrial pathway, where arachidonic and linolenic acids were by far the most potent inducers of PTP opening, mitochondrial depolarization, and cell death (Scorrano *et al.*, 2001) (Fig. 4). In summary, palmitate (and presumably other saturated fatty acids) can activate both mitochondria-dependent and -independent pathways, and the relative contribution of each should be addressed in specific models of cell death. It must be stressed, however, that the two pathways can communicate through downstream mediators: thus, C_{16}-ceramide generated from palmitate can feed back into mitochondria (Perry *et al.*, 2000; Kroesen *et al.*, 2001) generating a loop that may recruit the mitochondrial pathway later in the apoptotic cascade.

C. UNSATURATED FATTY ACIDS AND CELL DEATH

As mentioned in the Introduction, polyunsaturated fatty acids can enter key metabolic pathways without activation to their CoA esters. The most important for apoptotic signaling is arachidonic acid, which is released by activated phospholipases A_2 and then converted into prostaglandins, prostacyclins, and tromboxanes by COX, and into leukotrienes by LOX (Ara and Teicher, 1996; Williams et al., 1999). The biological properties of arachidonic acid have recently been covered in an excellent review by Brash (2001), to which the reader is referred for further reference. Here we will specifically focus on aspects that may be of relevance to cell death in the context of tumor formation/progression, which may involve mitochondria. It must be appreciated that the issue is complex because two opposing effects may overlap. Arachidonic acid may promote cell death in its nonesterified form (Wenzel and Hale, 1978b; Chan et al., 1998), and favor instead cell survival through its COX and LOX metabolites (Hong et al., 2001). Thus, the outcome may vary depending on the specific pathway(s) available for production and metabolism of arachidonic acid.

Experiments with animal models have suggested a link between fat content in the diet and risk of metastatic prostate cancer, which would largely depend on the supply of arachidonic acid and its transformation in prostaglandins and other COX metabolites (Gann et al., 1994). Consistently, arachidonic acid stimulates the growth of prostate cancer cells (Ghosh and Myers, 1998), possibly through overproduction of 5- and 15-LOX metabolites (Ikawa et al., 1999), which could act as antiapoptotic autocrine factors. The second important pathway in cancer development is mediated by arachidonic acid metabolism through COX-2, which is specifically upregulated in a variety of cancer cells (Eberhart et al., 1994; Oshima et al., 1996; Sheng et al., 1997; Liu et al., 1998, 2000; Shao et al., 2000) and causes mammary tumors when overexpressed under the control of the murine mammary tumor virus promoter (Liu et al., 2001). Thus, in cells in which LOX and COX are very active arachidonic acid may represent the source of powerful *antiapoptotic* agents, and thus favor tumor progression.

On the other hand, the importance of nonesterified arachidonic acid as a potential *proapoptotic* agent is supported by a variety of *in vitro* experimental models. Cells deficient in cytosolic phospholipase A_2 (cPLA$_2$), a major source of arachidonic acid, became resistant to cell death induced by TNF-α (Hayakawa et al., 1993) and, conversely, overexpression of cPLA$_2$ protected against cell death (Sapirstein et al., 1996); a cell line deficient in Δ6-desaturase (a key enzyme in the biosynthesis of arachidonic acid) was resistant to cell death induced by TNF-α (Reid et al., 1991); and inhibition of CoA-independent transacylase, which allows remodeling of arachidonate within membrane phospholipids, caused accumulation of nonesterified fatty acids and apoptosis (Surette et al., 1996, 1999). The role of arachidonic acid in apoptosis is also supported by a recent study in which its intracellular levels could be modulated by overexpression of COX-2 or of fatty acyl-CoA ligase, two of the major intracellular pathways that dispose of free arachidonic acid. Cell death was reduced by induction of either enzyme, which also protected from the killing effects of added arachidonic acid (Cao et al., 2000). An important link with the mitochondrial pathway is suggested by the findings that apoptosis induction by arachidonic acid involved activation of caspase-3, a process that is amplified by release of mitochondrial cytochrome c (Liu et al., 1996) and Smac/DIABLO (Du et al., 2000); and that Bax, which kills cells via the mitochondrial pathway (Wolter et al., 1997; Jürgensmeier et al., 1998; Gross et al., 1998; Narita et al., 1998; Pastorino et al., 1998; Więckowski et al., 2001), is essential for the apoptotic response of cells to nonsteroidal antiinflammatory drugs (NSAIDs) (Zhang et al., 2000). Our own results are consistent with the idea that mitochondria are key targets for added and possibly endogenous arachidonic acid released by cPLA$_2$ upon its activation by TNF-α (Scorrano et al., 2001). Indeed, in our model arachidonic acid demonstrably caused cell death through induction of PTP opening, which was followed by cytochrome c release (Scorrano et al., 2001). Importantly, arachidonic acid also stimulates the production of ceramide (Müller et al., 1995; Jayadev et al., 1994, 1997; Chan et al., 1998; Surette et al., 1999), which may feed back on mitochondria causing or potentiating PTP opening (Arora et al., 1997; Scorrano et al., 1999; Pacher and Hajnóczky, 2001).

Despite the complexity of the issue, inhibition of LOX and COX (particularly of COX-2) is a very viable pharmacological strategy for treatment of malignancies; and it is reassuring that NSAIDs unequivocally reduce the risk of colon cancer (Thun et al., 1991) irrespective of whether the effect depends more on the elevation of intracellular nonesterified arachidonic acid (Chan et al., 1998; Cao et al., 2000) or on the decreased production of its COX and LOX metabolites (Hong et al., 2001).

IV. CONCLUSIONS AND PERSPECTIVES

In this article we have discussed the basis for the complex effects of fatty acids and related compounds on mitochondria, and how these mechanisms may contribute to the regulation of the mitochondrial pathways to cell death. There is no doubt that manipulation of the dietary contents of fatty acids (Gann et al., 1994; Petrik et al., 2000) and treatment with NSAIDs (Thun et al., 1991; Liu et al., 2000) may reduce the incidence of gastrointestinal and prostate tumors. We hope that the identification of mitochondria as targets of the proapoptotic effects of fatty acids, and an increased understanding of the mechanism that serves as the basis for their tissue-specific effects, will help lead to the design of new pharmacological strategies to fight neoplastic diseases.

ACKNOWLEDGMENTS

Research in our laboratories is supported by the Italian Ministry for the University and Scientific Research, the National Research Council, Telethon-Italy (Grant 1141), the Armenise Harvard

Foundation and the NIH-PHS (USA) (to P.B.), and the Polish State Committee for Scientific Research (Grants KBN 6 P04A 057 09 and 6 P04A 005 16) (to L.W.).

REFERENCES

Andreyev, A. Y., Bondareva, T. O., Dedukhova, V. I., Mokhova, E. N., Skulachev, V. P., Tsofina, L. M., Volkov, N. I., and Vygodina, T. V. (1989). The ATP/ADP-antiporter is involved in the uncoupling effect of fatty acids on mitochondria. *Eur. J. Biochem.* **182,** 585–592.

Aquila, H., Link, T. A., and Klingenberg, M. (1987). Solute carriers involved in energy transfer of mitochondria form a homologous protein family. *FEBS Lett.* **212,** 1–9.

Ara, G., and Teicher, B. A. (1996). Cyclooxygenase and lipooxygenase inhibitors in cancer therapy. *Prostaglandins Leukot. Essent. Fatty Acids* **54,** 3–16.

Arora, A. S., Jones, B. J., Patel, T. C., Bronk, S. F., and Gores, G. J. (1997). Ceramide induces hepatocyte cell death through disruption of mitochondrial function in the rat. *Hepatology* **25,** 958–963.

Azzone, G. F., and Azzi, A. (1965). Volume changes in liver mitochondria. *Proc. Natl. Acad. Sci. USA* **53,** 1084–1089.

Bannerman, P., Nichols, W., Puhalla, S., Oliver, T., Berman, M., and Pleasure, D. (2000). Early migratory rat neural crest cells express functional gap junctions: Evidence that neural crest cell survival requires gap junction function. *J. Neurosci. Res.* **61,** 605–615.

Barakat, H. A., Kasperek, G. J., Dohm, G. L., Tapscott, E. B., and Snider, R. D. (1982). Fatty acid oxidation by liver and muscle preparations of exhaustively exercised rats. *Biochem. J.* **208,** 419–424.

Bernardi, P. (1992). Modulation of the mitochondrial cyclosporin A-sensitive permeability transition pore by the proton electrochemical gradient. Evidence that the pore can be opened by membrane depolarization. *J. Biol. Chem.* **267,** 8834–8839.

Bernardi, P. (1999). Mitochondrial transport of cations: Channels, exchangers and permeability transition. *Physiol. Rev.* **79,** 1127–1155.

Bernardi, P., Broekemeier, K. M., and Pfeiffer, D. R. (1994). Recent progress on regulation of the mitochondrial permeability transition pore; a cyclosporin-sensitive pore in the inner mitochondrial membrane. *J. Bioenerg. Biomembr.* **26,** 509–517.

Bernardi, P., Scorrano, L., Colonna, R., Petronilli, V., and Di Lisa, F. (1999). Mitochondria and cell death. Mechanistic aspects and methodological issues. *Eur. J. Biochem.* **264,** 687–701.

Bernardi, P., Petronilli, V., Di Lisa F., and Forte, M. (2001). A mitochondrial perspective on cell death. *Trends Biochem. Sci.* **26,** 112–117.

Beutner, G., Rück, A., Riede, B., Welte, W., and Brdiczka, D. (1996). Complexes between kinases, mitochondrial porin and adenylate translocator in rat brain resemble the permeability transition pore. *FEBS Lett.* **396,** 189–195.

Blazquez, C., Galve-Roperh, I., and Guzman, M. (2000). De novo-synthesized ceramide signals apoptosis in astrocytes via extracellular signal-regulated kinase. *FASEB J.* **14,** 2315–2322.

Bode, A. M., Byrd, S., and Klug, G. A. (1990). The relationship between plasma free fatty acids and liver mitochondrial function in vivo. *Biochim. Biophys. Acta* **1047,** 161–167.

Boss, O., Samec, S., Paoloni-Giacobino, A., Rossier, C., Dulloo, A., Seydoux, J., Muzzin, P., and Giacobino, J. P. (1997). Uncoupling protein-3: A new member of the mitochondrial carrier family with tissue-specific expression. *FEBS Lett.* **408,** 39–42.

Boyer, C. S., Moore, G. A., and Moldeus, P. (1993). Submitochondrial localization of the NAD^+ glycohydrolase. Implications for the role of pyridine nucleotide hydrolysis in mitochondrial calcium fluxes. *J. Biol. Chem.* **268,** 4016–4020.

Brash, A. R. (2001). Arachidonic acid as a bioactive molecule. *J. Clin. Invest.* **107,** 1339–1345.

Broekemeier, K. M., and Pfeiffer, D. R. (1995). Inhibition of the mitochondrial permeability transition by cyclosporin A during long time frame experiments: Relationship between pore opening and the activity of mitochondrial phospholipases. *Biochemistry* **34,** 16440–16449.

Broekemeier, K. M., Dempsey, M. E., and Pfeiffer, D. R. (1989). Cyclosporin A is a potent inhibitor of the inner membrane permeability transition in liver mitochondria. *J. Biol. Chem.* **264,** 7826–7830.

Brustovetsky, N., and Klingenberg, M. (1994). The reconstituted ADP/ATP carrier can mediate H^+ transport by free fatty acids, which is further stimulated by mersalyl. *J. Biol. Chem.* **269,** 27329–27336.

Cai, J., and Jones, D. P. (1998). Superoxide in apoptosis. Mitochondrial generation triggered by cytochrome c loss. *J. Biol. Chem.* **273,** 11401–11404.

Cao, Y., Pearman, A. T., Zimmerman, G. A., McIntyre, T. M., and Prescott, S. M. (2000). Intracellular unesterified arachidonic acid signals apoptosis. *Proc. Natl. Acad. Sci. USA* **97,** 11280–11285.

Chan, T. A., Morin, P. J., Vogelstein, B., and Kinzler, K. W. (1998). Mechanisms underlying nonsteroidal antiinflammatory drug-mediated apoptosis. *Proc. Natl. Acad. Sci. USA* **95,** 681–686.

Costantini, P., Petronilli, V., Colonna, R., and Bernardi, P. (1995). On the effects of paraquat on isolated mitochondria. Evidence that paraquat causes opening of the cyclosporin A-sensitive permeability transition pore synergistically with nitric oxide. *Toxicology* **99,** 77–88.

Costantini, P., Chernyak, B. V., Petronilli, V., and Bernardi, P. (1996). Modulation of the mitochondrial permeability transition pore by pyridine nucleotides and dithiol oxidation at two separate sites. *J. Biol. Chem.* **271,** 6746–6751.

Crompton, M. (1999). The mitochondrial permeability transition pore and its role in cell death. *Biochem. J.* **341,** 233–249.

Crompton, M., Ellinger, H., and Costi, A. (1988). Inhibition by cyclosporin A of a Ca^{2+}-dependent pore in heart mitochondria activated by inorganic phosphate and oxidative stress. *Biochem. J.* **255,** 357–360.

Davidson, A. M., and Halestrap, A. P. (1990). Partial inhibition by cyclosporin A of the swelling of liver mitochondria in vivo and in vitro induced by sub-micromolar $[Ca^{2+}]$, but not by butyrate. Evidence for two distinct swelling mechanisms. *Biochem. J.* **268,** 147–152.

Dedukhova, V. I., Mokhova, E. N., Skulachev, V. P., Starkov, A. A., Arrigoni Martelli, E., and Bobyleva, V. A. (1991). Uncoupling effect of fatty acids on heart muscle mitochondria and submitochondrial particles. *FEBS Lett.* **295,** 51–54.

de Pablo, M. A., Susin, S. A., Jacotot, E., Larochette, N., Costantini, P., Ravagnan, L., Zamzami, N., and Kroemer, G. (1999). Palmitate induces apoptosis via a direct effect on mitochondria. *Apoptosis* **4,** 81–87.

De Petrocellis, L., Melck, D., Palmisano, A., Bisogno, T., Laezza, C., Bifulco, M., and Di Marzo, V. (1998). The endogenous cannabinoid anandamide inhibits human breast cancer cell proliferation. *Proc. Natl. Acad. Sci. USA* **95,** 8375–8380.

Devane, W. A., Hanuš, L., Breuer, A., Pertwee, R. G., Stevenson, L. A., Griffin, G., Gibson, D., Mandelbaum, A., Etinger, A., and Mechoulam, R. (1992). Isolation and structure of a brain constituent that binds to the cannabinoid receptor. *Science* **258,** 1946–1949.

de Vries, J. E., Vork, M. M., Roemen, T. H., de Jong, Y. F., Cleutjens, J. P., van der Vusse, G. J., and van Bilsen, M. (1997). Saturated but not mono-unsaturated fatty acids induce apoptotic cell death in neonatal rat ventricular myocytes. *J. Lipid Res.* **38,** 1384–1394.

Di Lisa, F., and Bernardi, P. (1998). Mitochondrial function as a determinant of recovery or death in cell response to injury. *Mol. Cell. Biochem.* **184,** 379–391.

Di Lisa, F., and Ziegler, M. (2001). Pathophysiological relevance of mitochondria in NAD^+ metabolism. *FEBS Lett.* **492,** 4–8.

Di Lisa, F., Menabò, R., and Siliprandi, N. (1989). L-Propionyl-carnitine protection of mitochondria in ischemic rat hearts. *Mol. Cell. Biochem.* **88,** 169–173.

Di Lisa, F., Menabò, R., Canton, M., Barile, M., and Bernardi, P. (2001). Opening of the mitochondrial permeability transition pore causes depletion of mitochondrial and cytosolic NAD^+ and is a causative event in the death of myocytes in postischemic reperfusion of the heart. *J. Biol. Chem.* **276,** 2571–2575.

Di Marzo, V., Fontana, A., Cadas, H., Schinelli, S., Cimino, G., Schwartz, J. C., and Piomelli, D. (1994). Formation and inactivation of endogenous cannabinoid anandamide in central neurons. *Nature* **372,** 686–691.

Du, C., Fang, M., Li, Y., Li, L., and Wang, X. (2000). Smac, a mitochondrial protein that promotes cytochrome c-dependent caspase activation by eliminating IAP inhibition. *Cell* **102**, 33–42.

Duszyński, J., and Wojtczak, L. (1975). Effect of metal cations on the inhibition of adenine nucleotide translocation by acyl-CoA. *FEBS Lett.* **50**, 74–78.

Eberhart, C. E., Coffey, R. J., Radhika, A., Giardiello, F. M., Ferrenbach, S., and DuBois, R. N. (1994). Up-regulation of cyclooxygenase 2 gene expression in human colorectal adenomas and adenocarcinomas. *Gastroenterology* **107**, 1183–1188.

Echtay, K. S., Winkler, E., and Klingenberg, M. (2000). Coenzyme Q is an obligatory cofactor for uncoupling protein function. *Nature* **408**, 609–613.

Echtay, K. S., Winkler, E., Frischmuth, K., and Klingenberg, M. (2001). Uncoupling proteins 2 and 3 are highly active H^+ transporters and highly nucleotide sensitive when activated by coenzyme Q (ubiquinone). *Proc. Natl. Acad. Sci. USA* **98**, 1416–1421.

Ekert, P. G., Silke, J., Connolly, L. M., Reid, G. E., Moritz, R. L., and Vaux, D. L. (2000). Identification of DIABLO, a mammalian protein that promotes apoptosis by binding to and antagonizing IAP Proteins. *Cell* **102**, 43–53.

Engstová, H., Žačková, M., Růžička, M., Meinhardt, A., Hanuš, J., Krämer, R., and Ježek, P. (2001). Natural and azido fatty acids inhibit phosphate transport and activate fatty acid anion uniport mediated by the mitochondrial phosphate carrier. *J. Biol. Chem.* **276**, 4683–4691.

Epps, D. E., Schmid, P. C., Natarajan, V., and Schmid, H. H. (1979). N-Acylethanolamine accumulation in infarcted myocardium. *Biochem. Biophys. Res. Commun.* **90**, 628–633.

Epps, D. E., Palmer, J. W., Schmid, H. H., and Pfeiffer, D. R. (1982). Inhibition of permeability-dependent Ca^{2+} release from mitochondria by N-acylethanolamines, a class of lipids synthesized in ischemic heart tissue. *J. Biol. Chem.* **257**, 1383–1391.

Eskes, R., Desagher, S., Antonsson, B., and Martinou, J. C. (2000). Bid induces the oligomerization and insertion of Bax into the outer mitochondrial membrane. *Mol. Cell. Biol.* **20**, 929–935.

Fleury, C., and Sanchis, D. (1999). The mitochondrial uncoupling protein-2: Current status. *Int. J. Biochem. Cell Biol.* **31**, 1261–1278.

Fontaine, E., and Bernardi, P. (1999). Progress on the mitochondrial permeability transition pore. Regulation by complex I and ubiquinone analogs. *J. Bioenerg. Biomembr.* **31**, 335–345.

Fontaine, E., Eriksson, O., Ichas, F., and Bernardi, P. (1998a). Regulation of the permeability transition pore in skeletal muscle mitochondria. Modulation by electron flow through the respiratory chain complex I. *J. Biol. Chem.* **273**, 12662–12668.

Fontaine, E., Ichas, F., and Bernardi, P. (1998b). A ubiquinone-binding site regulates the mitochondrial permeability transition pore. *J. Biol. Chem.* **273**, 25734–25740.

Frey, T. G., and Mannella, C. A. (2000). The internal structure of mitochondria. *Trends. Biochem. Sci.* **25**, 319–324.

Gann, P. H., Hennekens, C. H., Sacks, F. M., Grodstein, F., Giovannucci, E. L., and Stampfer, M. J. (1994). Prospective study of plasma fatty acids and risk of prostate cancer. *J. Natl. Cancer Inst.* **86**, 281–286.

Garlid, K. D., Orosz, D. E., Modrianský, M., Vassanelli, S., and Ježek, P. (1996). On the mechanism of fatty acid-induced proton transport by mitochondrial uncoupling protein. *J. Biol. Chem.* **271**, 2615–2620.

Ghosh, J., and Myers, C. E. (1998). Inhibition of arachidonate 5–lipoxygenase triggers massive apoptosis in human prostate cancer cells. *Proc. Natl. Acad. Sci. USA* **95**, 13182–13187.

Gross, A., Jockel, J., Wei, M. C., and Korsmeyer, S. J. (1998). Enforced dimerization of Bax results in its translocation, mitochondrial dysfunction and apoptosis. *EMBO J.* **17**, 3878–3885.

Gross, A., Yin, X. M., Wang, K., Wei, M. C., Jockel, J., Milliman, C., Erdjument, B. H., Tempst, P., and Korsmeyer, S. J. (1999). Caspase cleaved Bid targets mitochondria and is required for cytochrome c release, while Bcl-X_L prevents this release but not tumor necrosis factor-R1/Fas death. *J. Biol. Chem.* **274**, 1156–1163.

Hansen, H. H., Hansen, S. H., Schousboe, A., and Hansen, H. S. (2000). Determination of the phospholipid precursor of anandamide and other N-acylethanolamine phospholipids before and after sodium azide-induced toxicity in cultured neocortical neurons. *J. Neurochem.* **75**, 861–871.

Hansen, H. S., Moesgaard, B., Hansen, H. H., Schousboe, A., and Petersen, G. (1999a). Formation of N-acyl-phosphatidylethanolamine and N-acylethanolamine (including anandamide) during glutamate-induced neurotoxicity. *Lipids* **34** (Suppl.) S327–S330.

Hansen, H. S., Hansen, H. H., Moesgaard, B., Petersen, G., and Hansen S. H. (1999b). H_2O_2-induced formation of N-acylethanolamine phospholipids (NAPE) and N-acylethanolamine (NAE) in cortical neurons. *Chem. Phys. Lipids* **101**, 156.

Hansen, H. S., Moesgaard, B., Hansen, H. H., and Petersen, G. (2000). N-Acylethanolamines and precursor phospholipids—relation to cell injury. *Chem. Phys. Lipids* **108**, 135–150.

Haworth, R. A., and Hunter, D. R. (1979). The Ca^{2+}-induced membrane transition in mitochondria. II. Nature of the Ca^{2+} trigger site. *Arch. Biochem. Biophys.* **195**, 460–467.

Hayakawa, M., Ishida, N., Takeuchi, K., Shibamoto, S., Hori, T., Oku, N., Ito, F., and Tsujimoto, M. (1993). Arachidonic acid-selective cytosolic phospholipase A_2 is crucial in the cytotoxic action of tumor necrosis factor. *J. Biol. Chem.* **268**, 11290–11295.

Hermesh, O., Kalderon, B., and Bar-Tana, J. (1998). Mitochondria uncoupling by a long chain fatty acyl analogue. *J. Biol. Chem.* **273**, 3937–3942.

Hickson-Bick, D. L., Buja, M. L., and McMillin, J. B. (2000). Palmitate-mediated alterations in the fatty acid metabolism of rat neonatal cardiac myocytes. *J. Mol. Cell. Cardiol.* **32**, 511–519.

Hillard, C. J., and Jarrahian, A. (2000). The movement of N-arachidonoylethanolamine (anandamide) across cellular membranes. *Chem. Phys. Lipids* **108**, 123–134.

Hong, K. H., Bonventre, J. C., O'Leary, E., Bonventre, J. V., and Lander, E. S. (2001). Deletion of cytosolic phospholipase A_2 suppresses Apc^{Min}-induced tumorigenesis. *Proc. Natl. Acad. Sci. USA* **98**, 3935–3939.

Hunter, D. R., and Haworth, R. A. (1979a). The Ca^{2+}-induced membrane transition in mitochondria. I. The protective mechanisms. *Arch. Biochem. Biophys.* **195**, 453–459.

Hunter, D. R., and Haworth, R. A. (1979b). The Ca^{2+}-induced membrane transition in mitochondria. III. Transitional Ca^{2+} release. *Arch. Biochem. Biophys.* **195**, 468–477.

Hunter, D. R., Haworth, R. A., and Southard, J. H. (1976). Relationship between configuration, function, and permeability in calcium-treated mitochondria. *J. Biol. Chem.* **251**, 5069–5077.

Ikawa, H., Kamitani, H., Calvo, B. F., Foley, J. F., and Eling, T. E. (1999). Expression of 15–lipoxygenase-1 in human colorectal cancer. *Cancer Res.* **59**, 360–366.

Jabůrek, M., Vařecha, M., Ježek, P., and Garlid, K. D. (2001). Alkylsulfonates as probes of uncoupling protein transport mechanism. Ion pair transport demonstrates that direct H^+ translocation by UCP1 is not necessary for uncoupling. *J. Biol. Chem.* **276**, 31897–31905.

Jacobs, E. E., and Sanadi, D. R. (1960). The reversible removal of cytochome c from mitochondria. *J. Biol. Chem.* **235**, 531–534.

Jarmuszkiewicz, W. (2001). Uncoupling proteins in mitochondria of plants and some microorganisms. *Acta Biochim. Pol.* **48**, 145–155.

Jarmuszkiewicz, W., Sluse-Goffart, C. M., Hryniewiecka, L., and Sluse, F. E. (1999). Identification and characterization of a protozoan uncoupling protein in Acanthamoeba castellanii. *J. Biol. Chem.* **274**, 23198–23202.

Jarmuszkiewicz, W., Milani, G., Fortes, F., Schreiber, A. Z., Sluse, F. E., and Vercesi, A. E. (2000). First evidence and characterization of an uncoupling protein in fungi kingdom: CpUCP of *Candida parapsilosis*. *FEBS Lett.* **467**, 145–149.

Jayadev, S., Linardic, C. M., and Hannun, Y. A. (1994). Identification of arachidonic acid as a mediator of sphingomyelin hydrolysis in response to tumor necrosis factor α. *J. Biol. Chem.* **269**, 5757–5763.

Jayadev, S., Hayter, H. L., Andrieu, N., Gamard, C. J., Liu, B., Balu, R., Hayakawa, M., Ito, F., and Hannun, Y. A. (1997). Phospholipase A_2 is necessary for tumor necrosis factor α-induced ceramide generation in L929 cells. *J. Biol. Chem.* **272**, 17196–17203.

Ježek, P. (1999). Fatty acid interaction with mitochondrial uncoupling proteins. *J. Bioenerg. Biomembr.* **31**, 457–466.

Ježek, P., Hanuš, J., Semrad, C., and Garlid, K. D. (1996). Photoactivated azido fatty acid irreversibly inhibits anion and proton transport through the mitochondrial incoupling protein. *J. Biol. Chem.* **271,** 6199–6205.

Ježek, P., Costa, A. D., and Vercesi, A. E. (1997a). Reconstituted plant uncoupling mitochondrial protein allows for proton translocation via fatty acid cycling mechanism. *J. Biol. Chem.* **272,** 24272–24278.

Ježek, P., Modrianský, M., and Garlid, K. D. (1997b). A structure-activity study of fatty acid interaction with mitochondrial uncoupling protein. *FEBS Lett.* **408,** 166–170.

Ježek, P., Modrianský, M., and Garlid, K. D. (1997c). Inactive fatty acids are unable to flip-flop across the lipid bilayer. *FEBS Lett.* **408,** 161–165.

Ježek, P., Engstová, H., Žačková, M., Vercesi, A. E., Costa, A. D., Arruda, P., and Garlid, K. D. (1998). Fatty acid cycling mechanism and mitochondrial uncoupling proteins. *Biochim. Biophys. Acta* **1365,** 319–327.

Jürgensmeier, J. M., Xie, Z., Deveraux, Q., Ellerby, L., Bredesen, D., and Reed, J. C. (1998). Bax directly induces release of cytochrome c from isolated mitochondria. *Proc. Natl. Acad. Sci. USA* **95,** 4997–5002.

Kamp, F., and Hamilton, J. A. (1992). pH gradients across phospholipid membranes caused by fast flip-flop of un-ionized fatty acids. *Proc. Natl. Acad. Sci. USA* **89,** 11367–11370.

Kerner, J., and Hoppel, C. (2000). Fatty acid import into mitochondria. *Biochim. Biophys. Acta* **1486,** 1–17.

Klingenberg, M. (1990). Mechanism and evolution of the uncoupling protein of brown adipose tissue. *Trends Biochem. Sci.* **15,** 108–112.

Klingenberg, M. (1999). Uncoupling protein—a useful energy dissipator. *J. Bioenerg. Biomembr.* **31,** 419–430.

Klingenberg, M., and Huang, S. G. (1999). Structure and function of the uncoupling protein from brown adipose tissue. *Biochim. Biophys. Acta* **1415,** 271–296.

Kong, J. Y., and Rabkin, S. W. (2000). Palmitate-induced apoptosis in cardiomyocytes is mediated through alterations in mitochondria: Prevention by cyclosporin A. *Biochim. Biophys. Acta* **1485,** 45–55.

Kowaltowski, A. J., Castilho, R. F., and Vercesi, A. E. (1996). Opening of the mitochondrial permeability transition pore by uncoupling or inorganic phosphate in the presence of Ca^{2+} is dependent on mitochondrial-generated reactive oxygen species. *FEBS Lett.* **378,** 150–152.

Kroemer, G., and Reed, J. C. (2000). Mitochondrial control of cell death. *Nat. Med.* **6,** 513–519.

Kroesen, B. J., Pettus, B., Luberto, C., Busman, M., Sietsma, H., de Leij, L., and Hannun, Y. A. (2001). Induction of apoptosis through B-cell receptor cross-linking occurs via de novo generated C16–ceramide and involves mitochondria. *J. Biol. Chem.* **276,** 13606–13614.

Laloi, M., Klein, M., Riesmeier, J. W., Müller-Röber, B., Fleury, C., Bouillaud, F., and Ricquier, D. (1997). A plant cold-induced uncoupling protein. *Nature* **389,** 135–136.

Lehninger, A. L., and Remmert, L. F. (1959). An endogenous uncoupling and swelling agent in liver mitochondria and its enzymic function. *J. Biol. Chem.* **234,** 2459–2464.

Lerner, E., Shug, A. L., Elson, C., and Shrago, E. (1972). Reversible inhibition of adenine nucleotide translocation by long chain fatty acyl coenzyme A esters in liver mitochondria of diabetic and hibernating animals. *J. Biol. Chem.* **247,** 1513–1519.

Li, H., Zhu, H., Xu, C. J., and Yuan, J. (1998). Cleavage of BID by caspase 8 mediates the mitochondrial damage in the Fas pathway of apoptosis. *Cell* **94,** 491–501.

Li, L. Y., Luo, X., and Wang, X. (2001). Endonuclease G is an apoptotic DNase when released from mitochondria. *Nature* **412,** 95–99.

Listenberger, L. L., Ory, D. S., and Schaffer, J. E. (2001). Palmitate-induced apoptosis can occur through a ceramide-independent pathway. *J. Biol. Chem.* **276,** 14890–14895.

Liu, C. H., Chang, H. S., Narko, K., Trifan, O. C., Wu, M. T., Smith, E., Haudenschild, C., Lane, T. F., and Hla, T. (2001). Overexpression of cyclooxygenase-2 is sufficient to induce tumorigenesis in transgenic mice. *J. Biol. Chem.* **276,** 18563–18569.

Liu, X., Kim, C. N., Yang, J., Jemmerson, R., and Wang, X. (1996). Induction of apoptotic program in cell-free extracts: Requirement for dATP and cytochrome c. *Cell* **86**, 147–157.

Liu, X. H., Yao, S., Kirschenbaum, A., and Levine, A. C. (1998). NS398, a selective cyclooxygenase-2 inhibitor, induces apoptosis and down-regulates bcl-2 expression in LNCaP cells. *Cancer Res.* **58**, 4245–4249.

Liu, X. H., Kirschenbaum, A., Yao, S., Lee, R., Holland, J. F., and Levine, A. C. (2000). Inhibition of cyclooxygenase-2 suppresses angiogenesis and the growth of prostate cancer in vivo. *J. Urol.* **164**, 820–825.

Luo, X., Budihardjo, I., Zou, H., Slaughter, C., and Wang, X. (1998). Bid, a Bcl2 interacting protein, mediates cytochrome c release from mitochondria in response to activation of cell surface death receptors. *Cell* **94**, 481–490.

Maccarrone, M., Salvati, S., Bari, M., and Finazzi-Agrò, A. (2000). Anandamide and 2-arachidonoyl-glycerol inhibit fatty acid amide hydrolase by activating the lipoxygenase pathway of the arachidonate cascade. *Biochem. Biophys. Res. Commun.* **278**, 576–583.

Mao, W., Yu, X. X., Zhong, A., Li, W., Brush, J., Sherwood, S. W., Adams, S. H., and Pan, G. (1999). UCP4, a novel brain-specific mitochondrial protein that reduces membrane potential in mammalian cells. *FEBS Lett.* **443**, 326–330.

Morel, F., Lauquin, G., Lunardi, J., Duszyński, J., and Vignais, P. V. (1974). An appraisal of the functional significance of the inhibitory effect of long chain acyl-CoAs on mitochondrial transports. *FEBS Lett.* **39**, 133–138.

Müller, G., Ayoub, M., Storz, P., Rennecke, J., Fabbro, D., and Pfizenmaier, K. (1995). PKC ξ is a molecular switch in signal transduction of TNF-α, bifunctionally regulated by ceramide and arachidonic acid. *EMBO J.* **14**, 1961–1969.

Mutomba, M. C., Yuan, H., Konyavko, M., Adachi, S., Yokoyama, C. B., Esser, V., McGarry, J. D., Babior, B. M., and Gottlieb, R. A. (2000). Regulation of the activity of caspases by L-carnitine and palmitoylcarnitine. *FEBS Lett.* **478**, 19–25.

Narita, M., Shimizu, S., Ito, T., Chittenden, T., Lutz, R. J., Matsuda, H., and Tsujimoto, Y. (1998). Bax interacts with the permeability transition pore to induce permeability transition and cytochrome c release in isolated mitochondria. *Proc. Natl. Acad. Sci. USA* **95**, 14681–14686.

Nicholls, D. G., and Locke, R. M. (1984). Thermogenic mechanisms in brown fat. *Physiol. Rev.* **64**, 1–64.

Nicolli, A., Basso, E., Petronilli, V., Wenger, R. M., and Bernardi, P. (1996). Interactions of cyclophilin with the mitochondrial inner membrane and regulation of the permeability transition pore, a cyclosporin A-sensitive channel. *J. Biol. Chem.* **271**, 2185–2192.

Oshima, M., Dinchuk, J. E., Kargman, S. L., Oshima, H., Hancock, B., Kwong, E., Trzaskos, J. M., Evans, J. F., and Taketo, M. M. (1996). Suppression of intestinal polyposis in $Apc^{\Delta 716}$ knockout mice by inhibition of cyclooxygenase 2 (COX-2). *Cell* **87**, 803–809.

Pacher, P., and Hajnóczky, G. (2001). Propagation of the apoptotic signal by mitochondrial waves. *EMBO J.* **20**, 4107–4121.

Pande, S. V., and Blanchaer, M. C. (1971). Reversible inhibition of mitochondrial adenosine diphosphate phosphorylation by long chain acyl coenzyme A esters. *J. Biol. Chem.* **246**, 402–411.

Pastorino, J. G., Snyder, J. W., Serroni, A., Hoek, J. B., and Farber, J. L. (1993). Cyclosporin and carnitine prevent the anoxic death of cultured hepatocytes by inhibiting the mitochondrial permeability transition. *J. Biol. Chem.* **268**, 13791–13798.

Pastorino, J. G., Snyder, J. W., Hoek, J. B., and Farber, J. L. (1995). Ca^{2+} depletion prevents anoxic death of hepatocytes by inhibiting mitochondrial permeability transition. *Am. J. Physiol.* **268**, C676–685.

Pastorino, J. G., Chen, S. T., Tafani, M., Snyder, J. W., and Farber, J. L. (1998). The overexpression of Bax produces cell death upon induction of the mitochondrial permeability transition. *J. Biol. Chem.* **273**, 7770–7775.

Paumen, M. B., Ishida, Y., Muramatsu, M., Yamamoto, M., and Honjo, T. (1997). Inhibition of carnitine palmitoyltransferase I augments sphingolipid synthesis and palmitate-induced apoptosis. *J. Biol. Chem.* **272**, 3324–3329.

Perry, D. K., Carton, J., Shah, A. K., Meredith, F., Uhlinger, D. J., and Hannun, Y. A. (2000). Serine palmitoyltransferase regulates de novo ceramide generation during etoposide-induced apoptosis. *J. Biol. Chem.* **275,** 9078–9084.

Petrik, M. B., McEntee, M. F., Chiu, C. H., and Whelan, J. (2000). Antagonism of arachidonic acid is linked to the antitumorigenic effect of dietary eicosapentaenoic acid in Apc(Min/+) mice. *J. Nutr.* **130,** 1153–1158.

Petronilli, V., Cola, C., and Bernardi, P. (1993a). Modulation of the mitochondrial cyclosporin A-sensitive permeability transition pore. II. The minimal requirements for pore induction underscore a key role for transmembrane electrical potential, matrix pH, and matrix Ca^{2+}. *J. Biol. Chem.* **268,** 1011–1016.

Petronilli, V., Cola, C., Massari, S., Colonna, R., and Bernardi, P. (1993b). Physiological effectors modify voltage sensing by the cyclosporin A-sensitive permeability transition pore of mitochondria. *J. Biol. Chem.* **268,** 21939–21945.

Petronilli, V., Costantini, P., Scorrano, L., Colonna, R., Passamonti, S., and Bernardi, P. (1994a). The voltage sensor of the mitochondrial permeability transition pore is tuned by the oxidation-reduction state of vicinal thiols. Increase of the gating potential by oxidants and its reversal by reducing agents. *J. Biol. Chem.* **269,** 16638–16642.

Petronilli, V., Nicolli, A., Costantini, P., Colonna, R., and Bernardi, P. (1994b). Regulation of the permeability transition pore, a voltage-dependent mitochondrial channel inhibited by cyclosporin A. *Biochim. Biophys. Acta* **1187,** 255–259.

Petronilli, V., Miotto, G., Canton, M., Colonna, R., Bernardi, P., and Di Lisa, F. (1999). Transient and long-lasting openings of the mitochondrial permeability transition pore can be monitored directly in intact cells by changes of mitochondrial calcein fluorescence. *Biophys. J.* **76,** 725–734.

Petronilli, V., Penzo, D., Scorrano, L., Bernardi, P., and Di Lisa F. (2001). The mitochondrial permeability transition, release of cytochrome *c* and cell death. Correlation with the duration of pore openings *in situ*. *J. Biol. Chem.* **276,** 12030–12034.

Pfaff, E., and Klingenberg, M. (1970). Unpublished observations cited by Weidemann, M. J., Erdelt, H., and Klingenberg, M. (1970). Adenine nucleotide translocation of mitochondria. Identification of carrier sites. *Eur. J. Biochem.* **16,** 313–335.

Polčic, P., Šabová, L., and Kolarov, J. (1997). Fatty acids induced uncoupling of Saccharomyces cerevisiae mitochondria requires an intact ADP/ATP carrier. *FEBS Lett.* **412,** 207–210.

Reid, T., Ramesha, C. S., and Ringold, G. M. (1991). Resistance to killing by tumor necrosis factor in an adipocyte cell line caused by a defect in arachidonic acid biosynthesis. *J. Biol. Chem.* **266,** 16580–16586.

Rial, E., and Gonzalez-Barroso, M. M. (2001). Physiological regulation of the transport activity in the uncoupling proteins UCP1 and UCP2. *Biochim. Biophys. Acta* **1504,** 70–81.

Ricquier, D., and Bouillaud, F. (2000). The uncoupling protein homologues: UCP1, UCP2, UCP3, StUCP and AtUCP. *Biochem. J.* **345** (Pt 2) 161–179.

Rottenberg, H., and Hashimoto, K. (1986). Fatty acid uncoupling of oxidative phosphorylation in rat liver mitochondria. *Biochemistry* **25,** 1747–1755.

Samartsev, V. N., Smirnov, A. V., Zeldi, I. P., Markova, O. V., Mokhova, E. N., and Skulachev, V. P. (1997). Involvement of aspartate/glutamate antiporter in fatty acid-induced uncoupling of liver mitochondria. *Biochim. Biophys. Acta* **1319,** 251–257.

Sapirstein, A., Spech, R. A., Witzgall, R., and Bonventre, J. V. (1996). Cytosolic phospholipase A_2(PLA$_2$), but not secretory PLA$_2$, potentiates hydrogen peroxide cytotoxicity in kidney epithelial cells. *J. Biol. Chem.* **271,** 21505–21513.

Sarker, K. P., Obara, S., Nakata, M., Kitajima, I., and Maruyama, I. (2000). Anandamide induces apoptosis of PC-12 cells: Involvement of superoxide and caspase-3. *FEBS Lett.* **472,** 39–44.

Scholz, R., Schwabe, U., and Soboll, S. (1984). Influence of fatty acids on energy metabolism. 1. Stimulation of oxygen consumption, ketogenesis and CO_2 production following addition of octanoate and oleate in perfused rat liver. *Eur. J. Biochem.* **141,** 223–230.

Schönfeld, P. (1990). Does the function of adenine nucleotide translocase in fatty acid uncoupling depend on the type of mitochondria? *FEBS Lett.* **264,** 246–248.

Schönfeld, P. (1992). Anion permeation limits the uncoupling activity of fatty acids in mitochondria. *FEBS Lett.* **303**, 190–192.

Schönfeld, P., and Bohnensack, R. (1997). Fatty acid-promoted mitochondrial permeability transition by membrane depolarization and binding to the ADP/ATP carrier. *FEBS Lett.* **420**, 167–170.

Schönfeld, P., Wojtczak, A. B., Geelen, M. J., Kunz, W., and Wojtczak, L. (1988). On the mechanism of the so-called uncoupling effect of medium- and short-chain fatty acids. *Biochim. Biophys. Acta* **936**, 280–288.

Schönfeld, P., Schild, L., and Kunz, W. (1989). Long-chain fatty acids act as protonophoric uncouplers of oxidative phosphorylation in rat liver mitochondria. *Biochim. Biophys. Acta* **977**, 266–272.

Schönfeld, P., Ježek, P., Belyaeva, E. A., Borecký, J., Slyshenkov, V. S., Więckowski, M. R., and Wojtczak, L. (1996). Photomodification of mitochondrial proteins by azido fatty acids and its effect on mitochondrial energetics. Further evidence for the role of the ADP/ATP carrier in fatty-acid-mediated uncoupling. *Eur. J. Biochem.* **240**, 387–393.

Schönfeld, P., Więckowski, M. R., and Wojtczak, L. (1997). Thyroid hormone-induced expression of the ADP/ATP carrier and its effect on fatty acid-induced uncoupling of oxidative phosphorylation. *FEBS Lett.* **416**, 19–22.

Schönfeld, P., Więckowski, M. R., and Wojtczak, L. (2000). Long-chain fatty acid-promoted swelling of mitochondria: Further evidence for the protonophoric effect of fatty acids in the inner mitochondrial membrane. *FEBS Lett.* **471**, 108–112.

Schwarz, H., Blanco, F. J., and Lotz, M. (1994). Anadamide an endogenous cannabinoid receptor agonist inhibits lymphocyte proliferation and induces apoptosis. *J. Neuroimmunol.* **55**, 107–115.

Scorrano, L., Petronilli, V., and Bernardi, P. (1997). On the voltage dependence of the mitochondrial permeability transition pore. A critical appraisal. *J. Biol. Chem.* **272**, 12295–12299.

Scorrano, L., Petronilli, V., Di Lisa, F., and Bernardi, P. (1999). Commitment to apoptosis by GD3 ganglioside depends on opening of the mitochondrial permeability transition pore. *J. Biol. Chem.* **274**, 22581–22585.

Scorrano, L., Penzo, D., Petronilli, V., Pagano, F., and Bernardi, P. (2001). Arachidonic acid is a potent inducer of cell death through the mitochondrial permeability transition. Implications for TNFα apoptotic signaling. *J. Biol. Chem.* **276**, 12035–12040.

Seitz, H. J., Müller, M. J., Krone, W., and Tarnowski, W. (1977). Coordinate control of intermediary metabolism in rat liver by the insulin/glucagon ratio during starvation and after glucose refeeding. Regulatory significance of long-chain acyl-CoA and cyclic AMP. *Arch. Biochem. Biophys.* **183**, 647–663.

Shao, J., Sheng, H., Inoue, H., Morrow, J. D., and DuBois, R. N. (2000). Regulation of constitutive cyclooxygenase-2 expression in colon carcinoma cells. *J. Biol. Chem.* **275**, 33951–33956.

Sheng, H., Shao, J., Kirkland, S. C., Isakson, P., Coffey, R. J., Morrow, J., Beauchamp, R. D., and DuBois, R. N. (1997). Inhibition of human colon cancer cell growth by selective inhibition of cyclooxygenase-2. *J. Clin. Invest.* **99**, 2254–2259.

Shug, A., Lerner, E., Elson, C., and Shrago, E. (1971). The inhibition of adenine nucleotide translocase activity by oleoyl CoA and its reversal in rat liver mitochondria. *Biochem. Biophys. Res. Commun.* **43**, 557–563.

Siliprandi, D., Biban, C., Testa, S., Toninello, A., and Siliprandi, N. (1992). Effects of palmitoyl CoA and palmitoyl carnitine on the membrane potential and Mg^{2+} content of rat heart mitochondria. *Mol. Cell. Biochem.* **116**, 117–123.

Skulachev, V. P. (1991). Fatty acid circuit as a physiological mechanism of uncoupling of oxidative phosphorylation. *FEBS Lett.* **294**, 158–162.

Skulachev, V. P. (1999). Anion carriers in fatty acid-mediated physiological uncoupling. *J. Bioenerg. Biomembr.* **31**, 431–445.

Sparagna, G. C., and Hickson-Bick, D. L. (1999). Cardiac fatty acid metabolism and the induction of apoptosis. *Am. J. Med. Sci.* **318**, 15–21.

Sparagna, G. C., Hickson-Bick, D. L., Buja, L. M., and McMillin, J. B. (2000). A metabolic role for mitochondria in palmitate-induced cardiac myocyte apoptosis. *Am. J. Physiol. Heart Circ. Physiol.* **279**, H2124–H2132.

Spinedi, A., Di Bartolomeo, S., and Piacentini, M. (1999). N-Oleoylethanolamine inhibits glucosylation of natural ceramides in CHP-100 neuroepithelioma cells: Possible implications for apoptosis. *Biochem. Biophys. Res. Commun.* **255,** 456–459.

Surette, M. E., Winkler, J. D., Fonteh, A. N., and Chilton, F. H. (1996). Relationship between arachidonate–phospholipid remodeling and apoptosis. *Biochemistry* **35,** 9187–9196.

Surette, M. E., Fonteh, A. N., Bernatchez, C., and Chilton, F. H. (1999). Perturbations in the control of cellular arachidonic acid levels block cell growth and induce apoptosis in HL-60 cells. *Carcinogenesis* **20,** 757–763.

Susin, S. A., Lorenzo, H. K., Zamzami, N., Marzo, I., Snow, B. E., Brothers, G. M., Mangion, J., Jacotot, E., Costantini, P., Loeffler, M., Larochette, N., Goodlett, D. R., Aebersold, R., Siderovski, D. P., Penninger, J. M., and Kroemer, G. (1999). Molecular characterization of mitochondrial apoptosis-inducing factor. *Nature* **397,** 441–446.

Szabó, I., and Zoratti, M. (1991). The giant channel of the inner mitochondrial membrane is inhibited by cyclosporin A. *J. Biol. Chem.* **266,** 3376–3379.

Szabó, I., and Zoratti, M. (1993). The mitochondrial permeability transition pore may comprise VDAC molecules. I. Binary structure and voltage dependence of the pore. *FEBS Lett.* **330,** 201–205.

Thun, M. J., Namboodiri, M. M., and Heath, C. W., Jr. (1991). Aspirin use and reduced risk of fatal colon cancer. *N. Engl. J. Med.* **325,** 1593–1596.

Uyemura, S. A., Luo, S., Moreno, S. N., and Docampo, R. (2000). Oxidative phosphorylation, Ca^{2+} transport, and fatty acid-induced uncoupling in malaria parasites mitochondria. *J. Biol. Chem.* **275,** 9709–9715.

Vercesi, A. E., Martins, I. S., Silva, M. A. P., Leite, H. M. F., Cuccovia, I. M., and Chaimovich, H. (1995). PUMPing plants. *Nature* **375,** 24.

Vinogradov, A., Scarpa, A., and Chance, B. (1972). Calcium and pyridine nucleotide interaction in mitochondrial membranes. *Arch. Biochem. Biophys.* **152,** 646–654.

Walter, L., Nogueira, V., Leverve, X., Bernardi, P., and Fontaine, E. (2000). Three classes of ubiquinone analogs regulate the mitochondrial permeability transition pore through a common site. *J. Biol. Chem.* **275,** 29521–29527.

Walter, L., Miyoshi, H., Leverve, X., Bernardi, P., and Fontaine, E. (2002). Regulation of the mitochondrial permeability transition pore by ubiquinone analogs. A progress report. *Free Radic. Res.* **36,** 405–412.

Wenzel, D. C., and Hale, T. W. (1978a). Toxicity of free fatty acids for cultured rat heart muscle and endothelioid cells. I. Saturated long-chain fatty acids. *Toxicology* **11,** 109–117.

Wenzel, D. C., and Hale, T. W. (1978b). Toxicity of free fatty acids for cultured rat heart muscle and endothelioid cells. II. Unsaturated long-chain fatty acids. *Toxicology* **11,** 119–125.

Więckowski, M. R., and Wojtczak, L. (1997). Involvement of the dicarboxylate carrier in the protonophoric action of long-chain fatty acids in mitochondria. *Biochem. Biophys. Res. Commun.* **232,** 414–417.

Więckowski, M. R., and Wojtczak, L. (1998). Fatty acid-induced uncoupling of oxidative phosphorylation is partly due to opening of the mitochondrial permeability transition pore. *FEBS Lett.* **423,** 339–342.

Więckowski, M. R., Brdiczka, D., and Wojtczak, L. (2000). Long-chain fatty acids promote opening of the reconstituted mitochondrial permeability transition pore. *FEBS Lett.* **484,** 61–64.

Więckowski, M. R., Vyssokikh, M., Dymkowska, D., Antonsson, B., Brdiczka, D., and Wojtczak, L. (2001). Oligomeric C terminal-truncated Bax preferentially releases cytochrome c but not adenylate kinase from mitochondria, outer membrane vesicles and proteoliposomes. *FEBS Lett.* **505,** 453–459.

Williams, C. S., Mann, M., and DuBois, R. N. (1999). The role of cyclooxygenases in inflammation, cancer, and development. *Oncogene* **18,** 7908–7916.

Winkler, E., and Klingenberg, M. (1994). Effect of fatty acids on H^+ transport activity of the reconstituted uncoupling protein. *J. Biol. Chem.* **269,** 2508–2515.

Wojtczak, L. (1976). Effect of long-chain fatty acids and acyl-CoA on mitochondrial permeability, transport, and energy-coupling processes. *J. Bioenerg. Biomembr.* **8,** 293–311.

Wojtczak, L., and Lehninger, A. L. (1961). Formation and disappearance of an endogenous uncoupling factor during swelling and contraction of mitochondria. *Biochim. Biophys. Acta* **51,** 442–456.

Wojtczak, L., and Schönfeld, P. (1993). Effect of fatty acids on energy coupling processes in mitochondria. *Biochim. Biophys. Acta* **1183,** 41–57.

Wojtczak, L., and Więckowski, M. R. (1999). The mechanisms of fatty acid-induced proton permeability of the inner mitochondrial membrane. *J. Bioenerg. Biomembr.* **31,** 447–455.

Wojtczak, L., and Załuska, H. (1967). The inhibition of translocation of adenine nucleotides through mitochondrial membranes by oleate. *Biochem. Biophys. Res. Commun.* **28,** 76–81.

Wojtczak, L., Adams, V., and Brdiczka, D. (1988). Effect of oleate on the apparent K_m of monoamine oxidase and the amount of membrane-bound hexokinase in isolated rat hepatocytes: Further evidence for the controlling role of the surface charge in hexokinase binding. *Mol. Cell Biochem.* **79,** 25–30.

Wojtczak, L., Więckowski, M. R., and Schönfeld, P. (1998). Protonophoric activity of fatty acid analogs and derivatives in the inner mitochondrial membrane: A further argument for the fatty acid cycling model. *Arch. Biochem. Biophys.* **357,** 76–84.

Wolter, K. G., Hsu, Y. T., Smith, C. L., Nechushtan, A., Xi, X. G., and Youle, R. J. (1997). Movement of Bax from the cytosol to mitochondria during apoptosis. *J. Cell Biol.* **139,** 1281–1292.

Žačková, M., Krämer, R., and Ježek, P. (2000). Interaction of mitochondrial phosphate carrier with fatty acids and hydrophobic phosphate analogs. *Int. J. Biochem. Cell Biol.* **32,** 499–508.

Zborowski, J., and Wojtczak, L. (1963). Induction of swelling of liver mitochondria by fatty acids of various chain length. *Biochim. Biophys. Acta* **70,** 596–598.

Zhang, L., Yu, J., Park, B. H., Kinzler, K. W., and Vogelstein, B. (2000). Role of Bax in the apoptotic response to anticancer agents. *Science* **290,** 989–992.

Zoratti, M., and Szabó, I. (1995). The mitochondrial permeability transition. *Biochim. Biophys. Acta* **1241,** 139–176.

Trinucleotide Repeat Disease

The Androgen Receptor in Spinal and Bulbar Muscular Atrophy

Jessica L. Walcott[*,†] and Diane E. Merry[*]

[*]*Department of Biochemistry and Molecular Pharmacology, Thomas Jefferson University, Philadelphia, Pennsylvania 19107, and*
[†]*Graduate Group in Pharmacological Sciences, University of Pennsylvania School of Medicine, Philadelphia, Pennsylvania 19104*

I. SBMA: A Clinical and Pathological View of The Disease
II. The CAG/Polyglutamine Diseases
III. Androgen Receptor: Overview and Domain Organization
IV. Effects of CAG Expansion on AR Function
   A. Function of the AR in the Nervous System
   B. Transactivation Capacity and Ligand Binding
   C. Levels of AR mRNA and Protein
V. Molecular Pathology
   A. Altered Conformation of Expanded Repeat AR
   B. The Role of Neuronal Intranuclear Inclusions in Disease
   C. Proteolysis
   D. Role of Ligand in SBMA
   E. Interacting Proteins
VI. Mouse Models of SBMA
VII. Polyglutamine Expansion and Neuronal Dysfunction
References

It has been more than 10 years since the discovery that the expansion of a simple CAG trinucleotide repeat within the coding region of the androgen receptor gene leads to the motor neuronopathy spinal and bulbar muscular atrophy (SBMA). A flurry of investigation into this and the other, more recently discovered, polyglutamine diseases has led to an understanding of many aspects of the molecular pathogenesis of this family of diseases. A characteristic pathological feature of the polyglutamine diseases is the occurrence in affected neurons of ubiquitinated aggregates; such aggregates also contain, among others, proteins involved in the folding and degradation of the mutant proteins. Aggregates themselves are likely not directly cytotoxic, but rather mark the accumulation of all or part of the mutant protein. Furthermore, aggregation occurs because of the inefficient clearance of the mutant protein by the ubiquitin–proteasome pathway for protein degradation. These findings are common to the polyglutamine diseases and reflect the general problem of folding/degrading expanded polyglutamines. In SBMA, the altered metabolism of the androgen receptor is ligand dependent. How the accumulation of the mutant protein causes neuronal dysfunction and disease is not well understood, but several cellular processes have been implicated. Although these findings provide insight into the toxic function of the expanded polyglutamine protein, additional investigations have led to the finding that intrinsic AR transactivational function is somewhat diminished in the presence of the expanded polyglutamine; this likely leads to the partial androgen insensitivity that characterizes patients with SBMA. The recent development of useful animal and cell models of SBMA will lead to increased understanding of disease pathogenesis, as well as to the development of new and better therapeutic strategies. © 2002, Elsevier Science (USA).

I. SBMA: A CLINICAL AND PATHOLOGICAL VIEW OF THE DISEASE

X-linked spinal and bulbar muscular atrophy (SBMA, Kennedy's disease) is a slowly progressive neuromuscular disorder characterized by the loss of motor neurons in the spinal cord and brainstem. The symptoms include the adult onset of symmetrical proximal muscle weakness, atrophy, and fasciculations. Bulbar muscle weakness manifests as difficulty with speech and swallowing (Kennedy *et al.*, 1968). Affected males often show signs of partial androgen insensitivity such as gynecomastia, reduced fertility, and testicular atrophy despite normal or increased serum testosterone levels (Arbizu *et al.*, 1983). The finding of partial androgen insensitivity along with mapping of the disease gene to the X chromosome made the androgen receptor (*AR*) gene a candidate gene for SBMA (Fischbeck *et al.*, 1986, 1991). The *AR* gene was subsequently found to contain a polymorphic trinucleotide CAG repeat that is expanded to pathogenic lengths in patients with SBMA (La Spada *et al.*, 1991). Expansion of the trinucleotide repeat in exon 1 of the *AR* gene leads to expression of an expanded polyglutamine repeat within the AR protein. The polyglutamine tract contains 40–62 residues in SBMA patients, but only 10–36 residues in normal individuals (La Spada *et al.*, 1991). The age of onset is inversely correlated with repeat length, such that longer repeats are found with earlier age of onset (Doyu *et al.*, 1992; La Spada *et al.*, 1992).

The main pathological finding in SBMA is motor neuron loss in the anterior horn of the spinal cord and in motor nuclei in the brainstem with a subclinical loss of sensory neurons in the dorsal root ganglia (Sobue *et al.*, 1989). Another pathological marker of the disease is the presence of AR-containing ubiquitinated neuronal intranuclear inclusions (NII) in motor neurons of the anterior horn of the spinal cord (Li *et al.*, 1998a). These inclusions have also been observed at a lower frequency in regions unaffected in SBMA including the dermis, scrotal skin, testis, heart, and kidney (Li *et al.*, 1998b).

II. THE CAG/POLYGLUTAMINE DISEASES

SBMA is one of a group of progressive neurodegenerative diseases resulting from a polyglutamine repeat expansion. The group includes Huntington's disease (HD), dentatorubral-pallidoluysian atrophy (DRPLA), and several spinocerebellar ataxias (SCA1, 2, 3, 6, 7, and 17) (reviewed by Nakamura *et al.*, 2001; Cummings and Zoghbi, 2000; Paulson and Fischbeck, 1996; Zoghbi and Orr, 2000). With the exception of X-linked SBMA, all share an autosomal dominant mode of inheritance. Despite widespread expression of the mutant protein, each disease is associated with the loss of a specific subset of neurons. Furthermore, each disease is marked not only by CAG repeat instability, but also by an inverse correlation between age of disease onset and CAG repeat length. Animal models, tissue culture, and patient tissue experiments support the theory that these diseases are caused by the acquisition of a toxic function conferred by the expanded polyglutamine. As with SBMA, a common pathological sign of abnormal metabolism is the presence of protein aggregation (Lin *et al.*, 1999; Paulson, 1999; Zoghbi and Orr, 2000). Expanded polyglutamine proteins form neuronal aggregates containing all or part of the mutant protein. In SCA2 and SCA6, the mutant proteins accumulate in the cytoplasm in aggregates that do not contain ubiquitin (Cummings and Zoghbi, 2000; Huynh *et al.*, 1999; Ishikawa *et al.*, 1999). More commonly, ubiquitin-positive intranuclear inclusions appear in affected neurons in SBMA, HD, SCA3 (MJD), SCA7, and DRPLA (Davies *et al.*, 1997; DiFiglia *et al.*, 1997; Hayashi *et al.*, 1998; Holmberg *et al.*, 1998; Li *et al.*, 1998a; Paulson *et al.*, 1997). The significance of these differences in aggregate localization is unknown. Regardless of cellular localization, the presence of neuronal inclusions signals the abnormal accumulation of all or part of the mutant protein.

III. ANDROGEN RECEPTOR: OVERVIEW AND DOMAIN ORGANIZATION

The androgen receptor (AR) is a member of the steroid/thyroid hormone receptor superfamily. Upon translation, the AR exists in an unliganded aporeceptor complex in the cytoplasm. Components of this complex include the heat shock proteins Hsp90, Hsp70, Hsp56, dnaJ, and p23 (Pratt and Welsh, 1994). Upon ligand binding, AR undergoes a conformational change, dissociates from the aporeceptor complex, and translocates into the nucleus where it binds as a dimer to hormone response elements (HRE) in the DNA. At the HRE, AR interacts with corepressors or coactivators to recruit basal transcription factors to stimulate or repress transcription of target genes.

The AR is a 919 amino acid protein containing the functional domains specific to the steroid/thyroid hormone receptor superfamily (see Fig. 1). The polyglutamine tract begins at position 58, within the amino-terminal region of the AR. More distal to the polyglutamine tract resides a rather poorly defined transcriptional activation motif, AF1, which resides between residues 142 and 337 and requires ligand binding for its function (He et al., 2001). The DNA-binding domain, found at amino acids 559–624, is highly conserved throughout the steroid hormone receptor superfamily and contains two zinc finger motifs responsible for DNA recognition and receptor dimerization (Tsai and O'Malley, 1994). The ligand-dependent nuclear localization signal (NLS) overlaps the carboxyl-terminal portion of the DNA-binding domain. This nuclear targeting sequence from amino acids 617–633 consists of two clusters of basic amino acids separated by 10 amino acids (Zhou et al., 1994a) and extends into the hinge region (Tsai and O'Malley, 1994). The carboxyl-terminus contains the ligand-binding domain (amino acids 676–919), which binds testosterone and dihydrotestosterone (DHT), both with high affinity. A separate transcriptional activation domain, AF2, is located within the ligand-binding domain (LBD) (Alen et al., 1999; Bevan et al., 1999). The AF2 domain in steroid hormone receptors is, in general, responsible for ligand-dependent recruitment of coactivators containing the LXXLL motif to a hydrophobic cleft in the LBD. However, the AF2 region alone only weakly binds LXXLL motif-containing coactivators from the p160 family (He et al., 1999). The AF2

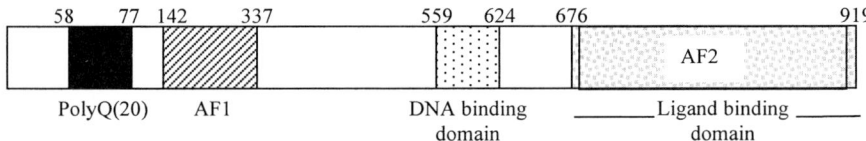

FIGURE 1 Schematic diagram of the androgen receptor. Numbers are given in the context of a 20 glutamine repeat, which begins at amino acid residue 57. AF1 and AF2 refer to activation function domains 1 and 2. The AF2 domain is roughly localized with the ligand-binding domain.

region interacts with the amino-terminus of AR (residues 1–503); this interaction is ligand dependent and recruits coactivators, including those of the p160 family, for transactivation (Alen *et al.*, 1999; He *et al.*, 1999, 2000, 2001; Langley *et al.*, 1995). The amino/carboxyl-terminal interaction that occurs during dimerization of the AR results from the interactions of the AF1 and AF2 transcriptional activation domains and produces an antiparallel organization (Langley *et al.*, 1995, 1998).

IV. EFFECTS OF CAG EXPANSION ON AR FUNCTION

A. FUNCTION OF THE AR IN THE NERVOUS SYSTEM

The normal function of AR with respect to male sexual development has been well documented (Brinkmann *et al.*, 1999; Jenster *et al.*, 1991; MacLean *et al.*, 1997; Tsai and O'Malley, 1994; Zhou *et al.*, 1994b). However, the role of AR in the nervous system, where it is widely expressed, is less well understood. Within the nervous system, the AR is expressed at relatively high levels in spinal and brainstem motor neurons (Sar and Stumpf, 1977). Several studies indicate that androgens play a trophic role in motor neurons of the spinal cord (Jones and Oblinger, 1994; Kujawa *et al.*, 1991, 1993; Nordeen *et al.*, 1985; Perez and Kelley, 1996; Yu, 1989). The AR was also shown to play a role in regulating structural plasticity of motor neurons (Garcia-Segura *et al.*, 1994; Kurz *et al.*, 1986; Matsumoto *et al.*, 1995). Furthermore, the overexpression of normal AR in a neuronal cell line was shown to increase soma and neurite size of differentiated neurons while leading to increased androgen-dependent resistance to serum withdrawal (Brooks *et al.*, 1998). How these neuronal roles for the AR are altered by expansion of the polyglutamine tract is an important area for investigation.

Loss of AR function results in complete androgen insensitivity with testicular feminization with no signs of motor neuron degeneration (Pinsky *et al.*, 1992; Quigley *et al.*, 1992). The absence of neurological symptoms in individuals with a loss of AR function indicates that the CAG expansion does not result in a full loss of AR function. However, as symptoms of androgen insensitivity occur in SBMA, it is possible that a partial loss of normal AR function contributes to disease pathogenesis.

B. TRANSACTIVATION CAPACITY AND LIGAND BINDING

Experiments addressing AR's normal function as a transcriptional transactivator have aimed at examining whether the partial loss of function seen in SBMA is due to reduced transactivation ability. Several studies indicate that the presence of an expanded repeat in AR leads to reduced transcriptional capacity with no altered androgen-binding properties (Butler et al., 1998; Chamberlain et al., 1994; Kazemi-Esfarjani et al., 1995; Mhatre et al., 1993; Nakajima et al., 1996). Expanding the CAG repeat from 52 to 212 resulted in a repeat length-dependent decrease in transcriptional transactivation ability in a mouse glioma and neuroblastoma hybrid cell line (Nakajima et al., 1996). The equilibrium binding affinity values, K_d, did not differ for normal versus expanded repeat AR in these experiments. However, Western blot analysis showed decreasing levels of AR expression with increasing repeat length, which may account for the reduced ability to transactivate reporter gene activity. In Cos-7 and CV-1 cells, progressive expansion of the repeat was also shown to result in a linear decrease in transactivation function (Chamberlain et al., 1994). AR77 retained only 60% of the transactivation function of AR25. Such decreases in transactivational capacity likely contribute to the partial androgen insensitivity of SBMA. In addition, the partial loss of AR transactivation may contribute to the neurological deficits of SBMA, perhaps by decreasing the ability of the AR to provide trophic support to motor neurons.

These results indicate that AR repeat expansion leads to altered transcriptional competence without an alteration in ligand-binding affinity. Studies using patient cells indicate that expansion of the repeat may alter the maximal binding capacity (B_{max}) of the AR (Danek et al., 1994; Warner et al., 1992), consistent with a reduction in steady-state AR levels. In contrast, others found that repeat expansion led to reduced equilibrium binding affinity (K_d) of the AR in a cohort of SBMA patients (MacLean et al., 1995). The basis for the differences in AR ligand binding is unknown. Variations in the effect of repeat expansion on ligand binding may contribute to differences in the extent of androgen insensitivity observed in SBMA patients.

C. LEVELS OF AR mRNA AND PROTEIN

Loss of AR mRNA and protein with repeat expansion may contribute to partial androgen insensitivity and/or to motor neuron cell loss, which along with age-related decreases in testosterone levels could contribute to the adult onset of SBMA. A linear inverse relationship has been found between CAG repeat number and AR mRNA and protein levels, without alterations in equilibrium binding affinity or intrinsic transcriptional capacity (Choong et al., 1996). Furthermore, reduced levels of AR65 compared to normal AR were observed in mouse motor neuron hybrid cell lines stably expressing AR24 and AR65 (Brooks et al., 1997). Additional support for the notion that AR transcription is reduced comes from the finding of reduced levels of AR mRNA in spinal cord from patients (Nakamura et al., 1994).

In addition to lower levels of AR protein caused by reduced transcription of the AR gene, it is possible that an altered metabolism of the mutant AR leads to lower levels of the soluble full-length protein. It is indeed a paradox that concomitant with reduced transcription of the expanded AR gene, the presence of intranuclear AR aggregates reflects the *accumulation* of at least a fragment of the AR. It may be that the symptoms of androgen insensitivity result from lower steady-state levels of the full-length protein, combined with its decreased transactivational capacity, whereas neurological deficits stem from the deficient clearance of the mutant protein.

V. MOLECULAR PATHOLOGY

A. ALTERED CONFORMATION OF EXPANDED REPEAT AR

Despite the finding that polyglutamine expansion leads to decreased transcriptional competence of the AR, the neurodegenerative diseases caused by polyglutamine expansions result *primarily* from the acquisition of a toxic protein function, rather than from the loss of intrinsic function. Expanded polyglutamine tracts alone, as well as truncated forms of polyglutamine proteins, have been shown to be toxic (Diamond *et al.*, 2000; Hackam *et al.*, 1998; Ikeda *et al.*, 1996; Mangiarini *et al.*, 1996; Merry *et al.*, 1998). Toxic properties might result from an altered protein conformation due to the presence of the expanded glutamine tract, or from accumulation of all or part of the AR, due to its inefficient clearance. Several theories exist to explain polyglutamine aggregation. In one, the expanded glutamine tract serves as a substrate for transglutaminase, linking it to other proteins by isodipeptide bonds (Green, 1993). Transglutaminase is a ubiquitous enzyme that stably cross-links glutamine residues in proteins with primary amines. Such activity could lead to polyglutamine protein accumulation, due to decreased efficiency in degrading such stably cross-linked protein species (Green, 1993). Alternatively, polyglutamine aggregation could occur through the formation of a "polar zipper," in which stable β-pleated sheets are held together by hydrogen bonding between main chain and side chain amides (Perutz *et al.*, 1994). The formation of such polar zippers would represent a phase change that occurs when short stretches of polyglutamine in a random coil structure transition to hydrogen-bonded hairpins when the glutamine repeat region is expanded (Perutz, 1996). Further evidence that the conformation of polyglutamine repeat-containing proteins changes with expansion comes from the development of antibodies that preferentially bind expanded polyglutamines (Persichetti *et al.*, 1999; Trottier *et al.*, 1995).

An altered conformation of the protein upon polyglutamine expansion might also lead to altered protein–protein interactions, which could then be involved in the aggregation process or in conferring toxicity. Normal AR binds Hsp90, Hsc70, Hsp56, dnaJ, and p23 when in the aporeceptor complex in its unliganded state. To date, no alterations of this poised configuration have been identified for expanded repeat AR. In addition to these known interactions involved in the androgen receptor's normal function as a steroid hormone receptor, various interacting proteins have been identified that may be involved in the molecular pathogenesis of SBMA. An up-to-date list of AR interactions can be found at The Androgen Receptor Gene

Mutations Database World Wide Web Server at http://www.mcgill.ca/androgendb/. Some of these will be reviewed here in the context of sequestration into intranuclear inclusions as well as their role as coactivators for AR function. The point in the metabolism of the AR that such altered conformations may occur remains unknown. The finding of only partial androgen insensitivity in patients affected with SBMA indicates that the normal function of the AR is generally spared. Such sparing of function suggests that any pathogenic altered conformational states of the mutant AR occur concomitant with or subsequent to DNA binding and the transcriptional activation of target genes.

B. THE ROLE OF NEURONAL INTRANUCLEAR INCLUSIONS IN DISEASE

Despite the common pathological finding of aggregation in cell culture, animal models, and patient tissue for SBMA and other polyglutamine diseases (Lin et al., 1999; Paulson, 1999; Zoghbi and Orr, 1999), the role of NII in the pathogenesis of SBMA is still unclear. Inclusions distribute to regions spared by disease in SBMA (Li et al., 1998b), HD (Gutekunst et al., 1999; Kuemmerle et al., 1999), SCA2 (Koyano et al., 1999), and SCA7 (Holmberg et al., 1998). Cell culture and transgenic mouse studies of SBMA and other polyglutamine diseases have also revealed a dissociation of inclusion formation from toxicity (Klement et al., 1998; Saudou et al., 1998; Simeoni et al., 2000).

Although NII may be neither necessary nor sufficient for neuronal dysfunction and death, their presence may contribute to the pathogenic process by sequestering proteins and thereby disrupting cellular homeostasis. In cells expressing full-length expanded AR treated with ligand, AR accumulates in cytoplasmic aggregates that contain Hsp70, Hsp90, the ubiquitin-like protein NEDD8, PA700 (26S proteasome cap), steroid receptor coactivator 1 (SRC-1), and mitochondria (Stenoien et al., 1999). Histological studies of mice expressing a truncated highly expanded repeat AR showed intranuclear inclusions positive for ubiquitin, the molecular chaperones HDJ1, HDJ-2/HSDJ, and Hsc70, components of the 26S proteasome, and CREB-binding protein (CBP) (Abel et al., 2001).

The sequestration of proteins in NII may contribute to disease pathogenesis. CBP is sequestered into polyglutamine inclusions found in SBMA patient tissue and in cell culture models of SBMA (McCampbell et al., 2000). Furthermore, CBP was found in NII in models and human tissue from other polyglutamine diseases (McCampbell et al., 2000; Nucifora et al., 2001). CBP is an important transcriptional coactivator for cAMP-mediated gene transcription (Chrivia et al., 1993), and several studies have shown that CBP-dependent transcription is reduced in the presence of expanded polyglutamines (McCampbell et al., 2000; Nucifora et al., 2001). Not only is CBP-dependent transcription restored by exogenous CBP, but expanded polyglutamine-specific cell toxicity is ameliorated by CBP. The importance of the sequestration of this and other transcriptional coactivators in the pathogenesis of polyglutamine disease is further revealed by the finding that histone deacetylase inhibitors have been shown to modulate neurological symptoms in a fly model of polyglutamine disease (Steffan et al., 2001).

Whether polyglutamine aggregates are themselves directly cytotoxic, the presence of NII in polyglutamine disease signals a failure of the cell to effectively degrade the mutant polyglutamine protein. Indeed, aggregates sequester several heat shock proteins and components of the proteasome (Abel et al., 2001; Chai et al., 1999; Cummings et al., 1998; Stenoien et al., 1999), proteins that are necessary for the normal folding and ultimate degradation of most proteins. Thus, the presence of proteins involved in the proteasome pathway in NII indicates a failed attempt to properly fold and degrade mutant expanded AR. Half-life studies of a truncated expanded AR revealed a substantially longer half-life than that of a normal repeat truncated AR (Bailey et al., 2002). Pharmacological inhibition of the proteasome promoted further aggregation of the mutant AR (Bailey et al., 2002) as well as of mutant ataxin-1 and ataxin-3 (Chai et al., 1999; Cummings et al., 1999). Mutant ataxin-1 has also been shown to resist ubiquitin-mediated degradation (Cummings et al., 1999). Furthermore, disruption of the ubiquitin proteasome pathway through the use of genetic mutations in the ubiquitination pathway (Cummings et al., 1999; Saudou et al., 1998) inhibited the formation of NII while increasing toxicity. These results implicate the proteasomal degradation of expanded polyglutamines in both aggregate formation and cellular toxicity.

Other proteins related to the ubiquitin proteasome degradation pathway, although not associated with NII, may also be involved in altered turnover of the mutant AR. E6AP, a ubiquitin protein ligase, is known to interact with the AR. Whether this protein is involved with the ubiquitination of the AR is, however, unknown.

Posttranslational modifications may also play a role in the altered metabolism of the mutant AR. The AR is a substrate for SUMO conjugation (Poukka et al., 2000), and interacts with a SUMO-conjugating protein, Ubc9 (Poukka et al., 1999). SUMO conjugation is enhanced by ligand binding; whether this modification is altered by the polyglutamine expansions is under investigation. In addition, the effects of repeat expansion on the phosphorylation of the AR have not been extensively investigated. The phosphorylation state of the AR influences ligand binding (Brinkmann, 1999); furthermore, upon ligand binding, the AR becomes hyperphosphorylated (Blok et al., 1998). The AR is a substrate for several protein kinases (reviewed by Blok et al., 1998; Brinkmann et al., 1999; Ikonen et al., 1994; Lin et al., 2001; Nazareth and Weigel, 1996), including DNA-dependent protein kinase, protein kinase A, protein kinase C, mitogen-activated kinase, casein kinase II, Akt kinase (Lin et al., 2001), and a novel nuclear protein kinase (Moilanen et al., 1998) whether the phosphorylation of AR by any one of these kinases is altered by polyglutamine expansion remains to be determined.

C. PROTEOLYSIS

In SBMA patient tissue, NII contain only epitopes for the amino-terminus of the AR (Li et al., 1998a,b), suggesting that the mutant AR within NII may represent a proteolyzed form. Several studies have identified truncated forms of the mutant AR (Abdullah *et al.*, 1998; Butler *et al.*, 1998; Ellerby *et al.*, 1999; Merry *et al.*, 1998); however, only recently has an *in vitro* system reproduced the amino-epitope inclusions found in patient material (Walcott, J. and Merry, D., unpublished observations). The relevance of such truncated products to disease is still undetermined. Nonetheless, the presence of expanded polyglutamine-containing fragments of AR has been correlated with increased toxicity and aggregate formation (Abdullah *et al.*, 1998; Butler *et al.*, 1998; Merry *et al.*, 1998), suggesting that such proteolysis may exacerbate the disease process.

The proteases responsible for the putative proteolysis of mutant AR in patients remain elusive. Candidate proteases have been described from cell culture studies and may play a role *in vivo*. The AR is a substrate for caspase-3 (Ellerby *et al.*, 1999; Kobayashi *et al.*, 1998; Wellington *et al.*, 1998). Furthermore, susceptibility of the AR to cleavage by caspase-3 was shown to increase with expansion of the polyglutamine repeat. The inefficient degradation of the expanded polyglutamine tract may also contribute to the formation and accumulation of proteasome-resistant fragments (I. Andriola and D. Merry, unpublished observations). Whatever protease is involved, the formation of AR fragments containing the expanded repeat would create highly toxic polyglutamine species that likely contribute to the pathogenesis of disease.

D. ROLE OF LIGAND IN SBMA

SBMA affects males almost exclusively, although mildly affected females have been described (Ferlini *et al.*, 1995). The fact that the polyglutamine expansion results in the acquisition of a toxic property, and that all other polyglutamine expansion diseases are inherited in a dominant fashion, makes it curious that female carriers of SBMA are relatively spared. The basis for this gender specificity may be due to the random inactivation of the X chromosome in affected neurons of carrier females. An alternative, but not exclusive, possibility is that female carriers of the SBMA mutation are protected by low circulating levels of androgenic steroids. Indeed, the recent development of a mouse model for SBMA indicates that females are protected by low circulating hormone levels (E. Chevalier-Larsen, C. O'Brien, and D. Merry, unpublished observations). Thus, understanding the role of ligand binding in the metabolic fate of the AR may lead to insights into disease pathogenesis.

AR binds the androgens testosterone (T) and dihydrotestosterone (DHT) with high affinity, but can also bind antiandrogens and other steroids. In the absence of ligand, the AR is found in the cytoplasm with a perinuclear distribution (Kemppainen *et al.*, 1992). Following treatment with testosterone, dihydrotestosterone, estradiol, progesterone, the synthetic progestin RU486, and the antiandrogens hydroxyflutamide and cyproterone acetate, the AR translocates into the nucleus. The nonandrogenic steroids and antiandrogen compounds elicit this response only at high concentrations. With the exception of the antagonist hydroxyflutamide, all of these compounds lead to the transcriptional activation of the AR

(Kemppainen et al., 1992). Studies on the kinetics of AR translocation following ligand treatment indicate that the AR is recycled and can move into and out of the nucleus following ligand treatment and withdrawal (Tyagi et al., 2000). Unlike many of the steroid hormone receptors, ligand binding stabilizes the AR, increasing it half-life from 1–1.5 h (Kemppainen et al., 1992; Zhou et al., 1995) to 6 h (Kemppainen et al., 1992). Nonandrogenic steroids and antiandrogen compounds do not stabilize AR even though they bind the receptor and can lead to its nuclear import (Kemppainen et al., 1992).

How does expansion of the polyglutamine alter the normal response of AR to ligand? In several cell culture models of SBMA, expanded repeat AR forms *cytoplasmic* aggregates with occasional nuclear aggregates following androgen treatment (Becker et al., 2000; Butler et al., 1998; Simeoni et al., 2000; Stenoien et al., 1999), although others found ligand to inhibit aggregation (Ellerby et al., 1999). The basis for these differences is unknown. A caveat in the use of cell models to understand AR aggregation is that in these models, aggregation occurs only following the growth of cells under stressed conditions (i.e., charcoal-stripped serum). Nevertheless, findings from such experiments can yield insights into the effects of different ligands on the fate of the mutant AR. Although androgen treatment led to aggregate formation, the antiandrogen compounds hydroxyflutamide and Casodex were able to inhibit aggregate formation (Becker et al., 2000). These results suggest that pharmacological manipulation of the mutant AR may lead to a better understanding of parameters involved in the acquisition of an aggregation-prone (and presumably toxic) state. Indeed, an inducible cell model of SBMA reveals a role for ligand in the abnormal metabolism of the AR (Walcott, J. and Merry, D., unpublished observations). The effects of agonists and antagonists will need to be studied further in animal models of SBMA to better understand how these compounds might be manipulated to develop treatments for SBMA.

E. INTERACTING PROTEINS

It has been hypothesized that the identification of interactors of polyglutamine proteins should contribute to an understanding of pathways affected by repeat expansion. A number of proteins have been found as coactivators for AR that interact with the receptor in a ligand-dependent manner and potentiate ligand-dependent transcriptional activation. The p160 family of coactivators, the 160-kDa proteins steroid receptor coactivator 1 (SRC1), transcription intermediary factor 2 (TIF2) (GRIP homologue), and amplified in breast cancer 1 (AIB1) protein interact with AR and other nuclear receptors (Bevan et al., 1999). The p160 coactivator GRIP1 interacts with the amino-terminal domain of AR in a repeat length-dependent manner such that increasing the polyglutamine length inhibits coactivation of AR by GRIP1 (Irvine et al., 2000). Another member of this family, SRC1, interacts with the AF1 and AF2 domains of AR, and as with GRIP1, the coactivation of AR by SRC1 is decreased by repeat expansion (Bevan et al., 1999).

CBP has also been shown to directly interact with AR and to be a coactivator for AR-dependent transactivation (Aarnisalo et al., 1998; Fronsdal et al., 1998).

It has also been shown to interact with other polyglutamine-containing proteins, including huntingtin, atrophin-1, and ataxin-3 (McCampbell *et al.*, 2000; Nucifora *et al.*, 2001; Steffan *et al.*, 2000). As previously discussed (Section V.B), CBP colocalizes with the inclusions found in cell culture models and transgenic mice made with truncated forms of expanded AR, in SBMA patient tissue, and in cell culture models of MJD (Abel *et al.*, 2001; McCampbell *et al.*, 2000). CBP is also found in aggregates in cell culture and mouse models of HD, human HD brain tissue, and DRPLA brain tissue (Nucifora *et al.*, 2001). Thus, CBP is an example of an interacting protein whose sequestration into NII plays a role in polyglutamine disease pathogenesis (McCampbell *et al.*, 2001; Steffan *et al.*, 2001).

Another AR interacting protein, Ran, is important for its role in nuclear transport. The nuclear localization of expanded repeat-containing proteins is important for disease pathogenesis (Klement *et al.*, 1998; Saudou *et al.*, 1998). Ran, a protein essential for most nucleocytoplasmic transport events, is a 25-kDa Ras-like GTP-binding protein that cycles between GTP and GDP bound states, coordinating nucleocytoplasmic transport by its nucleotide-bound state and cellular localization. Ran was found to interact with the amino-terminal domain of AR (Hsiao *et al.*, 1999). The AR/Ran interaction decreases as the length of the AR polyglutamine tract increases. Although the nuclear import of the expanded AR appears to be normal (Brooks *et al.*, 1997), it is possible that a weakened interaction of Ran with expanded polyglutamine-containing AR might result in subtle alterations in the shuttling of AR (Hsiao *et al.*, 1999).

In addition to the use of interacting proteins to understand the molecular pathogenesis of SBMA, the identification of genes whose expression is altered by the polyglutamine expansion in the AR will likely lead to the understanding of additional pathways disrupted by the expansion. The identification of such genes with altered transcription in other polyglutamine disease models is providing insights in these diseases (Li *et al.*, 1999; Lin *et al.*, 2000; Wyttenbach *et al.*, 2001). Specific tubulin isoforms have been shown to be transcriptionally regulated by the AR (Butler *et al.*, 2001). It will be important to determine in mouse models of SBMA if an alteration of this target gene contributes to disease pathogenesis, perhaps by causing transport-dependent motor neuron dysfunction.

VI. MOUSE MODELS OF SBMA

SBMA has been difficult to model in transgenic mice. In 1995, Bingham and co-workers created transgenic mice expressing mutant AR with 45 repeats under control of the ubiquitous inducible Mx, myosin light chain, and neuron-specific enolase promoters. No overt phenotype was detected in these mice, nor were subclinical symptoms found despite analysis of grip strength and ventral root studies. The absence of pathology may have been due to low levels of transgene expression in regions affected in SBMA. However, transgenic mice created using AR with 65 repeats with the neuron-specific enolase (NSE) and neurofilament light chain (NFL) promoters, with levels of transgene expression two to five times that of endogenous AR expression, were also unable to elicit a neuronal phenotype (Merry et al., 1996). These mice showed no evidence of pathology or behavioral abnormalities up to 2 years of age. Also, transgenic mice created using large yeast artificial chromosomes (YACs) containing human AR with 45 repeats failed to show any neurological phenotype, although meiotic instability was documented (La Spada et al., 1998). The results of these early studies suggested that very high expression levels or extremely long repeat lengths were necessary to model SMBA in a mouse.

The finding that protein truncation enhanced polyglutamine-dependent toxicity in cell (Hackam et al., 1998; Ikeda et al., 1996; Merry et al., 1998) and animal models (Ikeda et al., 1996; Mangiarini et al., 1996) indicated that the toxic properties of an expanded polyglutamine might be more easily achieved in the context of a truncated form of the polyglutamine-containing protein. Mice expressing high levels of the amino-terminal portion of the AR protein bearing a highly expanded repeat (112) developed neurological disease (Abel et al., 2001). The phenotypes of these mice created with truncated expanded AR depended on the distribution of protein governed by the promoter. Pan-neuronal expression from the prion protein (PrP) promoter resulted in widespread neurological disease consisting of gait abnormalities, resting tremor, hypoactivity, handling-induced seizures, and early death. Restricted expression of the same transgene construct led to a more specific phenotype: NFL promoter-driven expression in motor neuron populations of the cortex, brainstem, and spinal cord resulting in a limited gait abnormality. Furthermore, no evidence of neuronal degeneration or muscle pathology was observed in truncated AR-expressing mice, indicating that neuronal *dysfunction,* rather than neuronal cell *death,* was responsible for the observed phenotypes. It is possible that the motor neuron specificity of SBMA depends on additional AR protein sequences, perhaps determining specificity through protein interactions and/or cellular processes associated with AR metabolism.

Recently, mice were created using the human androgen receptor promoter to drive expression of a highly expanded polyglutamine region containing 239 repeats without the context of the AR protein (Adachi et al., 2001). These mice developed neurological symptoms of muscle weakness and ataxia. As with the mice developed by Abel et al. (2001), nuclear inclusions were found in more areas than typically seen in SBMA and there were no signs of degeneration or neuronal cell death observed in neurons affected in the disease (Adachi et al., 2001).

Transgenic mice created with truncated forms of AR driven by heterologous promoters, or with pure polyglutamine driven by the AR promoter, reproduced some aspects of SBMA, particularly the histological findings of NII (Abel et al., 2001; Adachi et al., 2001). However, the gender and the motor neuron specificity of SBMA remained difficult to model. Our recent findings with highly expanded full-length AR driven by the PrP promoter indicate not only that the symptoms of SBMA are ligand dependent, but that expression of the full-length AR protein is likely necessary to achieve the motor neuron specificity of SBMA (C. J. O'Brien, E. Chevalier-Larsen, and D. Merry, unpublished results).

VII. POLYGLUTAMINE EXPANSION AND NEURONAL DYSFUNCTION

Much of the work in understanding the molecular pathogenesis of spinal and bulbar muscular atrophy within the past 11 years has centered on the development of valid cell and animal models of disease. Experiments using these models have brought an increased understanding of the effects of polyglutamine expansion on AR function and its metabolic fate. These studies reveal that the mutant AR accumulates in the form of neuronal intranuclear inclusions, but that these aggregates likely *mark* the accumulation of the mutant protein and are not themselves directly pathogenic. The sequestration of proteins, either in aggregates or in a preaggregated AR state, may contribute to disease pathogenesis, as likely occurs with CREB-binding protein (CBP), but whether such a mechanism is important to the onset or progression of disease remains to be seen. What then is responsible for the dysfunction of motor neurons in SBMA in response to the accumulated polyglutamine-expanded AR? Studies in the next few years should illuminate the cellular processes that are responsible for the motor neuron deficits and ultimate cell death. Indeed, alterations in T-type Ca^{2+} channel activity have been observed in a cell model of SBMA (Sculptoreanu *et al.*, 2000); such changes may contribute to neuronal dysfunction as well as cell death. Other alterations may include the transcription of AR-responsive genes, or other secondary targets. Pathogenic mechanisms may include proteasomal dysfunction, as has been observed in Huntington's disease (Bence *et al.*, 2001; Jana *et al.*, 2001), as well as in our cellular model of SBMA (I. Andriola and D. Merry, unpublished observations). In addition, polyglutamine expansion may affect other systems involved with protein folding and degradation, including the function of molecular chaperones, as well as the activity of ubiquitinating enzymes. The elucidation of any one of these dysfunctional processes should lead to the development of novel therapeutics. In addition, the further understanding of the inefficient clearance of the mutant polyglutamine-expanded AR should lead to therapies designed to aid the cell in degrading this toxic protein.

REFERENCES

Aarnisalo, P., Palvimo, J. J., and Janne, O. A. (1998). CREB-binding protein in androgen receptor-mediated signaling. *Proc. Natl. Acad. Sci. USA* **95,** 2122–2127.

Abdullah, A. A. R., Trifiro, M. A., Panet-Raymond, V., Alvarado, C., de Tourreil, S., Frankel, D., Schipper, H. M., and Pinsky, L. (1998). Spinobulbar muscular atrophy: Polyglutamine-expanded androgen receptor is proteolytically resistant in vitro and processed abnormally in transfected cells. *Hum. Mol. Genet.* **7,** 379–384.

Abel, A., Walcott, J., Woods, J., Duda, J., and Merry, D. E. (2001). Expression of expanded repeat androgen receptor produces neurologic disease in transgenic mice. *Hum. Mol. Genet.* **10,** 107–116.

Adachi, H., Kume, A., Li, M., Nakagomi, Y., Niwa, H., Do, J., Sang, C., Kobayashi, Y., Doyu, M., and Sobue, G. (2001). Transgenic mice with an expanded CAG repeat controlled by the human AR promoter show polyglutamine nuclear inclusions and neuronal dysfunction without neuronal cell death. *Hum. Mol. Genet.* **10**, 1039–1048.

Alen, P., Claessens, F., Verhoeven, G., Rombauts, W., and Peeters, B. (1999). The androgen receptor amino-terminal domain plays a key role in p160 coactivator-stimulated gene transcription. *Mol. Cell. Biol.* **19**, 6085–6097.

Arbizu, T., Santamaria, J., Gomez, J. M., Quilez, A., and Serra, J. P. (1983). A family with adult spinal and bulbar muscular atrophy, X-linked inheritance and associated testicular failure. *J. Neurol. Sci.* **59**, 371–382.

Bailey, C. K., Andriola, I. F. M., Kampinga, H. H., and Merry, D. E. (2002). Molecular chaperones enhance the degradation of expanded polyglutamine repeat androgen receptor in a cellular model of spinal and bulbar muscular atrophy. *Hum. Mol. Genet.* **11**, 515–523.

Becker, M., Elke, M., Schneikert, J., Krug, H. F., and Cato, A. C. B. (2000). Cytoplasmic localization and the choice of ligand determine aggregate formation by androgen receptor with amplified polyglutamine stretch. *J. Cell Biol.* **149**, 255–262.

Bence, N. F., Sampat, R. M., and Kopito, R. R. (2001). Impairment of the ubiquitin-proteasome system by protein aggregation. *Science* **292**, 1552–1555.

Bevan, C., Hoare, S., Claessens, F., Heery, D., and Parker, M. (1999). The AF1 and AF2 domains of the androgen receptor interact with distinct regions of SRC1. *Mol. Cell. Biol.* **19**, 8383–8392.

Bingham, P. M., Scott, M. O., Wang, S., McPhaul, M. J., Wilson, E. M., Garbern, J. Y., Merry, D. E., and Fischbeck, K. H. (1995). Stability of an expanded trinucleotide repeat in the androgen receptor gene in transgenic mice. *Nat. Genet.* **9**, 191–196.

Blok, L. J., de Ruiter, P. E., and Brinkmann, A. O. (1998). Forskolin-induced dephosphorylation of the androgen receptor impairs ligand binding. *Biochemistry* **37**, 3850–3857.

Brinkmann, A. O., Blok, L. J., de Ruiter, P. E., Doesburg, P., Steketee, K., Berrevoets, C. A., and Trapman, J. (1999). Mechanisms of androgen receptor activation and function. *J. Steroid Biochem. Mol. Biol.* **69**, 307–313.

Brooks, B. P., Paulson, H. L., Merry, D. E., Salazar-Grueso, E. F., Brinkmann, A. O., Wilson, E. M., and Fischbeck, K. H. (1997). Characterization of an expanded glutamine repeat androgen receptor in a neuronal cell culture system. *Neurobiol. Dis.* **4**, 313–323.

Brooks, B. P., Merry, D. E., Paulson, H. L., Lieberman, A., Kolson, D. L., and Fischbeck, K. H. (1998). A cell culture model for androgen-inducible responses in motor neurons. *J. Neurochem.* **70**, 1054–1060.

Butler, R., Leigh, P. N., McPhaul, M. J., and Gallo, J. M. (1998). Truncated forms of the androgen receptor are associated with polyQ expansion in X-linked spinal and bulbar muscular atrophy. *Hum. Mol. Genet.* **7**, 121–127.

Butler, R., Leigh, P. N., and Gallo, J. M. (2001). Androgen-induced up-regulation of tubulin isoforms in neuroblastoma cells. *J. Neurochem.* **78**, 854–861.

Chai, Y., Koppenhafer, S. L., Shoesmith, S. J., Perez, M. K., and Paulson, H. L. (1999). Evidence for proteasome involvement in polyglutamine disease: Localization to nuclear inclusions in SCA3/MJD and suppression of polyglutamine aggregation in vitro. *Hum. Mol. Genet.* **8**, 673–682.

Chamberlain, N. L., Driver, E. D., and Miesfeld, R. L. (1994). The length and location of CAG trinucleotide repeats in the androgen receptor N-terminal domain affect transactivation function. *Nucleic Acids Res.* **22**, 3181–3186.

Choong, C. S., Kamppainen, J. A., Zhou, Z. X., and Wilson, E. M. (1996). Reduced androgen receptor gene expression with first exon CAG repeat expansion. *Mol. Endocrinol.* **10**, 1527–1535.

Chrivia, J. C., Kwok, R. P., Lamb, N., Hagiwara, M., Montminy, M. R., and Goodman, R. H. (1993). Phosphorylated CREB binds specifically to the nuclear protein CBP. *Nature* **365**, 855–859.

Cummings, C., and Zoghbi, H. (2000). Fourteen and counting: Unraveling trinucleotide repeat diseases. *Hum. Mol. Genet.* **9**, 909–916.

Cummings, C. J., Mancini, M. A., Antalffy, B., DeFranco, D. B., Orr, H. T., and Zoghbi, H. Y. (1998). Chaperone suppression of aggregation and altered subcellular proteasome localization imply protein misfolding in SCA1. *Nat. Genet.* **19**, 148–154.

Cummings, C. J., Reinstein, E., Jiang, Y.-H., Ciechanover, A., Orr, H. T., Beaudet, A. L., and Zoghbi, H. Y. (1999). Mutation of the E6-AP ubiquitin ligase reduces nuclear inclusion frequency while accelerating polyglutamine-induced pathology in SCA1 transgenic mice. *Am. J. Hum. Genet.* **65** (Suppl.), A29.

Danek, A., Witt, T. N., Mann, K., Schweikert, H. U., Romalo, G., LaSpada, A. R., and Fischbeck, K. H. (1994). Decrease in androgen binding and effect of androgen treatment in a case of X-linked bulbospinal neuronopathy. *Clin. Invest.* **72**, 892–897.

Davies, S. W., Turmaine, M., Cozens, B. A., DiFiglia, M., Sharp, A. H., Ross, C. A., Scherzinger, E., Wanker, E. E., Mangiarini, L., and Bates, G. P. (1997). Formation of neuronal intranuclear inclusions underlies the neurological dysfunction in mice transgenic for the HD mutation. *Cell* **90**, 537–548.

Diamond, M. I., Robinson, M. R., and Yamamoto, K. R. (2000). Regulation of expanded polyglutamine protein aggregation and nuclear localization by the glucocorticoid receptor. *Proc. Natl. Acad. Sci. USA* **97**, 657–661.

DiFiglia, M., Sapp, E., Chase, K. O., Davies, S. W., Bates, G. P., Vonsattel, J. P., and Aronin, N. (1997). Aggregation of huntingtin in neuronal intranuclear inclusions and dystrophic neurites in brain. *Science* **277**, 1990–1993.

Doyu, M., Sobue, G., Mukai, E., Kachi, T., Ysauda, T., Mitsuma, T., and Takahashi, A. (1992). Severity of X-linked recessive bulbospinal neuronopathy correlates with size of the tandem CAG repeat in androgen receptor gene. *Ann. Neurol.* **32**, 707–710.

Ellerby, L. M., Hackam, A. S., Propp, S. S., Ellerby, H. M., Rabizadeh, S., Cashman, N. R., Trifiro, M. A., Pinsky, L., Wellington, C. L., Salvesen, G. S., Hayden, M. R., and Bredesen, D. E. (1999). Kennedy's disease: Caspase cleavage of the androgen receptor is a crucial event in cytotoxicity. *J. Neurochem.* **72**, 185–195.

Ferlini, A., Patrosso, M. C., Guidetti, D., Merlini, L., Uncini, A., Ragno, M., Plasmati, R., Fini, S., Repetto, M., Vezzoni, P., and Forabosco, A. (1995). Androgen receptor gene (CAG)n repeat analysis in the differential diagnosis between Kennedy disease and other motoneuron disorders. *Am. J. Med. Genet.* **55**, 105–111.

Fischbeck, K. H., Ionasescu, V., Ritter, A. W., Ionasescu, R., Davies, K., Ball, S., Bosch, P., Burns, T., Hausmanowa-Petrusewicz, I., Borkowska, J., Ringel, S. P., and Stern, L. Z. (1986). Localization of the gene for X-linked spinal muscular atrophy. *Neurology* **36**, 1595–1598.

Fischbeck, K. H., Souders, D., and La Spada, A. R. (1991). A candidate gene for X-linked spinal muscular atrophy. *Adv. Neurol.* **56**, 209–213.

Fronsdal, K., Engedal, N., Slagsvold, T., and Saatcioglu, F. (1998). CREB binding protein is a coactivator for the androgen receptor and mediates cross-talk with AP-1. *J. Biol. Chem.* **273**, 31853–31859.

Garcia-Segura, L. M., Chowen, J. A., Parducz, A., and Naftolin, F. (1994). Gonadal hormones as promoters of structural synaptic plasticity: Cellular mechanisms. *Prog. Neurobiol.* **44**, 279–307.

Green, H. (1993). Human genetic diseases due to codon reiteration: Relationship to an evolutionary mechanism. *Cell* **74**, 955–956.

Gutekunst, C.-A., Li, S.-H., Yi, H., Mulroy, J. S., Kuemmerle, S., Jones, R., Rye, D., Ferrante, R. J., Hersch, S. M., and Li, X.-J. (1999). Nuclear and neuropil aggregates in Huntington's disease: Relationship to neuropathology. *J. Neurosci.* **19**, 2522–2534.

Hackam, A. S., Singaraja, R., Wellington, C. L., Metzler, M., McCutcheon, K., Zhang, T., Kalchman, M., and Hayden, M. R. (1998). The influence of huntingtin protein size on nuclear localization and cellular toxicity. *J. Cell Biol.* **141**, 1097–1105.

Hayashi, Y., Kakita, A., Yamada, M., Koide, R., Igarashi, S., Takano, H., Ikeuchi, T., Wakabayashi, K., Egawa, S., Tsuji, S., and Takahashi, H. (1998). Hereditary dentatorubral-pallidoluysian atrophy: Detection of widespread ubiquitinated neuronal and glial intranuclear inclusions in the brain. *Acta Neuropathol.* **96**, 547–552.

He, B., Kemppainen, J., Voegel, J., Gronemeyer, H., and Wilson, E. M. (1999). Activation function 2 in the human androgen receptor ligand binding domain mediates interdomain communication with the NH(2)-terminal domain. *J. Biol. Chem.* **274**, 37219–37225.

He, B., Kemppainen, J., and Wilson, E. M. (2000). FXXLF and WXXLF sequences mediate the NH2-terminal interaction with the ligand binding domain of the androgen receptor. *J. Biol. Chem.* **275,** 22986–22994.

He, B., Bowen, N. T., Minges, J. T., and Wilson, E. M. (2001). Androgen-induced NH2- and COOH-terminal interaction inhibits p160 coactivator recruitment by activation function 2. *J. Biol. Chem.* **276,** 42293–42301.

Holmberg, M., Duyckaerts, C., Durr, A., Cancel, G., Gourfinkel-An, I., Damier, P., Faucheux, B., Trottier, Y., Hirsch, E. C., Agid, Y., and Brice, A. (1998). Spinocerebellar ataxia type 7 (SCA7): A neurodegenerative disorder with neuronal intranuclear inclusions. *Hum. Mol. Genet.* **7,** 913–918.

Hsiao, P. W., Lin, D. L., Nakao, R., and Chang, C. (1999). The linkage of Kennedy's neuron disease to ARA24, the first identified androgen receptor polyglutamine region-associated coactivator. *J. Biol. Chem.* **274,** 20229–20234.

Huynh, D., Del Bigio, M., Ho, D., and Pulst, S. (1999). Expression of ataxin-2 in brains from normal individuals and patients with Alzheimer's disease and spinocerebellar ataxia 2. *Ann. Neurol.* **45,** 232–241.

Ikeda, H., Yamaguchi, M., Sugai, S., Aze, Y., Naruiya, S., and Kakizuka, A. (1996). Expanded polyglutamine in the Machado-Joseph disease protein induces cell death in vitro and in vivo. *Nat. Genet.* **13,** 196–202.

Ikonen, T., Palvimo, J. J., Kallio, P. J., Reinikainen, P., and Janne, O. A. (1994). Stimulation of androgen-regulated transactivation by modulators of protein phosphorylation. *Endocrinology* **135,** 1359–1366.

Irvine, R., Ma, H., Yu, M., Ross, R., Stallcup, M., and Coetzee, G. (2000). Inhibition of p160-mediated coactivation with increasing androgen receptor polyglutamine length. *Hum. Mol. Genet.* **9,** 267–274.

Ishikawa, K., Fujigasaki, H., Saegusa, H., Ohwada, K., Fujita, T., Iwamoto, H., Komatsuzaki, Y., Toru, S., Toriyama, H., Watanabe, M., Ohkoshi, N., Shoji, S., Kanazawa, I., Tanabe, T., and Mizusawa, H. (1999). Abundant expression and cytoplasmic aggregations of a1A voltage-dependent calcium channel protein associated with neurodegeneration in spinocerebellar ataxia type 6. *Hum. Mol. Genet.* **8,** 1185–1193.

Jana, N. R., Zemskov, E. A., Wang, G.-h., and Nukina, N. (2001). Altered proteasomal function due to the expression of polyglutamine-expanded truncated N-terminal huntingtin induces apoptosis by caspase activation through mitochondrial cytochrome c release. *Hum. Mol. Genet.* **10,** 1049–1059.

Jenster, G., van der Korput, H. A., van Vroonhoven, C., van der Kwast, T. H., Trapman, J., and Brinkmann, A. O. (1991). Domains of the human androgen receptor involved in steroid binding, transcriptional activation, and subcellular localization. *Mol. Endocrinol.* **5,** 1396–1404.

Jones, K. J., and Oblinger, M. M. (1994). Androgenic regulation of tubulin gene expression in axotomized hamster facial motoneurons. *J. Neurosci.* **14,** 3620–3627.

Kazemi-Esfarjani, P., Trifiro, M. A., and Pinsky, L. (1995). Evidence for a repressive function of the long polyglutamine tract in the human androgen receptor: Possible pathogenetic relevance for the $(CAG)_n$-expanded neuronopathies. *Hum. Mol. Genet.* **4,** 523–527.

Kemppainen, J. A., Lane, M. V., Sar, M., and Wilson, E. M. (1992). Androgen receptor phosphorylation, turnover, nuclear transport and transcriptional activation: Specificity for steroids and antihormones. *J. Biol. Chem.* **267,** 968–974.

Kennedy, W. R., Alter, M., and Sung, J. H. (1968). Progressive proximal spinal and bulbar muscular atrophy of late onset: A sex-linked recessive trait. *Neurology* **18,** 671–680.

Klement, I. A., Skinner, P. J., Kaytor, M. D., Yi, H., Hersch, S. M., Clark, H. B., Zoghbi, H. Y., and Orr, H. T. (1998). Ataxin-1 nuclear localization and aggregation: Role in polyglutamine-induced disease in SCA1 transgenic mice. *Cell* **95,** 41–53.

Kobayashi, Y., Miwa, S., Merry, D. E., Kume, A., Mei, L., Doyu, M., and Sobue, G. (1998). Caspase-3 cleaves the expanded androgen receptor protein of spinal and bulbar muscular atrophy in a polyglutamine repeat length-dependent manner. *Biochem. Biophys. Res. Commun.* **252,** 145–150.

Koyano, S., Uchihara, T., Fujigasaki, H., Nakamura, A., Yagishita, S., and Iwabuchi, K. (1999). Neuronal intranuclear inclusions in spinocerebellar ataxia type 2: Triple-labeling immunofluorescent study. *Neurosci. Lett.* **273**, 117–120.

Kuemmerle, S., Gutekunst, C., Klein, A., Li, X., Li, S., Beal, M., Hersch, S., and Ferrante, R. (1999). Huntington aggregates may not predict neuronal death in Huntington's disease. *Ann. Neurol.* **46**, 232–241.

Kujawa, K. A., Emeric, E., and Jones, K. J. (1991). Testosterone differentially regulates the regenerative properties of injured hamster facial motor neurons. *J. Neurosci.* **11**, 3898–3908.

Kujawa, K. A., Jacob, J. M., and Jones, K. J. (1993). Testosterone regulation of the regenerative proerties of injured rat sciatic motor neurons. *J. Neurosci. Res.* **35**, 268–273.

Kurz, E. M., Sengelaub, D. R., and Arnold, A. P. (1986). Androgens regulate the dendritic length of mammalian motoneruons in adulthood. *Science* **232**, 395–398.

Langley, E., Zhou, Z.-x., and Wilson, E. M. (1995). Evidence for an anti-parallel orientation of the ligand-activated human androgen receptor dimer. *J. Biol. Chem.* **270**, 29983–29990.

Langley, E., Kemppainen, J. A., and Wilson, E. M. (1998). Intermolecular NH2-/carboxyl-terminal interactions in androgen receptor dimerization revealed by mutations that cause androgen insensitivity. *J. Biol. Chem.* **273**, 92–101.

La Spada, A. R., Wilson, E. M., Lubahn, D. B., Harding, A. E., and Fischbeck, K. H. (1991). Androgen receptor gene mutations in X-linked spinal and bulbar muscular atrophy. *Nature* **353**, 77–79.

La Spada, A. R., Roling, D., Harding, A. E., Warner, C. L., Speigel, R., Hausmanowa-Petrusewicz, I., Yee, W.-C., and Fischbeck, K. H. (1992). Meiotic stability and genotype-phenotype correlation of the expanded trinucleotide repeat in X-linked spinal and bulbar muscular atrophy. *Nat. Genet.* **2**, 301–304.

La Spada, A. R., Peterson, K. R., Meadows, S. A., McClain, M. E., Jeng, G., Chmelar, R. S., Haugen, H. A., Chen, K., Singer, M. J., Moore, D., Trask, B. J., Fischbeck, K. H., Clegg, C. H., and McKnight, G. S. (1998). Androgen receptor YAC transgenic mice carrying CAG 45 alleles show trinucleotide repeat instability. *Hum. Mol. Genet.* **7**, 959–967.

Li, M., Miwa, S., Kobayashi, Y., Merry, D. E., Yamamoto, M., Tanaka, F., Doyu, M., Hashizume, Y., Fischbeck, K. H., and Sobue, G. (1998a). Nuclear inclusions of the androgen receptor protein in spinal and bulbar muscular atrophy. *Ann. Neurol.* **44**, 249–254.

Li, M., Nakagomi, Y., Kobayashi, Y., Merry, D. E., Tanaka, F., Doyu, M., Mitsuma, T., Hashizume, Y., Fischbeck, K. H., and Sobue, G. (1998b). Nonneural nuclear inclusions of androgen receptor protein in spinal and bulbar muscular atrophy. *Am. J. Pathol.* **153**, 695–701.

Li, S.-H., Cheng, A. L., Li, H., and Li, X.-J. (1999). Cellular defects and altered gene expression in PC12 cells stably expressing mutant huntingtin. *J. Neurosci.* **19**, 5159–5172.

Lin, H.-K., Yeh, S., Kang, H.-Y., and Chang, C. (2001). Akt suppresses androgen-induced apoptosis by phosphorylating and inhibiting androgen receptor. *Proc. Natl. Acad. Sci. USA* **98**, 7200–7205.

Lin, X., Cummings, C., and Zoghbi, H. (1999). Expanding our understanding of polyglutamine diseases through mouse models. *Neuron* **24**, 499–502.

Lin, X., Antalffy, B., Kang, D., Orr, H. T., and Zoghbi, H. Y. (2000). Polyglutamine expansion down-regulates specific neuronal genes before pathologic changes in SCA1. *Nat. Neurosci.* **3**, 157–163.

MacLean, H. E., Choi, W.-T., Rekaris, G., Warne, G. L., and Zajac, J. D. (1995). Abnormal androgen receptor binding affinity in subjects with Kennedy's disease (spinal and bulbar muscular atrophy). *J. Clin. Endocrinol. Metab.* **80**, 508–516.

MacLean, H. E., Warne, G. L., and Zajac, J. D. (1997). Localization of functional domains in the androgen receptor. *J. Steroid Biochem. Mol. Biol.* **62**, 233–242.

Mangiarini, L., Sathasivam, K., Seller, M., Cozens, B., Harper, A., Hetherington, C., Lawton, M., Trottier, Y., Lehrach, H., Davies, S. W., and Bates, G. P. (1996). Exon 1 of the HD gene with an expanded CAG repeat is sufficient to cause a progressive neurological phenotype in transgenic mice. *Cell* **87**, 493–506.

Matsumoto, A., Arai, Y., Urano, A., and Hyodo, S. (1995). Molecular basis of neuronal plasticity to gonadal steroids. *Funct. Neurol.* **10**, 59–76.

McCampbell, A., Taylor, J. P., Taye, A. A., Robitschek, J., Li, M., Walcott, J., Merry, D., Chai, Y., Paulson, H., Sobue, G., and Fischbeck, K. H. (2000). CREB-binding protein sequestration by expanded polyglutamine. *Hum. Mol. Genet.* **9,** 2197–2202.

McCampbell, A., Taye, A. A., Whitty, L., Penney, E., Steffan, J. S., and Fischbeck, K. H. (2001). Histone deacetylase inhibitors reduce polyglutamine toxicity. *Proc. Natl. Acad. Sci. USA* **98,** 15179–15184.

Merry, D. E., McCampbell, A., Taye, A. A., Winston, R. L., and Fischbeck, K. H. (1996). Toward a mouse model for spinal and bulbar muscular atrophy: Effect of neuronal expression of androgen receptor in transgenic mice. *Am. J. Hum. Genet.* **59** (Suppl.), A271.

Merry, D. E., Kobayashi, Y., Bailey, C. K., Taye, A. A., and Fischbeck, K. H. (1998). Cleavage, aggregation, and toxicity of the expanded androgen receptor in spinal and bulbar muscular atrophy. *Hum. Mol. Genet.* **7,** 693–701.

Mhatre, A. N., Trifiro, M. A., Kaufman, M., Kazemi-Esfarjani, P., Figlewicz, D., Rouleau, G., and Pinsky, L. (1993). Reduced transcriptional regulatory competence of the androgen receptor in X-linked spinal and bulbar muscular atrophy. *Nat. Genet.* **5,** 184–188.

Moilanen, A.-M., Karvonen, U., Poukka, H., Janne, O. A., and Palvimo, J. J. (1998). Activation of androgen receptor function by a novel nuclear protein kinase. *Mol. Biol. Cell* **9,** 2527–2543.

Nakajima, H., Kimura, F., Nakagawa, T., Furutama, D., Shinoda, K., Shimizu, A., and Ohsawa, N. (1996). Transcriptional activation by the androgen receptor in X-linked spinal and bulbar muscular atrophy. *J. Neurol. Sci.* **142,** 12–16.

Nakamura, K., Jeong, S.-Y., Uchihara, T., Anno, M., Nagashima, K., Nagashima, T., Ikeda, S.-I., Tsuji, S., and Kanazawa, I. (2001). SCA17, a novel autosomal dominant cerebellar ataxia caused by an expanded polyglutamine in TATA-binding protein. *Hum. Molec. Genet.* **10,** 1441–1448.

Nakamura, M., Mita, S., Murakami, T., Uchino, M., Watanabe, S., Tokunaga, M., Kumamoto, T., and Ando, M. (1994). Exonic trinucleotide repeats and expression of androgen receptor gene in spinal cord from X-linked spinal and bulbar muscular atrophy. *J. Neurol. Sci.* **122,** 74–79.

Nazareth, L. V., and Weigel, N. L. (1996). Activation of the human androgen receptor through a protein kinase A signaling pathway. *J. Biol. Chem.* **271,** 19900–19907.

Nordeen, E. J., Nordeen, K. W., Sengelaub, D. R., and Arnold, A. P. (1985). Androgens prevent normally occurring cell death in a sexually dimorphic spinal nucleus. *Science* **229,** 671–673.

Nucifora, J. F. C., Sasaki, M., Peters, M. F., Huang, H., Cooper, J. K., Yamada, M., Takahashi, H., Tsuji, S., Troncoso, J., Dawson, V. L., Dawson, T. M., and Ross, C. A. (2001). Interference by huntingtin and atrophin-1 with CBP-mediated transcription leading to cellular toxicity. *Science* **291,** 2423–2428.

Paulson, H. (1999). Protein fate in neurodegenerative proteinopathies; polyglutamine diseases join the misfold. *Am. J. Hum. Genet.* **64,** 339–345.

Paulson, H. L., and Fischbeck, K. H. (1996). Trinucleotide repeats in neurogenetic disorders. *Annu. Rev. Neurosci.* **19,** 79–107.

Paulson, H. L., Perez, M. K., Trottier, Y., Trojanowski, J. Q., Subramony, S. H., Das, S. S., Vig, P., Mandel, J.-L., Fischbeck, K. H., and Pittman, R. N. (1997). Intranuclear inclusions of expanded polyglutamine protein in spinocerebellar ataxia type 3. *Neuron* **19,** 1–20.

Perez, J., and Kelley, D. B. (1996). Trophic effects of androgen: Receptor expression and the survival of laryngeal motor neurons after axotomy. *J. Neurosci.* **16,** 6625–6633.

Persichetti, F., Trettel, F., Huang, C. C., Fraefel, C., Timmers, H. T., Gusella, J. F., and MacDonald, M. E. (1999). Mutant huntingtin forms in vivo complexes with distinct context-dependent conformations of the polyglutamine segment. *Neurobiol. Dis.* **6,** 364–375.

Perutz, M. (1996). Glutamine repeats and inherited neurodegenerative diseases: Molecular aspects. *Curr. Opin. Struct. Biol.* **6,** 848–858.

Perutz, M. F., Johnson, T., Suzuki, M., and Finch, J. T. (1994). Glutamine repeats as polar zippers: Their possible role in inherited neurodegenerative diseases. *Proc. Natl. Acad. Sci. USA* **91,** 5355–5358.

Pinsky, L., Trifiro, M., Kaufman, M., Beitel, L. K., Mhatre, A., Kazemi-Esfarjani, P., Sabbaghian, N., Lumbroso, R., Alvarado, C., Vasiliou, M., and Gottlieb, B. (1992). Androgen resistance due to mutation of the androgen receptor. *Clin. Invest. Med.* **15,** 456–472.

Poukka, H., Aarnisalo, P., Karvonen, U., Palvimo, J. J., and Janne, O. A. (1999). Ubc9 interacts with the androgen receptor and activates receptor-dependent transcription. *J. Biol. Chem.* **274,** 19441–19446.

Poukka, H., Karvonen, U., Janne, O. A., and Palvimo, J. J. (2000). Covalent modification of the androgen receptor by small ubiquitin-like modifier 1 (SUMO-1). *Proc. Natl. Acad. Sci. USA* **97,** 14145–14150.

Pratt, W. B., and Welsh, M. J. (1994). Chaperone functions of the heat shock proteins associated with steroid receptors. *Cell Biol.* **5,** 83–93.

Quigley, C. A., Friedman, K. J., Johnson, A., Lafreniere, R. G., Silverman, L. M., Lubahn, D. B., Brown, T. R., Wilson, E. M., Willard, H. F., and French, F. S. (1992). Complete deletion of the androgen receptor gene: Definition of the null phenotype of the androgen insensitivity syndrome and determination of carrier status. *J. Clin. Endocrinol. Metab.* **74,** 927–933.

Sar, M., and Stumpf, W. E. (1977). Androgen concentration in motor neurons of cranial nerves and spinal cord. *Science* **19,** 77–79.

Saudou, F., Finkbeiner, S., Devys, D., and Greenberg, M. E. (1998). Huntingtin acts in the nucleus to induce apoptosis but death does not correlate with the formation of intranuclear inclusions. *Cell* **95,** 55–66.

Sculptoreanu, A., Abramovici, H., Abdullah, A. A. R., Bibikova, A., Panet-Raymond, V., Frankel, D., Schipper, H. M., Pinsky, L., and Trifiro, M. A. (2000). Increased T-type Ca2+ channel activity as a determinant of cellular toxicity in neuronal cell lines expressing polyglutamine-expanded human androgen receptors. *Mol. Cell. Biochem.* **203,** 23–31.

Simeoni, S., Mancini, M. A., Stenoien, D. L., Marcelli, M., Weigel, N. L., Zanisi, M., Martini, L., and Poletti, A. (2000). Motoneuronal cell death is not correlated with aggregate formation of androgen receptors containing an elongated polyglutamine tract. *Hum. Mol. Genet.* **9,** 133–144.

Sobue, G., Hashizume, Y., Mukai, E., Hirayama, M., Mitsuma, T., and Takahashi, A. (1989). X-linked recessive bulbospinal neuronopathy: A clinicopathological study. *Brain* **112,** 209–232.

Steffan, J., Kazantsev, A., Spasic-Boskovic, O., Greenwald, M., Zhu, Y. Z., Gohler, H., Wanker, E., Bates, G., Housman, D., and Thompson, L. (2000). The Huntington's disease protein interacts with p53 and CREB-binding protein and represses transcription. *Proc. Natl. Acad. Sci. USA* **97,** 6763–6768.

Steffan, J. S., Bodai, L., Pallos, J., Poelman, M., McCampbell, A., Apostol, B. L., Kazantsev, A., Schmidt, E., Zhu, Y.-Z., Greenwald, M., Kurokawa, R., Housman, D. E., Jackson, G. R., Marsh, J. L., and Thompson, L. M. (2001). Histone deacetylase inhibitors arrest polyglutamine-dependent neurodegeneration in Drosophila. *Nature* **413,** 739–743.

Stenoien, D. L., Cummings, C. J., Adams, H. P., Mancini, M. G., Patel, K., DeMartino, G. N., Marcelli, M., Weigel, N. L., and Mancini, M. A. (1999). Polyglutamine-expanded androgen receptors form aggregates that sequester heat shock proteins, proteasome components and SRC-1, and are suppressed by the HDJ-2 chaperone. *Hum. Mol. Genet.* **8,** 731–741.

Trottier, Y., Lutz, Y., Stevanin, G., Imbert, G., Devys, D., Cancel, G., Saudou, F., Weber, C., David, G., Tora, L., Agid, Y., Brice, A., and Mandel, J.-L. (1995). Polyglutamine expansion as a pathological epitope in Huntington's disease and four dominant cerebellar ataxias. *Nature* **378,** 403–406.

Tsai, M.-J., and O'Malley, B. W. (1994). Molecular mechanisms of action of steroid/thryoid receptor superfamily members. *Annu. Rev. Biochem.* **63,** 451–486.

Tyagi, R. K., Lavrovsky, Y., Ahn, S. C., Song, C. S., Chatterjee, B., and Roy, A. K. (2000). Dynamics of intracellular movement of nucleocytoplasmic recycling of the ligand-activated androgen receptor in living cells. *Mol. Endocrinol.* **14,** 1162–1174.

Warner, C. L., Griffen, J. E., Wilson, J. D., Jacobs, L. D., Murray, K. R., Fischbeck, K. H., Dickoff, D., and Griggs, R. C. (1992). X-linked spinomuscluar atrophy; a kindred with associated abnormal androgen binding. *Neurology* **42,** 2181–2184.

Wellington, C. L., Ellerby, L. M., Hackam, A. S., Margolis, R. L., Trifiro, M. A., Sangaraja, R., McCutcheon, K., Salvesen, G. S., Propp, S. S., Bromm, M., Rowland, K. J., Zhang, T., Rasper, D., Roy, S., Thornberry, N., Pinsky, L., Kakizuka, A., Ross, C. A., Nicholson, D. W., Bredesen, D. E., is evident from PGI_2-mediated inhibition of smooth muscle proliferation (Zucker *et al.*, 1998) and modulation of both reverse cholesterol transport from vascular cells (Morishita *et al.*, 1990) and cellular adhesion to the vessel wall (Sinzinger *et al.*, 1987) *in vitro*. Delivery of PGI_2 synthase *in vivo* can prevent proliferation and migration of smooth muscle cell (SMC), key features of restenosis and atherosclerosis (Harada *et al.*, 1999; Numaguchi *et al.*, 1999b). In addition, the response to a thrombotic stimulus was reduced in mice lacking the PGI_2 receptor (Murata *et al.*, 1997), confirming the role of PGI_2 as an endogneous antithrombotic agent. PGI_2-mediated protection of cardiomyocytes *in vitro* (Adderley and Fitzgerald, 1999) and *in vivo* (Xiao *et al.*, 2001) has also been reported. Outside the cardiovascular system, PGI_2 is thought to be important in the regulation of gastric acid secretion and renal blood flow, renin release, and glomerular filtration rate in the kidney (Komhoff *et al.*, 1998). More recently attention has focused on PGI_2 as a regulator of gene transcription with roles in inflammation, remodeling, and angiogenesis (Jones *et al.*, 1999; Gupta *et al.*, 2000; Tong *et al.*, 2000; Hoshikawa *et al.*, 2001b). Clinically, chronic PGI_2 treatment can reduce pulmonary vascular resistance in patients with primary pulmonary hypertension (McLaughlin *et al.*, 1998) and has had some usage in patients with peripheral vascular disease.

PGI_2 transduces its effects through a membrane-associated receptor termed the IP. Studies of this receptor have been hampered by the lack of an IP antagonist, highly selective agonists, or specific anti-IP antibodies. Cloning of the cDNA for the IP from several species, including human [h(Boie *et al.*, 1994)] and mouse [m(Namba *et al.*, 1994)], indicated it membership of the G protein-coupled receptor (GPCR) superfamily (Fig. 1) and facilitated detailed examination of IP's signaling, regulation, and function.

II. EXPRESSION AND DISTRIBUTION OF IP

The hIP gene encodes a protein consisting of 386 amino acid residues with a deduced molecular weight of 40,956 kDa ((Ogawa *et al.*, 1995), Fig. 1). Spanning approximately 7 kb and consisting of three exons and two introns, the gene maps to chromosome bands 19q13.3. The hIP promoter region contains several regulatory motifs including AP1 and AP2 binding sites and a glucocorticoid response element but no conventional TATA and CCAAT boxes (Ogawa *et al.*, 1995). Studies directly examining the regulation of IP gene expression have not been carried out, although in one study IP expression was up-regulated by proinflammatory and megakaryocytopoietic cytokines (Sasaki *et al.*, 1997). Control of IP gene regulation may, however, be relevant in settings of vascular disease; in both animal models of hypertension (Numaguchi *et al.*, 1999a) and human pulmonary hypertension (Hoshikawa *et al.*, 2001b) IP expression was decreased. IP-mediated control of gene expression may also modulate mitogen-activated protein (MAP) kinase pathways, and antiproliferative/apoptopic gene expression, responsible for prostacyclin-dependent modulation of pulmonary vascular remodeling in hypoxia (Hoshikawa *et al.*, 2001a).

Human IP mRNA is abundantly expressed in the kidney (Boie et al., 1994) where PGI_2, the principal eicosanoid synthesized, is thought to regulate renal blood flow, renin release, and glomerular filtration rate (Komhoff et al., 1998). Receptor mRNA is also present in the lung where PGI_2 can modulate vascular tone (McLaughlin et al., 1998) and directed overexpression of PGI_2 synthase reduces elevated pulmonary blood pressure (Geraci et al., 1999). The IP is also expressed in both the brain and the liver where the role of PGI_2 is unknown. Expression within the cardiovascular system is most abundant in the aorta (Nakagawa et al., 1994), consistent with the major biological role of PGI_2 in platelet and macrovascular homeostasis. IP is also expressed in the atrium and ventricle of the heart indicating a possible role for PGI_2 in cardiac tissue. Indeed, increased PGI_2 biosynthesis was implicated in cytoprotection against oxidant-induced cardiac injury (Adderley and Fitzgerald, 1999; Dowd et al., 2001) as a homeostatic response to thromboxane-mediated cardiac decompensation (Zhang et al., 2001b).

Examination of IP expression in other species has yielded results similar to humans, with abundant expression in the kidney, lung, liver, and heart (Namba et al., 1994; Sasaki et al., 1994). A major difference between humans and rodents, however, is the marked expression of IP in the thymus, although the relevance is not clear. In addition, IP is widely distributed in the brain, although again the function of PGI_2 in the central nervous system remains unknown. Despite its expression in murine thymus, IP deletion has not been reported to result in disordered T cell function and the IP appears to play a minor role, if any, in murine T cell maturation (Rocca et al., 1999).

III. LIGAND BINDING AND SIGNAL TRANSDUCTION

Prostanoids are part of the eicosanoid family of 20-carbon fatty acid arachidonic acid derivatives (Narumiya et al., 1999). Two structural features of the ligand, a cyclopentane ring and its β side chain, primarily determine the binding affinities of the prostanoid receptors. The binding pocket of the IP can evidently accommodate the cyclopentane rings of the I and E type prostanoids. PGI_2 analogues such as iloprost and cicaprost bind the IP with an affinity equal to PGI_2, as does PGE_1, but the affinity of PGE_2 is much lower. The rank order of affinity of ligands (agonists) for the cloned hIP was iloprost = cicaprost > PGE_1 > carbacyclin >> PGE_2 > PGD_2, $PGF_{2\alpha}$ (Boie et al., 1994; Nakagawa et al., 1994). Currently no antagonist for the IP is available. Cross-reactivity of IP ligands with receptors for PGE_2 (EP) has been reported. IP ligands including cicaprost (generally considered to be the most selective IP agonist available) bind to EP3 at relatively low concentrations, whereas iloprost, a commonly used IP agonist, also demonstrates significant cross-reactivity with EP1 (Narumiya et al., 1999).

IP couples to its intracellular signaling cascades via activation of heterotrimeric G proteins. The best characterized of these reactions is activation of G_s with a concomitant increase in cellular cAMP (Boie *et al.*, 1994). Indeed, this signaling cascade is thought to be responsible for many of the established biological actions of PGI_2. It is clear, however, that, similar to other GPCRs, IP may activate more than one G protein. Ligation of IP induces increased inositol phosphate generation/intracellular calcium mobilization in overexpression cell models (Smyth *et al.*, 1996; Lawler *et al.*, 2001), presumably via activation of Gq, whereas inhibition of cAMP generation, via Gi, has also been reported for the mouse (m) receptor (Lawler *et al.*, 2001). Indeed, mIP-Gq and mIP-Gi coupling may depend on a protein kinase A (PKA)-mediated switch in effector coupling (Lawler *et al.*, 2001), such as has been described for β_2-adrenoreceptor (AR)-mediated activation of MAP kinases (Daaka *et al.*, 1997). The physiological relevance of multiple potential IP-G protein combinations is not known. It appears likely, however, that a reductionist view of IP-Gs activation alone may simplify biological reality. Reports of GPCR compartmentalization in microdomains of the cell membrane underscore this possibility. In rat cardiomyocytes the β1-AR and β2-AR show distinct cellular targeting with the latter confined totally to calveolae (Rybin *et al.*, 2000). This differential distribution colocalizes the receptors with different components of the cAMP signaling pathway and may direct differences in their signaling characteristics. Although the IP itself has not been examined colocalization of the components of the PGI_2 biosynthetic pathway with caveolin-1 in endothelial cells has been reported (Liou *et al.*, 2000; Spisni *et al.*, 2001).

IV. IP REGULATION

The effects of PGI_2 are limited as a result of its short half-life (2 min) *in vivo* and by agonist-dependent regulation of its receptor (Smyth *et al.*, 1998, 2000). It has been known for some time that elevated PGI_2 levels are accompanied by decreased IP binding and responsiveness in human vascular disease, for example, myocardial infarction (Jaschonek *et al.*, 1986) and preeclampsia (Klockenbusch *et al.*, 1996), and during administration of PGI_2 and its analogues (Modesti *et al.*, 1987). Furthermore, agonist-induced regulation of IP may be important in settings of plaque rupture, such as unstable angina, where biosynthesis of PGI_2 is increased during ischemic episodes but vasoconstrictor signals predominate (Fitzgerald *et al.*, 1986a,b; Oates *et al.*, 1988; FitzGerald, 1991).

Agonist-induced regulation of GPCR is a multistep process (Ferguson and Caron, 1998). In the "classic" pathway, agonist-induced GPCR phosphorylation is mediated by second messenger-activated kinases, protein kinase C (PKC) or PKA, and/or GPCR kinases (GRKs). Binding of an adapter protein, arrestin, leads to uncoupling of the receptor from the G protein and receptor sequestration. The internalized receptor may be recycled to the cell surface for another round of signal transduction, may be down-regulated via lysosomal degradation, or may direct activation of additional signaling systems. A complex interplay of second messenger kinases and GRKs has emerged from recent studies. The β2-AR, a prototypic

GPCR, is phosphorylated and regulated in a GRK2-dependent manner, which is, in turn, dependent on PKA-mediated phosphorylation of GRK2, facilitating its association with the receptor (Cong et al., 2001). Indeed a so-called A-kinase anchoring protein (AKAP79) acts as a scaffold in the formation of a signaling complex comprising the β2-AR, PKA, PKC, and a protein phosphatase (Fraser et al., 2000).

Studies of homologous IP regulation have frequently focused on changes in receptor sensitivity and abundance resulting from exposure to agonist for several hours. Preincubation of platelets with iloprost resulted in a dose-dependent down-regulation of [^{3}H]iloprost binding accompanied by a loss of IP-mediated inhibition of platelet-endothelial cell adhesion (Darius et al., 1995). Internalization of a high-affinity binding site for iloprost, presumed to be the prostacyclin receptor, has also been reported in platelets (Giovanazzi et al., 1997). Iloprost-induced inhibition of smooth muscle proliferation was attenuated by iloprost pretreatment of cells for 24 h, an effect that did not extend to the antiproliferative actions of PGE$_1$, suggesting homologous regulation of IP (Grosser et al., 1995; Zucker et al., 1998).

The availability of the cDNA for IP has allowed the generation of overexpression cell models for direct investigation of IP regulation. Agonist activation of hIP induced a rapid (seconds) receptor PKC-mediated phosphorylation and uncoupling from adenylyl cyclase (Smyth et al., 1996, 1998). Treatment of cells for longer periods of time (up to 1 h) induced a phosphorylation-independent sequestration of hIP via at least two endocytotic pathways (Smyth et al., 2000). In addition, mIP may be phosphorylation by PKA (Lawler et al., 2001) and/or GRKs (Smyth et al., 1996, 1998) with little or no impact on homologous regulation, whereas an uncharacterized phosphorylation- and sequestration-independent regulatory pathway (Nilius et al., 2000) may also contribute to agonist-mediated receptor responsiveness. Whatever the mechanism, rapid uncoupling of the receptor from its effector(s), together with the short half-life of its autocoidal ligand at physiological pH, predicts a tight regulation of PGI$_2$ responsiveness *in vivo*.

It has become increasingly apparent that heterologous regulation of prostanoid receptors may play a role in receptor responsiveness. The thromboxane receptor (TP) α isoform, but not the β isoform, is heterologously regulated by IP- (Walsh et al., 2000) or PGD$_2$ receptor (DP)- (Foley et al., 2001) mediated activation of PKA, whereas both TP isoforms are subject to cross-regulation by EP1-mediated PKC activation (Walsh and Kinsella, 2000). The potential for cross-regulation of IP has not been formally studied, although PKC-dependent hIP phosphorylation was evident during activation of the thrombin receptor (Smyth et al., 1996). Contrary to expectations, however, activation of TP but not the thrombin receptor, both of which activate PKC, led to enhanced IP signaling in platelets (Murray et al., 1990), suggesting that heterologous regulation of IP may be more complex than expected.

The formation of GPCR homodimers and heterodimers may underlie this complexity. Mutant bradykinin B2 receptors that do not form homodimers do not undergo agonist-induced phosphorylation, desensitization, or down-regulation (AbdAlla et al., 1999), whereas heterodimers of the angiotensin II AT-1 with B2 receptors are sequestered via a distinct pathway not utilized by either receptor alone (AbdAlla *et al.*, 2000). Indeed B2 receptor expression enhanced angiotensin II signaling whereas AT-1 receptor expression reduced B2 signaling, suggesting that heterodimerization, and not necessarily an interplay of kinase pathways, may direct heterologous receptor regulation. Furthermore, the AT-2 receptor can act as an AT-1 receptor-specific antagonist as a direct result of heterodimerization of the receptor subtypes (AbdAlla *et al.*, 2001a). Although not yet examined, such direct receptor–receptor interactions may contribute to heterologous regulation of TPs by IP, DP, or EP1 (Walsh *et al.*, 2000; Walsh and Kinsella, 2000; Foley *et al.*, 2001). IP homodimers form in transfected HEK 293 cells (Smyth *et al.*, unpublished observations). Whether this association, or IP interaction with other GPCRs, is important for its functional regulation is currently under investigation.

V. MUTATION STUDIES

Homology between IP and other prostaglandin receptors (EP subtypes 1–4 for PGE_2, DP1 and DP2 for PGD_2, TPα and TPβ for TxA_2, and FP for $PGF_{2\alpha}$) is less than 50% and is greatest in the transmembrane regions, especially the seventh (Narumiya *et al.*, 1999). Phylogenetic comparison of the prostanoid receptor family reveals three subclusters: the first consists of the relaxant relaxant receptors EP2, EP4, IP, and DP, which increase cAMP generation; the second consists of the contractile receptors EP1, FP, and TP, which increase intracellular calcium levels; and the third consists of EP3, which can couple to both elevation of intracellular calcium and a decrease in cAMP (Boie *et al.*, 1995; Narumiya *et al.*, 1999).

Several potentially interesting features, common to GPCR, are present in the IP sequence (Boie *et al.*, 1994; Fig. 1). There are two potential sites for N-linked glycosylation, located in the N-terminus and first extracellular loop. Cysteine residues, in the extraceullar regions, which may form an intramolecular disulfide bridge, are conserved in all prostanoid receptors (Narumiya *et al.*, 1999). A basic arginine residue, a DPWXY motif, and an isoleucine residue, located in the seventh transmembrane domain, are also conserved among all prostanoid receptors. In addition to these regions, a pair of cysteine residues, which may be palmitoylated, is found in the C-terminus along with several potential sites for phosphorylation by serine/threonine kinases. Interestingly, the presence of a CAAX (CSCL) motif for isoprenylation, at the C-terminal end of the protein, makes IP unique among the prostanoid receptors (Hayes *et al.*, 1999). Mutational studies have directly addressed the regions important for prostanoid receptor functional regulation.

All of the prostanoid receptors that have been cloned so far, DP1, DP2, FP, both TP isoforms, EP subtypes 1–4 and IP, have N-glycosylation sites (Coleman *et al.*, 1994; Narumiya *et al.*, 1999). The addition of carbohydrate moieties to form a glycosylated GPCR is believed to be important for intracellular trafficking, stability, protein folding, membrane localization, ligand binding, and signal transduction with varying effects depending on the receptor being studied. Glycosylation of CXCR-4 (Zhou and Tai, 1999), TPα (Walsh et al., 1998), and EP3α (Huang and Tai, 1998) are necessary both for correct sorting to the plasma membrane and for normal ligand binding and/or signal transduction. In contrast, although plasma membrane localization of glycosylation-deficient β-AR (Rands et al., 1990), platelet-activating receptor (Garcia Rodriguez et al., 1995), or EP3β receptors (Boer et al., 2000) was reduced, ligand binding and signal transduction were preserved. Disruption of hIP glycosylation at either Asn-7 or Asn-78 residues (Fig. 1) reduced membrane localization, ligand binding, and both hIP–adenylyl cyclase and –phospholipase coupling (Zhang et al., 2001a). Single substitution of Asn-78 resulted in significantly greater deficiency in hIP–adenylyl cyclase coupling compared to Asn-7, whereas hIP carrying both mutations was completely unable to bind its ligand or transduce the cAMP signal, despite substantial residual receptor at the plasma membrane. Interestingly, activation of phospholipase C activation was completely dependent on the integrity of both sites, suggesting that hIP signal transduction through multiple G protein cascades may, in part, be related to the glycosylation of the receptor.

Several GPCRs utilize the C terminal region to direct their regulation. Disruption of a consensus sites for PKC-mediated phosphorylation of hIP, serine 328, abolished rapid agonist-induced phosphorylation and desensitization mediated by this kinase (Smyth et al., 1998) with no affect on receptor sequestration (Smyth et al., 2000). Other consensus sites for PKC phosphorylation of hIP, at serine 374 (Smyth et al., 1998, 2000), or mIP at serine 357 (Lawler et al., 2001), have been disrupted with no effect on regulation. However, the loss of serine 357 from the mIP prevented agonist-induced PKA-mediated phosphorylation and abolished mIP-Gi and mIP-Gq activation. Truncation of the entire C terminal region of hIP generated a receptor in which ligand binding, adenylyl cyclase, and partial PLC activation were preserved (Smyth et al., 1998), but in which all agonist-induced regulatory responses were abolished (Smyth et al., 1998, 2000), indicating the critical role played by this region in IP regulation.

Mutational studies have also addressed the regions important for IP to bind its ligand. Neither mutation of C terminally located phosphorylation sites nor the CAAX motif had any effect on ligand binding (Hayes et al., 1999), in agreement with the current understanding that the ligand-binding pocket of the prostanoid receptors is formed mainly by the first, second, and seventh transmembrane domains. In a study with chimeric receptors Ser-50, in the first transmembrane domain of mIP, was identified as important for binding of the cyclopentane ring of D, E, and I type prostanoids (Kobayashi et al., 2000). Recognition of the α-side chain of IP ligands was determined by the sixth and/or seventh transmembrane region, which may confer the ability of IP to discriminate a structural difference between PGE_1 and PGE_2 side chains. In another study, point mutation of a single amino acid Leucine 304 in the seventh transmembrane region conferred iloprost binding on the EP2, again demonstrating the importance of this domain for prostanoid binding (Kedzie et al., 1998).

Although not addressed directly for the IP, mutational studies have determined the relevance of (1) the conserved extracellular cysteine residues 105 and 183 to ligand binding to the α isoform of TP (Chiang *et al.*, 1996; D'Angelo, 1996); (2) the conserved seventh transmembrane arginine residue for ligand binding to EP2 (Huang and Tai, 1995), EP3 (Kedzie *et al.*, 1998), and FP (Rehwald *et al.*, 1999); and (3) a possible role for the conserved DPWXY motif in G-protein activation mediated via EP3 (Satoh *et al.*, 1999). Signal transduction was also impaired in TPα in which a third intracellular loop cysteine 223 was replaced with an alanine (D'Angelo *et al.*, 1996) or a second intracellular loop phenylalanine 138 was replaced with an aspartic acid or tyrosine reside (Zhou *et al.*, 1999). The exact determinants of ligand binding and effector coupling of IP remain to be established. However, as mentioned earlier the presence of a CAAX (CSLC) motif at the C terminal end of IP makes it unique. Isoprenylation of IP has been demonstrated and disruption of this modification, by mutation of the CSLC motif to SSLC, generates a receptor with normal ligand binding, reduced inositol phosphate generation, and absent adenylyl cyclase activation (Hayes *et al.*, 1999), suggesting that this novel aspect of IP biology may contribute to its coupling to a downstream effector. In addition, switching of the β2-AR from the Gs-adenylyl cyclase cascade to MAP kinase activation is dependent on prior activation of the primary signaling cascade (Daaka *et al.*, 1997). A similar switching mechanism has been reported for mIP in which IP-adenylyl cyclase activation induced PKA-mediated phosphorylation of serine 357 and directed subsequent IP–Gq and IP–Gi coupling (Lawler *et al.*, 2001).

VI. KNOCKOUT AND TRANSGENE STUDIES

Generation of mice in which the gene for IP was disrupted (Murata *et al.*, 1997) permitted evaluation of the importance of this eicosanoid *in vivo*. IP null mice were viable and reproductive, had normal basal blood pressure and heart rate, and had no increased occurrence of vascular disorders, suggesting that IP may not act as a homeostatic regulator in the systemic vasculature. However, complete removal of the authentic ligand for the IP, by targeted deletion of the PGI$_2$ synthase, generated mice that were hypertensive and exhibited renal and aortic abnormalities including arterial sclerosis (Yokoyama *et al.*, 2001). There was, however, a concomitant increase in biosynthesis of other eicosanoids, including PGE$_2$ and TxA2, probably as a result of shunting to other synthetic pathways, which could account for the distinct phenotype from the IP null animals.

IP deficiency enhanced the thrombotic response to vascular injury (Murata *et al.*, 1997; Chen *et al.*, 2002), the development of atherosclerosis (Egan *et al.*, 2000), and exaggerated the platelet activation response to vascular injury (Chen *et al.*, 2002). Furthermore, increased PGI$_2$ generation in the vascular wall, by gene transfer of PGIS into rat carotid arteries, significantly inhibited neointimal formation resulting from balloon injury (Todaka *et al.*, 1999). These data confirm the proposed role of PGI_2 as an endogneous antiplatelet, antithrombotic, and antimitogenic mediator in settings of vascular injury.

A second consequence of IP deficiency was a dramatic reduction in inflammation comparable to wild-type mice in which COX-derived eicosanoid production was prevented with indomethacin (Murata et al., 1997). Furthermore, the hyperalgesic response to acetic acid was equivalent in IP-deficient mice and indomethacin-treated controls, although spinal and supraspinal nociceptive responses were not altered. These observation suggest, somewhat surprisingly, that PGI_2 may act as a principal mediator of vascular changes in inflammation and peripheral nociception. However, these consequences seem conditioned by genetic background.

Clinically, PGI_2 has been used successfully in the treatment of pulmonary hypertension (McLaughlin et al., 1998). IP knockout mice developed more severe pulmonary hypertension and vascular remodeling after chronic hypoxic exposure when compared to wild-type mice (Takeuchi et al., 2001), whereas transgenic mice specifically overexpressing PGIS in the lung were protected from the development of pulmonary hypertension after exposure to chronic hypobaric hypoxia (Geraci et al., 1999). Finally, the cardioprotective effects of PGI_2, which have been reported *in vitro,* are also evident *in vivo*; IP-deficient mice developed significantly greater myocardial infarct compared to wild types due to the loss of a direct protective effect of PGI_2 on cardiomyocytes during ischemia reperfusion (Xiao et al., 2001).

VII. MULTIPLE PGI_2 RECEPTORS?

The GPCRs for many mediators, including the eicosanoids, exist as multiple isoforms. These can be individual gene products, such as the EPs 1 through 4, or splice variants generated from a single gene, such as the EP3, TP, FP, and DP isoforms (Narumiya et al., 1999). Despite much research variants of the IP have not been cloned. However, pharmacological data consistent with multiple IPs has been reported. Evidence for a novel IP, distinct from the cloned receptor, has been reported in brain (Takechi et al., 1996; Watanabe et al., 1999) and kidney (Hebert et al., 1995). Indeed, the generation of IP knockout mice suggested *in vivo* activity of PGI_2 at sites distinct from the cloned IP as these mice were normotensive whereas PGI_2 synthase knockout mice were hypertensive (Yokoyama et al., 2001), although altered eicosanoid formation in the PGIS-deficient animals cannot be ruled out as a possible explanation for these inconsistencies. PGI_2 activity at peroxisome proliferator-activated receptors (PPARs) has been demonstrated and may be important for blastocyst implantation. However, although PGI_2 and iloprost activated PPARα and PPAR$\delta(\beta)$, another high-affinity PGI_2 analogue, cicaprost, did not. Furthermore, although the loss of PGI_2-mediated PPAR$\delta(\beta)$ activation was thought to underlie the implantation defect in COX-2-deficient mice, no implantation defect was evident in PPAR$\delta(\beta)$-deficient mice (Peters et al., 2000). Other potential PGI_2-mediated PPARδ-mediated actions include a role in colorectal cancer

(Gupta et al., 2000) and angiogenesis (Lim et al., 1999; Spisni et al., 2001), although the functional consequences remain undefined.

The question of a second GPCR for PGI_2 remains open and indeed the answer may not lie in the identification of a new gene product or splice variant. The formation of heterodimers of both highly homologous and divergent GPCR can generate "new" receptor sites with signaling and ligand-binding properties distinct from the individual receptors studies (Jordan and Devi, 1999; AbdAlla et al., 2000; George et al., 2000; Rocheville et al., 2000a,b) and these complexes may form in disease states (AbdAlla et al., 2001b). The formation of heterodimeric complexes in which IP is a partner has never been addressed, however, IP is frequently coexpressed with other members of the prostanoid receptor family. IP and its closest relative in the eicosanoid receptor family, the EP2 receptor (39%), both activate a common effector (Gs-adenylyl cyclase), two features that predict heterodimerization (AbdAlla et al., 2000; Rocheville et al., 2000a), and a putative dimerization motif (Hebert et al., 1996) is located in the first transmembrane domain of IP (Fig. 1).

GPCR can also interact with modifying proteins that are not related to the G-protein signaling cascade. RAMPS (receptor activity modifying proteins) interact with the calcitonin and calcitonin receptor-like receptor to substantially alter their pharmacology (Sexton et al., 2001), whereas the interaction of $\beta 2$-AR with EBP50 modulates its membrane trafficking (Cao et al., 1999). Novel IP interacting proteins may explain the apparent inconsistencies between the pharmacological and molecular data and reveal the presence of multiple receptor "subtypes."

VIII. CONCLUSIONS

PGI_2 biological effects are mediated through its receptor, the IP. The recent cloning of the IP gene and generation of IP knockout mice have confirmed the antithrombotic actions of PGI_2 *in vivo*. The widespread distribution of IP in the cardiovascular system as well the kidney, lung, liver, and central nervous system, and the possibility of novel PGI_2 receptors, suggest that PGI_2 may play roles beyond those currently identified.

REFERENCES

AbdAlla, S., Zaki, E., Lother, H., and Quitterer, U. (1999). Involvement of the amino terminus of the B(2) receptor in agonist-induced receptor dimerization. *J. Biol. Chem.* **274**(37), 26079–26084.

AbdAlla, S., Lother, H., and Quitterer, U. (2000). AT1-receptor heterodimers show enhanced G-protein activation and altered receptor sequestration. *Nature* **407**(6800), 94–98.

AbdAlla, S., Lother, H., Abdel-Tawab, A. M., and Quitterer, U. (2001a). The angiotensin II AT2 receptor is an AT1 receptor antagonist. *J. Biol. Chem.* **276**(43), 39721–39726.

AbdAlla, S., Lother, H., el Massiery, A., and Quitterer, U. (2001b). Increased AT(1) receptor heterodimers in preeclampsia mediate enhanced angiotensin II responsiveness. *Nat. Med.* **7**(9), 1003–1009.

Adderley, S. R., and Fitzgerald, D. J. (1999). Oxidative damage of cardiomyocytes is limited by extracellular regulated kinases 1/2-mediated induction of cyclooxygenase-2. *J. Biol. Chem.* **274**(8), 5038–5046.

Boer, U., Neuschafer-Rube, F., Moller, U., and Puschel, G. P. (2000). Requirement of N-glycosylation of the prostaglandin E2 receptor EP3beta for correct sorting to the plasma membrane but not for correct folding. *Biochem. J.* **350**(Pt. 3), 839–847.

Boie, Y., Rushmore, T. H., Darmon-Goodwin, A., Grygorczyk, R., Slipetz, D. M., Metters, K. M., and Abramovitz, M. (1994)). Cloning and expression of a cDNA for the human prostanoid IP receptor. *J. Biol. Chem.* **269**(16), 12173–12178.

Boie, Y., Sawyer, N., Slipetz, D. M., Metters, K. M., and Abramovitz, M. (1995). Molecular cloning and characterization of the human prostanoid DP receptor. *J. Biol. Chem.* **270**(32), 18910–18916.

Cao, T. T., Deacon, H. W., Reczek, D., Bretscher, A., and von Zastrow, M. (1999). A kinase-regulated PDZ-domain interaction controls endocytic sorting of the beta2-adrenergic receptor. *Nature* **401**(6750), 286–290.

Catella-Lawson, F., McAdam, B., Morrison, B. W., Kapoor, S., Kujubu, D., Antes, L., Lasseter, K. C., Quan, H., Gertz, B. J., and FitzGerald, G. A. (1999). Effects of specific inhibition of cyclooxygenase-2 on sodium balance, hemodynamics, and vasoactive eicosanoids. *J. Pharmacol. Exp. Ther.* **289**(2), 735–741.

Cheng, Y., Austin, S. C., Rocca, B., Koller, B. H., Coffman, T. M., Grosser, T., Lawson, J. A., and FitzGerald, G. A. (2002). Role of prostacyclin in the cardiovascular response to thromboxane A2. *Science* **296**(5567), 539–541.

Chiang, N., Kan, W. M., and Tai, H. H. (1996). Site-directed mutagenesis of cysteinyl and serine residues of human thromboxane A2 receptor in insect cells. *Arch. Biochem. Biophys.* **334**(1), 9–17.

Coleman, R. A., Smith, W. L., and Narumiya, S. (1994). International Union of Pharmacology classification of prostanoid receptors: Properties, distribution, and structure of the receptors and their subtypes. *Pharmacol. Rev.* **46**(2), 205–229.

Cong, M., Perry, S. J., Lin, F. T., Fraser, I. D., Hu, L. A., Chen, W., Pitcher, J. A., Scott, J. D., and Lefkowitz, R. J. (2001). Regulation of membrane targeting of the G protein-coupled receptor kinase 2 by protein kinase A and its anchoring protein AKAP79. *J. Biol. Chem.* **276**(18), 15192–15199.

Daaka, Y., Luttrell, L. M., and Lefkowitz, R. J. (1997). Switching of the coupling of the beta2-adrenergic receptor to different G proteins by protein kinase A. *Nature* **390**(6655), 88–91.

D'Angelo, D. D., Eubank, J. J., Davis, M. G., and Dorn, G. W., 2nd. (1996). Mutagenic analysis of platelet thromboxane receptor cysteines. Roles in ligand binding and receptor-effector coupling. *J. Biol. Chem.* **271**(11), 6233–6240.

Darius, H., Binz, C., Veit, K., Fisch, A., and Meyer, J. (1995). Platelet receptor desensitization induced by elevated prostacyclin levels causes platelet-endothelial cell adhesion. *J. Am. Coll. Cardiol.* **26**(3), 800–806.

Dowd, N. P., Scully, M., Adderley, S. R., Cunningham, A. J., and Fitzgerald, D. J. (2001). Inhibition of cyclooxygenase-2 aggravates doxorubicin-mediated cardiac injury in vivo. *J. Clin. Invest.* **108**(4), 585–590.

Egan, K., Austin, S., Smyth, E. M., and FitzGerald, G. A. (2000). Accelerated atherogenesis in prota-cyclin receptor deficient mice. *Circulation* **102**, 234.

Ferguson, S. S., and Caron, M. G. (1998). G protein-coupled receptor adaptation mechanisms. *Semin. Cell. Dev. Biol.* **9**(2), 119–127.

Fitzgerald, D. J., Doran, J., Jackson, E., and FitzGerald, G. A. (1986a). Coronary vascular occlusion mediated via thromboxane A2-prostaglandin endoperoxide receptor activation in vivo. *J. Clin. Invest.* **77**(2), 496–502.

Fitzgerald, D. J., Roy, L., Catella, F., and FitzGerald, G. A. (1986b). Platelet activation in unstable coronary disease. *N. Engl. J. Med.* **315**(16), 983–989.

FitzGerald, G. A. (1991). Mechanisms of platelet activation: Thromboxane A2 as an amplifying signal for other agonists. *Am. J. Cardiol.* **68**(7), 11B–15B.

Foley, J. F., Kelley, L. P., and Kinsella, B. T. (2001). Prostaglandin D(2) receptor-mediated desensitization of the alpha isoform of the human thromboxane A(2) receptor. *Biochem. Pharmacol.* **62**(2), 229–239.

Fraser, I. D., Cong, M., Kim, J., Rollins, E. N., Daaka, Y., Lefkowitz R. J., and Scott, J. D. (2000). Assembly of an A kinase-anchoring protein-beta(2)-adrenergic receptor complex facilitates receptor phosphorylation and signaling. *Curr. Biol.* **10**(7), 409–412.

Garcia Rodriguez, C., Cundell, D. R., Tuomanen, E. I., Kolakowski, L. F., Jr., Gerard, C., and Gerard, N. P. (1995). The role of N-glycosylation for functional expression of the human platelet-activating factor receptor. Glycosylation is required for efficient membrane trafficking. *J. Biol. Chem.* **270**(42), 25178–25184.

George, S. R., Fan, T., Xie, Z., Tse, R., Tam, V., Varghese, G., and O'Dowd, B. F. (2000). Oligomerization of mu- and delta-opioid receptors. Generation of novel functional properties. *J. Biol. Chem.* **275**(34), 26128–26135.

Geraci, M. W., Gao, B., Shepherd, D. C., Moore, M. D., Westcott, J. Y., Fagan, K. A., Alger, L. A., Tuder, R. M., and Voelkel, N. F. (1999). Pulmonary prostacyclin synthase overexpression in transgenic mice protects against development of hypoxic pulmonary hypertension. *J. Clin. Invest.* **103**(11), 1509–1515.

Giovanazzi, S., Accomazzo, M. R., Letari, O., Oliva, D., and Nicosia, S. (1997). Internalization and down-regulation of the prostacyclin receptor in human platelets. *Biochem. J.* **325**(Pt. 1), 71–77.

Grosser, T., Bonisch, D., Zucker, T. P., and Schror, K. (1995). Iloprost-induced inhibition of proliferation of coronary artery smooth muscle cells is abolished by homologous desensitization. *Agents Actions Suppl.* **45**, 85–91.

Gupta, R. A., Tan, J., Krause, W. F., Geraci, M. W., Willson, T. M., Dey, S. K., and DuBois, R. N. (2000). Prostacyclin-mediated activation of peroxisome proliferator-activated receptor delta in colorectal cancer. *Proc. Natl. Acad. Sci. USA* **97**(24), 13275–13280.

Harada, M., Toki, Y., Numaguchi, Y., Osanai, H., Ito, T., Okumura, K., and Hayakawa, T. (1999). Prostacyclin synthase gene transfer inhibits neointimal formation in rat balloon-injured arteries without bleeding complications. *Cardiovasc. Res.* **43**(2), 481–491.

Hayes, J. S., Lawler, O. A., Walsh, M. T., and Kinsella, B. T. (1999). The prostacyclin receptor is isoprenylated. Isoprenylation is required for efficient receptor-effector coupling. *J. Biol. Chem.* **274**(34), 23707–23718.

Hebert, R. L., Regnier, L., and Peterson, L. N. (1995). Rabbit cortical collecting ducts express a novel prostacyclin receptor. *Am. J. Physiol.* **268**(1, Pt. 2), F145–F154.

Hebert, T. E., Moffett, S., Morello, J. P., Loisel, T. P., Bichet, D. G., Barret, C., and Bouvier, M. (1996). A peptide derived from a beta2-adrenergic receptor transmembrane domain inhibits both receptor dimerization and activation. *J. Biol. Chem.* **271**(27), 16384–16392.

Herschman, H. R. (1998). Recent progress in the cellular molecular biology of prostaglandin synthesis. *Trends Cardiovasc. Med.* **8**, 145–150.

Hoshikawa, Y., Voelkel, N. F., Gesell, T. L., Moore, M. D., Golpon, H. A., Keith, R. L., Arcaroli, J. J., Narumiya, S., and Geraci, M. W. (2001a). Alteration of pulmonary gene expression due to prostacyclin receptor targeted disruption. Eicosanoids and Other Bioactive Lipids in Cacer, Inflammation and Related Diseases. 7th International Conference, Nashville, Tennessee, October 14–17, 2001.

Hoshikawa, Y., Voelkel, N. F., Gesell, T. L., Moore, M. D., Morris, K. G., Alger, L. A., Narumiya, S., and Geraci, M. W. (2001b). Prostacyclin receptor-dependent modulation of pulmonary vascular remodeling. *Am. J. Respir. Crit. Care Med.* **164**(2), 314–318.

Huang, C., and Tai, H. H. (1995). Expression and site-directed mutagenesis of mouse prostaglandin E2 receptor EP3 subtype in insect cells. *Biochem. J.* **307**(Pt. 2), 493–498.

Huang, C., and Tai, H. H. (1998). Prostaglandin E2 receptor EP3alpha subtype: The role of N-glycosylation in ligand binding as revealed by site-directed mutagenesis. *Prostaglandins Leukot. Essent. Fatty Acids.* **59**(4), 265–271.

Jaschonek, K., Karsch, K. R., Weisenberger, H., Tidow, S., Faul, C., and Renn, W. (1986). Platelet prostacyclin binding in coronary artery disease. *J. Am. Coll. Cardiol.* **8**(2), 259–266.

Jones, M. K., Wang, H., Peskar, B. M., Levin, E., Itani, R. M., Sarfeh, I. J., and Tarnawski, A. S. (1999). Inhibition of angiogenesis by nonsteroidal anti-inflammatory drugs: Insight into mechanisms and implications for cancer growth and ulcer healing. *Nat. Med.* **5**(12), 1418–1423.

Jordan, B. A., and Devi, L. A. (1999). G-protein-coupled receptor heterodimerization modulates receptor function. *Nature* **399**(6737), 697–700.

Kedzie, K. M., Donello, J. E., Krauss, H. A., Regan, J. W., and Gil, D. W. (1998). A single amino-acid substitution in the EP2 prostaglandin receptor confers responsiveness to prostacyclin analogs. *Mol. Pharmacol.* **54**(3), 584–590.

Klockenbusch, W., Hohlfeld, T., Wilhelm, M., Somville, T., and Schror, K. (1996). Platelet PGI2 receptor affinity is reduced in pre-eclampsia. *Br. J. Clin. Pharmacol.* **41**(6), 616–618.

Kobayashi, T., Ushikubi, F., and Narumiya, S. (2000). Amino acid residues conferring ligand binding properties of prostaglandin I and prostaglandin D receptors. Identification by site-directed mutagenesis. *J. Biol. Chem.* **275**(32), 24294–24303.

Komhoff, M., Lesener, B., Nakao, K., Seyberth, H. W., and Nusing, R. M. (1998). Localization of the prostacyclin receptor in human kidney. *Kidney Int.* **54**(6), 1899–1908.

Lawler, O. A., Miggin, S. M., and Kinsella, B. T. (2001a). Protein kinase A mediated phosphorylation of serine 357 of the mouse prostacyclin receptor regulates its coupling to Gs-, to Gi- and to Gq-coupled effector signaling. *J. Biol. Chem.* **6,** 6.

Lim, H., Gupta, R. A., Ma, W. G., Paria, B. C., Moller, D. E., Morrow, J. D., DuBois, R. N., Trzaskos, J. M., and Dey, S. K. (1999). Cyclo-oxygenase-2-derived prostacyclin mediates embryo implantation in the mouse via PPARdelta. *Genes. Dev.* **13**(12), 1561–1574.

Liou, J. Y., Shyue, S. K., Tsai, M. J., Chung, C. L., Chu, K. Y., and Wu, K. K. (2000). Colocalization of prostacyclin synthase with prostaglandin H synthase-1 (PGHS-1) but not phorbol ester-induced PGHS-2 in cultured endothelial cells. *J. Biol. Chem.* **275**(20), 15314–15320.

McAdam, B. F., Catella-Lawson, F., Mardini, I. A., Kapoor, S., Lawson, J. A., and FitzGerald, G. A. (1999). Systemic biosynthesis of prostacyclin by cyclooxygenase (COX)-2: The human pharmacology of a selective inhibitor of COX-2. *Proc. Natl. Acad. Sci. USA* **96**(1), 272–277.

McLaughlin, V. V., Genthner, D. E., Panella, M. M., and Rich, S. (1998). Reduction in pulmonary vascular resistance with long-term epoprostenol (prostacyclin) therapy in primary pulmonary hypertension. *N. Engl. J. Med.* **338**(5), 273–277.

Modesti, P. A., Fortini, A., Poggesi, L., Boddi, M., Abbate, R., and Gensini, G. F. (1987). Acute reversible reduction of PGI2 platelet receptors after iloprost infusion in man. *Thromb. Res.* **48**(6), 663–669.

Moncada, S., and Vane, J. R. (1981). Prostacyclin: Homeostatic regulator or biological curiosity? *Clin. Sci. (Colch.)* **61**(4), 369–372.

Morishita, H., Yui, Y., Hattori, R., Aoyama, T., and Kawai, C. (1990). Increased hydrolysis of cholesteryl ester with prostacyclin is potentiated by high density lipoprotein through the prostacyclin stabilization. *J. Clin. Invest.* **86**(6), 1885–1891.

Murata, T., Ushikubi, F., Matsuoka, T., Hirata, M., Yamasaki, A., Sugimoto, Y., Ichikawa, A., Aze, Y., Tanaka, T., Yoshida, N., Ueno, A., Oh-ishi, S., and Narumiya, S. (1997). Altered pain perception and inflammatory response in mice lacking prostacyclin receptor. *Nature* **388**(6643), 678–682.

Murray, R., Shipp, E., and FitzGerald, G. A. (1990). Prostaglandin endoperoxide/thromboxane A2 receptor desensitization. Cross-talk with adenylate cyclase in human platelets. *J. Biol. Chem.* **265**(35), 21670–21675.

Nakagawa, O., Tanaka, I., Usui, T., Harada, M., Sasaki, Y., Itoh, H., Yoshimasa, T., Namba, T., Narumiya, S., and Nakao, K. (1994). Molecular cloning of human prostacyclin receptor cDNA and its gene expression in the cardiovascular system. *Circulation* **90**(4), 1643–1647.

Namba, T., Oida, H., Sugimoto, Y., Kakizuka, A., Negishi, M., Ichikawa, A., and Narumiya, S. (1994). cDNA cloning of a mouse prostacyclin receptor. Multiple signaling pathways and expression in thymic medulla. *J. Biol. Chem.* **269**(13), 9986–9992.

Narumiya, S., Sugimoto, Y., and Ushikubi, F. (1999). Prostanoid receptors: Structures, properties and functions. *Physiol. Rev.* **79**(4), 1193–1226.

Nilius, S. M., Hasse, A., Kuger, P., Schror, K., and Meyer-Kirchrath, J. (2000). Agonist-induced long-term desensitization of the human prostacyclin receptor. *FEBS Lett.* **484**(3), 211–216.

Numaguchi, Y., Harada, M., Osanai, H., Hayashi, K., Toki, Y., Okumura, K., Ito, T., and Hayakawa, T. (1999a). Altered gene expression of prostacyclin synthase and prostacyclin receptor in the thoracic aorta of spontaneously hypertensive rats. *Cardiovasc. Res.* **41**(3), 682–688.

Numaguchi, Y., Naruse, K., Harada, M., Osanai, H., Mokuno, S., Murase, K., Matsui, H., Toki, Y., Ito, T., Okumura, K., and Hayakawa, T. (1999b). Prostacyclin synthase gene transfer accelerates reendothelialization and inhibits neointimal formation in rat carotid arteries after balloon injury. *Arterioscler. Thromb. Vasc. Biol.* **19**(3), 727–733.

Oates, J. A., FitzGerald, G. A., Branch, R. A., Jackson, E. K., Knapp, H. R., and Roberts, L. J., 2nd. (1988). Clinical implications of prostaglandin and thromboxane A2 formation (1). *N. Engl. J. Med.* **319**(11), 689–698.

Ogawa, Y., Tanaka, I., Inoue, M., Yoshitake, Y., Isse, N., Nakagawa, O., Usui, T., Itoh, H., Yoshimasa, T., Narumiya, S., et al. (1995). Structural organization and chromosomal assignment of the human prostacyclin receptor gene. *Genomics* **27**(1), 142–148.

Peters, J. M., Lee, S. S., Li, W., Ward, J. M., Gavrilova, O., Everett, C., Reitman, M. L., Hudson, L. D., and Gonzalez, F. J. (2000). Growth, adipose, brain, and skin alterations resulting from targeted disruption of the mouse peroxisome proliferator-activated receptor beta(delta). *Mol. Cell. Biol.* **20**(14), 5119–5128.

Rands, E., Candelore, M. R., Cheung, A. H., Hill, W. S., Strader, C. D., and Dixon, R. A. (1990). Mutational analysis of beta-adrenergic receptor glycosylation. *J. Biol. Chem.* **265**(18), 10759–10764.

Rehwald, M., Neuschafer-Rube, F., de Vries, C., and Puschel, G. P. (1999). Possible role for ligand binding of histidine 81 in the second transmembrane domain of the rat prostaglandin F2alpha receptor. *FEBS Lett.* **443**(3), 357–362.

Rocca, B., Spain, L. M., Pure, E., Langenbach, R., Patrono, C., and FitzGerald, G. A. (1999). Distinct roles of prostaglandin H synthases 1 and 2 in T-cell development. *J. Clin. Invest.* **103**(10), 1469–1477.

Rocheville, M., Lange, D. C., Kumar, U., Patel, S. C., Patel, R. C., and Patel, Y. C. (2000a). Receptors for dopamine and somatostatin: Formation of hetero-oligomers with enhanced functional activity. *Science* **288**(5463), 154–157.

Rocheville, M., Lange, D. C., Kumar, U., Sasi. R., Patel, R. C., and Patel, Y. C. (2000b). Subtypes of the somatostatin receptor assemble as functional homo- and heterodimers. *J. Biol. Chem.* **275**(11), 7862–7869.

Rybin, V. O., Xu, X., Lisanti, M. P., and Steinberg, S. F. (2000). Differential targeting of beta-adrenergic receptor subtypes and adenylyl cyclase to cardiomyocyte caveolae. A mechanism to functionally regulate the cAMP signaling pathway. *J. Biol. Chem.* **275**(52), 41447–41457.

Sasaki, Y., Usui, T., Tanaka, I., Nakagawa, O., Sando, T., Takahashi, T., Namba, T., Narumiya, S., and Nakao, K. (1994). Cloning and expression of a cDNA for rat prostacyclin receptor. *Biochim. Biophys. Acta.* **1224**(3), 601–605.

Sasaki, Y., Takahashi, T., Tanaka, I., Nakamura, K., Okuno, Y., Nakagawa, O., Narumiya, S., and Nakao, K. (1997). Expression of prostacyclin receptor in human megakaryocytes. *Blood* **90**(3), 1039–1046.

Satoh, S., Chang, C., Katoh, H., Hasegawa, H., Nakamura, K., Aoki, J., Fujita, H., Ichikawa, A., and Negishi, M. (1999). The key amino acid residue of prostaglandin EP3 receptor for governing G protein association and activation steps. *Biochem. Biophys. Res. Commun.* **255**(1), 164–168.

Sexton, P. M., Albiston, A., Morfis, M., and Tilakaratne, N. (2001). Receptor activity modifying proteins. *Cell. Signal.* **13**(2), 73–83.

Sinzinger, H., Zidek, T., Fitscha, P., O'Grady, J., Wagner, O., and Kaliman, J. (1987). Prostaglandin I2 reduces activation of human arterial smooth muscle cells in-vivo. *Prostaglandins* **33**(6), 915–918.

Smyth, E. M., Nestor, P. V., and FitzGerald, G. A. (1996). Agonist-dependent phosphorylation of an epitope-tagged human prostacyclin receptor. *J. Biol. Chem.* **271**(52), 33698–33704.

Smyth, E. M., Li, W. H., and FitzGerald, G. A. (1998). Phosphorylation of the prostacyclin receptor during homologous desensitization. A critical role for protein kinase c. *J. Biol. Chem.* **273**(36), 23258–23266.

Smyth, E. M., Austin, S. C., Reilly, M. P., and FitzGerald, G. A. (2000). Internalization and sequestration of the human prostacyclin receptor. *J. Biol. Chem.* **275**(41), 32037–32045.

Spisni, E., Griffoni, C., Santi, S., Riccio, M., Marulli, R., Bartolini, G., Toni, M., Ullrich, V., and Tomasi, V. (2001). Colocalization prostacyclin (PGI2) synthase—caveolin-1 in endothelial cells and new roles for PGI2 in angiogenesis. *Exp. Cell Res.* **266**(1), 31–43.

Takechi, H., Matsumura, K., Watanabe, Y., Kato, K., Noyori, R., and Suzuki, M. (1996). A novel subtype of the prostacyclin receptor expressed in the central nervous system. *J. Biol. Chem.* **271**(10), 5901–5906.

Takeuchi, K., Kato, S., Ogawa, Y., Kanatsu, K., and Umeda, M. (2001). Role of endogenous prostacyclin in gastric ulcerogenic and healing responses—a study using IP-receptor knockout mice. *J. Physiol. Paris* **95**(1–6), 75–80.

Todaka, T., Yokoyama, C., Yanamoto, H., Hashimoto, N., Nagata, I., Tsukahara, T., Hara, S., Hatae, T., Morishita, R., Aoki, M., Ogihara, T., Kaneda, Y., and Tanabe, T. (1999). Gene transfer of human prostacyclin synthase prevents neointimal formation after carotid balloon injury in rats. *Stroke* **30**(2), 419–426.

Tong, B. J., Tan, J., Tajeda, L., Das, S. K., Chapman, J. A., DuBois, R. N., and Dey, S. K. (2000). Heightened expression of cyclooxygenase-2 and peroxisome proliferator-activated receptor-delta in human endometrial adenocarcinoma. *Neoplasia.* **2**(6), 483–490.

Topper, J. N., Cai, J., Falb, D., and Gimbrone, M. A, Jr. (1996). Identification of vascular endothelial genes differentially responsive to fluid mechanical stimuli: Cyclooxygenase-2, manganese superoxide dismutase., and endothelial cell nitric oxide synthase are selectively up-regulated by steady laminar shear stress. *Proc. Natl. Acad. Sci. USA* **93**(19), 10417–10422.

Vane, J. R. (1969). The release and fate of vaso-active hormones in the circulation. The Second Gaddum Memorial Lecture. *Br. J. Pharmacol.* **35**, 209–242.

Vane, J. R., and Botting, R. M. (1995). Pharmacodynamic profile of prostacyclin. *Am. J. Cardiol.* **75**(3), 3A–10A.

Walsh, M. T., and Kinsella, B. T. (2000). Regulation of the human prostanoid TPalpha and TPbeta receptor isoforms mediated through activation of the EP(1) and IP receptors. *Br. J. Pharmacol.* **131**(3), 601–609.

Walsh, M. T., Foley, J. F., and Kinsella, B. T. (1998). Characterization of the role of N-linked glycosylation on the cell signaling and expression of the human thromboxane A2 receptor alpha and beta isoforms. *J. Pharmacol. Exp. Ther.* **286**(2), 1026–1036.

Walsh, M. T., Foley, J. F., and Kinsella, B. T. (2000). The alpha, but not the beta, isoform of the human thromboxane A2 receptor is a target for prostacyclin-mediated desensitization. *J. Biol. Chem.* **275**(27), 20412–20423.

Watanabe, Y., Matsumura, K., Takechi, H., Kato, K., Morii, H., Bjorkman, M., Langstrom, B., Noyori, R., and Suzuki, M. (1999). A novel subtype of prostacyclin receptor in the central nervous system. *J. Neurochem.* **72**(6), 2583–2592.

Xiao, C. Y., Hara, A., Yuhki, K., Fujino, T., Ma, H., Okada, Y., Takahata, O., Yamada, T., Murata, T., Narumiya, S., and Ushikubi, F. (2001). Roles of prostaglandin I2 and thromboxane A2 in cardiac ischemia-reperfusion injury: A study using mice lacking their respective receptors. *Circulation* **104**, 2210–2215.

Yokoyama, C., Yabuki, T., Shimonishi, M., Wada, M., Hatae, T., Takeda, J., Okabe, M., and Tanabe, T. (2001). Prostacyclin deficiency in mice induces vascular disorders in kidney. *Ist Takeda Science Foundation Symposium on Pharma Sciences—Lipids in signaling and related diseases,* Tokyo, Japan. October 1–3.

Zhang, Z., Austin, S. C., and Smyth, E. M. (2001a). Glycosylation of the human prostacyclin receptor: Role in ligand binding and signal transduction. *Mol. Pharmacol.* **60**(3), 480–487.

Zhang, Z. B., Vezza, R., Plappert, T., McNamara, P., Tang, L. X., Austin, S., Pratico, D., St. John-Sutton, M., and FitzGerald, G. A. (2001b). COX-2 dependent thromboxane formation mediates cardiac failure in Gh transgenic mice. *Arterioscler. Thromb. Vas. Biol.* **21,** 686.

Zhou, H., and Tai, H. H. (1999). Characterization of recombinant human CXCR4 in insect cells: Role of extracellular domains and N-glycosylation in ligand binding. *Arch. Biochem. Biophys.* **369**(2), 267–276.

Zhou, H., Yan, F., Yamamoto, S., and Tai, H. H. (1999). Phenylalanine 138 in the second intracellular loop of human thromboxane receptor is critical for receptor-G-protein coupling. *Biochem. Biophys. Res. Commun.* **264**(1), 171–175.

Zucker, T. P., Bonisch, D., Hasse, A., Grosser, T., Weber, A. A., and Schror, K. (1998). Tolerance development to antimitogenic actions of prostacyclin but not of prostaglandin E1 in coronary artery smooth muscle cells. *Eur. J. Pharmacol.* **345**(2), 213–220.

Sterol Regulatory Element-Binding Protein Family as Global Regulators of Lipid Synthetic Genes in Energy Metabolism

Hitoshi Shimano

Department of Internal Medicine, Institute of Clinical Medicine, University of Tsukuba, Tsukuba, Ibaraki 305-8575, Japan

I. Introduction
II. Molecular Aspects of SREBPs
    A. *Structure of SREBPs*
    B. *Cleavage of SREBPs*
    C. *DNA-Binding Sites of SREBPs*
    D. *Target Genes of SREBPs*
    E. *Modification of Gene Expression by Cofactors for SREBPs*
III. Members of the SREBP Family
    A. *SREBP-1a, -1c, and -2*
    B. *Differences among SREBP Isoforms in Target DNA Specificity*
IV. *In Vivo* Functions of SREBPs

    A. *SREBP Transgenic Mice*
    B. *SREBP Knockout Mice*
    C. *Nutritional Regulation of Lipogenic Enzyme Genes by SREBP-1*
    D. *The Effects of Insulin/Glucose on SREBP-1c Expression*
 V. Transcriptional Regulation of Lipid Synthesis by SREBP-2 and -1c in the Liver
VI. Promoter Analysis of SREBP-1c
    A. *Presence of SRE and an Oxysterol-Inducible Region in the SREBP-1c Promoter*
    B. *LXR/RXR Activation of SREBP-1c: A New Link between Cholesterol and Fatty Acid Regulation*
VII. ADD1/SREBP-1c in Adipogenesis
VIII. Clinical Aspects of SREBPs
    A. *SREBP-1c and Insulin Resistance*
    B. *SREBP-1 and Fatty Livers*
    C. *SREBP-1c and Atherogenic Remnant Lipoproteins*
IX. Cross-Talk of Transcription Factors for Lipid Metabolism
X. Future Aspects of SREBPs
    References

Sterol regulatory element-binding proteins (SREBPs) have been established as lipid synthetic transcription factors for cholesterol and fatty acid synthesis. SREBPs are synthesized as membrane-bound precursors with their N-terminal active portions entering the nucleus to activate target genes after proteolytic cleavage in a sterol-regulated manner. This cleavage step is regulated by a putative sterol-sensing molecule, SREBP-activating protein (SCAP), that forms a complex with SREBPs and traffics between the rough endoplasmic reticulum and Golgi. DNA cis-elements that SREBPs bind, originally identified as sterol-regulatory elements (SREs), now expands to a variety of SRE-like sequences and some of E-boxes, which makes SREBPs eligible to regulate a wide range of lipid genes. Animal experiments including transgenic and knockout mice suggest that three isoforms, SREBP-1a, -1c, and -2, have different roles in lipid synthesis. In differentiated tissues and organs, SREBP-1c is involved in fatty acid, whereas SREBP-2 plays a major role in regulation of cholesterol synthesis. SREBP-1a is expressed in growing cells, providing both cholesterol and fatty acids that are required for membrane synthesis. SREBP-1c seems to be a mediator for insulin/glucose signaling to lipogenesis, and could be involved in insulin resistance, remnant lipoproteins, and fatty livers. Future studies in this field will certainly focus on understanding the molecular mechanisms sensing cellular sterol and energy states leading to the activation of SREBP-mediated gene transcription. © 2002, Elsevier Science (USA).

I. INTRODUCTION

Long-term regulation of genes involved in metabolism is controlled at the transcriptional level. Good examples are the biosynthetic pathways for cholesterol and fatty acids, both necessary for cellular membrane structure and function. Growing cells need to synthesize both cholesterol and fatty acids according to their demand for growth. The cholesterol biosynthetic pathway is linked to the synthesis of bile acids and steroid hormones in differentiated tissues such as the liver and steroidogenic organs. In contrast, lipogenesis, the synthesis of fatty acids and triglycerides, is necessary for energy storage in lipogenic organs such as liver and adipose tissue. Cholesterol synthesis is tightly regulated to maintain cellular cholesterol levels by a fine-tuned feedback system whereas fatty acid synthesis is driven primarily by the availability of carbohydrates and the actions of hormones such as insulin (Levanon *et al.,* 1990; Ridgway *et al.,* 1992). Despite these different patterns of regulation, both lipid biosynthetic pathways are controlled at the transcriptional level. Recent evidence suggests that they are controlled by a common family of transcription factors designated sterol regulatory element-binding proteins (SREBPs) (Brown and Goldstein, 1997). Here we review recent progress in determining the molecular basis of physiological functions of SREBPs. SREBPs have now been established as dominant lipid synthetic regulators. However, the physiological roles of SREBPs need to be understood in light of global lipid and energy metabolism in which nuclear receptors such as peroxisome proliferator-activated receptor (PPAR) α and γ and liver X receptors (LXRs) are also involved.

II. MOLECULAR ASPECTS OF SREBPs

A. STRUCTURE OF SREBPs

SREBPs were purified and cloned as nuclear factors that bind to the sterol-regulatory element (SRE) common to low-density lipoprotein (LDL) receptor and hydroxymethylglutaryl-coenzyme A (HMG-CoA) synthase genes (Briggs *et al.,* 1993; Wang *et al.,* 1993; Yokoyama *et al.,* 1993). Their unique aspect is that they are produced as membrane-bound transcription factors. SREBP precursors, referred to as membrane forms, are bound to rough endoplasmic reticulum (rER) or nuclear envelope with two membrane-span regions. Both the amino- and carboxyl-terminal portions of the proteins project into the cytoplasm. The N-terminal cytosolic domain contains an acidic amino acid-rich transactivation domain and a basic helix–loop–helix leucine zipper (bHLH-zip) domain that is a key for homodimerization and specific binding to SREs. The transactivation domain and bHLH-zip region flank a glycine, serine, proline-rich region the function of which is unknown. The carboxyl-terminal portion in the cytosol is designated as a regulatory domain. As described in the next section, this domain is essential for sterol regulation

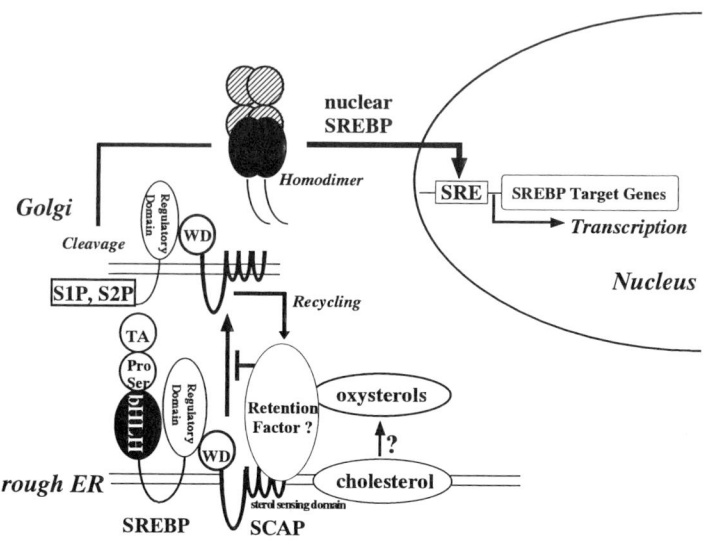

FIGURE 1 Intracellular SREBP–SCAP complex itinerary and sterol regulation.

because SREBP forms a complex with SREBP cleavage activating protein (SCAP) through this domain (Fig. 1) (Hua *et al.*, 1996; Sakai *et al.*, 1997).

B. CLEAVAGE OF SREBPs

Membrane-bound SREBP precursors must undergo a two-step proteolytic cleavage to release their active forms (nuclear SREBPs) containing bHLH-zip into the nucleus to activate their target genes (Sakai *et al.*, 1996; Wang *et al.*, 1994) (Fig. 1). The molecular mechanism responsible for cleavage of SREBPs has been elegantly and extensively analyzed and shown to be a key event in sterol regulation by Brown and Goldstein (1997, 1998, 1999; Brown *et al.*, 2000). SREBP precursor and SCAP form a complex in the rough ER membrane (Sakai *et al.*, 1997). SCAP is a multiple membrane-spanning protein containing a consensus sequence for a sterol-sensing domain that is shared by HMG-CoA reductase, Niemann-Pick C1 disease gene product (NPC1), and Patched (Carstea *et al.*, 1997; Hua *et al.*, 1996). SCAP functions as a sterol sensor and is a prerequisite for SREBP cleavage (Rawson *et al.*, 1999; Sakai *et al.*, 1998). When cellular (and presumably rER) sterol is abundant, SREBP–SCAP stays in the rER. When cellular cholesterol is depleted, the SREBP–SCAP complex targets to the Golgi where site-1 and site-2 proteases await. Site-1 protease, activated by autocatalysis, cleaves SREBP at the loop region. The first cleavage is immediately followed by the second cleavage by site-2 protease in the first membrane-spanning region to liberate the N-terminal of SREBP (Duncan *et al.*, 1997, 1998; Espenshade *et al.*, 1999; Rawson *et al.*, 1997; Sakai *et al.*, 1998). SCAP presumably recycles to rER to meet another newly synthesized SREBP precursor. Escorted by Importin β (Nagoshi et al., 1999), the liberated homodimer of nuclear SREBP enters the nucleus and activates the transcription of genes involved in cholesterol and fatty acid synthesis by binding to SREs or to palindromic sequences called E-boxes within their promoter regions. Therefore, cellular sterol regulation is entirely controlled by sterol-regulated interorganelle trafficking of the SREBP–SCAP complex between the rER and Golgi, and subsequent cleavage of SREBP. The molecular mechanism of this sterol-sensing system is currently unknown. Some oxysterols such as 25-hydroxy cholesterol efficiently causes the SREBP–SCAP complex to be retained at the rER, thereby inhibiting cleavage. It is tempting to speculate that an unknown oxysterol acceptor may bind to the sterol-sensing domain of SCAP as a retention factor. An oxysterol-binding protein has been cloned as an acceptor for 25-hydroxy cholesterol, and was interestingly shown to move to the Golgi when bound to oxysterols (Levanon et al., 1990; Ridgway et al., 1992). However, its contribution to this sterol regulation mechanism is another enigma.

The fate of nuclear SREBP seems to be degradation by the ubiquitin/proteasome-dependent pathway, which could be another potential level of SREBP regulation (Hirano et al., 2001).

In contrast to this sterol regulation, sterol-independent cleavage of SREBP has been also suggested. Caspase-3 (CPP32) can cut out nuclear SREBP during apoptosis (Wang et al., 1996). Tumor necrosis factor-α (TNF-α), activating neutral shpingomyelinase and ceramide, was also reported to activate SREBP, presumably through the same cleavage process (Lawler et al., 1998; Scheek et al., 1997). The physiological relevance of this sterol-independent cleavage is currently unknown.

In another report, unsaturated fatty acids regulate cleavage of SREBP-1 but not SREBP-2 (Hannah et al., 2000). These data might reveal a new regulation of fatty acid metabolism mediated by SREBP-1 at the cleavage level in contrast to sterol regulation by SREBP-2. This distinct regulation by SREBP-1 and -2 in the lipid-sensing mechanism could be due to the structure difference in the regulatory domain of SREBP-1 and -2.

C. DNA-BINDING SITES OF SREBPs

The DNA-binding specificity of SREBPs is another unique feature of this transcription factor family. The cis elements are designated as SREs because they were originally identified as elements common to HMG-CoA synthase and reductase and LDL receptor gene promoters in sterol regulation. The original SRE sequence is ATCACCCCAC (Briggs et al., 1993). Spiegelman's group independently cloned a rat homologue of SREBP-1c, designated adipocyte determination and differentiation factor 1 (ADD1), as a specific factor for adipose tissue differentiation (Tontonoz et al., 1993). In the screening of ADD1-binding sites from random DNA sequences, ADD1 preferentially bound to an E-box sequence: ATCACGTGAT (Kim et al., 1995). Therefore SREBP-1 has a dual binding specificity for both SRE and E-box. A tyrosine residue in the basic region of SREBPs where other bHLH proteins have an arginine residue is a key to this dual binding specificity of SREBPs (Kim et al., 1995). Mutant SREBPs where arginine was substituted for the tyrosine lost binding specificity to the SRE, but retained binding to E-boxes (Kim et al., 1995). Presumably this dual binding specificity makes SREBPs eligible for a broad spectrum of target genes covering both cholesterogenic and lipogenic genes. The 2.3-Å resolution cocrystal structure of the DNA-binding portion of SREBP-1a bound to an SRE revealed a quasisymmetric homodimer with an asymmetric DNA–protein interface (Parraga et al., 1998). Our recent data on binding specificity of SREBP to SRE and E-box from transfection studies indicate that SREBPs have a higher affinity and lower capacity for SRE than E-box (Amemiya-Kudo et al., 2002).

D. TARGET GENES OF SREBPs

The current list of SREBP target genes is shown in Table I. They can be categorized as cholesterol synthetic genes and lipogenic enzyme genes. Genes involved as cholesterol synthesis contain SRE(s) that are identical or very similar to the

TABLE I SREBP Target Genes

Cholesterogenic enzymes
  HMG-CoA synthase (Smith et al., 1988)
  HMG-CoA reductase (Vallett et al., 1996)
  Farnesyl diphosphate synthase (Ericsson et al., 1996b)
  Squalene synthase (Guan et al., 1997)
  7-Dehydrocholesterol reductase (Kim et al., 2001)
  Lanosterol 14α-demethylase (Cyp51) (Rozman et al., 1999)
  SREBP-2 (Sato et al., 1996)

Lipogenic enzymes
  Acetyl-CoA carboxylase (Magana et al., 1997)
  Fatty acid synthase (Magana and Osborne, 1996)
  Stearoyl CoA desaturase-1 and 2 (Tabor et al., 1998, 1999)
  ATP citrate lyase (Moon et al., 2000; Sato et al., 2000)
  Acetyl-CoA synthase (Ikeda et al., 2001; Sone et al., 2002)
  Glycerol-3-phosphate acyltransferase (Ericsson et al., 1997)
  SREBP-1 (Amemiya-Kudo, 2000)

Other lipoprotein and lipid metabolism
  LDL receptor (Smith et al., 1990)
  Lipoprotein lipase (Schoonjans et al., 2000)
  HDL receptor (scavenger receptor type BI) (Lopez and McLean, 1999)
  Acyl-CoA binding protein (Diazepam) (Tabor et al., 1998)
  PPAR γ (Fajas et al., 1999)
  StARP (Christenson et al., 2001; Shea-Eaton et al., 2001b)
  Cholesterol ester transfer protein (Gauthier et al., 1999)
  Neutral cholesterol ester hydrolase (Natarajan et al., 1998)

Genes repressed by SREBP
  MTP (Sato et al., 1999)
  Caveolin (Bist et al., 1997)

Lipogenic enzymes, which are involved in energy storage through synthesis of fatty acids and triglycerides, are coordinately regulated at the transcriptional level during different metabolic states (Goodridge, 1987; Hillgartner *et al.*, 1995). Recent *in vivo* studies demonstrated that SREBP-1c plays a crucial role in the dietary regulation of most hepatic lipogenic genes. These studies include the effects of overexpression of SREBP-1 on hepatic lipogenic gene expression (Shimano *et al.*, 1997a, b) and physiological changes of SREBP-1c protein in normal mice after dietary manipulation such as placement on high carbohydrate diets, polyunsaturated fatty acid-enriched diets, and a fasting–refeeding regimen (Horton *et al.*, 1998a; Kim *et al.*, 1999; Thewke *et al.*, 1998; Worgall *et al.*, 1998; Yahagi *et al.*, 1999). Similar coordinated changes in SREBP-1c and lipogenic gene expression upon fasting and refeeding were also observed in adipose tissue, another lipogenic organ (Kim *et al.*, 1998a). Finally, we established a role for SREBP-1 in nutritional regulation of lipogenic genes by showing a severely impaired nutritional induction of lipogenic gene expression in SREBP-1 null mice (Shimano *et al.*, 1999). All the previous reports suggest that the SREBP-1c gene is regulated at the mRNA level. Up-regulation of hepatic SREBP-1 mRNA levels was observed in rodent livers on a fasting–refeeding regimen, a chronic high carbohydrate diet, and in primary hepatocytes with a high glucose medium (Foretz *et al.*, 1999b; Hasty *et al.*, 2000; Horton *et al.*, 1998a; Yahagi *et al.*, 1999). In contrast, down-regulation of SREBP-1c was observed in livers from fasted rodents, insulin-depleted diabetic rats with streptzotocin treatment, and mice on a diet containing polyunsaturated fatty acids (Horton *et al.*, 1998a; Kim *et al.*, 1999; Shimomura *et al.*, 1999b; Thewke *et al.*, 1998; Worgall *et al.*, 1998; Yahagi *et al.*, 1999). This nutritional regulation of SREBP-1c expression can result in changes in the amount of nuclear SREBP-1c because sterol regulation of the SREBP cleavage system is leaky to SREBP-1c as will be discussed later.

D. THE EFFECTS OF INSULIN/GLUCOSE ON SREBP-1c EXPRESSION

It is known that both glucose and insulin are required for the production of fatty acids via the induction of hepatic lipogenic enzymes. Cumulative lines of evidence suggest that this action of glucose and insulin could be mediated by induction of SREBP-1c. Different groups have shown that SREBP-1c is up-regulated by insulin *in vivo* and in primary hepatocyte cultures (Azzout-Marniche *et al.*, 2000; Foretz *et al.*, 1999a; Shimomura *et al.*, 1999b). These observations raised the possibility that SREBP-1c could be a metabolic mediator of the insulin effect in the liver (Flier and Hollenberg, 1999). It is interesting to explore the relationship between insulin receptor signaling and SREBP-1c induction. PI3 kinase inhibitors decrease nuclear and precursor SREBP-1c proteins (Hasty *et al.*, 2000). In IRS-2 knockout mice, hepatic SREBP-1c expression is increased associated with leptin resistance (Tobe *et al.*, 2001). In contrast, we and other groups have shown that glucose can induce SREBP-1c expression whereas glucose analogue, which cannot be metabolized, decreases SREBP-1c expression (Foretz et al., 1999b; Hasty et al., 2000). This is consistent with previous data that glucose must be metabolized for this effect on lipogenic gene induction and that insulin could only be permissive to this glucose action (Doiron et al., 1994; Mourrieras et al., 1997). Dietary fructose can activate SREBP-1c expression in the mouse liver to the same or an even higher extent than glucose, suggesting that a sugar metabolite can mediate SREBP-1c expression (A. Takahashi, unpublished observation). Further studies are required to clarify the mechanism by which insulin/glucose induces SREBP-1c expression.

V. TRANSCRIPTIONAL REGULATION OF LIPID SYNTHESIS BY SREBP-2 AND -1c IN THE LIVER

Figure 2 depicts a current diagram for different functions of SREBP-2 and SREBP-1c in the regulation of lipid synthesis in the liver. Both SREBP-2 and -1c are subjected to the cleavage system by SCAP, S1P, and S2P. If SCAP is disrupted, no cleavage of SREBP-1 or -2 occurs in the cultured cells or in liver (Matsuda et al., 2001; Rawson et al., 1999), suggesting that there is no other cleavage system. Then, in the presence of a sufficient amount of cholesterol, is lipogenesis shut down because cleavage of SREBP-1 is cancelled? Our recent

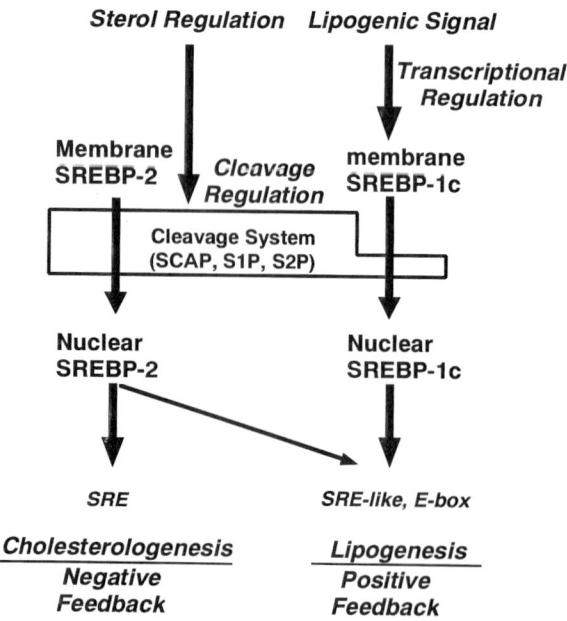

FIGURE 2 Differential regulation of SREBP-2 and -1c for cholesterol and fatty acid synthesis.

findings using a well-differentiated liver cell line showed that cleavage of SREBP-1 was less sensitive to sterol suppression than SREBP-2, suggesting that SREBP-1 could have a chance to liberate its active form into the nucleus even in the presence of sterols, thereby activating lipogenic genes by increasing its own amount (Hasty *et al.*, 2000). In another report, unsaturated fatty acids can regulate cleavage of SREBP-1, but not SREBP-2 (Hannah *et al.*, 2000). These data might suggest a new regulation of fatty acid metabolism mediated by SREBP-1 in contrast to sterol regulation by SREBP-2. The key to understanding this distinct regulation by SREBP-1 and -2 is the lipid-sensing mechanism of SCAP.

VI. PROMOTER ANALYSIS OF SREBP-1c

A. PRESENCE OF SRE AND AN OXYSTEROL-INDUCIBLE REGION IN THE SREBP-1c PROMOTER

Transcriptional regulation of lipogenic enzymes is controlled by the amount of SREBP-1c mRNA. This led us to analyze the promoter of SREBP-1c in an effort to understand the regulation of SREBP-1c and lipogenic enzymes (Amemiya-Kudo *et al.*, 2000) (Fig. 3). A cluster of putative binding sites of several transcription

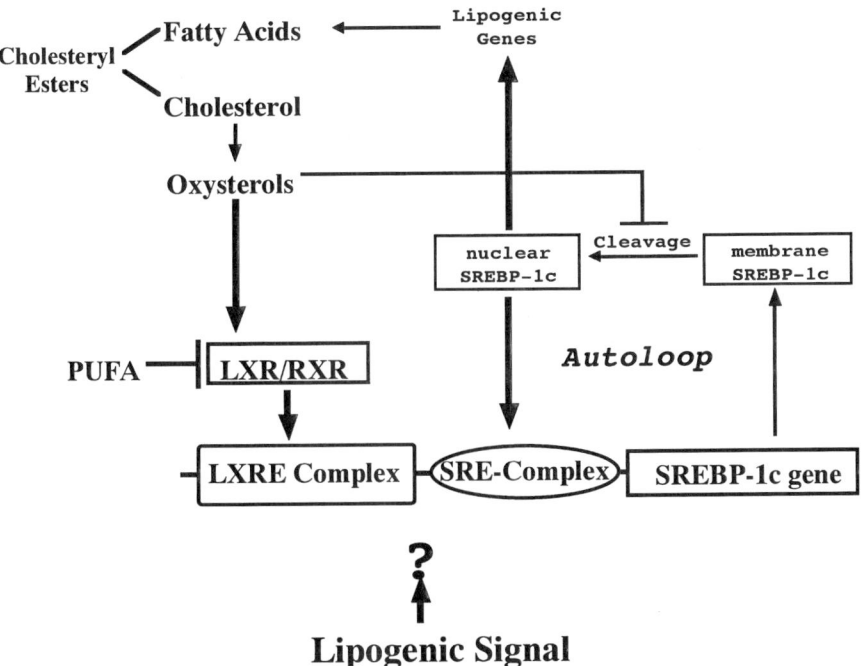

FIGURE 3 Regulation of SREBP-1c promoter.

factors composed of an NF-Y site, E-box, sterol-regulatory element, and Sp1 site was located at −90 bp of the SREBP-1c promoter. Luciferase reporter gene and gel shift assays indicated that the NF-Y site and the SRE in the SRE complex are responsible for SREBP activation. This suggests an autoloop production of SREBP-1c through the SRE complex, possibly explaining an overshoot phenomenon in induction of SREBP-1c and its downstream genes in the livers of refed mice (Horton *et al.*, 1998a). We also found another region that is up-regulated by oxysterols upstream of the SRE complex. This region was later identified as a liver X receptor/retinoid X receptor (LXR/RXR) binding site.

B. LXR/RXR ACTIVATION OF SREBP-1c: A NEW LINK BETWEEN CHOLESTEROL AND FATTY ACID REGULATION

Seeking dominant factors that activate the SREBP-1c promoter, an expression cloning strategy was undertaken. We cloned and identified LXR α and β as strong activators of the SREBP-1c promoter (Yoshikawa *et al.*, 2001). Deletion and mutation analysis of the SREBP-1c promoter identified two sets of new LXR-binding sites (LXR elements) that are very similar to an LXRE recently identified in the adenosine triphosphate (ATP)-binding cassette (ABC1) gene promoter (Costet *et al.*, 2000; Repa *et al.*, 2000b; Schwartz *et al.*, 2000). Either LXR, RXR, or their respective ligands can synergistically activate the SREBP-1c promoter. Recent reports on the effects of pharmacological LXR ligands and mice deficient in LXR α, β, or both also indicated that LXR/RXR activates SREBP-1c expression (Repa *et al.*, 2000a; Schultz *et al.*, 2000). This activation of SREBP-1c by LXR/RXR is accompanied by a substantial increase in nuclear SREBP-1c, leading to activation of its target lipogenic genes (Repa *et al.*, 2000a; Yoshikawa *et al.*, 2001). LXR/RXR seems to be a dominant regulator for SREBP-1c expression. LXR has recently been established as an oxysterol receptor and plays a key role in dealing with an excess amount of free cholesterol in the cells by enhancing cholesterol efflux or bile acid synthesis. Because SREBP-1c can activate fatty acid synthesis, activation of SREBP-1c by the oxysterol receptor could be another cellular adaptation for excess cholesterol by enhancing formation of cholesterol esters. This might suggest a new link between cholesterol and fatty acid metabolism. As shown in Fig. 3, the SREBP-1c promoter can be activated regardless of sterol levels, ensuring lipogenesis or cholesterol ester formation.

Statins and polyunsaturated fatty acids can suppress SREBP-1c expression. This suppression is mediated through decreased availability of endogenous LXR ligand(s), decreased production of ligand by statin (DeBose-Boyd *et al.*, 2001), and antagonizing ligand activation and binding to LXR by PUFA (Ou *et al.*, 2001; Yoshikawa *et al.*, 2002). PUFA have also been shown to inhibit SREBP-1c by a posttranscriptional mechanism and by inhibiting cleavage (Hannah *et al.*, 2000; Xu *et al.*, 2001; Yahagi *et al.*, 1999).

VII. ADD1/SREBP-1c IN ADIPOGENESIS

SREBP-1c was also cloned as a transcription factor for genes involved in adipogenesis, designated ADD1 (Tontonoz et al., 1993). It has been suggested that ADD1/SREBP-1c is involved in adipogenesis by activating PPAR γ, a master gene for adipogenesis, by two mechanisms (Tontonoz et al., 1994). First, ADD1 directly activates promoters of PPAR $\gamma1$ and $\gamma3$ and induces gene expression of PPAR γ (Fajas et al., 1999). Second and more interestingly, ADD1 induces production or secretion of an unidentified lipid molecule, which can be a ligand for PPAR γ (Kim et al., 1998b). This lipid molecule could be an endogenous ligand for PPAR γ and a key for adipogenesis. However, the roles of ADD1/SREBP-1c in adipose tissue seem to be very complex. SREBP-1 knockout mice have slightly less fat, but adipogenesis and lipogenesis were apparently normal, although this could be explained by a compensatory activation of SREBP-2 (Shimano et al., 1997b, 1999). Overexpression of nuclear SREBP-1c in a fat-specific manner using aP2 promoter in transgenic mice impaired adipose tissue differentiation, causing a decreased amount of fat and severe insulin resistance and hyperinsulinemia (Shimomura et al., 1998). Overexpression of nuclear SREBP-1c might have caused a disturbance of coordinated transcriptional machinery for adipogenesis or differentiation of adipocytes. These apparently opposing observations lead us to speculate that the timing and amount of ADD1/SREBP-1c expression might be important for adipogenesis. Likewise, overexpression of SREBP-1a seems to disturb the normal turnover of adipose tissue. Phosphoenolpyruvate carboxykinase (PEPCK)–SREBP-1a transgenic mice gradually lose adipose tissues as they grow fatty livers (Shimano et al., 1996).

VIII. CLINICAL ASPECTS OF SREBPs

A. SREBP-1c AND INSULIN RESISTANCE

The relationship between SREBP-1c and insulin resistance was first described in the report that adipose tissue-specific overexpression of SREBP-1c caused marked general insulin resistance and hyperinsulinemia in transgenic mice. Because these aP2–SREBP1c mice presented with a decrease in mass of adipose tissue with disturbance of adipocyte differentiation, the mice were considered a model of lipodystrophic diabetic animals as well as fatless mice produced by overexpression of a specific dominant negative form of C/EBPs (AZip/F1 mice). Insulin resistance and hyperinsulinemia in aP2–SREBP-1c mice were reversed by treatment with a low dose of leptin, suggesting that leptin deficiency is a causative factor of insulin resistance in this lipodystrophic animal (Shimomura et al., 1999c). Because thiazolidinedione is also effective in amelioration of insulin resistance of this animal, impaired adipocyte differentiation could also be involved in this pathological process. In contrast to lipodystrophy, an intensively studied obesity model, ob/ob mice, is also associated with severe insulin resistance and hyperinsulinemia. However, the mechanism of insulin resistance by leptin deficiency is different. We recently established ob/ob SREBP-1 knockout doubly mutant mice (Yahagi et al., 2002). The absence of SREBP-1 on an ob/ob background ameliorated fatty livers, but sustained hyperinsulinemia, suggesting that SREBP-1 is not involved in insulin resistance in ob/ob mice. Further investigation on the involvement of SREBP-1c in insulin resistance is needed as fatty acids have been thought to be deeply involved in insulin resistance.

B. SREBP-1 AND FATTY LIVERS

The roles of SREBP-1 in hepatic lipogenesis have been well established. Consequently it has been suggested that SREBP-1 could be involved in fatty livers. Forced activation of SREBP-1a or -1c resulted in fatty liver due to hepatic overproduction of triglycerides (Shimano et al., 1996, 1997a). In addition, other murine models for obesity of insulin resistance also show a similar hepatic phenotype. ob/ob mice or fatty Zucker rats, which are mutated in leptin and leptin receptor, respectively, show tremendous obesity and insulin resistance. Hyperinsulinemia and later onset of diabetes are associated with fatty livers. Hepatic expression of SREBP-1c in these rodents is dysregulated and showed a higher level of nuclear SREBP-1c protein than wild-type animals (Shimomura et al., 1999a; N. Yahagi et al., submitted). Together with ameriolation of fatty livers in ob/ob: SREBP-1 knockout doubly mutant mice (Yahagi et al., 2002), these observations suggest that activation of SREBP-1c is deeply involved in hyperinsulinemic fatty livers.

C. SREBP-1c AND ATHEROGENIC REMNANT LIPOPROTEINS

In mice, the plasma level of remnant particles is not elevated because they possess active LDL receptors that highly participate in plasma clearance of these particles in mice. In transgenic mice overexpressing nuclear SREBP-1a, hepatic production of lipids is extremely high and causes accumulation of a huge amount of triglycerides and cholesterol esters in livers, but plasma levels of these lipids are substantially lower than in control mice (Shimano et al., 1996). When these SREBP-1a transgenic mice are mated to LDL receptor knockout mice, the plasma levels of both cholesterol and triglycerides are markedly increased (Horton et al., 1999). These data suggest that the normolipidemia in SREBP-1a transgenic mice is a result of hepatic secretion and recapture of lipoproteins owing to sustained expression of LDL receptors, rather than hepatic lipoprotein secretion defect. A similar situation is applicable to ob/ob mice. Despite a state of excess energy, ob/ob mice do not show severe hyperlipidemia although associated with slightly high plasma triglyceride and HDL cholesterol levels (Silver et al., 1999). When LDL receptors are disrupted, ob/ob LDLRKO mice showed remarkably severe hypertriglyceridemia and hypercholesterolemia with a tremendous accumulation

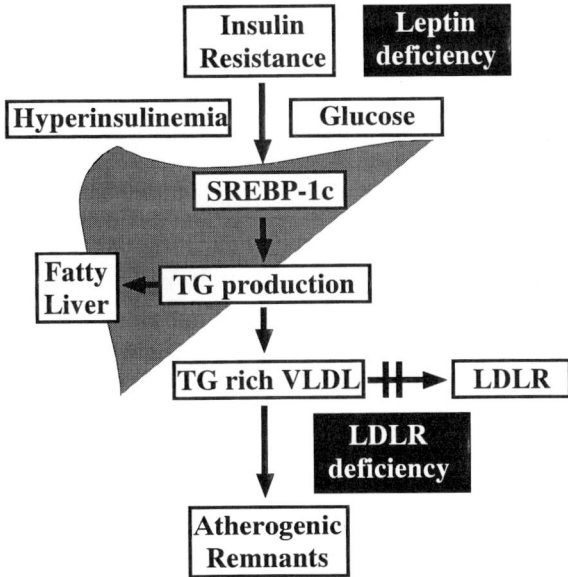

FIGURE 4 Production of fatty livers and remnant lipoproteins in ob/ob:LDL receptor doubly mutant mice.

of remnant particles (Hasty *et al.*, 2001). These hyperlipidemic mice developed marked atherosclerosis in the aortic sinus and arch even on a regular chow diet, indicating that accumulated remnants are very atherogenic. Hypertriglyceridemia was ameliorated by food restriction or short-term letin treatment, suggesting a link between excess caloric intake and overproduction of remnant particles. Figure 4 shows a sequence of potential events from excess energy state to atherosclerosis. Hepatic overproduction of triglycerides can result in production and secretion of triglyceride-rich very low-density lipoproteins. These particles are converted in circulation to remnant lipoproteins that are atherogenic. It also suggests an important clinical possibility: caloric restriction may be effective for reducing atherogenic lipoproteins.

IX. CROSS-TALK OF TRANSCRIPTION FACTORS FOR LIPID METABOLISM

Lipid metabolism has been nutritionally regulated at the transcriptional level for genes that are engaged in metabolic pathways. Recent progress in understanding transcription factors involved in lipid metabolism uncovered major players of cellular energy metabolism. These are SREBP-1, -2, PPAR α, PPAR γ, and LXR. Their roles are summarized in Fig. 5. Biosynthesis of cholesterol is

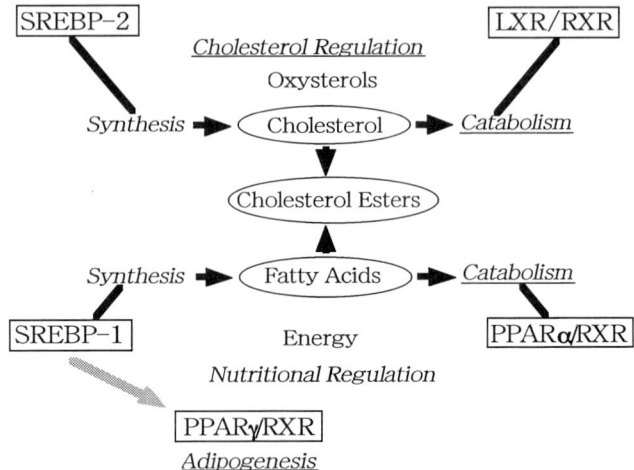

FIGURE 5 Roles of major transcription factors for lipid metabolism.

regulated by SREBP-2. Cellular adaptation for excess amounts of cholesterol such as cholesterol efflux regulated by ABC1 and hepatic bile acid synthesis is regulated by the nuclear receptor LXR/RXR. Fatty acid synthesis is regulated by SREBP-1 as discussed in this article. Catabolism of fatty acid is controlled by PPAR α. In different nutritional states, corresponding nutritional factors are regulated by changing amounts of active forms or recruitment of their specific ligands, and play a role in controlling appropriate gene expression for adaptic nutritional regulation. However, it seems that they do not work independently, but play their specified roles very efficiently by interacting with each other. These nutritional transcription factors form a diverse and complex network. Figure 6 shows the paradigm of cross-talk of these nutritional transcription factors from recent reports and our new findings. Although the roles of SREBP-1 and -2 are specialized to fatty acid and cholesterol synthesis, respectively, their functions can overlap because they share DNA-binding specificity and the cleavage system composed of SCAP-S1P, and S2P is indispensable for active forms of SREBP. The coordination of cholesterol and fatty acid synthesis is necessary for processes such as production of cellular membranes or lipoproteins. The presence of an overlapping mechanism of cholesterol and fatty acid synthesis through SREBP-1 and -2 is not surprising in those cases.

As described in the promoter analysis of SREBP-1c, LXR/RXR plays a dominant role in the expression of SREBP-1c. The oxysterol receptor controls lipogenic transcription factors, showing a novel link between cholesterol and fatty acid synthesis, the physiological relevance of which is currently unknown. It has also been reported that PPAR γ activates LXR α expression (Chawla *et al.*, 2001).

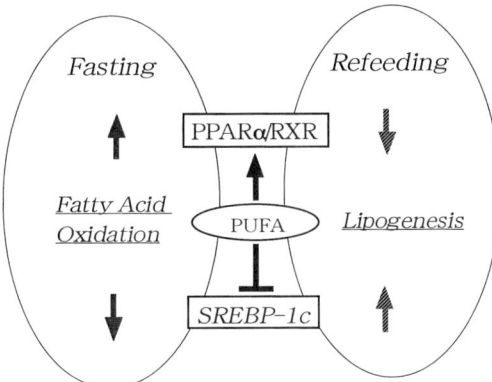

FIGURE 6 Reciprocal regulation of fatty acid metabolism by PPAR α and SREBP-1c.

Regulation of fatty acid metabolism depends entirely upon energy states (Fig. 6). In an energy-depleted state such as fasting, PPAR α and thus fatty acid degradation is induced to produce ketones as alternative energy sources. Simultaneously, SREBP-1c and lipogenesis are completely suppressed. Conversely, in a refed state, PPAR α is suppressed and SREBP-1c is activated to convert excess energy into triglycerides. This reciprocal regulation of PPAR α and SREBP-1c plays a crucial role in nutritional regulation of fatty acid metabolism. Interestingly, the regulation of PPAR α and SREBP-1c is also connected. Polyunsaturated fatty acids are shown to be ligands for PPAR α whereas PUFA suppress SREBP-1c by different mechanisms.

LXR and PPAR are nuclear receptors and share RXR as a heterodimer partner. They require RXR to bind to and activiate their specific DNA sequence in the promoters of their target genes. Understanding the kinetics of nuclear receptors requires detailed studies on heterodimerization, ligand specificity, and protein–protein interaction. However, depending upon RXR availability, these two nuclear receptors could compete with each other and modify the opposite function. We recently found in transfection studies with cultured cells that PPARs and LXR suppress each other by competing for RXR and, more interestingly, forming a nonfunctional PPAR–LXR heterodimer. For cross-talk of these three nuclear receptors, the concentrations of receptors and ligands in the cells are crucial. The physiological relevance to these interactions of LXR and PPAR needs to be tested *in vivo*. Our findings also suggest that PPAR α and nuclear receptors can regulate their target genes by changing their own local levels as well as the amounts of their ligands. SREBP-1c has been shown to regulate lipogenic genes in an autoregulatory fashion by changing its own amount.

As shown in Fig. 7, these mutual interactions of nuclear receptors and SREBPs suggest the presence of a complex and fine-tuned network of transcription factors and open up a new paradigm of nutritional regulation of gene expression in energy metabolism. By modifying each other, these factors can regulate their target gene

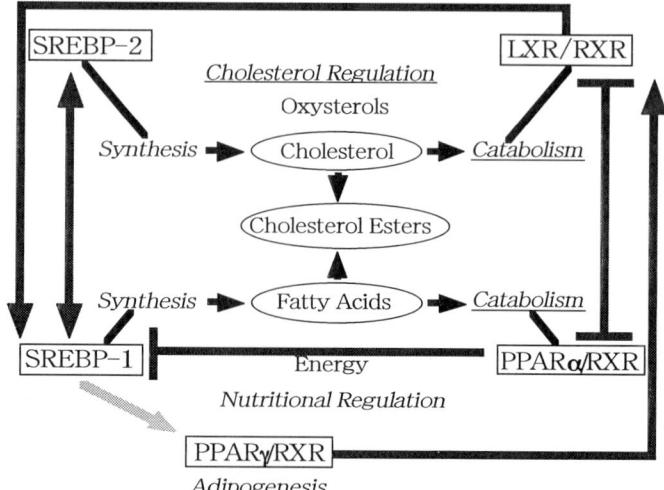

FIGURE 7 Cross-talk of major transcription factors for lipid metabolism.

expression more efficiently. This network also suggests difficulty in developing new therapeutics in this field because a substance that modifies one factor might cause modification of other connected factors, which makes it difficult to predict consequences in an entire system *in vivo*. All agonists and antagonists for nuclear receptors that are currently candidates for drugs for metabolic diseases should be carefully tested for this.

X. FUTURE ASPECTS OF SREBPs

The list of SREBP target genes is still expanding. More detailed analysis on unknown target genes using DNA microarray or DNA chip technology will help complete our understanding of SREBP functions. It is necessary to elucidate a sterol-sensing system and identify signaling molecules for cellular cholesterol amount. The SCAP–SREBP-2 complex should play a key role in this mechanism. Identification of energy-sensing molecules is another interesting topic. It is likely that SREBP-1 is involved in this mechanism and plays a role as a mediator for energy storage.

ACKNOWLEDGMENTS

I greatly thank my co-workers Naoya Yahagi, Michiyo Amemiya-Kudo, Alyssa H. Hasty, Tomohiro Yoshikawa, Takashi Matsuzaka, and Hirohito Sone for their contribution to our work presented in this review, and Nobuhiro Yamada, Shun Ishibashi, J. L. Goldstein, and M. S. Brown for continuous support.

REFERENCES

Amemiya-Kudo, M., Shimano, H., Yoshikawa, T., Yahagi, N., Hasty, A. H., Okazaki, H., Tamura, Y., Shionoiri, F., Iizuka, Y., Ohashi, K., et al. (2000). Promoter analysis of the mouse sterol regulatory element-binding protein-1c gene. *J. Biol. Chem.* **275,** 31078–31085.

Amemiya-Kudo, M., Shimano, H., Hasty, A. H., Yahagi, N., Yoshikawa, T., Matsuzaka, T., Okazaki, H., Tamura, Y., Iizuka, Y., Ohashi, K., Osuga, J. I., Harada, K., Gotoda, T., Sato, R., Kimura, S., Ishibashi, S., and Yamada, N. (2002). Transcriptional activities of nuclear SREBP-1a, -1c, and -2 to different target promoters of lipogenic and cholesterogenic genes. *J. Lipid Res.* **43**(8), 1220–1235.

Athanikar, J. N., Sanchez, H. B., and Osborne, T. F. (1997). Promoter selective transcriptional synergy mediated by sterol regulatory element binding protein and Sp1: A critical role for the Btd domain of Sp1. *Mol. Cell. Biol.* **17,** 5193–5200.

Azzout-Marniche, D., Becard, D., Guichard, C., Foretz, M., Ferre, P., and Foufelle, F. (2000). Insulin effects on sterol regulatory-element-binding protein-1c (SREBP-1c) transcriptional activity in rat hepatocytes. *Biochem. J.* **350**(Pt. 2), 389–393.

Bennett, M. K., and Osborne, T. F. (2000). Nutrient regulation of gene expression by the sterol regulatory element binding proteins: Increased recruitment of gene-specific coregulatory factors and selective hyperacetylation of histone H3 in vivo. *Proc. Natl. Acad. Sci. USA* **97,** 6340–6344.

Bennett, M. K., Lopez, J. M., Sanchez, H. B., and Osborne, T. F. (1995). Sterol regulation of fatty acid synthase promoter. Coordinate feedback regulation of two major lipid pathways. *J. Biol. Chem.* **270,** 25578–25583.

Bennett, M. K., Ngo, T. T., Athanikar, J. N., Rosenfeld, J. M., and Osborne, T. F. (1999). Co-stimulation of promoter for low density lipoprotein receptor gene by sterol regulatory element-binding protein and Sp1 is specifically disrupted by the yin yang 1 protein. *J. Biol. Chem.* **274,** 13025–13032.

Bist, A., Fielding, P. E., and Fielding, C. J. (1997). Two sterol regulatory element-like sequences mediate up-regulation of caveolin gene transcription in response to low density lipoprotein free cholesterol. *Proc. Natl. Acad. Sci. USA* **94,** 10693–10698.

Briggs, M. R., Yokoyama, C., Wang, X., Brown, M. S., and Goldstein, J. L. (1993). Nuclear protein that binds sterol regulatory element of low density lipoprotein receptor promoter. I. Identification of the protein and delineation of its target nucleotide sequence. *J. Biol. Chem.* **268,** 14490–14496.

Brown, M. S., and Goldstein, J. L. (1997). The SREBP pathway: Regulation of cholesterol metabolism by proteolysis of a membrane-bound transcription factor. *Cell.* **89,** 331–340.

Brown, M. S., and Goldstein, J. L. (1998). Sterol regulatory element binding proteins (SREBPs): Controllers of lipid synthesis and cellular uptake. *Nutr. Rev.* **56,** S1–3; discussion S54–75.

Brown, M. S., and Goldstein, J. L. (1999). A proteolytic pathway that controls the cholesterol content of membranes, cells, and blood. *Proc. Natl. Acad. Sci. USA* **96,** 11041–11048.

Brown, M. S., Ye, J., Rawson, R. B., and Goldstein, J. L. (2000). Regulated intramembrane proteolysis: A control mechanism conserved from bacteria to humans. *Cell* **100,** 391–398.

Carstea, E. D., Morris, J. A., Coleman, K. G., Loftus, S. K., Zhang, D., Cummings, C., Gu, J., Rosenfeld, M. A., Pavan, W. J., Krizman, D. B., et al. (1997). Niemann-Pick C1 disease gene: Homology to mediators of cholesterol homeostasis [see comments]. *Science* **277,** 228–231.

Chawla, A., Boisvert, W. A., Lee, C. H., Laffitte, B. A., Barak, Y., Joseph, S. B., Liao, D., Nagy, L., Edwards, P. A., Curtiss, L. K., et al. (2001). A PPAR gamma-LXR-ABCA1 pathway in macrophages is involved in cholesterol efflux and atherogenesis. *Mol. Cell.* **7,** 161–171.

Chouinard, R. A., Jr., Luo, Y., Osborne, T. F., Walsh, A., and Tall, A. R. (1998). Sterol regulatory element binding protein-1 activates the cholesteryl ester transfer protein gene in vivo but is not required for sterol up-regulation of gene expression. *J. Biol. Chem.* **273,** 22409–22414.

Christenson, L. K., Osborne, T. F., McAllister, J. M., and Strauss, J. F., 3rd (2001). Conditional response of the human steroidogenic acute regulatory protein gene promoter to sterol regulatory element binding protein-1a. *Endocrinology* **142,** 28–36.

Costet, P., Luo, Y., Wang, N., and Tall, A. R. (2000). Sterol-dependent transactivation of the human ABC1 promoter by LXR/RXR. *J. Biol. Chem.* **275,** 28240–28245.

DeBose-Boyd, R. A., Ou, J., Goldstein, J. L., and Brown, M. S. (2001). Expression of sterol regulatory element-binding protein 1c (SREBP-1c) mRNA in rat hepatoma cells requires endogenous LXR ligands. *Proc. Natl. Acad. Sci. USA* **98,** 1477–1482.

Doiron, B., Cuif, M. H., Kahn, A., and Diaz-Guerra, M. J. (1994). Respective roles of glucose, fructose, and insulin in the regulation of the liver-specific pyruvate kinase gene promoter. *J. Biol. Chem.* **269,** 10213–10216.

Dooley, K. A., Millinder, S., and Osborne, T. F. (1998). Sterol regulation of 3-hydroxy-3-methylglutaryl-coenzyme A synthase gene through a direct interaction between sterol regulatory element binding protein and the trimeric CCAAT-binding factor/nuclear factor Y. *J. Biol. Chem.* **273,** 1349–1356.

Dooley, K. A., Bennett, M. K., and Osborne, T. F. (1999). A critical role for cAMP response element-binding protein (CREB) as a co-activator in sterol-regulated transcription of 3-hydroxy-3-methylglutaryl coenzyme A synthase promoter. *J. Biol. Chem.* **274,** 5285–5291.

Duncan, E. A., Brown, M. S., Goldstein, J. L., and Sakai, J. (1997). Cleavage site for sterol-regulated protease localized to a Leu-Ser bond in the lumenal loop of sterol regulatory element-binding protein-2. *J. Biol. Chem.* **272,** 12778–12785.

Duncan, E. A., Dave, U. P., Sakai, J., Goldstein, J. L., and Brown, M. S. (1998). Second-site cleavage in sterol regulatory element-binding protein occurs at transmembrane junction as determined by cysteine panning. *J. Biol. Chem.* **273,** 17801–17809.

Ericsson, J., Jackson, S. M., and Edwards, P. A. (1996a). Synergistic binding of sterol regulatory element-binding protein and NF-Y to the farnesyl diphosphate synthase promoter is critical for sterol-regulated expression of the gene. *J. Biol. Chem.* **271,** 24359–24364.

Ericsson, J., Jackson, S. M., Lee, B. C., and Edwards, P. A. (1996b). Sterol regulatory element binding protein binds to a cis element in the promoter of the farnesyl diphosphate synthase gene. *Proc. Natl. Acad. Sci. USA* **93,** 945–950.

Ericsson, J., Jackson, S. M., Kim, J. B., Spiegelman, B. M., and Edwards, P. A. (1997). Identification of glycerol-3-phosphate acyltransferase as an adipocyte determination and differentiation factor 1- and sterol regulatory element-binding protein-responsive gene. *J. Biol. Chem.* **272,** 7298–7305.

Ericsson, J., Usheva, A., and Edwards, P. A. (1999). YY1 is a negative regulator of transcription of three sterol regulatory element-binding protein-responsive genes. *J. Biol. Chem.* **274,** 14508–14513.

Espenshade, P. J., Cheng, D., Goldstein, J. L., and Brown, M. S. (1999). Autocatalytic processing of site-1 protease removes propeptide and permits cleavage of sterol regulatory element-binding proteins. *J. Biol. Chem.* **274,** 22795–22804.

Fajas, L., Schoonjans, K., Gelman, L., Kim, J. B., Najib, J., Martin, G., Fruchart, J. C., Briggs, M., Spiegelman, B. M., and Auwerx, J. (1999). Regulation of peroxisome proliferator-activated receptor gamma expression by adipocyte differentiation and determination factor 1/sterol regulatory element binding protein 1: Implications for adipocyte differentiation and metabolism. *Mol. Cell. Biol.* **19,** 5495–5503.

Flier, J. S., and Hollenberg, A. N. (1999). ADD-1 provides major new insight into the mechanism of insulin action [letter; comment]. *Proc. Natl. Acad. Sci. USA* **96,** 14191–14192.

Foretz, M., Guichard, C., Ferre, P., and Foufelle, F. (1999a). Sterol regulatory element binding protein-1c is a major mediator of insulin action on the hepatic expression of glucokinase and lipogenesis-related genes [see comments]. *Proc. Natl. Acad. Sci. USA* **96,** 12737–12742.

Foretz, M., Pacot, C., Dugail, I., Lemarchand, P., Guichard, C., Le Liepvre, X., Berthelier-Lubrano, C., Spiegelman, B., Kim, J. B., Ferre, P., and Foufelle, F. (1999b). ADD1/SREBP-1c is required in the activation of hepatic lipogenic gene expression by glucose. *Mol. Cell. Biol.* **19,** 3760–3768.

Gauthier, B., Robb, M., Gaudet, F., Ginsburg, G. S., and McPherson, R. (1999). Characterization of a cholesterol response element (CRE) in the promoter of the cholesteryl ester transfer protein gene: Functional role of the transcription factors SREBP-1a, -2, and YY1. *J. Lipid. Res.* **40,** 1284–1293.

Goodridge, A. G. (1987). Dietary regulation of gene expression: Enzymes involved in carbohydrate and lipid metabolism. *Annu. Rev. Nutr.* **7,** 157–185.

Guan, G., Dai, P. H., Osborne, T. F., Kim, J. B., and Shechter, I. (1997). Multiple sequence elements are involved in the transcriptional regulation of the human squalene synthase gene. *J. Biol. Chem.* **272**, 10295–10302.

Gunther, M., Laithier, M., and Brison, O. (2000). A set of proteins interacting with transcription factor Sp1 identified in a two-hybrid screening. *Mol. Cell. Biochem.* **210**, 131–142.

Hannah, V. C., Ou, J., Luong, A., Goldstein, J. L., and Brown, M. S. (2001). Unsaturated fatty acids down-regulate SREBP isoforms 1a and 1c by two mechanisms in HEK-293 cells. *J. Biol. Chem.* **276**, 4365–4372.

Hasty, A. H., Shimano, H., Yahagi, N., Amemiya-Kudo, M., Perrey, S., Yoshikawa, T., Osuga, J., Okazaki, H., Tamura, Y., Iizuka, Y., *et al.* (2000). Sterol regulatory element-binding protein-1 is regulated by glucose at the transcriptional level. *J. Biol. Chem.* **275**, 31069–31077.

Hasty, A. H., Shimano, H., Osuga, J.-i., Namatame, I., Takahashi, A., Yahagi, N., Perrey, S., Iizuka, Y., Tamura, Y., Amemiya-Kudo, M., *et al.* (2001). Severe hypercholesterolemia, hypertriglyceridemia and atherosclerosis in mice lacking both leptin and the low density lipoprotein receptor. *J. Biol. Chem.* **276**, 37402–37408.

Hillgartner, F. B., Salati, L. M., and Goodridge, A. G. (1995). Physiological and molecular mechanisms involved in nutritional regulation of fatty acid synthesis. *Physiol. Rev.* **75**, 47–76.

Hirano, Y., Yoshida, M., Shimizu, M., and Sato, R. (2001). Direct demonstration of rapid degradation of nuclear sterol regulatory element-binding proteins by the ubiquitin-proteasome pathway. *J. Biol. Chem.* **27**, 27.

Horton, J. D., Bashmakov, Y., Shimomura, I., and Shimano, H. (1998a). Regulation of sterol regulatory element binding proteins in livers of fasted and refed mice. *Proc. Natl. Acad. Sci. USA* **95**, 5987–5992.

Horton, J. D., Shimomura, I., Brown, M. S., Hammer, R. E., Goldstein, J. L., and Shimano, H. (1998b). Activation of cholesterol synthesis in preference to fatty acid synthesis in liver and adipose tissue of transgenic mice overproducing sterol regulatory element-binding protein-2. *J. Clin. Invest.* **101**, 2331–2339.

Horton, J. D., Shimano, H., Hamilton, R. L., Brown, M. S., and Goldstein, J. L. (1999). Disruption of LDL receptor gene in transgenic SREBP-1a mice unmasks hyperlipidemia resulting from production of lipid-rich VLDL. *J. Clin. Invest.* **103**, 1067–1076.

Hua, X., Yokoyama, C., Wu, J., Briggs, M. R., Brown, M. S., Goldstein, J. L., and Wang, X. (1993). SREBP-2, a second basic-helix-loop-helix-leucine zipper protein that stimulates transcription by binding to a sterol regulatory element. *Proc. Natl. Acad. Sci. USA* **90**, 11603–11607.

Hua, X., Wu, J., Goldstein, J. L., Brown, M. S., and Hobbs, H. H. (1995). Structure of the human gene encoding sterol regulatory element binding protein-1 (SREBF1) and localization of SREBF1 and SREBF2 to chromosomes 17p11.2 and 22q13. *Genomics* **25**, 667–673.

Hua, X., Nohturfft, A., Goldstein, J. L., and Brown, M. S. (1996). Sterol resistance in CHO cells traced to point mutation in SREBP cleavage-activating protein. *Cell* **87**, 415–426.

Ikeda, Y., Yamamoto, J., Okamura, M., Fujino, T., Takahashi, S., Takeuchi, K., Osborne, T. F., Yamamoto, T. T., Ito, S., and Sakai, J. (2001). Transcriptional regulation of the murine acetyl CoA synthetase 1 gene through multiple clustered binding sites for SREBPs and a single neighboring site for Sp1. *J. Biol. Chem.* **2**, 2.

Inoue, J., Sato, R., and Maeda, M. (1998). Multiple DNA elements for sterol regulatory element-binding protein and NF-Y are responsible for sterol-regulated transcription of the genes for human 3-hydroxy-3-methylglutaryl coenzyme A synthase and squalene synthase. *J. Biochem. (Tokyo)* **123**, 1191–1198.

Jackson, S. M., Ericsson, J., Mantovani, R., and Edwards, P. A. (1998). Synergistic activation of transcription by nuclear factor Y and sterol regulatory element binding protein. *J. Lipid Res.* **39**, 767–776.

Kast, H. R., Nguyen, C. M., Anisfeld, A. M., Ericsson, J., and Edwards, P. A. (2001). CTP: Phosphocholine cytidylyltransferase, a new sterol- and SREBP- responsive gene. *J. Lipid Res.* **42**, 1266–1272.

Kim, H. J., Takahashi, M., and Ezaki, O. (1999). Fish oil feeding decreases mature sterol regulatory element-binding protein 1 (SREBP-1) by down-regulation of SREBP-1c mRNA in mouse liver. A possible mechanism for down-regulation of lipogenic enzyme mrnas. *J. Biol. Chem.* **274,** 25892–25898.

Kim, J. B., Spotts, G. D., Halvorsen, Y. D., Shih, H. M., Ellenberger, T., Towle, H. C., and Spiegelman, B. M. (1995). Dual DNA binding specificity of ADD1/SREBP1 controlled by a single amino acid in the basic helix-loop-helix domain. *Mol. Cell. Biol.* **15,** 2582–2588.

Kim, J. B., Sarraf, P., Wright, M., Yao, K. M., Mueller, E., Solanes, G., Lowell, B. B., and Spiegelman, B. M. (1998a). Nutritional and insulin regulation of fatty acid synthetase and leptin gene expression through ADD1/SREBP1. *J. Clin. Invest.* **101,** 1–9.

Kim, J. B., Wright, H. M., Wright, M., and Spiegelman, B. M. (1998b). ADD1/SREBP1 activates PPARgamma through the production of endogenous ligand. *Proc. Natl. Acad. Sci. USA* **95,** 4333–4337.

Kim, J. H., Lee, J. N., and Paik, Y. K. (2001). Cholesterol biosynthesis from lanosterol. A concerted role for Sp1 and NF-Y-binding sites for sterol-mediated regulation of rat 7-dehydrocholesterol reductase gene expression. *J. Biol. Chem.* **276,** 18153–18160.

Korn, B. S., Shimomura, I., Bashmakov, Y., Hammer, R. E., Horton, J. D., Goldstein, J. L., and Brown, M. S. (1998). Blunted feedback suppression of SREBP processing by dietary cholesterol in transgenic mice expressing sterol-resistant SCAP(D443N). *J. Clin. Invest.* **102,** 2050–2060.

Lawler, J. F., Jr., Yin, M., Diehl, A. M., Roberts, E., and Chatterjee, S. (1998). Tumor necrosis factor-alpha stimulates the maturation of sterol regulatory element binding protein-1 in human hepatocytes through the action of neutral sphingomyelinase. *J. Biol. Chem.* **273,** 5053–5059.

Levanon, D., Hsieh, C. L., Francke, U., Dawson, P. A., Ridgway, N. D., Brown, M. S., and Goldstein, J. L. (1990). cDNA cloning of human oxysterol-binding protein and localization of the gene to human chromosome 11 and mouse chromosome 19. *Genomics* **7,** 65–74.

Lopez, D., and McLean, M. P. (1999). Sterol regulatory element-binding protein-1a binds to cis elements in the promoter of the rat high density lipoprotein receptor SR-BI gene. *Endocrinology* **140,** 5669–5681.

Lopez, J. M., Bennett, M. K., Sanchez, H. B., Rosenfeld, J. M., and Osborne, T. E. (1996). Sterol regulation of acetyl coenzyme A carboxylase: A mechanism for coordinate control of cellular lipid. *Proc. Natl. Acad. Sci. USA* **93,** 1049–1053.

Magana, M. M., and Osborne, T. F. (1996). Two tandem binding sites for sterol regulatory element binding proteins are required for sterol regulation of fatty-acid synthase promoter. *J. Biol. Chem.* **271,** 32689–32694.

Magana, M. M., Lin, S. S., Dooley, K. A., and Osborne, T. F. (1997). Sterol regulation of acetyl coenzyme A carboxylase promoter requires two interdependent binding sites for sterol regulatory element binding proteins. *J. Lipid Res.* **38,** 1630–1638.

Magana, M. M., Koo, S. H., Towle, H. C., and Osborne, T. F. (2000). Different sterol regulatory element-binding protein-1 isoforms utilize distinct co-regulatory factors to activate the promoter for fatty acid synthase. *J. Biol. Chem.* **275,** 4726–4733.

Mater, M. K., Thelen, A. P., Pan, D. A., and Jump, D. B. (1999). Sterol response element-binding protein 1c (SREBP1c) is involved in the polyunsaturated fatty acid suppression of hepatic S14 gene transcription. *J. Biol. Chem.* **274,** 32725–32732.

Matsuda, M., Korn, B. S., Hammer, R. E., Moon, Y. A., Komuro, R., Horton, J. D., Goldstein, J. L., Brown, M. S., and Shimomura, I. (2001). SREBP cleavage-activating protein (SCAP) is required for increased lipid synthesis in liver induced by cholesterol deprivation and insulin elevation. *Genes Dev.* **15,** 1206–1216.

Matsuzaka, T., Shimano, H., Yahagi, N., Amemiya-Kudo, M., Yoshikawa, T., Hasty, A. H., Tamura, Y., Osuga, J., Okazaki, H., Iizuka, Y., Takahashi, A., Sone, H., Gotoda, T., Ishibashi, S., and Yamada, N. (2002). Dual regulation of mouse Delta(5)- and Delta(6)-desaturase gene expression by SREBP-1 and PPARalpha. *J. Lipid Res.* **43**(1), 107–114.

Miserez, A. R., Cao, G., Probst, L. C., and Hobbs, H. H. (1997). Structure of the human gene encoding sterol regulatory element binding protein 2 (SREBF2). *Genomics* **40,** 31–40.

Moon, Y. A., Lee, J. J., Park, S. W., Ahn, Y. H., and Kim, K. S. (2000). The roles of sterol regulatory element-binding proteins in the transactivation of the rat ATP citrate-lyase promoter. *J. Biol. Chem.* **275**, 30280–30286.

Mourrieras, F., Foufelle, F., Foretz, M., Morin, J., Bouche, S., and Ferre, P. (1997). Induction of fatty acid synthase and S14 gene expression by glucose, xylitol and dihydroxyacetone in cultured rat hepatocytes is closely correlated with glucose 6-phosphate concentrations. *Biochem. J.* **326**, 345–349.

Naar, A. M., Beaurang, P. A., Robinson, K. M., Oliner, J. D., Avizonis, D., Scheek, S., Zwicker, J., Kadonaga, J. T., and Tjian, R. (1998). Chromatin, TAFs, and a novel multiprotein coactivator are required for synergistic activation by Sp1 and SREBP-1a in vitro. *Genes Dev.* **12**, 3020–3031.

Nagoshi, E., Imamoto, N., Sato, R., and Yoneda, Y. (1999). Nuclear import of sterol regulatory element-binding protein-2, a basic helix-loop-helix-leucine zipper (bHLH-Zip)-containing transcription factor, occurs through the direct interaction of importin beta with HLH-Zip. *Mol. Biol. Cell* **10**, 2221–2233.

Nakajima, T., Hamakubo, T., Kodama, T., Inazawa, J., and Emi, M. (1999). Genomic structure and chromosomal mapping of the human sterol regulatory element binding protein (SREBP) cleavage-activating protein (SCAP) gene *J. Hum. Genet.* **44**, 402–407.

Natarajan, R., Ghosh, S., and Grogan, W. M. (1998). Molecular cloning of the promoter for rat hepatic neutral cholesterol ester hydrolase: Evidence for transcriptional regulation by sterols. *Biochem. Biophys. Res. Commun.* **243**, 349–355.

Ou, J., Tu, H., Shan, B., Luk, A., DeBose-Boyd, R. A., Bashmakov, Y., Goldstein, J. L., and Brown, M. S. (2001). Unsaturated fatty acids inhibit transcription of the sterol regulatory element-binding protein-1c (SREBP-1c) gene by antagonizing ligand- dependent activation of the LXR. *Proc. Natl. Acad. Sci. USA* **98**, 6027–6032.

Pai, J. T., Guryev, O., Brown, M. S., and Goldstein, J. L. (1998). Differential stimulation of cholesterol and unsaturated fatty acid biosynthesis in cells expressing individual nuclear sterol regulatory element-binding proteins. *J. Biol. Chem.* **273**, 26138–26148.

Parraga, A., Bellsolell, L., Ferre-D'Amare, A. R., and Burley, S. K. (1998). Co-crystal structure of sterol regulatory element binding protein 1a at 2.3 A resolution. *Structure* **6**, 661–672.

Rawson, R. B., Zelenski, N. G., Nijhawan, D., Ye, J., Sakai, J., Hasan, M. T., Chang, T. Y., Brown, M. S., and Goldstein, J. L. (1997). Complementation cloning of S2P, a gene encoding a putative metalloprotease required for intramembrane cleavage of SREBPs. *Mol. Cell* **1**, 47–57.

Rawson, R. B., DeBose-Boyd, R., Goldstein, J. L., and Brown, M. S. (1999). Failure to cleave sterol regulatory element-binding proteins (SREBPs) causes cholesterol auxotrophy in Chinese hamster ovary cells with genetic absence of SREBP cleavage-activating protein. *J. Biol. Chem.* **274**, 28549–28556.

Repa, J. J., Liang, G., Ou, J., Bashmakov, Y., Lobaccaro, J. M., Shimomura, I., Shan, B., Brown, M. S., Goldstein, J. L., and Mangelsdorf, D. J. (2000a). Regulation of mouse sterol regulatory element-binding protein-1c gene (SREBP-1c) by oxysterol receptors, LXRalpha and LXRbeta. *Genes Dev.* **14**, 2819–2830.

Repa, J. J., Turley, S. D., Lobaccaro, J. A., Medina, J., Li, L., Lustig, K., Shan, B., Heyman, R. A., Dietschy, J. M., and Mangelsdorf, D. J. (2000b). Regulation of absorption and ABC1-mediated efflux of cholesterol by RXR heterodimers [see comments]. *Science* **289**, 1524–1529.

Ridgway, N. D., Dawson, P. A., Ho, Y. K., Brown, M. S., and Goldstein, J. L. (1992). Translocation of oxysterol binding protein to Golgi apparatus triggered by ligand binding. *J. Cell Biol.* **116**, 307–319.

Rozman, D., Fink, M., Fimia, G. M., Sassone-Corsi, P., and Waterman, M. R. (1999). Cyclic adenosine 3',5'-monophosphate(cAMP)/cAMP-responsive element modulator (CREM)-dependent regulation of cholesterogenic lanosterol 14alpha-demethylase (CYP51) in spermatids. *Mol. Endocrinol.* **13**, 1951–1962.

Sakai, J., Duncan, E. A., Rawson, R. B., Hua, X., Brown, M. S., and Goldstein, J. L. (1996). Sterol-regulated release of SREBP-2 from cell membranes requires two sequential cleavages, one within a transmembrane segment. *Cell* **85**, 1037–1046.

Sakai, J., Nohturfft, A., Cheng, D., Ho, Y. K., Brown, M. S., and Goldstein, J. L. (1997). Identification of complexes between the COOH-terminal domains of sterol regulatory element-binding proteins (SREBPs) and SREBP cleavage-activating protein. *J. Biol. Chem.* **272,** 20213–20221.

Sakai, J., Nohturfft, A., Goldstein, J. L., and Brown, M. S. (1998). Cleavage of sterol regulatory element-binding proteins (SREBPs) at site-1 requires interaction with SREBP cleavage-activating protein. Evidence from in vivo competition studies. *J. Biol. Chem.* **273,** 5785–5793.

Sakakura, Y., Shimano, H., Sone, H., Takahashi, A., Inoue, K., Toyoshima, H., Suzuki, S., and Yamada, N. (2001). Sterol Regulatory Element-Binding Proteins Induce an Entire Pathway of Cholesterol Synthesis. *Biochem. Biophys. Res. Commun.* **286,** 176–183.

Sanchez, H. B., Yieh, L., and Osborne, T. F. (1995). Cooperation by sterol regulatory element-binding protein and Sp1 in sterol regulation of low density lipoprotein receptor gene. *J. Biol. Chem.* **270,** 1161–1169.

Sato, R., Inoue, J., Kawabe, Y., Kodama, T., Takano, T., and Maeda, M. (1996). Sterol-dependent transcriptional regulation of sterol regulatory element-binding protein-2. *J. Biol. Chem.* **271,** 26461–26464.

Sato, R., Miyamoto, W., Inoue, J., Terada, T., Imanaka, T., and Maeda, M. (1999). Sterol regulatory element-binding protein negatively regulates microsomal triglyceride transfer protein gene transcription. *J. Biol. Chem.* **274,** 24714–24720.

Sato, R., Okamoto, A., Inoue, J., Miyamoto, W., Sakai, Y., Emoto, N., Shimano, H., and Maeda, M. (2000). Transcriptional regulation of the ATP citrate-lyase gene by sterol regulatory element-binding proteins. *J. Biol. Chem.* **275,** 12497–12502.

Scheek, S., Brown, M. S., and Goldstein, J. L. (1997). Sphingomyelin depletion in cultured cells blocks proteolysis of sterol regulatory element binding proteins at site 1. *Proc. Natl. Acad. Sci. USA* **94,** 11179–11183.

Schoonjans, K., Gelman, L., Haby, C., Briggs, M., and Auwerx, J. (2000). Induction of LPL gene expression by sterols is mediated by a sterol regulatory element and is independent of the presence of multiple E boxes. *J. Mol. Biol.* **304,** 323–334.

Schultz, J. R., Tu, H., Luk, A., Repa, J. J., Medina, J. C., Li, L., Schwendner, S., Wang, S., Thoolen, M., Mangelsdorf, D. J., et al. (2000). Role of LXRs in control of lipogenesis. *Genes Dev.* **14,** 2831–2838.

Schwartz, K., Lawn, R. M., and Wade, D. P. (2000). ABC1 Gene expression and apoa-I-mediated cholesterol efflux are regulated by LXR. *Biochem. Biophys. Res. Commun.* **274,** 794–802.

Shea-Eaton, W., Lopez, D., and McLean, M. P. (2001a). Yin Yang 1 protein negatively regulates high-density lipoprotein receptor gene transcription by disrupting binding of sterol regulatory element binding protein to the sterol regulatory element. *Endocrinology* **142,** 49–58.

Shea-Eaton, W. K., Trinidad, M. J., Lopez, D., Nackley, A., and McLean, M. P. (2001b). Sterol regulatory element binding protein-1a regulation of the steroidogenic acute regulatory protein gene. *Endocrinology* **142,** 1525–1533.

Sheng, Z., Otani, H., Brown, M. S., and Goldstein, J. L. (1995). Independent regulation of sterol regulatory element-binding proteins 1 and 2 in hamster liver. *Proc. Natl. Acad. Sci. USA* **92,** 935–938.

Shimano, H., Horton, J. D., Hammer, R. E., Shimomura, I., Brown, M. S., and Goldstein, J. L. (1996). Overproduction of cholesterol and fatty acids causes massive liver enlargement in transgenic mice expressing truncated SREBP-1a [see comments]. *J. Clin. Invest.* **98,** 1575–1584.

Shimano, H., Horton, J. D., Shimomura, I., Hammer, R. E., Brown, M. S., and Goldstein, J. L. (1997a). Isoform 1c of sterol regulatory element binding protein is less active than isoform 1a in livers of transgenic mice and in cultured cells. *J. Clin. Invest.* **99,** 846–854.

Shimano, H., Shimomura, I., Hammer, R. E., Herz, J., Goldstein, J. L., Brown, M. S., and Horton, J. D. (1997b). Elevated levels of SREBP-2 and cholesterol synthesis in livers of mice homozygous for a targeted disruption of the SREBP-1 gene [see comments]. *J. Clin. Invest.* **100,** 2115–2124.

Shimano, H., Yahagi, N., Amemiya-Kudo, M., Hasty, A. H., Osuga, J., Tamura, Y., Shionoiri, F., Iizuka, Y., Ohashi, K., Harada, K., et al. (1999). Sterol regulatory element-binding protein-1 as a

key transcription factor for nutritional induction of lipogenic enzyme genes. *J. Biol. Chem.* **274**, 35832–35839.
Shimomura, I., Shimano, H., Horton, J. D., Goldstein, J. L., and Brown, M. S. (1997). Differential expression of exons 1a and 1c in mRNAs for sterol regulatory element binding protein-1 in human and mouse organs and cultured cells. *J. Clin. Invest.* **99**, 838–845.
Shimomura, I., Hammer, R. E., Richardson, J. A., Ikemoto, S., Bashmakov, Y., Goldstein, J. L., and Brown, M. S. (1998). Insulin resistance and diabetes mellitus in transgenic mice expressing nuclear SREBP-1c in adipose tissue: Model for congenital generalized lipodystrophy. *Genes Dev.* **12**, 3182–3194.
Shimomura, I., Bashmakov, Y., and Horton, J. D. (1999a). Increased levels of nuclear SREBP-1c associated with fatty livers in two mouse models of diabetes mellitus. *J. Biol. Chem.* **274**, 30028–30032.
Shimomura, I., Bashmakov, Y., Ikemoto, S., Horton, J. D., Brown, M. S., and Goldstein, J. L. (1999b). Insulin selectively increases SREBP-1c mRNA in the livers of rats with streptozotocin-induced diabetes [see comments]. *Proc. Natl. Acad. Sci. USA* **96**, 13656–13661.
Shimomura, I., Hammer, R. E., Ikemoto, S., Brown, M. S., and Goldstein, J. L. (1999c). Leptin reverses insulin resistance and diabetes mellitus in mice with congenital lipodystrophy. *Nature* **401**, 73–76.
Silver, D. L., Jiang, X.-C., and Tall, A. R. (1999). Increased high density lipoprotein (HDL) defective hepatic catabolism of apoA-I and ApoA-II, and decreased apoA-I mRNA in ob/ob mice. *J. Biol. Chem.* **274**, 4140–4146.
Smith, J. R., Osborne, T. F., Brown, M. S., Goldstein, J. L., and Gil, G. (1988). Multiple sterol regulatory elements in promoter for hamster 3-hydroxy-3-methylglutaryl-coenzyme A synthase. *J. Biol. Chem.* **263**, 18480–18487.
Smith, J. R., Osborne, T. F., Goldstein, J. L., and Brown, M. S. (1990). Identification of nucleotides responsible for enhancer activity of sterol regulatory element in low density lipoprotein receptor gene. *J. Biol. Chem.* **265**, 2306–2310.
Sone, H., Shimano, H., Sakakura, Y., Inoue, N., Amemiya-Kudo, M., Yahagi, N., Osawa, M., Suzuki, H., Yokoo, T., Takahashi, A., Iida, K., Toyoshima, H., Iwama, A., and Yamada, N. (2002). Acetyl-coenzyme A synthetase is a lipogenic enzyme controlled by SREBP-1 and energy status. *Am. J. Physiol. Endocrinol. Metab.* **282**(1), E222–230.
Swinnen, J. V., Alen, P., Heyns, W., and Verhoeven, G. (1998). Identification of diazepam-binding inhibitor/acyl-CoA-binding protein as a sterol regulatory element-binding protein-responsive gene. *J. Biol. Chem.* **273**, 19938–19944.
Tabor, D. E., Kim, J. B., Spiegelman, B. M., and Edwards, P. A. (1998). Transcriptional activation of the stearoyl-CoA desaturase 2 gene by sterol regulatory element-binding protein/adipocyte determination and differentiation factor 1. *J. Biol. Chem.* **273**, 22052–22058.
Tabor, D. E., Kim, J. B., Spiegelman, B. M., and Edwards, P. A. (1999). Identification of conserved cis-elements and transcription factors required for sterol-regulated transcription of stearoyl-CoA desaturase 1 and 2. *J. Biol. Chem.* **274**, 20603–20610.
Thewke, D. P., Panini, S. R., and Sinensky, M. (1998). Oleate potentiates oxysterol inhibition of transcription from sterol regulatory element-1-regulated promoters and maturation of sterol regulatory element-binding proteins. *J. Biol. Chem.* **273**, 21402–21407.
Tobe, K., Suzuki, R., Aoyama, M., Yamauchi, T., Kamon, J., Kubota, N., Terauchi, Y., Matsui, J., Akanuma, Y., Kimura, S., *et al.* (2001). Increased expression of the sterol regulatory element-binding protein (SREBP)-1 gene in insulin receptor substrate-2-/- mouse liver. *J. Biol. Chem.* **276**, 38337–38340.
Tontonoz, P., Kim, J. B., Graves, R. A., and Spiegelman, B. M. (1993). ADD1: A novel helix-loop-helix transcription factor associated with adipocyte determination and differentiation. *Mol. Cell. Biol.* **13**, 4753–4759.
Tontonoz, P., Hu, E., and Spiegelman, B. M. (1994). Stimulation of adipogenesis in fibroblasts by PPAR gamma 2, a lipid-activated transcription factor. *Cell.* **79**, 1147–1156.

Vallett, S. M., Sanchez, H. B., Rosenfeld, J. M., and Osborne, T. F. (1996). A direct role for sterol regulatory element binding protein in activation of 3-hydroxy-3-methylglutaryl coenzyme A reductase gene. *J. Biol. Chem.* **271,** 12247–12253.

Wang, X., Briggs, M. R., Hua, X., Yokoyama, C., Goldstein, J. L., and Brown, M. S. (1993). Nuclear protein that binds sterol regulatory element of low density lipoprotein receptor promoter. II. Purification and characterization. *J. Biol. Chem.* **268,** 14497–14504.

Wang, X., Sato, R., Brown, M. S., Hua, X., and Goldstein, J. L. (1994). SREBP-1, a membrane-bound transcription factor released by sterol-regulated proteolysis [see comments]. *Cell* **77,** 53–62.

Wang, X., Zelenski, N. G., Yang, J., Sakai, J., Brown, M. S., and Goldstein, J. L. (1996). Cleavage of sterol regulatory element binding proteins (SREBPs) by CPP32 during apoptosis. *EMBO J.* **15,** 1012–1020.

Worgall, T. S., Sturley, S. L., Seo, T., Osborne, T. F., and Deckelbaum, R. J. (1998). Polyunsaturated fatty acids decrease expression of promoters with sterol regulatory elements by decreasing levels of mature sterol regulatory element-binding protein. *J. Biol. Chem.* **273,** 25537–25540.

Xiong, S., Chirala, S. S., and Wakil, S. J. (2000). Sterol regulation of human fatty acid synthase promoter I requires nuclear factor-Y- and Sp-1-binding sites. *Proc. Natl. Acad. Sci. USA* **97,** 3948–3953.

Xu, J., Teran-Garcia, M., Park, J. H., Nakamura, M. T., and Clarke, S. D. (2001). Polyunsaturated fatty acids suppress hepatic sterol regulatory element-binding protein-1 expression by accelerating transcript decay. *J. Biol. Chem.* **276,** 9800–9807.

Yahagi, N., Shimano, H., Hasty, A. H., Amemiya-Kudo, M., Okazaki, H., Tamura, Y., Iizuka, Y., Shionoiri, F., Ohashi, K., Osuga, J., *et al.* (1999). A crucial role of sterol regulatory element-binding protein-1 in the regulation of lipogenic gene expression by polyunsaturated fatty acids. *J. Biol. Chem.* **274,** 35840–35844.

Yahagi, N., Shimano, H., Hasty, A. H., Matsuzaka, T., Ide, T., Yoshikawa, T., Amemiya-Kudo, M., Tomita, S., Okazaki, H., Tamura, Y., Iizuka, Y., Ohashi, K., Osuga, J., Harada, K., Gotoda, T., Nagai, R., Ishibashi, S., and Yamada, N. (2002). Absence of sterol regulatory element-binding protein-1 (SREBP-1) ameliorates fatty livers but not obesity or insulin resistance in Lep(ob)/Lep(ob) mice. *J. Biol. Chem.* **277**(22), 19353–19357.

Yang, W. S., and Deeb, S. S. (1998). Sp1 and Sp3 transactivate the human lipoprotein lipase gene promoter through binding to a CT element: Synergy with the sterol regulatory element binding protein and reduced transactivation of a naturally occurring promoter variant. *J. Lipid Res.* **39,** 2054–2064.

Yieh, L., Sanchez, H. B., and Osborne, T. F. (1995). Domains of transcription factor Sp1 required for synergistic activation with sterol regulatory element binding protein 1 of low density lipoprotein receptor promoter. *Proc. Natl. Acad. Sci. USA* **92,** 6102–6106.

Yokoyama, C., Wang, X., Briggs, M. R., Admon, A., Wu, J., Hua, X., Goldstein, J. L., and Brown, M. S. (1993). SREBP-1, a basic-helix-loop-helix-leucine zipper protein that controls transcription of the low density lipoprotein receptor gene. *Cell* **75,** 187–197.

Yoshikawa, T., Shimano, H., Amemiya-Kudo, H., Yahagi, N., Hasty, A., Matsuzaka, T., Okazaki, H., Tamura, Y., Iizuka, Y., Ohashi, K., *et al.* (2001). Identification of liver X receptor-retinoid X receptor as an activator of the sterol regulatory element-binding protein 1c gene promoter. *Mol. Cell. Biol.* **21,** 2991–3000.

Yoshikawa, T., Shimano, H., Yahagi, N., Ide, T., Amemiya-Kudo, M., Matsuzaka, T., Nakakuki, M., Tomita, S., Okazaki, H., Tamura, Y., Iizuka, Y., Ohashi, K., Takahashi, A., Sone, H., Osuga Ji, J., Gotoda, T., Ishibashi, S., and Yamada, N. (2002). Polyunsaturated fatty acids suppress sterol regulatory element-binding protein 1c promoter activity by inhibition of liver X receptor (LXR) binding to LXR response elements. *J. Biol. Chem.* **277**(3), 1705–1711.

BRASSINOSTEROIDS

PLANT COUNTERPARTS TO ANIMAL STEROID HORMONES?

STEVEN D. CLOUSE

Department of Horticultural Science, North Carolina State University, Raleigh, North Carolina 27695

I. Introduction
II. Biosynthesis
   A. Mevalonate to Squalene 2,3-Epoxide
   B. Cycloartenol to Campesterol
   C. Campesterol to Brassinolide
III. Physiological Responses
   A. Cell Expansion
   B. Cell Division
   C. Cell Differentiation and Organ Morphogenesis
IV. Signal Transduction
   A. BR Receptor
   B. Downstream Components of BR Signal Transduction
V. Conclusion and Future Prospects
   References

Brassinosteroids are polyhydroxylated derivatives of common plant membrane sterols such as campesterol. They occur throughout the plant kingdom and have been shown by genetic and biochemical analyses to be essential for normal plant growth and development. Numerous reviews have detailed the recent progress in our understanding of the biosynthesis, physiological responses, and molecular modes of action of brassinosteroids. It is clear that like their animal steroid counterparts, brassinosteroids have a defined receptor, can regulate the expression of specific genes, and can orchestrate complex physiological responses involved in growth. This review summarizes the current status of BR research, pointing out where appropriate the similarities and differences between the mechanism of action of brassinosteroids and the more thoroughly studied animal steroid hormones. © 2002, Elsevier Science (USA).

I. INTRODUCTION

Plants contain a variety of compounds derived from diverse biosynthetic pathways that are critical signals in regulating reproduction, embryogenesis, seed germination, vegetative growth, morphogenesis, and adaptation to the environment. By analogy with mammalian systems, these plant compounds are referred to as "hormones" because of their ability to profoundly affect physiological processes at concentrations well below nutrients and vitamins, and by their ability to initiate altered expression of specific genes and the posttranslational modification of enzymes and structural proteins (Davies, 1995). Contrary to the norm in animals, plants do not have special endocrine glands for the production of these hormones, and often plant hormones are synthesized in the same tissue or cell in which the hormone acts. Until recently plant biologists believed that the growth regulatory effects of plant hormones could be accounted for by five classes of compounds consisting of abscisic acid, auxins, cytokinins, ethylene, and gibberellins. During the past 5 years, however, molecular genetic and biochemical analysis of a number of dwarf mutants in *Arabidopsis thaliana,* pea, tomato, and rice (Clouse and Feldmann, 1999; Yamamuro *et al.,* 2000), has provided convincing evidence that a sixth class of plant hormones, termed brassinosteroids (BRs), is just as essential as the original five in promoting normal plant growth and development.

A number of plants have been shown to contain some of the same steroids as animal systems, including ecdysteroids, androgens, estrogens, and corticosteroids (Geuns, 1978). However, among plant steroids only BRs appear to be ubiquitously distributed in the plant kingdom while playing essential roles in modulating the expansion, division, and differentiation of cells at nanomolar to micromolar concentrations (Clouse and Sasse, 1998). BR research sprang from work beginning in the 1940s at the United States Department of Agriculture (USDA), in which organic extracts of pollen from over 60 species were assayed for new plant growth-promoting properties (Steffens, 1991). An extract from *Brassica napus* pollen proved to have particularly pronounced effects on cell elongation and division in bioassays and promoted overall growth when young seedlings were sprayed in greenhouse experiments (Mitchell *et al.,* 1970). The active component of the *B. napus* extract was isolated from 227 kg of *Brassica* pollen and determined to be a novel plant steroid (Grove *et al.,* 1979), which was named brassinolide (BL).

BL structure was determined by single-crystal X-ray analysis to be a polyhydroxylated derivative of 5α-cholestane, namely $(22R,23R,24S)$-$2\alpha,3\alpha,22,23$-tetrahydroxy-24-methyl-B-homo-7-oxa-5α-cholestan-6-one (Fig. 1). Thus plants

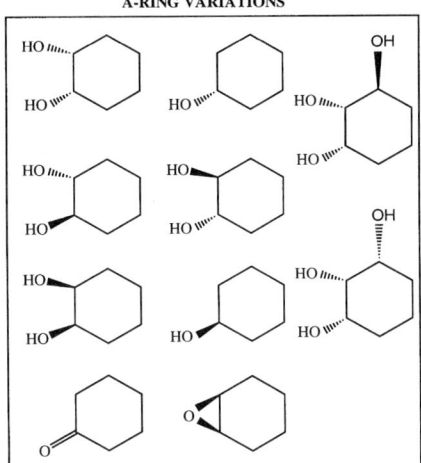
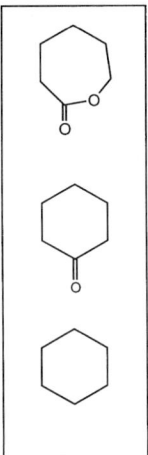

FIGURE 1 Structure of naturally occurring brassinosteroids. The structure of brassinolide is presented with possible variations in the steroid ring and side chain shown in boxes. Over 40 compounds have been identified in plants with different combinations of the variations shown.

produce a growth-promoting steroid with structural similarity to animal steroid hormones, including androgens, estrogens, and corticosteroids from vertebrates, and ecdysteroids from insects and crustacea. In addition to BL there are over 40 related BRs that have been characterized in tissues of at least 37 angiosperms (28 dicots and 9 monocots), 5 gymnosperms, 1 alga, and 1 pteridophyte (Fujioka, 1999). Structural variations in BRs occur primarily at C-2 and C-3 in the A ring, at C-6 in the B ring (where either a lactone, ketone, or deoxo function is found), in the side chain stereochemistry, and by the presence or absence of a methyl (methylene) or ethyl (ethylene) group at C-24. Many of the characterized BRs are biosynthetic precursors or metabolic products of BL, although some have independent biological activity without further conversion. BL, however, shows the highest biological activity in numerous bioassays suggesting that a lactone function at C-6/C-7, *cis*-vicinal hydroxyls at C-2 and C-3, *R* configuration of the hydroxyls at C-22/C-23, and a methyl substitution at C-24 are required for optimal BR activity (Mandava, 1988).

Endogenous levels of BRs vary from less than 0.1 to more than 100 ng/g fresh weight, depending on plant organ type, tissue age, and species (Fujioka, 1999). Pollen and immature seeds contain the highest levels and young, growing shoots have greater BR concentrations than mature tissue, which is not surprising considering the greater physiological response of immature tissue to BR (Mandava, 1988). Exogenous BR application to a variety of plant tissues, including epicotyls, hypocotyls, and peduncles in dicots, and coleoptiles and mesocotyls of monocots, has a dramatic positive effect on stem elongation (Mandava, 1988). BR treatment also stimulates tracheary element differentiation (Clouse and Zurek, 1991; Fukuda, 1997), increases rates of cell division (Hu *et al.,* 2000; Oh and Clouse, 1998), accelerates senescence, causes hyperpolarization of membranes, stimulates ATPase activity, alters the orientation of cortical microtubules, and mediates the effect of abiotic and biotic stresses (Sasse, 1999).

Many of these physiological responses, which were first observed by exogenous application of BRs to plant model systems, were later verified *in planta* by examination of the phenotypes of BR-deficient and -insensitive mutants. *Arabidopsis* BR mutants show a characteristic phenotype in the light, which includes dwarf stature, dark green, rounded leaves, prolonged life-span, reduced male fertility, and altered vascular development. In the dark, they exhibit some of the features of light-grown plants including shortened hypocotyls and open cotyledons. In BR-deficient mutants, these phenotypic aberrations are rescued to wild type only by exogenous application of BRs (Altmann, 1999; Clouse and Feldmann, 1999).

II. BIOSYNTHESIS

A. MEVALONATE TO SQUALENE 2,3-EPOXIDE

BRs, like their animal steroid counterparts, are products of the isoprenoid biosynthetic pathway originating with acetyl-Coenzyme A (CoA) and proceeding via mevalonate, isopentenyl pyrophosphate, geranyl pyrophosphate, and farnesyl

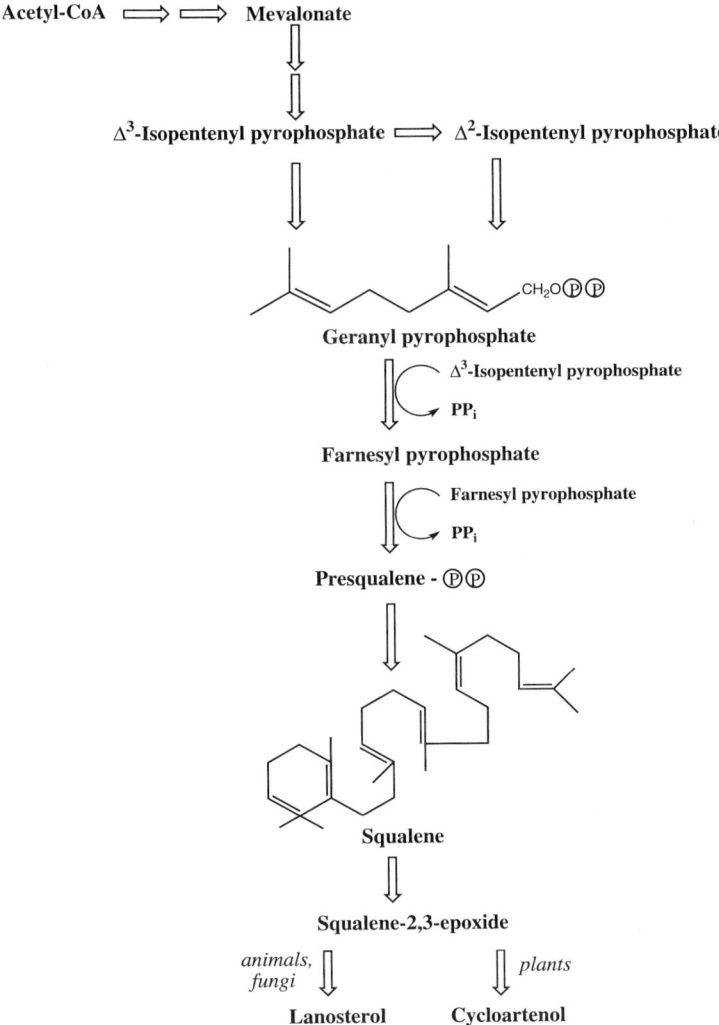

FIGURE 2 Biosynthetic pathway of the plant sterol precursor, cycloartenol, from mevalonate. Animals, fungi, and plants share a common pathway up to squalene-2,3-epoxide.

pyrophosphate (Fig. 2). Squalene is produced by condensation of two farnesyl pyrophosphate molecules, which is then converted to squalene-2,3-epoxide. It is generally accepted that the steps from mevalonate to squalene-2,3-epoxide are conserved between animals and plants, but the conversion of squalene-2,3-epoxide to sterol progenitors differs. In animals and fungi squalene-2,3-epoxide is converted to lanosterol, the precursor of cholesterol and the animal steroid hormones, whereas in plants it is converted to cycloartenol, the parent compound of all plant sterols, including campesterol, stigmasterol, and sitosterol (Benveniste, 1986).

Initial studies of BR biosynthesis employed cell suspension cultures of *Catharanthus roseus* fed with deuterated and tritiated putative biosynthetic intermediates (Fujioka and Sakurai, 1997; Sakurai, 1999). Conversion of the labeled compounds was monitored with sensitive techniques of gas chromatography–mass spectrometry (GC-MS) and a skeletal pathway was deduced in which BL is derived from campesterol via ring reduction, epimerization and oxidation, and side chain hydroxylations. Moreover, it was found that C-6 oxidation could occur before (early C-6 oxidation pathway) or after (late C-6 oxidation pathway) hydroxylation of the side chain. The discovery of BR and sterol-deficient mutants in *Arabidopsis* greatly refined the level of understanding of BL biosynthesis and its relatedness to animal steroid hormone biosynthesis. The general strategy of this approach combined genetic and biochemical analysis in which endogenous BR and sterol levels were measured in mutant and wild-type plants, followed by feeding experiments with labeled intermediates to identify putative positions in the pathway that were affected by the mutation. Cloning and sequencing revealed that the mutations were lesions in genes encoding proteins with significant homology to enzymes involved in animal sterol or steroid biosynthesis, and the use of recombinant proteins and labeled substrates confirmed that the respective enzymes were capable of catalyzing the step in the pathway that was blocked in that mutant.

B. CYCLOARTENOL TO CAMPESTEROL

BL biosynthesis can be divided into two major subgroups: general sterol synthesis (cycloartenol to campesterol) and a BR-specific pathway leading from campesterol to BL. Beyond their role as BR precursors, plant sterols serve as integral membrane components, regulating the fluidity and permeability of membranes and directly affecting the activity of membrane-associated proteins (Hartmann, 1998). Furthermore, recent evidence from animal studies suggests that sterols can themselves serve as ligands for specific nuclear receptors and can also interact with other transcription complexes to modulate signal transduction pathways (Edwards and Ericsson, 1999).

The conversion of cycloartenol to 24-methylenecycloartenol is accomplished via *S*-adenosylmethionine-dependent transmethylation, catalyzed by sterol methyltransferase 1 (SMT1). The methyl (and sometimes ethyl) substitutuents at C-24 are the only sterol carbons not derived from the acetate–mevalonate pathway. SMT1 shares up to 43% sequence identity with fungal sterol methyl transferases, including Erg6p, a C-24 methylase from *Saccharomyces cerevisiae* involved in the conversion of lanosterol to ergosterol (Diener *et al.*, 2000). Arabidopsis SMT1 suppresses the defects of the *erg6* mutant, suggesting a functional equivalence for the yeast and plant reductases (Husselstein *et al.*, 1996).

SMT1 expression is strongest in actively growing regions of the plant, particularly the shoot apex, and also is observed in specific stages and tissues of embryogenesis (Diener *et al.*, 2000). *Arabidopsis* mutants affecting SMT1 show some of the characteristics of BR-deficient mutants in adult plants, but cannot be rescued to wild type by BR treatment. Cholesterol accumulates in the *smt*1 mutant at the expense of sitosterol, whereas other sterol levels are relatively normal. Moreover, *smt*1 shows unique defects in embryogenesis not previously seen in BR mutants or in sterol biosynthetic mutants later in the pathway (Diener *et al.*, 2000).

The next step in plant sterol biosynthesis for which a cloned gene is available is the C-14 reduction of 4α-methyl-5α-ergosta-8,14,24(28)-trien-3β-ol to 4α-methylfecosterol, catalyzed by an integral membrane sterol C-14 reductase, which in *Arabidopsis* is encoded by the *FACKEL* gene (Jang *et al.*, 2000; Schrick *et al.*, 2000). Plant C-14 sterol reductases are 34–38% identical (53–56% similar) to the sterol reductase domain of lamin B receptors in human, *Xenopus*, rat, and chicken, and 28–39% identical (43–54% similar) to a putative sterol reducatase in human (SR-1) and a variety of C-14 sterol reductases from fungi (Schrick *et al.*, 2000). The catalytic domain of the sterol reductases appears to be conserved among eukaryotes and expression of the *FACKEL* gene in the *erg*24 mutant of *S. cerevisiae*, complements the defect in C-14 sterol reductase activity. Furthermore, *fackel* accumulates three unusual 8,14-diene sterols, similar to plant cell suspension cultures treated with the known C-14 reductase inhibitor, 15-azasterol (Schrick *et al.*, 2000).

The *fackel* mutant was first identified in a screen for mutations affecting seedling body organization (Mayer *et al.*, 1991), and further alleles were uncovered in a genetic screen for constitutive cytokinin response mutants (Jang *et al.*, 2000). The *fackel* mutant shares many of the phenotypic charactersitics of *smt*1, including severely disrupted embryogenesis, BR deficiency, and lack of rescue by exogenous BR application. An intriguing explanation for the observation that BR cannot rescue these early sterol biosynthetic mutants, although it does so with mutants later in the sterol pathway, is that specific sterols are generated early in the pathway that themselves act as signaling molecules regulating embryogenesis (Jang *et al.*, 2000; Schrick *et al.*, 2000). These putative sterol signals would be disrupted in the *fackel* and *smt*1 mutants, but not those downstream in the sterol pathway. Arguments supporting this hypothesis have been summarized in a recent review (Clouse, 2000).

The next step in the biosynthetic pathway to campesterol for which cloned genes and mutants are available involves the introduction of a C-5 double bond in the B ring of episterol to yield 5-dehydroepisterol (Fig. 3). The reaction involves a Δ^7-C-5-desaturase, NADH, cytochrome b_5 reductase, and cytochrome b_5, and has been documented in plants (Rahier *et al.*, 1997), fungi (Osumi *et al.*, 1979), and mammals (Kawata *et al.*, 1985). In Arabidopsis the Δ^7-C-5-desaturase is encoded by the *DWF7/STE*1 gene whereas in yeast it is encoded by *ERG*3. Mutations in *DWF7/STE*1 result in many of the phenotypic features of BR-deficient mutants (although not as severe), which are rescued by BR treatment (Choe *et al.*, 1999b). These mutants also accumulate Δ^7 sterols with a decrease in the corresponding Δ^5 sterols such as 24-methylenecholesterol, campesterol, campestanol, and all downstream BRs. Transgenic mutant plants expressing the yeast *ERG*3 gene regain the normal sterol profile (Gachotte *et al.*, 1995). Conversely, the *erg*3 mutant of

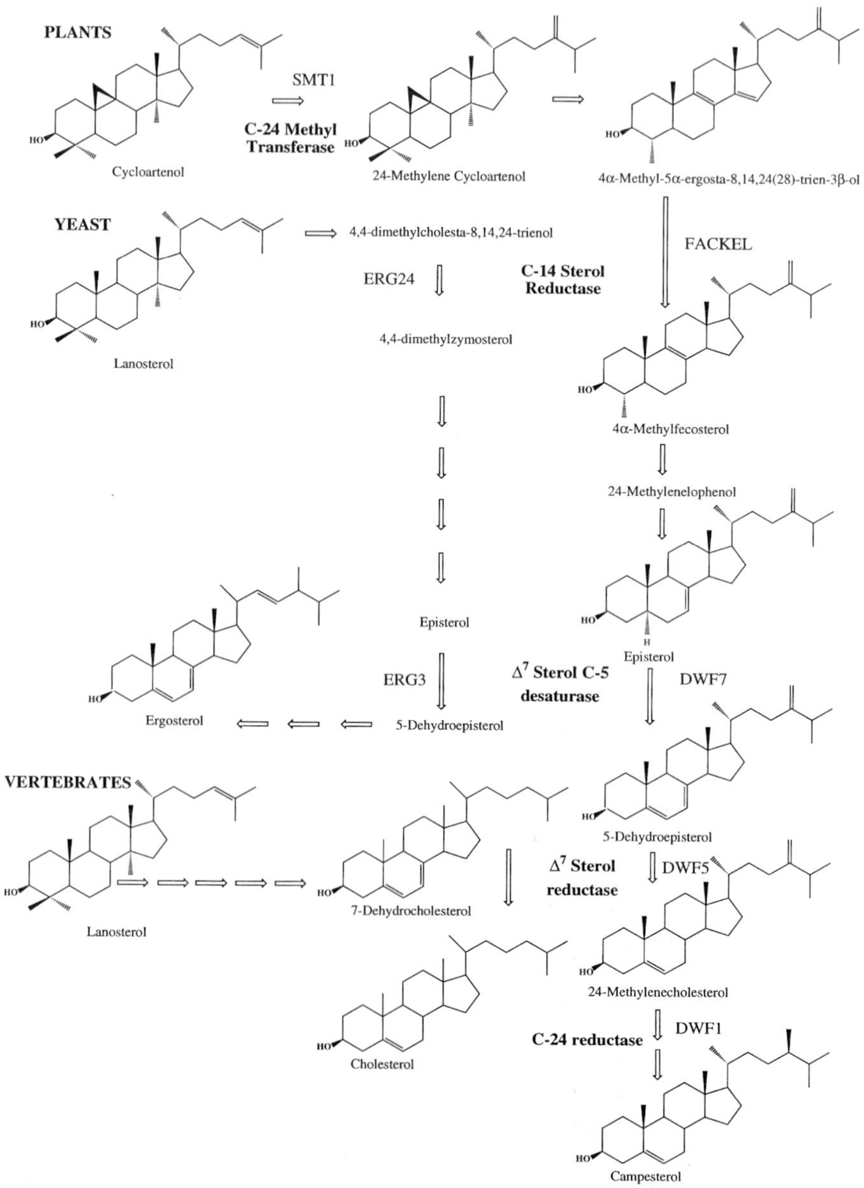

FIGURE 3 Biosynthetic pathways of sterols from cycloartenol and lanosterol. Protein names for steps in which cloned genes and mutants are available in *Arabidopsis* and yeast are indicated in capital letters, and the corresponding enzyme functions are presented in bold. Functional equivalence of several of the enzymes among plants, yeast, and vertebrates has been demonstrated.

yeast, which is defective in Δ^7-C-5-desaturase activity, regains the capacity to synthesize ergosterol when transformed with the *DWF7/STE1* gene (Husselstein *et al.,* 1999). The *STE1/DWF7* gene is 38% identical (50% similar) to the yeast *ERG3* gene and 35% identical (47% similar) to a human ortholog, which also rescues the *erg3* mutant (Husselstein *et al.,* 1999). All of these Δ^7-C-5-desaturases have conserved transmembrane domains and histidine clusters that are required for activity and they also share sequence similiarity with fatty acid desaturases involved in phospholipid and galactolipid biosynthesis (Ohlrogge and Browse, 1995). A putative DWF7/STE1 homolog with 80% amino acid identity to the original gene has been cloned in *Arabidopsis,* suggesting that duplicate genes might regulate this step. Such redundancy could account for the less severe phenotype of the *dwf7/ste1* mutants when compared to mutants in the BR-specific pathway (Choe *et al.,* 1999b).

The next two steps in the pathway involve reductions in the B ring and the side chain. Conversion of 5-dehydroepisterol to 24-methylenecholesterol is catalyzed by a Δ^7-sterol reductase, encoded by the *dwf5* gene of *Arabidopsis* (Choe *et al.,* 2000). The phenotypes of *dwf5* mutants are similar to *dwf7/ste1* and other BR-deficient mutants, but have some unique characteristics. The *dwf5-1* allele has increased fertility compared to other BR and sterol-deficient mutants, but also shows abnormal seeds with reduced germination rates. Moreover, *dwf5-1* does not show the prolonged life cycle typical of most BR mutants (Choe *et al.,* 2000). Like the sterol C-14 reductase discussed above, DWF5 shows many of the conserved amino acid motifs found in yeast, vertebrate, and chicken sterol reductases. For example, the human Δ^7-sterol reductase responsible for the conversion of 7-dehydrocholesterol to cholesterol shares 35% identity and 60% similarity to DWF5 (Moebius *et al.,* 1998). However, overexpression of the human reductase in the *dwf5* mutant did not result in complete rescue to wild type, althought the transgenic plants were somewhat larger than the mutant (Choe *et al.,* 2000).

The completion of campesterol biosynthesis involves the isomerization and reduction of 24-methylenecholesterol to campesterol by a C-24 sterol reductase. In *Arabidopsis* this is encoded by *DWF1*, an oxidoreductase with a flavin adenine dinucleotide-binding domain and a membrane anchoring region (Choe *et al.,* 1999a; Takahashi *et al.,* 1995). A majority of the known *dwf1* mutant alleles contain lesions in the FAD domain, suggesting a critical role for FAD-binding in DWF1 function. Subsequent experiments utilizing a green fluorescent protein–DWF1 fusion confirmed that DWF1 is an integral membrane protein, as expected for most sterol and steroid biosynthetic enzymes (Klahre *et al.,* 1998).

C. CAMPESTEROL TO BRASSINOLIDE

Conversion of campesterol to BL begins via a series of four reactions that result in the reduction of the C-5 double bond, yielding campestanol (Fig. 4). One of these steps, the conversion of (24*R*)-24-methylcholest-4-en-3-one to (24*R*)-24-methyl-5α-choestan-3-one has been characterized in some detail in *Arabidopsis*

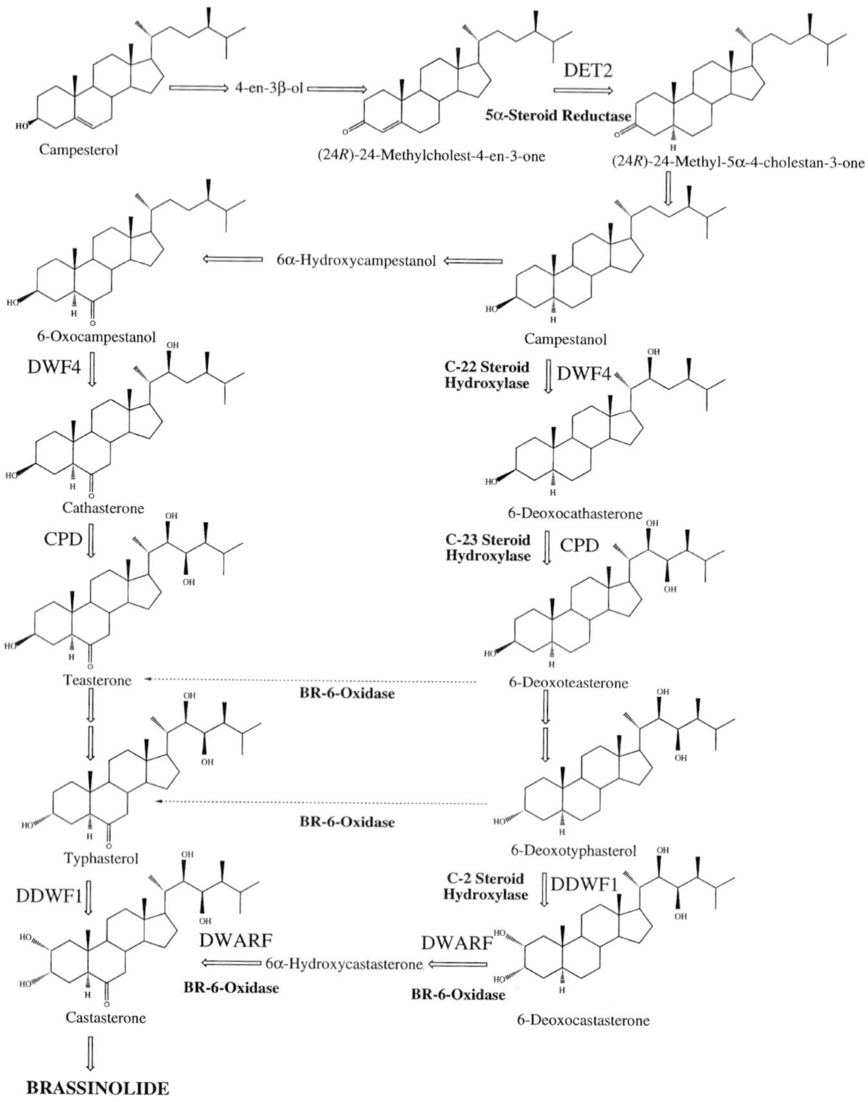

FIGURE 4 The biosynthetic pathway from campesterol to brassinolide. Protein names for steps in which cloned genes and mutants are available in *Arabidopsis* (DET2, DWF4, CPD), tomato (DWARF), and pea (DDWF1) are indicated in capital letters, and the corresponding enzyme functions are presented in bold. The pathway on the left is termed early C-6 oxidation and that on the right, late C-6 oxidation. BR-6-oxidase has been shown to convert 6-deoxocastasterone and 6-deoxotyphasterol to the corresponding 6-oxo compounds in *Arabidopsis,* whereas the conversion of 6-Deoxoteasterone has been demonstrated so far in recombinant yeast cells only.

(Noguchi et al., 1999b). The reaction is catalyzed by the *DET2* gene product, a protein with significant sequence identity (38–42%) to mammalian 5α-steroid reductases involved in the NADPH-dependent reduction of testosterone to dihydrotestosterone (Li *et al.*, 1996). The mammalian and *Arabidopsis* reductases share 80% of the known conserved residues among 5α-reductases, including a glutamate that is required for mammalian enzyme activity and that is changed to a lysine in the *det*2-1 and *det*2-6 mutant alleles. The full functional equivalence of the plant and mammalian enzymes has been convincingly demonstrated. When expressed in human embryonic kidney cells, recombinant DET2 protein reduced 3-oxo, $\Delta^{4,5}$ mammalian steroids, such as testosterone and progesterone, and this activity was inhibited by 4-MA, a competitive inhibitor of mammalian 5α-reductases (Li *et al.*, 1997). Moreover, overexpression of human 5α-reductases in transgenic *det*2 plants resulted in wild-type phenotype without BR application, a phenomenon again inhibited by 4-MA.

Although *det*2 has a typical BR mutant phenotype that can be rescued by BR treatment, it is not as severely affected as the *cpd* and *bri*1 mutants described below. This might result from about 10% residual campestanol present in the *det*2 mutant (Fujioka *et al.*, 1997), which, in turn, could arise from a redundant reductase catalyzing the same reaction, as has been proposed for *dwf*7 (Choe *et al.*, 1999b).

After campestanol, the BL biosynthetic pathway diverges into parallel pathways that differ only by whether C-6 oxidation occurs before or after side chain hydroxylation (Yokota, 1997). The endogenous co-occurrence of 6-deoxo and 6-oxo-BRs has been demonstrated in many plants where BL is also present including bean, wheat, rice, rye, Arizona cypress, *C. roseus,* and *Arabidopsis* (Choi *et al.*, 1997; Fujioka *et al.*, 1996; Fujioka and Sakurai, 1997; Griffiths *et al.*, 1995), suggesting that these pathways operate simultaneously. Feeding deuterium-labeled substrates to seedlings followed by GC-MS verified that both of these pathways are fully functional in *Arabidopsis* (Noguchi *et al.*, 2000), although 6-deoxo intermediates occur at much higher levels than the corresponding 6-oxo intermediates. Some plants, such as tomato, apparently do not synthesize 6-oxo intermediates at all (Bishop *et al.*, 1999; Koka *et al.*, 2000). Interconversion of intermediates at several points in the two pathways has been demonstrated in *Arabidopsis* seedlings (Noguchi *et al.*, 2000; Shimada *et al.*, 2001), suggesting BL biosynthesis is more of a metabolic grid than two independent linear pathways of early and late C-6 oxidation.

Side chain hydroxylation at C-22 and C-23 during BL biosynthesis is accomplished successively by the products of the *DWF*4 and *CPD* genes, which encode cytochrome P-450 oxidases with sequence similarity to mammalian steroid hydroxylases and each other (Choe *et al.*, 1998; Szekeres *et al.*, 1996). The phenotype of *dwf*4 mutants resembles that of other BR-deficient mutants and only 22α-hydroxylated intermediates in BL biosynthesis, and synthetic compounds such as 22-hydroxycampesterol, rescue *dwf*4 to wild type (Azpiroz *et al.*, 1998; Choe *et al.*, 1998). The *cpd* mutant is an extreme dwarf, is male sterile, and has disorganized vascular tissue. These defects are rescued only by 23α-hydroxylated BRs, indicating that CPD acts as a C-23 steroid hydroxylase (Szekeres *et al.*, 1996).

Assuming that levels of intermediates should be elevated immediately prior to a control point, then both the reactions catalyzed by DWF4 and CPD are good candidates for rate-limiting steps in BL biosynthesis. The DWF4 substrate, campestanol, is present in 500-fold excess over the product 6-deoxocathasterone, which in turn occurs at 20-fold higher levels than the CPD product 6-deoxoteasterone (Nomura *et al.*, 2001). Regulation of transcript and protein levels of enzymes involved in rate-limiting steps of BL biosynthesis would be expected to have the greatest impact on modulation of BL levels in response to environmental and internal signals. Indeed, transcription of the *CPD* gene is specifically down-regulated by BL in both dark and light (Mathur *et al.*, 1998) and *DWF4* transcript levels are low in wild type but increase dramatically in the BR-deficient *dwf*1 and *cpd* mutants, suggesting derepression in the absence of BR (Choe *et al.*, 1998). Increased *DWF4* expression is also seen in the BR-insensitive mutant, *bri*1, indicating that active BR signaling is required for repression of *DWF4* in wild type *Arabidopsis*. The importance of BR signaling in BL homeostasis is confirmed by the observation that *bri*1 mutants accumulate much higher levels of BRs than wild type (Noguchi *et al.*, 1999a), accompanied by increased metabolic flow through the pathway (Noguchi *et al.*, 2000).

Mutants affecting steps after CPD in the *Arabidopsis* BL biosynthetic pathway have not been definitively identified, although it is possible that the *dwf*8 mutant blocks the hydroxylation of typhasterol to castasterone (Clouse and Feldmann, 1999). In pea, the product of the *ddwf*1 (dark-induced *dwf*-like 1) gene, another membrane-bound cytochrome P-450 monooxygenase, has been shown to catalyze this conversion (Kang *et al.*, 2001). The function of DDWF1 as a C-2 hydroxylase involved in BR biosynthesis was suggested by the ability of recombinant DDWF1 to hydroxylate typhasterol to castasterone *in vitro* and by *in vivo* experiments involving sense and antisense transgenic plants in tobacco and *Arabidopsis* that either over or underexpressed *ddwf*1. Measurement of endogenous BR levels and feeding experiments with BR biosynthetic intermediates strongly suggested that DDWF1 was the C-2 hydroxylase responsible for conversion of typhasterol to castasterone and 6-deoxotyphasterol to 6-deoxocastasterone (Kang *et al.*, 2001).

The penultimate step in BL biosynthesis has been characterized in tomato. The tomato DWARF gene encodes a cytochrome P-450 monooxygenase that catalyzes the oxidation of 6-deoxocastasterone to castasterone via 6-hydroxycastasterone (Bishop *et al.*, 1999). A putative homologue of the tomato gene, *AtBR6ox*, has been identified in *Arabidopsis*. The 466 amino acid protein shares 68% identity (81% similarity) to the tomato DWARF cytochrome P-450, and is more closely related to the tomato gene than other *Arabidopsis* P-450s in the BL pathway such as DWF4 and CPD. Recombinant AtBR6ox is a steroid-6-oxidase that converts not only 6-deoxocastasterone to castasterone, but also 6-deoxotyphasterol to typhasterol, 3-dehydro-6-deoxoteasterone to 3-dehydroteasterone, and 6-deoxoteasterone to teasterone (Shimada *et al.*, 2001).

III. PHYSIOLOGICAL RESPONSES

A. CELL EXPANSION

Cell growth in plants and animals differs dramatically due to the presence in plants of a rigid, complex cell wall that must transiently yield to allow expansion of the protoplast. Plant cell expansion is critical for growth and differentiation in all organs and is the result of changes in gene expression and biochemical processes that lead to alterations in wall mechanical properties, cell hydraulics, and osmotic potential (Cosgrove, 1997). Regulation of the synthesis and activity of wall-modifying enzymes is one mechanism by which plant hormones such as BRs, auxins, and gibberellins promote cell elongation (Clouse, 1997; Clouse and Feldmann, 1999). Many of the known BR-regulated genes encode proteins that affect cell walls (Catala *et al.*, 1997; Koka *et al.*, 2000; Oh *et al.*, 1998; Zurek and Clouse, 1994) and BRs have been shown by biophysical measurements to alter wall mechanical properties in a number of stem tissues (Tominaga *et al.*, 1994; Wang *et al.*, 1993; Zurek *et al.*, 1994). Such temporary "loosening" of the rigid wall allows turgor-driven expansion to proceed and is accompanied by synthesis of new wall components to prevent thinning and weakening of the wall in the expanding cell.

The dwarf stature of BR-deficient mutants and the specific ability of BRs to rescue this defect provide convincing genetic evidence that BRs are essential for normal plant cell expansion. Moreover, light and electron microscopic examination of cell files in wild-type *Arabidopsis* and *cbb*, *dwf*4, *cpd*, and *dwf*1 BR-deficient mutants provides direct physical evidence that longitudinal cell expansion is greatly reduced in the absence of BR (Azpiroz *et al.*, 1998; Kauschmann *et al.*, 1996; Szekeres *et al.*, 1996; Takahashi *et al.*, 1995). Further *in vivo* evidence that BRs promote cell elongation comes from overexpression studies of BR biosynthetic enzymes in transgenic plants. Transgenic *Arabidopsis* overexpressing *DWF4* had higher levels of BR and longer hypocotyls than wild-type plants (Choe *et al.*, 2001).

The arrangement of cortical microtubules surrounding the plant cell is known to be important in determining the direction of cell expansion. BRs have been shown to effectively reconfigure microtubules to a transverse orientation, thus allowing longitudinal growth to proceed (Mayumi and Shibaoka, 1995). The *bul*1 mutant of *Arabidopsis* is a BR-deficient dwarf that is allelic to *dwf*7, which, as described above, is blocked in the Δ^7-C-5-desaturation step of campesterol biosynthesis (Catterou *et al.*, 2001b). Indirect immunofluorescence of α-tubulin in *bul*1 revealed that microtubule levels and organization were severely disrupted in the mutant. Treatment with BR restored microtubule architecture to that of wild-type *Arabidopsis* and rescued the cell elongation defects without activating tubulin gene expression (Catterou *et al.*, 2001a). Thus, a direct, nongenomic effect of BR on tubulin organization and cell elongation can be inferred.

Besides changes in cell wall properties, alterations in ion and water transport also affect elongation and BRs may be involved in these processes as well. Vacuolar

H^+-ATPases (V-ATPases) are proton pumps that acidify endomembrane compartments and affect solute uptake into the vacuole (Forgac, 1999). Transgenic carrot cell lines expressing antisense transcripts of subunit A of the V-ATPase are defective in cell expansion, suggesting a functional role for the complex in elongation (Gogarten et al., 1992). The det3 mutant of Arabidopsis exhibits organ-specific defects in elongation and a reduced response to BRs (Schumacher et al., 1999). The DET3 gene encodes subunit C of the V-ATPase, which is required for proper assembly of the complex. DET3 expression is tightly correlated with tissues undergoing rapid expansion, and it has been proposed that the BR signal transduction pathway regulates V-ATPase assembly by activating the C subunit (Schumacher et al., 1999).

Aquaporins contribute to water permeability in both plant and animal cells and facilitate water transport across membranes (Chrispeels and Agre, 1994). Measurements of osmotic permeability in protoplasts derived from wild-type Arabidopsis and BR-deficient and insensitive mutants, suggested that BL may control aquaporin activity (Morillon et al., 2001). Thus BRs affect all of the major processes involved in plant cell expansion, including osmotic potential, cell wall properties, and microtubule organization.

B. CELL DIVISION

Like their animal steroid counterparts, BRs are also known to modulate cell division. Auxins and cytokinins are required for normal cell division both *in planta* and in cell and tissue culture. BRs have also been shown to stimulate cell division in cultured parenchyma cells of *Helianthus tuberosus* (Clouse and Zurek, 1991) and in protoplasts of Chinese cabbage and petunia (Nakajima et al., 1996; Oh and Clouse, 1998). We have recently shown that BL affects the kinetics of the cell cycle in synchronized cell cultures of tobacco and also regulates the expression of genes associated with specific phases of the cell cycle, including histone H_2B and High Mobility Group-1 protein (J. Jiang and S. D. Clouse, unpublished).

The eukaryotic cell cycle can be divided into four phases, consisting of G_1, S (DNA replication), G_2, and M (karyo- and cytokinesis). Regulation of progression through the cell cycle has been highly conserved between mammals and plants and involves cyclins and their associated cyclin-dependent kinases, as well as the phosphorylation status of the retinoblastoma protein (Stals and Inze, 2001). D-type cyclins are critical for progression from G_1 to S, and estrogen promotes cell proliferation in ovary tissue in part by increasing transcript levels of cyclin D2 (Robker and Richards, 1998). Interestingly, a promotive role for BRs in *Arabidopsis* cell division has been implicated by the discovery (Hu et al., 2000) that BR treatment of det2 cell suspension cultures increases transcript levels of the gene encoding cyclin D3 (CycD3). CycD3 is also regulated by cytokinins, but BR effectively substitutes for cytokinin in the growth of *Arabidopsis* callus and cell suspension cultures (Hu et al., 2000). Thus, both plant and animal steroid hormones appear to regulate the cell cycle in part by affecting cyclin gene expression.

C. CELL DIFFERENTIATION AND ORGAN MORPHOGENESIS

In addition to elongation and division, differentiation into specific cell types is the hallmark of multicellular eukaryotic development, and again steroids play a pivotal role in signaling these events. Vascular plants have specific water- and nutrient-conducting tissues called the xylem and the phloem. Differentiation of xylem tissue involves a highly complex program of altered gene expression and biochemical events leading to secondary wall modifications and loss of cytoplasm. The resulting cells are hollow, water-conducting tracheary elements (Fukuda, 1997).

Auxins and cytokinins are required for vascular differentiation in plants, but substantial evidence also supports the role of BRs in this process. For example, nanomolar levels of BL stimulate tracheary element formation in *H. tuberosus* explants and isolated mesophyll cells of *Zinnia elegans* (Clouse and Zurek, 1991; Iwasaki and Shibaoka, 1991). In *Zinnia,* BRs also regulate the expression of several genes required for xylem formation (Yamamoto *et al.,* 1997), and measurements of endogenous hormone levels by GC-MS reveals dramatic increases in BR levels immediately prior to tracheary element differentiation (Yamamoto *et al.,* 2001). High levels of expression of a BR-regulated gene in paratracheary parenchyma cells surrounding vessel elements in soybean stems also suggest a role for BRs in xylem formation (Oh *et al.,* 1998). In *Arabidopsis,* the BR-deficient mutant *cpd* exhibits unequal division of the cambium, producing extranumerary phloem cell files at the expense of xylem cells (Szekeres *et al.,* 1996), whereas the sterol and BR-deficient mutant *dwf7* shows a similar increase in phloem versus xylem cells with a reduction in the total number of vascular bundles (Choe *et al.,* 1999b).

In animals the overall body plan and specific organs types are established in embryogenesis. In contrast, plants continue to form organs such as leaves throughout their life cycle from groups of proliferating cells termed meristems (Irish and Jenik, 2001). Specific cells within the shoot apical meristem are responsible for leaf initiation and defects in leaf development are characteristic of BR mutants in *Arabidopsis* and tomato (Clouse and Feldmann, 1999). Interestingly, recent work in tomato has shown that transcripts of the *DWARF* gene, encoding the cytochrome P-450 responsible for 6-deoxocastasterone to castasterone conversion during BL biosynthesis, accumulate in the meristem specifically within cells initiating the new leaf (Pien *et al.,* 2001). Although not yet established by direct measurements, this suggests the possibility of cell-specific BL biosynthesis in the leaf primordium.

Light quality, duration, and intensity profoundly influence plant morphogenesis. Dark-grown seedlings have elongated stems with a pronounced apical hook and contain undifferentiated chloroplast precursors. Upon exposure to light, the rate of stem elongation slows dramatically, the apical hook opens, and true leaves with mature chloroplasts develop (Chory *et al.,* 1996). The integration of light and hormone signaling pathways is critical for proper seedling development. Numerous BR-deficient mutants in *Arabidopsis,* pea, and tomato show defects in cell expansion in the dark, and depending on the severity of the mutant allele and the species, some of these mutants also show other defects in photomorphogenesis (Clouse and Feldmann, 1999). Recent work in pea has shown that the dark-inducible, light-repressible small G protein, Pra2, interacts with and activates Ddwf1, the cytochrome P-450 C-2 hydroxylase involved in BL biosynthesis (Kang et al., 2001). Thus, a novel link between light signal transduction and the endogenous levels of BR has likely been established.

BRs have been shown to influence numerous other physiological processes in plants, including carbohydrate partitioning, adaptation to various stresses, and affects on reproductive biology and senescence. These topics have been recently reviewed elsewhere (Altmann, 1999; Clouse and Sasse, 1998; Sasse, 1999) and will not be considered further here.

IV. SIGNAL TRANSDUCTION

A. BR RECEPTOR

The critical importance of steroid hormones in animal development and adult homeostasis and the resultant pathology ensuing from defects in biosynthesis and signal transduction of these hormones have promoted decades of research on steroid physiology, biochemistry, and molecular biology. In the classic model of steroid hormone action, an intracellular receptor, localized either in the nucleus or in the cytoplasm (complexed with heat shock proteins and immunophilins), binds the steroid ligand and becomes competent to recognize palindromic sequences in the promoters of steroid-responsive genes (Aranda and Pascual, 2001). The steroid receptor–ligand complex binds to DNA as a homodimer and interacts with coactivators or cosuppressors to regulate the transcriptional state of the target gene (Robyr et al., 2000).

Besides steroid receptors, which recognize androgens, estrogens, progesterone, ecdysones, glucocorticoids, and mineralocorticoids, a large superfamily of related nuclear receptors binds ligands such as retinoic acid, vitamin D, thyroid hormone, cholesterol derivatives, and pregnanes. All members of the superfamily have a related structure that consists of a variable N-terminal domain often associated with transcriptional activation, a highly conserved DNA-binding domain with two zinc fingers, a linker region, and a multifunctional domain that mediates ligand-binding, dimerization, and ligand-dependent transcriptional activation (Evans, 1988). Members of the nuclear receptor superfamily are widely found in vertebrates and invertebrates (Whitfield et al., 1999), but screening plant DNA with conserved animal sequences (S. D. Clouse, unpublished; P. Krishna, personal communication) and examination of the completed genome sequence of *Arabidopsis*, suggest that plants do not have proteins with all of the characteristic features of animal nuclear receptors. Thus, plants may have conserved the steroid signal without the associated nuclear receptor signaling system.

The majority of what is currently known about BR signaling has come from the study of a single genetic locus in *Arabidopsis, BRASSINOSTEROID*

INSENSITIVE 1(BRI1), originally identified by screening mutated *Arabidopsis* seedlings for their ability to elongate primary roots in the presence of normally inhibitory concentrations of BR (Clouse *et al.*, 1996). Numerous alleles of *bri* 1 have been isolated in several independent genetic screens (Kauschmann *et al.*, 1996; Li and Chory, 1997; Noguchi *et al.*, 1999a). The phenotype of null *bri* 1 mutants (extreme dwarfs with dark green, curled leaves, delayed development, and male sterility) appears to be identical to the most severe BR-deficient mutants, such as *cpd*, suggesting that the BRI1 protein is a critical component of the primary BR signal transduction pathway.

BRI1 encodes a leucine-rich repeat receptor kinase with significant sequence similarity to numerous plant receptor kinases involved in regulation of plant growth, morphogenesis, diseases resistance, and responses to environmental factors (Lease *et al.*, 1998). Both animal and plant receptor kinases exhibit similar functional domains including an extracellular ligand-binding domain, a single-pass transmembrane sequence, and an intracellular kinase domain. In animals both tyrosine receptor kinases, such as the epidermal growth factor receptor (Heldin, 1995), and serine/threonine receptor kinases, represented by the transforming growth factor-β (TGF-β) family of receptors (Massague, 1998), are found. All plant receptor kinases identified to date, however, are of the serine/threonine class. Plant receptor kinases can be further catergorized by the structural motifs represented in their extracellular domains, with leucine-rich repeats occurring quite commonly. Leucine-rich repeats create a surface mediating protein–protein interaction by forming solvent-exposed paralellel β-sheets. In mammals, receptors for some peptide hormones, such as nerve growth factor and gonadotropin, contain leucine-rich repeats (Braun *et al.*, 1991; Kobe and Deisenhofer, 1994).

The mechanism of action of many animal receptor kinases is known and generally involves ligand-mediated homo- or heterodimerization of the receptor followed by autophosphorylation of the intracellular domain. The activated kinase then becomes competent to recognize and phosphorylate downstream components of signal transduction, such as transcription factors, or in the case of direct, nongenomic effects, other cytoplasmic proteins involved in cellular metabolism (Heldin, 1995). Although plant receptor kinases have not yet been characterized to the same extent as their animal counterparts, evidence is accumulating that these plant receptors are likely to follow the same general paradigm of receptor kinase action (Trotochaud *et al.*, 1999; Wang *et al.*, 2001). Biochemical and genetic analyses have shown that both the extracellular and kinase domains of BRI1 are essential for proper function (Noguchi *et al.*, 1999a; Wang *et al.*, 2001).

Sequence analysis of BRI1 (Fig. 5) reveals that the protein product begins with a putative signal peptide, followed by a leucine zipper and 25 leucine-rich repeats, flanked by short sequences containing paired cysteines. Between repeat 21 and 22 lies a unique 70 amino acid island that is critical for biological function since the *bri* 1–113 mutant, which has a point mutation at Gly-611 within this region, exhibits a severely dwarfed phenotype (Friedrichsen *et al.*, 2000; Li and Chory, 1997). Following the extracellular domain is first a short hydrophobic sequence functioning

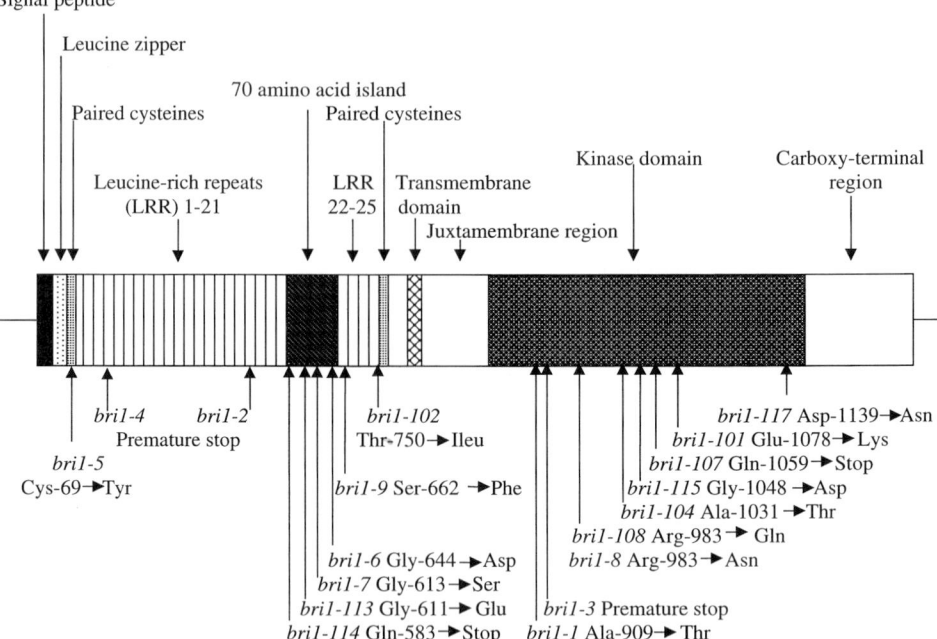

FIGURE 5 Structure of the BRI1 receptor kinase. Major functional domains are shown and the effect on protein structure of the known *bri*1 mutant alleles is indicated. Data derived from Li and Chory (1997), Noguchi *et al.*(1999a), and Friedrichsen *et al.* (2000).

as a transmembrane domain, then the juxtamembrane region, and finally the kinase domain containing all 12 conserved subdomains found in eukaryotic kinases. Numerous *bri* 1 mutant alleles with severe phenotypes result from point mutations in the kinase domain, including lesions in subdomain II at an invariant Ala residue found in all kinases and in subdomain VIII, which contains the highly conserved activation loop required for kinase activity (Friedrichsen *et al.*, 2000; Lease *et al.*, 1998; Noguchi *et al.*, 1999a). Recombinant BRI1 kinase domain, expressed either in *Escherichia coli* or mammalian cell cultures, is capable of autophosphorylation *in vitro* in the presence of $[\gamma\text{-}^{32}\text{P}]\text{ATP}$, confirming the biochemical function of this segment of the BRI1 protein. Moreover, phosphoamino acid analyses revealed that BRI1 phosphorylates on Ser and Thr residues only, as expected for a plant receptor kinase (Friedrichsen *et al.*, 2000; Oh *et al.*, 2000).

Matrix-assisted laser desorption/ionization mass spectrometry (MALDI/MS) of tryptic peptides was used to identify at least 12 sites of *in vitro* autophosphorylation in the BRI1 cytoplasmic domain, including five in the juxtamembrane region (N-terminal to the catalytic kinase domain), five in the kinase domain itself (one each in subdomain I and VIa and three in subdomain VIII), and two in the carboxy-terminal region. Some of the Ser and Thr residues autophosphorylated in BRI1 are conserved at corresponding positions in related plant Ser/Thr kinases, with the region of greatest Ser/Thr conservation occuring in the peptide 1038-DTHLSVSTLAGTPGYVPPEYYQSFR-1062, which lies in the highly conserved activation loop of kinase subdomain VIII (Johnson *et al.*, 1996). Outside of subdomain VIII, the only other strongly conserved sites for Ser or Thr in related kinases were in the positions equivalent to T-872 in the juxtamembrane region and T-982 in kinase subdomain VIa of the BRI1(Oh *et al.*, 2000). Whether these conserved sites are also autophosphorylated in other plant receptor kinases is currently not known, since BRI1 is the first plant receptor kinase to be examined in detail for specific phosphorylation sites. If the identified BRI1 sites are also affected *in vivo,* autophosphorylation in subdomain VIII would likely lead to kinase activation, whereas autophosphorylation of multiple sites in the juxtamembrane and carboxy-terminal regions would suggest multiple interacting cytoplasmic partners for BRI1, each with a specific phosphorylated target sequence within these regions.

The phenotype of the *bri*1 mutants and the structure of the BRI1 protein clearly indicate that this receptor kinase is a critical component of BR signal transduction, and may in fact be the primary, or only, receptor for BR in the cell. BRI1-green fluorescent protein fusions visualized with confocal microscopy clearly showed that BRI1 is localized in the plasma membrane (Friedrichsen *et al.*, 2000). Thus, if BRI1 is the BR receptor, perception of BR occurs at the cell surface and not in the cytoplasm or nucleus as in the traditional animal steroid receptor model. Evidence that the extracellular domain of BRI1 is required for perception of BR first came from studies of chimeric receptor kinases consisting of the extracellular, transmembrane, and juxtamembrane segments of BRI1 fused to the kinase domain of rice XA21, a leucine-rich repeat receptor kinase that confers resistance of rice to the pathogen *Xanthomonas oryzae* pv. *oryzae* (He *et al.*, 2000). In an incompatible interaction of rice with this pathogen, a hypersensitive response occurs that includes an oxidative burst, rapid activation of defense gene expression, and eventually cell death. This hypersensitive response also is conserved in rice cell suspension cultures expressing a full-length *XA21* construct when treated with an incompatible strain of *X. oryzae*. When rice cells expressing the chimeric *BRI1/XA21* construct were treated with 2 μM BR, a similar hypersensitive response was induced in the cells, indicating that the BRI1 extracellular domain perceived BR, and that this perception resulted in the activation of a signal transduction pathway involving the kinase domain from a different receptor kinase. Interestingly, a mutant construct in which Gly-611 of the 70 amino acid island of BRI1 was changed to Glu (as in *bri*1–113) was unable to activate the rice defense response in cells treated with BR, providing further functional evidence of the importance of this motif in BR perception (He *et al.*, 2000).

In a direct confirmation that BRI1 is either the BR receptor or part of a BR receptor complex, transgenic plants overexpressing a BRI1/green fluorescent protein were found to have increased binding of tritiated BR in their plasma membranes when compared to wild-type plants (Wang *et al.*, 2001). The BR-binding activity could be specifically immunoprecipitated by antibodies to green fluorescent protein, and binding was competitively inhibited by active BRs but not by biologically

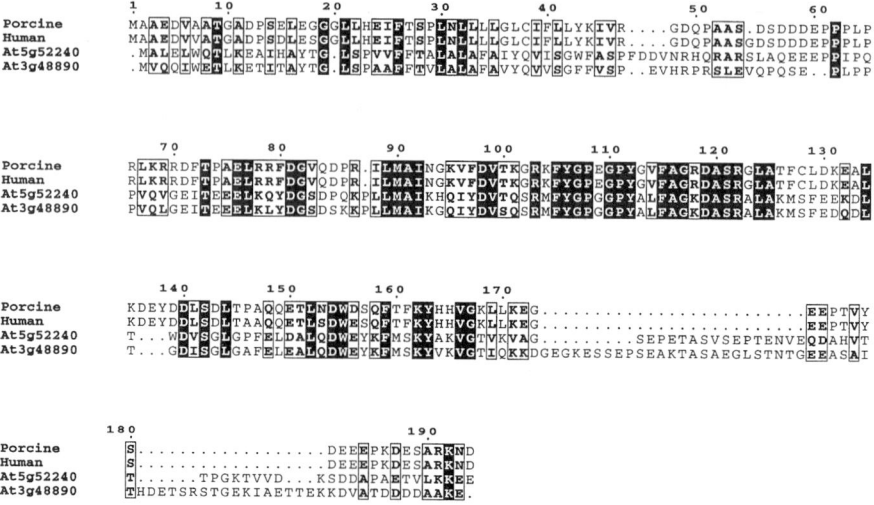

FIGURE 6 Sequence alignment of putative membrane-bound steroid receptors. Putative progesterone membrane receptors from pig (accession number X99714) and human (accession number Y12711) are aligned to two predicted proteins from the *Arabidopsis* genome sequence using CLUSTAL W. Amino acids shaded in black are identical between all three species.

inactive structural analogues such as ecdysone. The BR binding was of high affinity ($K_d = 7.4 \pm 0.9$ nM), which is consistent with physiological concentrations of BR and those concentrations giving typical responses in numerous bioassays. BR binding was abolished by mutations in the extracellular domain but not the kinase domain of BRI1. The specificity and high affinity of BR binding suggest that BRI1 is either the BR receptor or an integral component of a receptor complex, and that binding of BR to this receptor is a primary mechanism by which BR-induced events are initiated.

Leucine-rich repeats, which comprise the majority of the BRI1 extracellular domain, are generally not known for binding small molecules such as steroids (Kobe and Deisenhofer, 1994). It is possible that BL either binds directly to the unique 70 amino acid island, or instead first binds to another steroid-binding protein, which then associates with the extracellular domain of BRI1 to form a complex. In animals, high-affinity binding proteins for sex steroids have been identified that are not in the classic superfamily of steroid receptors. These include the membrane progesterone receptors (Falkenstein *et al.*, 1996; Gerdes *et al.*, 1998) and the soluble sex hormone-binding globulins (Forest and Pugeat, 1986). The *Arabidopsis* genome encodes several predicted proteins that have significant sequence identitiy to these mammalian sequences (Fig. 6). It would be of great interest to determine if recombinant proteins derived from these sequences bind labeled BR directly, and if these proteins interact with the extracellular domain of BRI1.

From what has been previously said, it is clear that BRs are perceived at the cell surface in plants, rather than intracellularly. However, it cannot be concluded that plant and animal steroid hormone signal transduction pathways have no similarities, since it is now widely accepted that in animals, membrane localized steroid receptors that are distinct from members of the nuclear receptor superfamily can mediate a variety of steroid-initiated responses (Wehling, 1997). These so called "nongenomic effects" differ from classic steroid hormone action by their rapid response (less than 10 min), their insensitivity to inhibitors of protein and RNA synthesis, and their ability to respond to hormones tethered to large molecules that prevent entry into the cell. All known animal steroid hormones exhibit these nongenomic effects and high-affinity steroid-binding sites in membranes from several tissue types have been identified (Watson and Gametchu, 1999). Nongenomic steroid effects include many responses more often associated with peptide hormones, and involve conventional second messengers such as phospholipase C, phosphoinositide turnover, and changes in intracellular pH and free calcium levels. In addition, membrane-initiated steroid events lead to activation of tyrosine kinases (Mendoza *et al.*, 1995; Tesarik *et al.*, 1993) and phosphorylation of a variety of proteins, including transcription factors (Falkenstein *et al.*, 2000).

In hepatocytes and sperm cells, treatment with progesterone results in a rapid influx of calcium. A high-affinity membrane-binding site ($K_d = 11$ nM) for progesterone was identified in porcine liver cells and the purified protein allowed antibody production and cloning of the corresponding cDNA (Falkenstein *et al.*, 1996, 2000). Heterologous expression of this cDNA in Chinese hamster ovary cells resulted in an increase in microsomal progesterone-binding capacity, and recombinant protein from the expressed cDNA showed similar progesterone specificity when compared to the native porcine liver protein. Moreover, treatment of human sperm cells with specific antibodies against the progesterone-binding protein resulted in a significant reduction in progesterone-induced calcium increases within the cell (Falkenstein *et al.*, 1999). Thus, the progesterone-binding protein appears to be a viable candidate for a membrane-bound steroid receptor mediating rapid nongenomic effects. Western blot analysis showed that a cross-reacting protein occurs in many mammalian species and a cDNA with high sequence identity was subsequently identified in humans (Gerdes *et al.*, 1998). The protein structure shows no similarity to intracellular progesterone receptors of the nuclear superfamily. As previously mentioned, two *Arabidopsis* genes encode predicted proteins with significant sequence identity to the putative mammalian receptors (Fig. 6) and it will be quite informative to determine if these *Arabidopsis* proteins play any role in plant steroid signaling.

B. DOWNSTREAM COMPONENTS OF BR SIGNAL TRANSDUCTION

Although the importance of BRI1 in BR perception has been clearly established, understanding downstream signaling events, including identification of cytoplasmic binding partners and kinase domain substrates for BRI1, remains an area of active research. Molecular genetic and biochemical approaches used in other receptor kinase studies, including yeast two-hybrid analysis (Bower et al., 1996; Gu et al., 1998), interaction cloning (Stone et al., 1994), and immunoprecipitation of receptor–protein complexes (Trotochaud et al., 1999), are currently being employed to identify putative *in vivo* binding partners for BRI1. Reports of two *in vitro* substrates of the BRI1 kinase domain have appeared, including a kinase-associated phosphatase (Schumacher and Chory, 2000) and the regulatory subunit of the V-ATPase previously mentioned (Schumacher et al., 1999) but no confirmation of these proteins as *in vivo* substrates has been published.

Another approach to identifying kinase substrates is the use of synthetic peptides to understand binding motifs and recognition consensus sequences (Kuriyan and Cowburn, 1997). Such a study has been undertaken for the BRI1 kinase domain and a putative consensus phosphorylation recognition sequences has been reported (Oh et al., 2000). The sequence has a similar configuration of basic and hydrophobic amino acids at $P-3$ through $P-6$ (relative to the phosphorylated Ser) as that found for target sequences of SNF1-related kinases (McMichael et al., 1993; Toroser and Huber, 1998). The optimum peptide for BRI1 phosphorylation was GRMKKIASVEMMKK, and using variations of this sequence it was determined that the positioning of residues both N- and C-terminal of the phosphorylated Ser was critical for optimal activity. Positive residues at $P-3$, $P-4$, $P+5$, and $P+6$ were essential with a preference for Lys over Arg. Based on these results, a preliminary consensus sequence of [RK]-[RK]-X(2)-S-X(2)-[LMVIFY]-X-[RK]-[RK] was proposed (Oh et al., 2000).

Using this consensus sequence to search the *Arabidopsis* nonredundant protein database returned a number of interesting proteins containing the recognition motif, some of which had obvious connections to signal transduction pathways (S. D. Clouse, unpublished). Presently, there is no direct evidence that these are true substrates *in vivo*, but the BRI1 substrate recognition sequence may provide a valuable molecular tool for further analysis. Interestingly, one of the proteins containing elements of the BRI recognition sequence was independently identified by subtractive hybridization in a screen for BR-regulated genes (Jiang and Clouse, 2001). This WD-domain protein is an apparent homologue of mammalian TRIP-1 (TGF-β receptor interacting protein), which is a kinase domain substrate of the TGF-β type II receptor (Chen et al., 1995). The TGF-β family of polypeptide growth factors regulates proliferation, lineage determination, differentiation, and death of cells, and plays a prominent role in the development and homeostasis of organisms from fruit fly to human (Massague, 1998). They are perceived at the cell surface by heterotetrameric type I and type II transmembrane Ser/Thr receptor kinases, which phosphorylate a number of cytoplasmic proteins. Among these substrates, TRIP-1 is unique in that it functions both as a modulator of TGF-β-regulated gene expression and as an essential subunit of the eukaryotic translation initiation factor, eIF3 (Choy and Derynck, 1998). Consisting of 10 or more subunits, eIF3 is the largest eukaryotic translation initiation factor and is required for ribosomal binding of mRNA to the 40 S subunit, thus playing an essential role in the initiation of eukaryotic protein synthesis. *Arabidopsis* TRIP-1 was recently shown to be a functional component of the plant eIF3 complex (Burks *et al.,* 2000).

BR treatment of bean, tobacco, and *Arabidopsis* cells results in a rapid increase in TRIP-1 transcript levels (Jiang and Clouse, 2001). Moreover, we have recently found that recombinant *Arabidopsis* TRIP-1 is strongly phosphorylated by the BRI kinase domain *in vitro,* as is a synthetic peptide derived from the TRIP-1 sequence (W. K. Ray and S. D. Clouse, unpublished). Regulation of gene expression and phosphorylation of eIF3 subunits both might be ways in which BR signal transduction could impinge directly on translation initiation. If phosphorylation of TRIP-1 by BRI1 enhances its competence to initiate assembly of the eIF3 complex, it would provide an effective method by which BR could promote cell growth.

We are currently using immunoprecipitation followed by mass spectrometry to determine if TRIP-1 is indeed phosphorylated by BRI1 *in vivo.* However, correlative evidence that TRIP-1 may play a role in BR signaling *in vivo* comes from the phenotype of transgenic *Arabidopsis* lines expressing antisense *TRIP-1* RNA. Several of the lines surviving beyond the seedling stage shared some of the characteristics of BR-deficient and -insensitive mutants, including dwarfism, altered leaf morphology, delayed development, reduced apical dominance, and reduced fertility (Jiang and Clouse, 2001). If TRIP-1 is a true *in vivo* substrate of BRI1, it would indicate that plants have evolved a unique mechanism to regulate cell growth by combining elements of steroid signaling and components of the TGF-β pathway.

V. CONCLUSION AND FUTURE PROSPECTS

We have seen that plants, like animals, use steroid signals to modulate a wide array of developmental programs. The biosynthetic pathways of plant and animal steroids share many common features, and several of the steps can be catalyzed equally well by enzymes of plant or animal origin. The apparent lack of nuclear steroid receptors in plants is a clear difference in signal transduction, but perception of steroids at the cell surface by animal cells has received much greater attention in the past decade. As more membrane steroid receptors are characterized in animal systems, it will be interesting to determine if any of these proteins, like BRI1, are themselves receptor kinases or if they form heterodimers with this class of signal transduction molecules. Preliminary results on the promotive effect of BRs on cell division is likely to lead to new areas of molecular research and will provide data for comparisons of the role of steroids in controlling cell proliferation in plants and animals. BRI1 analysis has comprised almost the entire body of BR signal transduction knowledge to date, but preliminary meeting reports of the isolation of new BR-insensitive mutants have appeared and publications should follow shortly. Further analysis of BRI1 will also be required to resolve some unanswered questions in BR signal transduction: (1) Does BRI1 form homo- or heterodimers? (2) Are accessory steroid-binding proteins required for BR binding? (3) What are the *in vivo* autophosphorylation sites? and (4) What are the number and nature of cytoplasmic-binding partners for the BRI1 kinase domain? Forthcoming results of molecular genetic and biochemical experiments from a number of laboratories should be illuminating and will establish further similarities and contrasts between plant and animal steroid signaling.

REFERENCES

Altmann, T. (1999). Molecular physiology of brassinosteroids revealed by the analysis of mutants. *Planta* **208,** 1–11.
Aranda, A., and Pascual, A. (2001). Nuclear hormone receptors and gene expression. *Physiol. Rev.* **81,** 1269–1304.
Azpiroz, R., Wu, Y., LoCascio, J., and Feldmann, K. (1998). An Arabidopsis brassinosteroid-dependent mutant is blocked in cell elongation. *Plant Cell* **10,** 219–230.
Benveniste, P. (1986). Sterol biosynthesis. *Annu. Rev. Plant Physiol.* **37,** 275–308.
Bishop, G. J., Nomura, T., Yokota, T., Harrison, K., Noguchi, T., Fujioka, S., Takatsuto, S., Jones, J. D. G., and Kamiya, Y. (1999). The tomato DWARF enzyme catalyses C-6 oxidation in brassinosteroid biosynthesis. *Proc. Natl. Acad. Sci. USA* **96,** 1761–1766.
Bower, M., Matias, D., Fernandes-Carvalho, E., Mazzurco, M., Gu, T., Rothstein, S., and Goring, D. (1996). Two members of the thiredoxin-h family interact with the kinase domain of a *Brassica* S locus receptor kinase. *Plant Cell* **8,** 1641–1650.
Braun, T., Schofield, P. R., and Sprengel, R. (1991). Amino-terminal leucine-rich repeats in gonadotropin receptors determine hormone selectivity. *EMBO J.* **10,** 1885–1890.
Burks, E. A., Bezerra, P. P., Le, H., Gallie, D. R., and Browning, K. S. (2000). Plant initiation factor 3 subunit composition resembles mammalian initiation factor 3 and has a novel subunit. *J. Biol. Chem.* **276,** 2122–2131.
Catala, C., Rose, J. K. C., and Bennett, A. (1997). Auxin-regulation and spatial localization of an endo-1,4-β-d-glucanase and a xyloglucan endotransglycosylase in expanding tomato hypocotyls. *Plant J.* **12,** 417–426.
Catterou, M., Dubois, F., Schaller, H., Aubanelle, L., Vilcot, B., Sangwan-Norreel, B. S., and Sangwan, R. S. (2001a). Brassinosteroids, microtubules and cell elongation in *Arabidopsis thaliana*. I. Molecular, cellular and physiological characterization of the Arabidopsis *bul*1 mutant, defective in the delta 7-sterol-C5-desaturation step leading to brassinosteroid biosynthesis. *Planta* **212,** 659–672.
Catterou, M., Dubois, F., Schaller, H., Aubanelle, L., Vilcot, B., Sangwan-Norreel, B. S., and Sangwan, R. S. (2001b). Brassinosteroids, microtubules and cell elongation in *Arabidopsis thaliana*. II. Effects of brassinosteroids on microtubules and cell elongation in the *bul*1 mutant. *Planta* **212,** 673–683.
Chen, R. H., Miettinen, P. J., Maruoka, E. M., Choy, L., and Derynck, R. (1995). A WD-domain protein that is associated with and phosphorylated by the type II TGF-beta receptor. *Nature* **377,** 548–552.
Choe, S., Dilkes, B. P., Fujioka, S., Takatsuto, S., Sakurai, A., and Feldmann, K. A. (1998). The *DWF*4 gene of Arabidopsis encodes a cytochrome P-450 that mediates multiple 22α-hydroxylation steps in brassinosteroid biosynthesis. *Plant Cell* **10,** 231–243.
Choe, S., Dilkes, B. P., Gregory, B. D., Ross, A. S., Yuan, H., Noguchi, T., Fujioka, S., Takatsuto, S., Tanaka, A., Yoshida, S., Tax, F., and Feldmann, K. A. (1999a). The Arabidopsis *dwarf*1 mutant is defective in the conversion of 24-methylenecholesterol to campesterol in brassinosteroid biosynthesis. *Plant Physiol.* **119,** 897–907.
Choe, S., Noguchi, T., Fujioka, S., Takatsuto, S., Tissier, C. P., Gregory, B. D., Ross, A. S., Tanaka, A., Yoshida, S., Tax, F. E., and Feldmann, K. A. (1999b). The Arabidopsis *dwf*7/*ste*1 mutant is defective in the Δ^7 sterol C-5 desaturation step leading to brassinosteroid biosynthesis. *Plant Cell* **11,** 207–221.

Choe, S., Tanaka, A., Noguchi, T., Fujioka, S., Takatsuto, S., Ross, A. S., Tax, F. E., Yoshida, S., and Feldmann, K. A. (2000). Lesions in the sterol delta reductase gene of Arabidopsis cause dwarfism due to a block in brassinosteroid biosynthesis. *Plant J.* **21,** 431–443.

Choe, S., Fujioka, S., Noguchi, T., Takatsuto, S., Yoshida, S., and Feldmann, K. A. (2001). Overexpression of DWARF4 in the brassinosteroid biosynthetic pathway results in increased vegetative growth and seed yield in Arabidopsis. *Plant J.* **26,** 573–582.

Choi, Y.-H., Fujioka, S., Nomura, T., Harada, A., Y., T., Takatsuto, S., and Sakurai, A. (1997). An alternative brassinolide biosynthetic pathway via late C-6 oxidation. *Phytochemistry* **44,** 609–613.

Chory, J., Catterjee, M., Cook, R., Elich, T., Fankhauser, C., Li, J., Nagpal, P., Neff, M., Pepper, A., Poole, D., Reed, J., and Vitart, V. (1996). From seed germination to flowering, light controls plant development via the pigment phytochrome. *Proc. Natl. Acad. Sci. USA* **93,** 12066–12071.

Choy, L., and Derynck, R. (1998). The type II transforming growth factor (TGF)-beta receptor-interacting protein TRIP-1 acts as a modulator of the TGF-beta response. *J. Biol. Chem.* **273,** 31455–31462.

Chrispeels, M. J., and Agre, P. (1994). Aquaporins: Water channel proteins of plant and animal cells. *Trends Biochem. Sci.* **19,** 421–425.

Clouse, S. D. (1997). Molecular genetic analysis of brassinosteroid action. *Physiol. Plant.* **100,** 702–709.

Clouse, S. D. (2000). Plant development: A role for sterols in embryogenesis. *Curr. Biol.* **10,** R601–R604.

Clouse, S. D., and Feldmann, K. A. (1999). Molecular genetics of brassinosteroid action. *In* "Brassinosteroids: Steroidal Plant Hormones" (A. Sakurai, T. Yokota, and S. D. Clouse, Eds.), pp. 163–190. Springer-Verlag, Tokyo.

Clouse, S. D., and Sasse, J. M. (1998). Brassinosteroids: Essential regulators of plant growth and development. *Annu. Rev. Plant Physiol. Plant Mol. Biol.* **49,** 427–451.

Clouse, S. D., and Zurek, D. (1991). Molecular analysis of brassinolide action in plant growth and development. *In* "Brassinosteroids Chemistry, Bioactivity, & Applications" (H. G. Cutler, T. Yokota, and G. Adam, Eds.), pp. 122–140. American Chemical Society, Washington, D.C.

Clouse, S. D., Langford, M., and McMorris, T. C. (1996). A brassinosteroid-insensitive mutant in *Arabidopsis thaliana* exhibits multiple defects in growth and development. *Plant Physiol.* **111,** 671–678.

Cosgrove, D. J. (1997). Relaxation in a high-stress environment: The molecular basis of extensible cell walls and enlargement. *Plant Cell* **9,** 1031–1041.

Davies, P. J. (1995). The plant hormones: Their nature, occurrence and functions. *In* "Plant Hormones: Physiology, Biochemistry and Molecular Biology" (P. J. Davies, Ed.), pp. 1–12. Kluwer Academic, Dordrecht.

Diener, A. C., Li, H., Zhou, W., Whoriskey, W. J., Nes, W. D., and Fink, G. R. (2000). Sterol methyltransferase 1 controls the level of cholesterol in plants. *Plant Cell* **12,** 853–870.

Edwards, P., and Ericsson, J. (1999). Signaling molecules derived from the cholesterol biosynthetic pathway. *Annu. Rev. Biochem.* **68,** 157–185.

Evans, R. (1988). The steroid and thyroid hormone receptor superfamily. *Science* **240,** 667–675.

Falkenstein, E., Meyer, C., Eisen, C., Scriba, P., and Wehling, M. (1996). Full-length cDNA sequence of a progesterone membrane-binding protein from porcine vascular smooth muscle cells. *Biochem. Biophys. Res. Commun.* **229,** 86–89.

Falkenstein, E., Heck, M., Gerdes, D., Grube, D., Christ, M., Weigel, M., Buddhikot, M., Meizel, and S., Wehling, M. (1999). Specific progesterone binding to a membrane protein and related nongenomic effects on Ca^{2+}-fluxes in sperm. *Endocrinology* **140,** 5999–6002.

Falkenstein, E., Tillmann, H. C., Christ, M., Feuring, M., and Wehling, M. (2000). Multiple actions of steroid hormones—a focus on rapid, nongenomic effects. *Pharmacol. Rev.* **52,** 513–556.

Forest, M., and Pugeat, M. (1986). "Binding Proteins of Seroid Hormones." John Libbey, London.

Forgac, M. (1999). Structure and properties of the vacuolar (H+)-ATPases. *J. Biol. Chem.* **274,** 12951–12954.

Friedrichsen, D. M., Joazeiro, C. A., Li, J., Hunter, T., and Chory, J. (2000). Brassinosteroid-insensitive-1 is a ubiquitously expressed leucine-rich repeat receptor serine/threonine kinase. *Plant Physiol.* **123,** 1247–1256.
Fujioka, S. (1999). Natural occurrence of brassinosteroids in the plant kingdom. *In* "Brassinosteroids: Steroidal Plant Hormones" (A. Sakurai, T. Yokota, and S. D. Clouse, Eds.), pp. 21–45. Springer-Verlag, Tokyo.
Fujioka, S., and Sakurai, A. (1997). Biosynthesis and metabolism of brassinosteroids. *Physiol. Plant.* **100,** 710–715.
Fujioka, S., Choi, Y.-H., Takatsuto, S., Yokota, T., Li, J., Chory, J., and Sakurai, A. (1996). Identification of castasterone, 6-deoxocastasterone, typhasterol and 6-deoxotyphasterol from the shoots of *Arabidopsis thaliana. Plant Cell Physiol.* **37,** 1201–1203.
Fujioka, S., Li, J., Choi, Y.-H., Seto, H., Takatsuto, S., Noguchi, T., Watanabe, T., Kuriyama, H., Yokota, T., Chory, J., and Sakurai, A. (1997). The Arabidopsis *deetiolated2* mutant is blocked early in brassinosteroid biosynthesis. *Plant Cell* **9,** 1951–1962.
Fukuda, H. (1997). Tracheary element differentiation. *Plant Cell* **9,** 1147–1156.
Gachotte, D., Meens, R., and Benveniste, P. (1995). An Arabidopsis mutant deficient in sterol biosynthesis: Heterologous complementation by *ERG* 3 encoding a delta 7-sterol-C-5-desaturase from yeast. *Plant J.* **8,** 407–416.
Gerdes, D., Wehling, M., Leube, B., and Falkenstein, E. (1998). Cloning and tissue expression of two putative steroid membrane receptors. *Biol. Chem.* **379,** 907–911.
Geuns, J. M. C. (1978). Steroid hormones and plant growth and development. *Phytochemistry* **17,** 1–14.
Gogarten, J. P., Fichmann, J., Braun, Y., Morgan, L., Styles, P., Taiz, S. L., DeLapp, K., and Taiz, L. (1992). The use of antisense mRNA to inhibit the tonoplast H+ ATPase in carrot. *Plant Cell* **4,** 851–864.
Griffiths, P. G., Sasse, J. M., Yokota, T., and Cameron, D. W. (1995). 6-Deoxotyphasterol and 3-dehydro-6-deoxoteasterone, possible precursors to brassinosteroids in the pollen of *Cupressus arizonica. Biosci. Biotech. Biochem.* **59,** 956–959.
Grove, M. D., Spencer, G. F., Rohwedder, W. K., Mandava, N. B., Worley, J. F., Warthen, J. D., Steffens, G. L., Flippen-Anderson, J. L., and Cook, J. C. (1979). Brassinolide, a plant growth-promoting steroid isolated from *Brassica napus* pollen. *Nature* **281,** 216–217.
Gu, T., Mazzurco, M., Sulaman, W., Matias, D. D., and Goring, D. R. (1998). Binding of an arm repeat protein to the kinase domain of the S-locus receptor kinase. *Proc. Natl. Acad. Sci. USA* **95,** 382–387.
Hartmann, M. (1998). Plant sterols and the membrane environment. *Trends Plant Sci.* **3,** 170–175.
He, Z., Wang, Z. Y., Li, J., Zhu, Q., Lamb, C., Ronald, P., and Chory, J. (2000). Perception of brassinosteroids by the extracellular domain of the receptor kinase BRI1. *Science* **288,** 2360–2363.
Heldin, C. (1995). Dimerization of cell surface receptors in signal transduction. *Cell* **80,** 213–224.
Hu, Y., Bao, F., and Li, J. (2000). Promotive effect of brassinosteroids on cell division involves a distinct CycD3-induction pathway in Arabidopsis. *Plant J.* **24,** 693–701.
Husselstein, T., Gachotte, D., Desprez, T., Bard, M., and Benveniste, P. (1996). Transformation of *Saccharomyces cerevisiae* with a cDNA encoding a sterol C-methyltransferase from *Arabidopsis thaliana* results in the synthesis of 24-ethyl sterols. *FEBS Lett.* **381,** 87–92.
Husselstein, T., Schaller, H., Gachotte, D., and Benveniste, P. (1999). Delta7-sterol-C5-desaturase: Molecular characterization and functional expression of wild-type and mutant alleles. *Plant Mol. Biol.* **39,** 891–906.
Irish, V. F., and Jenik, P. D. (2001). Cell lineage, cell signaling and the control of plant morphogenesis. *Curr. Opin. Genet. Dev.* **11,** 424–430.
Iwasaki, T., and Shibaoka, H. (1991). Brassinosteroids act as regulators of tracheary-element differentiation in isolated Zinnia mesophyll cells. *Plant Cell Physiol.* **32,** 1007–1014.
Jang, J. C., Fujioka, S., Tasaka, M., Seto, H., Takatsuto, S., Ishii, A., Aida, M., Yoshida, S., and Sheen, J. (2000). A critical role of sterols in embryonic patterning and meristem programming revealed by the fackel mutants of *Arabidopsis thaliana. Genes Dev.* **14,** 1485–1497.

Jiang, J., and Clouse, S. D. (2001). Expression of a plant gene with sequence similarity to animal TGF-β receptor interacting protein is regulated by brassinosteroids and required for normal plant development. *Plant J.* **26**, 35–45.
Johnson, L. N., Noble, M. E., and Owen, D. J. (1996). Active and inactive protein kinases: Structural basis for regulation. *Cell* **85**, 149–158.
Kang, J. G., Yun, J., Kim, D. H., Chung, K. S., Fujioka, S., Kim, J. I., Dae, H. W., Yoshida, S., Takatsuto, S., Song, P. S., and Park, C. M. (2001). Light and brassinosteroid signals are integrated via a dark-induced small G protein in etiolated seedling growth. *Cell* **105**, 625–636.
Kauschmann, A., Jessop, A., Koncz, C., Szekeres, M., Willmitzer, L., and Altmann, T. (1996). Genetic evidence for an essential role of brassinosteroids in plant development. *Plant J.* **9**, 701–713.
Kawata, S., Trzaskos, J. M., and Gaylor, J. L. (1985). Microsomal enzymes of cholesterol biosynthesis from lanosterol. Purification and characterization of delta 7-sterol 5-desaturase of rat liver microsomes. *J. Biol. Chem.* **260**, 6609–6617.
Klahre, U., Noguchi, T., Fujioka, S., Takatsuto, S., Yokota, T., Nomura, T., Yoshid, S., and Chua, N.-H. (1998). The Arabidopsis *DIMINUTO/DWARF* 1 gene encodes a protein involved in steroid synthesis. *Plant Cell* **10**, 1677–1690.
Kobe, B., and Deisenhofer, J. (1994). The leucine-rich repeat: A versatile binding motif. *Trends Biochem. Sci.* **19**, 415–421.
Koka, C. V., Cerny, R. E., Gardner, R. G., Noguchi, T., Fujioka, S., Takatsuto, S., Yoshida, S., and Clouse, S. D. (2000). A putative role for the tomato genes *DUMPY* and *CURL-3* in brassinosteroid biosynthesis and response. *Plant Physiol.* **122**, 85–98.
Kuriyan, J., and Cowburn, D. (1997). Molecular peptide recognition domains in eukaryotic signaling. *Annu. Rev. Biophys. Biomol. Struct.* **26**, 259–288.
Lease, K., Ingham, E., and Walker, J. C. (1998). Challenges in understanding RLK function. *Curr. Opin. Plant Biol.* **1**, 388–392.
Li, J., and Chory, J. (1997). A putative leucine-rich repeat receptor kinase involved in brassinosteroid signal transduction. *Cell* **90**, 929–938.
Li, J., Nagpal, P., Vitart, V., McMorris, T. C., and Chory, J. (1996). A role for brassinosteroids in light-dependent development of *Arabidopsis*. *Science* **272**, 398–401.
Li, J., Biswas, M. G., Chao, A., Russell, D. W., and Chory, J. (1997). Conservation of function between mammalian and plant steroid 5alpha-reductases. *Proc. Natl. Acad. Sci. USA* **94**, 3554–3559.
Mandava, N. B. (1988). Plant growth-promoting brassinosteroids. *Annu. Rev. Plant Physiol. Plant Mole. Biol.* **39**, 23–52.
Massague, J. (1998). TGF-beta signal transduction. *Annu. Rev. Biochem.* **67**, 753–791.
Mathur, J., Molnar, G., Fujioka, S., Takatsuto, S., Sakurai, A., Yokota, T., Adam, G., Voigt, B., Nagy, F., Maas, C., Schell, J., Koncz, C., and Szekeres, M. (1998). Transcription of the Arabidopsis *CPD* gene, encoding a steroidogenic cytochrome P-450, is negatively controlled by brassinosteroids. *Plant J.* **14**, 593–602.
Mayer, U., Torres-Ruiz, R., Berleth, T., Misera, S., and Jurgens, G. (1991). Mutations affecting the body organization in the *Arabidopsis* embryo. *Nature* **353**, 402–407.
Mayumi, K., and Shibaoka, H. (1995). A possible double role for brassinolide in the re-orientation of cortical microtubules in the epidermal cells of Azuki bean epicotyls. *Plant Cell Physiol.* **36**, 173–181.
McMichael, R. J., Klein, R., Salvucci, M., and Huber, S. (1993). Identification of the major regulatory phosphorylation site in sucrose-phosphate synthase. *Arch. Biochem. Biophys.* **307**, 248–252.
Mendoza, C., Soler, A., and Tesarik, J. (1995). Nongenomic steroid action: Independent targeting of a plasma membrane calcium channel and a tyrosine kinase. *Biochem. Biophys. Res. Commun.* **210**, 518–523.
Mitchell, J. W., Mandava, N. B., Worley, J. F., Plimmer, J. R., and Smith, M. V. (1970). Brassins: A new family of plant hormones from rape pollen. *Nature (London)* **225**, 1065–1066.
Moebius, F. F., Fitzky, B. U., Lee, J. N., Paik, Y. K., and Glossmann, H. (1998). Molecular cloning and expression of the human delta7-sterol reductase. *Proc. Natl. Acad. Sci. USA* **95**, 899–902.

Morillon, R., Catterou, M., Sangwan, R. S., Sangwan, B. S., and Lassalles, J. P. (2001). Brassinolide may control aquaporin activities in *Arabidopsis thaliana*. *Planta* **212**, 199–204.

Nakajima, N., Shida, A., and Toyama, S. (1996). Effects of brassinosteroid on cell division and colony formation of Chinese cabbage mesophyll protoplasts. *Jpn. J. Crop Sci.* **65**, 114–118.

Noguchi, T., Fujioka, S., Choe, S., Takatsuto, S., Yoshida, S., Yuan, H., Feldmann, K. A., and Tax, F. E. (1999a). Brassinosteroid-insensitive dwarf mutants of Arabidopsis accumulate brassinosteroids. *Plant Physiol.* **121**, 743–752.

Noguchi, T., Fujioka, S., Takatsuto, S., Sakurai, A., Yoshida, S., Li, J., and Chory, J. (1999b). Arabidopsis *det2* is defective in the conversion of (24R)-24-methylcholest-4-en-3-one to (24R)-24-methyl-5α-cholestan-3-one in brassinosteroid biosynthesis. *Plant Physiol.* **120**, 833–839.

Noguchi, T., Fujioka, S., Choe, S., Takatsuto, S., Tax, F. E., Yoshida, S., and Feldmann, K. A. (2000). Biosynthetic pathways of brassinolide in Arabidopsis. *Plant Physiol.* **124**, 201–209.

Nomura, T., Sato, T., Bishop, G. J., Kamiya, Y., Takatsuto, S., and Yokota, T. (2001). Accumulation of 6-deoxocathasterone and 6-deoxocastasterone in Arabidopsis, pea and tomato is suggestive of common rate-limiting steps in brassinosteroid biosynthesis. *Phytochemistry* **57**, 171–178.

Oh, M.-H., and Clouse, S. D. (1998). Brassinolide affects the rate of cell division in isolated leaf protoplasts of *Petunia hybrida*. *Plant Cell Rep.* **17**, 921–924.

Oh, M.-H., Romanov, W., Smith, R., Sasse, J., and Clouse, S. (1998). *BRU1* encodes a xyloglucan endo-transglycosylase that is expressed in inner and outer tissues of elongating soybean epicotyls. *Plant Cell Physiol.* **39**, 124–130.

Oh, M. H., Ray, W. K., Huber, S. C., Asara, J. M., Gage, D. A., and Clouse, S. D. (2000). Recombinant brassinosteroid insensitive 1 receptor-like kinase autophosphorylates on serine and threonine residues and phosphorylates a conserved peptide motif *in vitro*. *Plant Physiol.* **124**, 751–766.

Ohlrogge, J., and Browse, J. (1995). Lipid biosynthesis. *Plant Cell* **7**, 957–970.

Osumi, T., Nishino, T., and Katsuki, H. (1979). Studies on the delta 5-desaturation in ergosterol biosynthesis in yeast. *J. Biochem. (Tokyo)* **85**, 819–826.

Pien, S., Wyrzykowska, J., and Fleming, A. J. (2001). Novel marker genes for early leaf development indicate spatial regulation of carbohydrate metabolism within the apical meristem. *Plant J.* **25**, 663–674.

Rahier, A., Smith, M., and Taton, M. (1997). The role of cytochrome b5 in 4alpha-methyl-oxidation and C5(6) desaturation of plant sterol precursors. *Biochem. Biophys. Res. Commun.* **236**, 434–437.

Robker, R. L., and Richards, J. S. (1998). Hormonal control of the cell cycle in ovarian cells: Proliferation versus differentiation. *Biol. Reprod.* **59**, 476–482.

Robyr, D., Wolffe, A. P., and Wahli, W. (2000). Nuclear hormone receptor coregulators in action: Diversity for shared tasks. *Mol. Endocrinol.* **14**, 329–347.

Sakurai, A. (1999). Biosynthesis. *In* "Brassinosteroids: Steroidal Plant Hormones" (A. Sakurai, T. Yokota, and S. D. Clouse, Eds.), pp. 91–112. Springer-Verlag, Tokyo.

Sasse, J. (1999). Physiological Actions of Brassinosteroids. *In* "Brassinosteroids: Steroidal Plant Hormones" (A. Sakurai, T. Yokota, and S. D. Clouse, Eds.), pp. 137–161. Springe-Verlag, Tokyo.

Schrick, K., Mayer, U., Horrichs, A., Kuhnt, C., Bellini, C., Dangl, J., Schmidt, J., and Jurgens, G. (2000). FACKEL is a sterol C-14 reductase required for organized cell division and expansion in Arabidopsis embryogenesis. *Genes Dev.* **14**, 1471–1484.

Schumacher, K., and Chory, J. (2000). Brassinosteroid signal transduction: Still casting the actors. *Curr. Opin. Plant Biol.* **3**, 79–84.

Schumacher, K., Vafeados, D., McCarthy, M., Sze, H., Wilkins, T., and Chory, J. (1999). The Arabidopsis *det3* mutant reveals a central role for the vacuolar H(+)-ATPase in plant growth and development. *Genes Dev.* **13**, 3259–3270.

Shimada, Y., Fujioka, S., Miyauchi, N., Kushiro, M., Takatsuto, S., Nomura, T., Yokota, T., Kamiya, Y., Bishop, G. J., and Yoshida, S. (2001). Brassinosteroid-6-oxidases from *Arabidopsis* and tomato catalyze multiple C-6 oxidations in brassinosteroid biosynthesis. *Plant Physiol.* **126**, 770–779.

Stals, H., and Inze, D. (2001). When plant cells decide to divide. *Trends Plant Sci.* **6**, 359–364.

Steffens, G. L. (1991). U.S. Department of Agriculture Brassins Project: 1970–1980. *In* "Brassinosteroids Chemistry, Bioactivity, & Applications" (H. G. Cutler, T. Yokota, and G. Adam, Eds.), pp. 2–17. American Chemical Society, Washington, D.C.

Stone, J., Collinge, M., Smith, R., Horn, M., and Walker, J. (1994). Interaction of a protein phosphatase with an *Arabidopsis* serine-threonine receptor kinase. *Science* **266,** 793–795.

Szekeres, M., Nemeth, K., Koncz-kalman, Z., Mathur, J., Kauschmann, A., Altmann, T., Redei, G. P., Nagy, F., Schell, J., and Koncz, C. (1996). Brassinosteroids rescue the deficiency of CYP90, a cytochrome P-450, controlling cell elongation and de-etiolation in Arabidopsis. *Cell* **85,** 171–182.

Takahashi, T., Gasch, A., Nishizawa, N., and Chua, N. (1995). The *DIMINUTO* gene of *Arabidopsis* is involved in regulating cell elongation. *Genes Dev.* **9,** 97–107.

Tesarik, J., Moos, J., and Mendoza, C. (1993). Stimulation of protein tyrosine phosphorylation by a progesterone receptor on the cell surface of human sperm. *Endocrinology* **133,** 328–335.

Tominaga, R., Sakurai, N., and Kuraishi, S. (1994). Brassinolide-induced elongation of inner tissues of segments of squash (*Cucurbita maxima* Duch.) hypocotyls. *Plant Cell Physiol.* **35,** 1103–1106.

Toroser, D., and Huber, S. (1998). 3-Hydroxy-3-methylglutaryl-coenzyme A reductase kinase and sucrose-phosphate synthase kinase activities in cauliflower florets: Ca^{2+} dependence and substrate specificities. *Arch. Biochem. Biophys.* **355,** 291–300.

Trotochaud, A., Hao, T., Wu, G., Yang, Z., and Clark, S. (1999). The CLAVATA1 receptor-like kinase requires CLAVATA3 for its assembly into a signalling complex that includes KAPP and a Rho-related protein. *Plant Cell* **11,** 393–405.

Wang, T.-W., Cosgrove, D. J., and Arteca, R. N. (1993). Brassinosteroid stimulation of hypocotyl elongation and wall relaxation in pakchoi (*Brassica chinensis* cv *Lei-choi*). *Plant Physiol.* **101,** 965–968.

Wang, Z. Y., Seto, H., Fujioka, S., Yoshida, S., and Chory, J. (2001). BRI1 is a critical component of a plasma-membrane receptor for plant steroids. *Nature* **410,** 380–383.

Watson, C. S., and Gametchu, B. (1999). Membrane-initiated steroid actions and the proteins that mediate them. *Proc. Soc. Exp. Biol. Med.* **220,** 9–19.

Wehling, M. (1997). Specific, nongenomic actions of steroid hormones. *Annu. Rev. Physiol.* **59,** 365–393.

Whitfield, G. K., Jurutka, P. W., Haussler, C. A., and Haussler, M. R. (1999). Steroid hormone receptors: Evolution, ligands, and molecular basis of biologic function. *J. Cell Biochem.* **Suppl. 32/33,** 110–122.

Yamamoto, R., Demura, T., and Fukuda, H. (1997). Brassinosteroids induce entry into the final stage of tracheary element differentiation in cultured *Zinnia* cells. *Plant Cell Physiol.* **38,** 980–983.

Yamamoto, R., Fujioka, S., Demura, T., Takatsuto, S., Yoshida, S., and Fukuda, H. (2001). Brassinosteroid levels increase drastically prior to morphogenesis of tracheary elements. *Plant Physiol.* **125,** 556–563.

Yamamuro, C., Ihara, Y., Wu, X., Noguchi, T., Fujioka, S., Takatsuto, S., Ashikari, M., Kitano, H., and Matsuoka, M. (2000). Loss of function of a rice brassinosteroid insensitive 1 homolog prevents internode elongation and bending of the lamina joint. *Plant Cell* **12,** 1591–1606.

Yokota, T. (1997). The structure, biosynthesis and function of brassinosteroids. *Trends Plant Sci.* **2,** 137–143.

Zurek, D. M., and Clouse, S. D. (1994). Molecular cloning and characterization of a brassinosteroid-regulated gene from elongating soybean (*Glycine max* L.) epicotyls. *Plant Physiol.* **104,** 161–170.

Zurek, D. M., Rayle, D. L., McMorris, T. C., and Clouse, S. D. (1994). Investigation of gene expression, growth kinetics, and wall extensibility during brassinosteroid-regulated stem elongation. *Plant Physiol.* **104,** 505–513.

Endocannabinoids and Their Actions

Mauro Maccarrone and Alessandro Finazzi-Agrò

Department of Experimental Medicine and Biochemical Sciences, University of Rome "Tor Vergata," I-00133 Rome, Italy

    I. Introduction
   II. The Endocannabinoids
      A. Synthesis
      B. Degradation
      C. Molecular Targets
  III. Actions in the Central Nervous System
      A. Multiple Sclerosis
      B. Parkinson's Disease
  IV. Actions in the Periphery
      A. Platelet Activation
      B. Fertility
   V. Summary and Conclusions
      References

Endocannabinoids are a new class of lipid mediators, which includes amides and esters of long-chain polyunsaturated fatty acids. Anandamide (**I**) and 2-arachidonoylglycerol (**II**) are the main endogenous agonists of cannabinoid receptors, able to mimic several pharmacological effects of Δ^9-tetrahydrocannabinol (**III**), the active principle of *Cannabis sativa* preparations such as hashish and marijuana. The pathways leading to the synthesis and release of anandamide and 2-arachidonoylglycerol from neuronal and nonneuronal cells are rather uncertain. Instead, evidence has accumulated showing that the activity of these compounds at their specific receptors is limited by cellular uptake through a specific membrane transporter, followed by intracellular degradation by a fatty acid amide hydrolase. Here, the endocannabinoids and the endocannabinoid-like compounds most relevant for human physiology will be discussed, along with the synthetic and degradative pathways of anandamide and 2-arachidonoylglycerol and their molecular targets on the cell surface. The main actions of the endocannabinoids in human cells and tissues will also be reviewed, focusing on the activities most recently discovered in the central nervous system and in the periphery. © 2002, Elsevier Science (USA).

I. INTRODUCTION

Two main molecular targets of Δ^9-tetrahydrocannabinol (THC, **III**), the psychoactive principle of *Cannabis sativa* (Gaoni and Mechoulam, 1964), are the central CB1 (Matsuda *et al.*, 1990) and the peripheral CB2 (Munro *et al.*, 1993) cannabinoid (CB) receptors. Both were discovered and characterized more than four millennia after the beneficial effects of cannabis extracts had been exploited in folklore medicine (Mechoulam, 1986; Mechoulam and Hanus, 2000). Soon afterward, an endogenous THC-like molecule, called anandamide (**I**) from "ananda," the Sanskrit word for "bliss," was isolated and found to activate CB receptors, thus mimicking the psychotropic effects of THC (Devane *et al.*, 1992). In a few years other endogenous agonists of CB receptors were characterized and were collectively called "endocannabinoids" (Di Marzo, 1998; Mechoulam *et al.*, 1998; Pop, 1999). This article will review the metabolism of the endocannabinoids and their actions in human cells and tissues.

II. THE ENDOCANNABINOIDS

Endocannabinoids are lipid mediators, isolated from brain and peripheral tissues (Devane *et al.*, 1992; Mechoulam *et al.*, 1998) and found also in milk (Di Marzo *et al.*, 1998a). They include amides and esters of long-chain polyunsaturated fatty acids, which exibit cannabimimetic activity (Fig. 1). A "cannabimimetic" compound should prove to be "THC-mimetic" in a long series of bioassays described in the literature (Pertwee, 1997), and in particular in a "tetrad" of mouse behavioral assays that, when performed together, are highly indicative of cannabinoid-like activity (Martin *et al.*, 1991). The tetrad includes (1) the "open field" test, which measures locomotor activity, (2) the "ring immobility" test, where the time spent motionless in a ring is measured, (3) the "hot plate" test, used to determine antinociceptive effects, and (4) induction of rectal hypothermia. Cannabimimetic activity can be further addressed by measuring cross-tolerance to THC in these

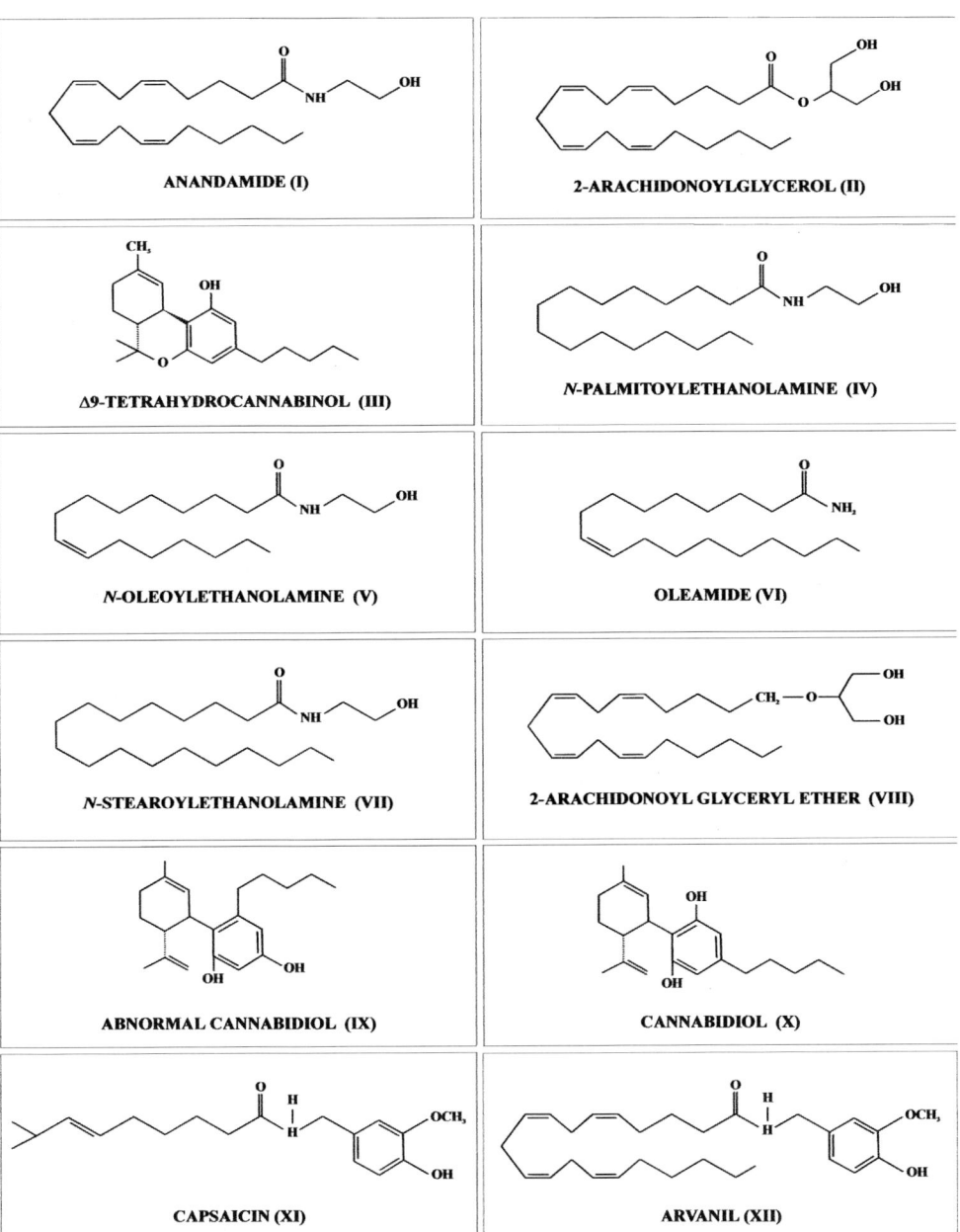

FIGURE 1 Structure of anandamide and related compounds.

assays, or by determining whether animals can distinguish between the effects of the compound under study and those of THC in a "drug discrimination" test (Di Marzo, 1998). The discovery of anandamide (*N*-arachidonoylethanolamine, AEA) in pig brain (Devane *et al.*, 1992), and the finding that this compound was cannabimimetic in the tetrad tests (Fride and Mechoulam, 1993) confirmed the hypothesis of the existence of endogenous ligands for the cannabinoid receptors. Although structurally different from plant cannabinoids, these metabolites were termed "endocannabinoids" in analogy with the "endorphins," i.e. the endogenous ligands of opiate receptors (Di Marzo and Fontana, 1995). Another arachidonate derivative, 2-arachidonoylglycerol (2-AG), found in both canine gut (Mechoulam *et al.*, 1995) and rat brain (Sugiura *et al.*, 1995), was shown to mimic THC by functionally activating CB receptors (Mechoulam *et al.*, 1995), and together with AEA is the most biologically active endocannabinoid described to date (Sugiura and Waku, 2000). The endocannabinoids inhibit gap junction communication in glial cells (Venance *et al.*, 1995), interact with dopaminergic neurotransmission (Giuffrida *et al.*, 1999), and are also involved in glutamate-induced neurotoxicity (Hampson *et al.*, 1998; Hansen *et al.*, 1999). In the periphery, AEA and 2-AG exert cardiovascular actions (Wagner *et al.*, 1998, 1999), playing a role as vasorelaxant (Randall *et al.*, 1996; Deutsch *et al.*, 1997; Sugiura *et al.*, 1998; Zygmunt *et al.*, 1999).

The content of *N*-acylethanolamines (NAE) and 2-AG in human brain is shown in Table I (Maccarrone *et al.*, 2001a). *N*-Palmitoylethanolamine (PEA, **IV**) is the

TABLE I Quantitative Analysis of *N*-Acylethanolamines and 2-Arachidonoylglycerol in Human Brain and Cells

Endocannabinoids [a] (nmol/mg protein)	Brain	Neuroblastoma CHP100 cells	Lymphoma U937 cells
AEA	0.16 ± 0.05 (7.7%)[b]	0.32 ± 0.06 (7.4%)	0.37 ± 0.07 (6.4%)
2-AG	0.10 ± 0.05 (4.8%)	0.25 ± 0.04 (5.8%)	0.22 ± 0.04 (3.8%)
OEA	0.49 ± 0.19 (23.6%)	1.26 ± 0.28 (29.2%)	1.91 ± 0.37 (33.0%)
PEA	1.04 ± 0.37 (50.0%)	2.00 ± 0.60 (46.5%)	2.70 ± 0.60 (46.6%)
SEA	0.29 ± 0.08 (13.9%)	0.48 ± 0.09 (11.1%)	0.59 ± 0.10 (10.2%)
Total amount	2.08 (100%)	4.31 (100%)	5.79 (100%)

[a] AEA, *N*-arachidonoylethanolamine (anandamide); 2-AG, 2-arachidonoylglycerol; OEA, *N*-oleoylethanolamine; PEA, *N*-palmitoyethanolamine; SEA, *N*-stearoylethanolamine.

[b] Values in parentheses represent percent of the total.

most abundant NAE, representing 50% of total endocannabinoid-like compounds, whereas AEA (7.7%) and 2-AG (4.8%) are the least abundant. PEA like cannabinoids has an antiinflammatory action in rat mastocytes and basophils (Facci et al., 1995), it protects cerebellar granule neurons from excitotoxic death (Skaper et al., 1996), and it has peripheral analgesic effects (Calignano et al., 1998). However, it does not bind to CB1 or CB2 receptors and its cannabimimetic properties are still under debate (Lambert and Di Marzo, 1999). N-Oleoylethanolamine (OEA, **V**) represented 24% of human brain NAEs and should be considered an "endocannabinoid-like" compound for which no pathophysiological roles, either cannabimimetic or noncannabimimetic, have been reported so far. Yet a close analogue of OEA, cis-9-octadecenoamide (oleamide, **VI**), was isolated from the cerebrospinal fluid of sleep-deprived mammals and was shown to induce normal sleep in rats (Cravatt et al., 1995). Oleamide was found to inhibit anandamide degradation in vitro, as well as in intact astrocytes in culture (di Tomaso et al., 1996), and to inhibit gap junction communication (Boger et al., 1998). Human brain also contains considerable amounts of N-stearoylethanolamine (SEA, **VII**), which amounts to 14% of total NAEs (Table I). To the best of our knowledge, no activities, either cannabimimetic or noncannabimimetic, have been reported so far for SEA. However, SEA has been also found in rat brain (Cadas et al., 1997), suggesting that it might be a structural lipid. Indeed, NAEs might affect lipid ordering in cell membranes (Bloom et al., 1997), and can have a more general "entourage effect," i.e. they might potentiate the effects of AEA or 2-AG by inhibiting their degradation (Mechoulam et al., 1998; Lambert and Di Marzo, 1999).

It seems noteworthy that AEA and 2-AG are present in comparable amounts in human brain (Table I). These amounts correspond to approximately 50 nmol AEA and 35 nmol 2-AG per gram fresh weight, and are consistent with the large amounts of endocannabinoids (up to 1 μM) recently found in human secretions such as milk (Di Marzo et al., 1998a). Interestingly, this is also the concentration range of γ-aminobutyric acid in human brain (Perry et al., 1971). Thus, these findings reinforce the idea that endocannabinoids might act in the central nervous system not only as modulatory substances (e.g., in an autocrine fashion) like eicosanoids and neuropeptides (Cadas et al., 1997), but also as neurotransmitters. Recent findings have indeed promoted AEA from candidate status to bona fide neurotransmitter (Giuffrida et al., 1999; Self, 1999). Finally, both human neuroblastoma (CHP100) and lymphoma (U937) cells in culture were found to have similar amounts and composition of endocannabinoid-like compounds (Table I). CHP100 and U937 cells are widely used as experimental models for neuronal and immune cells (Galiègue et al., 1995; Maccarrone et al., 1998a), and quantitation of NAEs and 2-AG in these systems might be indicative of the in vivo concentrations in neurons and peripheral lymphocytes. Moreover, endocannabinoid quantitation in human neuronal and immune cells supports the hypothesis that these compounds might be suitable signals to elicit common responses in the neuroimmune axis. Recently, a new ether-type endocannabinoid has been added the cohort of these lipid mediators, i.e., 2-arachidonoylglyceryl ether (noladin ether, **VIII**). This compound has been isolated from porcine brain and has been shown to bind to the CB1 and very weakly to the CB2 receptors, thus causing sedation, hypothermia, intestinal immobility, and mild antinociception in mice (Hanus *et al.,* 2001). Because ethers are generally stable *in vivo,* whereas AEA (an amide) and 2-AG (an ester) are rapidly hydrolyzed, noladin ether might lead to drug development.

A. SYNTHESIS

Unlike classical neurotransmitters and neuropeptides, AEA and 2-AG are not stored in intracellular compartments, but are produced on demand by receptor-stimulated cleavage of lipid precursors. The AEA precursor is an *N*-arachidonoyl-phosphatidylethanolamine (NArPE), which is believed to originate from the transfer of arachidonic acid from the *sn*-1 position of 1,2-*sn*-di-arachidonoylphosphatidylcholine to phosphatidylethanolamine, catalyzed by a calcium-dependent *N*-acyltransferase (*trans*-acylase) (Fig. 2). NArPE is then cleaved by a yet uncharacterized *N*-acylphosphatidylethanolamine (NAPE)-specific phospholipase D, which releases AEA and phosphatidic acid (Fig. 2). At present, it is not yet clear whether the *N*-acyltransferase or the (NAPE-specific) phospholipase D controls the ratelimiting step of AEA synthesis (Hansen *et al.,* 2000). However, a similar route can be operational also for the synthesis of the other cannabimimetic NAEs, since their precursors, *N*-acylethanolamine phospholipids, are ubiquitous constituents of animal and human cells, tissues, and body fluids (Schmid, 2000). An illustration of the receptor-dependent release of AEA from membrane phospholipids has been provided by microdialysis studies, which suggest that AEA may be released in the brain striatum upon activation of D_2-type dopamine receptors and that such release may be involved in counterbalancing the stimulatory effects of dopamine on motor activity (Giuffrida *et al.,* 1999; Beltramo *et al.,* 2000). An alternate synthesis of AEA can occur through the direct enzymatic condensation between the free arachidonic acid and ethanolamine (Fig. 2). This pathway is catalyzed by an anandamide synthase, which is most active at alkaline pH (8.5–10) and is independent of ATP and coenzyme A. Yet AEA synthase requires high micromolar concentrations of arachidonic acid and high millimolar concentrations of ethanolamine, i.e., much higher than the normal intracellular amounts of these compounds (Kruszka and Gross, 1994; Devane and Axelrod, 1994). Furthermore, an AEA synthase partially purified from porcine brain (Ueda *et al.,* 1995) was found to exhibit the same pH and temperature dependence, and the same inhibition profiles, of fatty acid amide hydrolase (FAAH), the enzyme catalyzing AEA hydrolysis (see below). Therefore, it is now believed that AEA synthase is instead an AEA hydrolase "working in reverse" under nonphysiological *in vitro* conditions (i.e., a "reverse" FAAH). In keeping with this concept, a cloned FAAH expressed in COS-7 cells was able to synthesize AEA when incubated with high millimolar concentrations of ethanolamine (Kurahashi *et al.,* 1997). To date, only another "AEA synthase" has been found in the mouse uterus (Paria *et al.,* 1996), and has been reported to show different profiles compared to the FAAH in the same tissue.

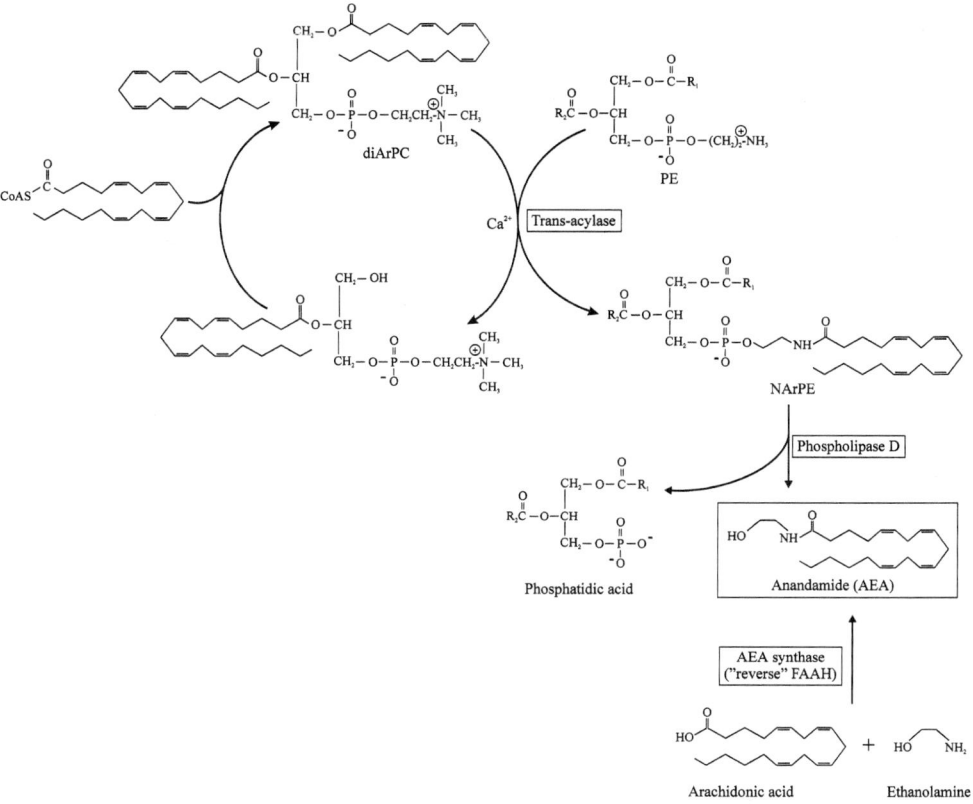

FIGURE 2 Possible pathways for anandamide biosynthesis in cells. In the upper part, membrane N-arachidonoylphosphatidylethanolamine (NArPE) is formed by the transfer of arachidonic acid from the sn-1 position of 1,2-sn-di-arachidonoylphosphatidylcholine (diArPC) to phosphatidylethanolamine (PE), catalyzed by a calcium-dependent N-acyltransferase (*trans*-acylase). Hydrolysis of NArPE by a yet uncharacterized phospholipase D releases anandamide (AEA) and phosphatidic acid. In the lower part, AEA is directly synthesized by an AEA synthase (probably an AEA hydrolase working in reverse, i. e., a "reverse" FAAH), which condensates arachidonic acid and ethanolamine.

Remarkably, AEA synthase and hydrolase activities in the uterus were regulated by sex hormones in the same way (Maccarrone *et al.*, 2000a). Whether a different enzyme or a "reverse" FAAH, the synthase activity may contribute to maintaining the equilibrium of AEA, at least in specific cell compartments where arachidonic acid and ethanolamine reach high concentrations.

The most likely route of 2-AG biosynthesis involves the same enzymatic cascade responsible for the generation of the second messengers 1, 2-diacylglycerol (DAG) and inositol trisphosphate (Fig. 3). Phospholipase C acting on phosphatidylinositol bisphosphate produces DAG, which is converted to 2-AG by an *sn*-1-DAG lipase activity (Stella *et al.*, 1997). An additional pathway for 2-AG formation may involve the hydrolysis of lysophospholipids or triacylglycerols (Fig. 3), and has

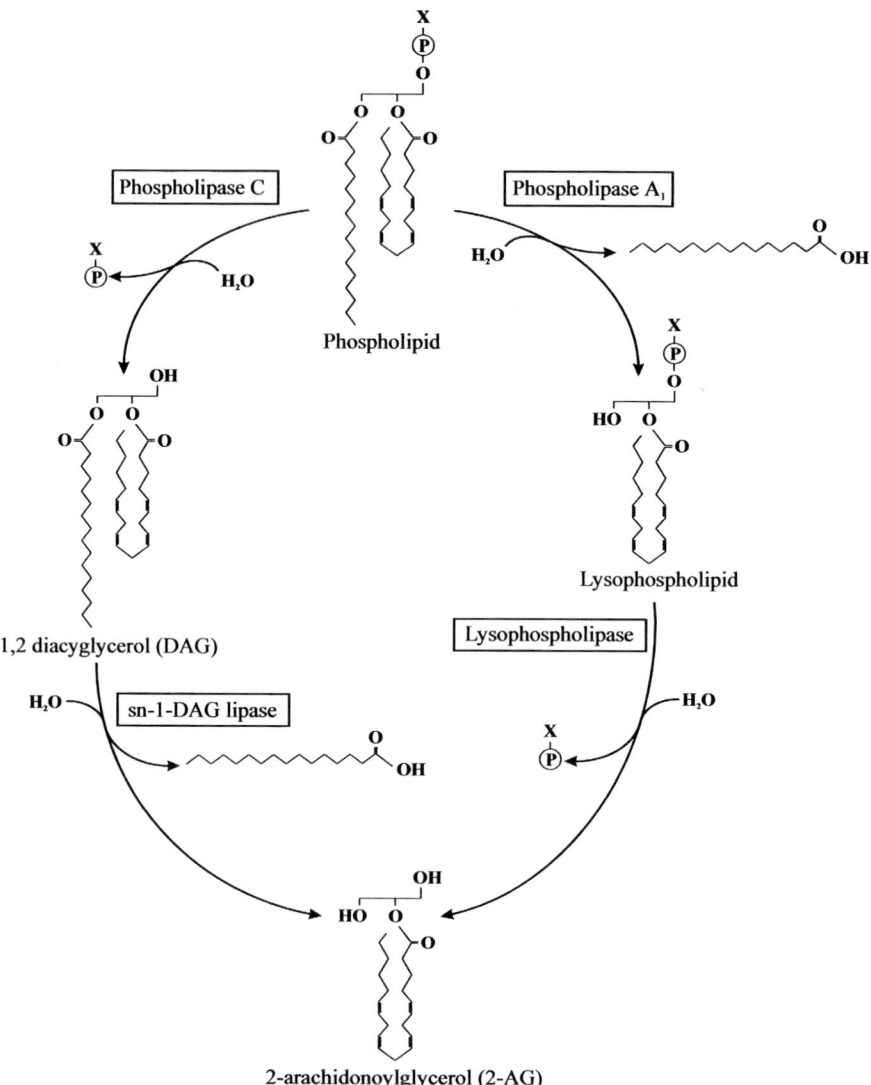

FIGURE 3 Possible pathways for 2-AG formation in cells. On the left, phospholipase C cleaves a phospholipid and generates 1,2-diacylglycerol (DAG). This diglyceride is then hydrolyzed by an *sn*-1-DAG lipase, which releases 2-AG. In the alternate route (on the right) a phospholipase A_1 converts a phospholipid into a lysophospholipid, which is further hydrolyzed by a lysophospholipase C to yield 2-AG.

been recently reviewed (Sugiura and Waku, 2000). Regardless of the route involved, 2-AG biosynthesis may be triggered by neural activity or by occupation of membrane receptors. For instance, in the rat hippocampus stimulation of the Schaffer collaterals, an excitatory fiber tract that utilizes glutamate as a neurotransmitter, strongly increases 2-AG levels (Stella *et al.*, 1997).

B. DEGRADATION

The biological activity of AEA and 2-AG is terminated by their removal from the extracellular space, which occurs through a two-step process: (1) cellular uptake by a high affinity transporter followed by (2) intracellular degradation by a fatty acid amide hydrolase (Fig. 4). Several properties of a selective AEA membrane transporter (AMT) have been characterized, though its molecular structure remains unknown (for a recent review see Hillard and Jarrahian, 2000). AMT has been shown to take up AEA according to a saturable process, which has the characteristics of a facilitated diffusion: it is bidirectional and independent on both energy and sodium, unlike amine and amino acid transporters (Hillard *et al.*, 1997). Moreover, the use of different analogues (Jarrahian *et al.*, 2000) has led to some generalizations on the properties of AMT: (1) at least one *cis* double bond in a long alkyl chain must be present for binding, (2) the AEA binding site can tolerate very bulky additions to the head group region, provided they are hydrophobic, (3) the carrier does not require a secondary amine in the carboxamido group, (4) the presence of an electron-donating group at the $2'$ position can stabilize the binding to the carrier but this interaction is very sensitive to the orientation and position of this group, and (5) aromatic substitutions in the head group region stabilize the binding to the carrier, possibly because of the introduction of aromatic stacking interactions (Hillard and Jarrahian, 2000).

Recently, AMT has been shown to be functionally coupled to CB1 receptors through nitric oxide (NO): activation of CB1 receptors by AEA releases NO, which in turn activates AMT (perhaps by nitrosylating a cysteine residue in the binding site) thus stimulating the removal of AEA from the extracellular space (Maccarrone *et al.*, 2000b). This regulatory loop represents a "timer," by which activation of CB1 receptors by AEA triggers the termination of the activity of AEA at the receptor itself. At present it is not yet clear whether 2-AG is transported into cells through AMT or through a different transporter (Fig. 4). Recent evidence, showing that in rat glioma (C6) cells NO stimulates the uptake of AEA and 2-AG to similar extents, and that the AMT inhibitor AM404 (Piomelli *et al.*, 1999) can block the uptake of both endocannabinoids with the same potency, suggests that a distinct 2-AG membrane transporter (AGMT) may not exist or it is very similar to AMT, though it has an efficacy for 2-AG lower than that for AEA (Bisogno *et al.*, 2001). Interestingly, human platelets have been recently shown to take up 2-AG through a selective AGMT (Maccarrone *et al.*, 2001c). AMT has been demonstrated in several cell types derived from brain and from the immune system (Hillard and Jarrahian, 2000). In particular, it has been characterized in many human cells like neuroblastoma (CHP100) and lymphoma (U937) (Maccarrone *et al.*, 1998b), platelets (Maccarrone *et al.*, 1999), endothelial cells (Maccarrone *et al.*, 2000b), mastocytes (Maccarrone *et al.*, 2000c), and peripheral lymphocytes (Maccarrone *et al.*, 2001b). It is noteworthy that the kinetic constants calculated for AMT in different human cells suggest that different levels of the same AMT, rather than different isoforms of it, are expressed on the cell surface. A feature of AMT that deserves further investigation is its ability to "work in reverse," i.e., to extrude AEA

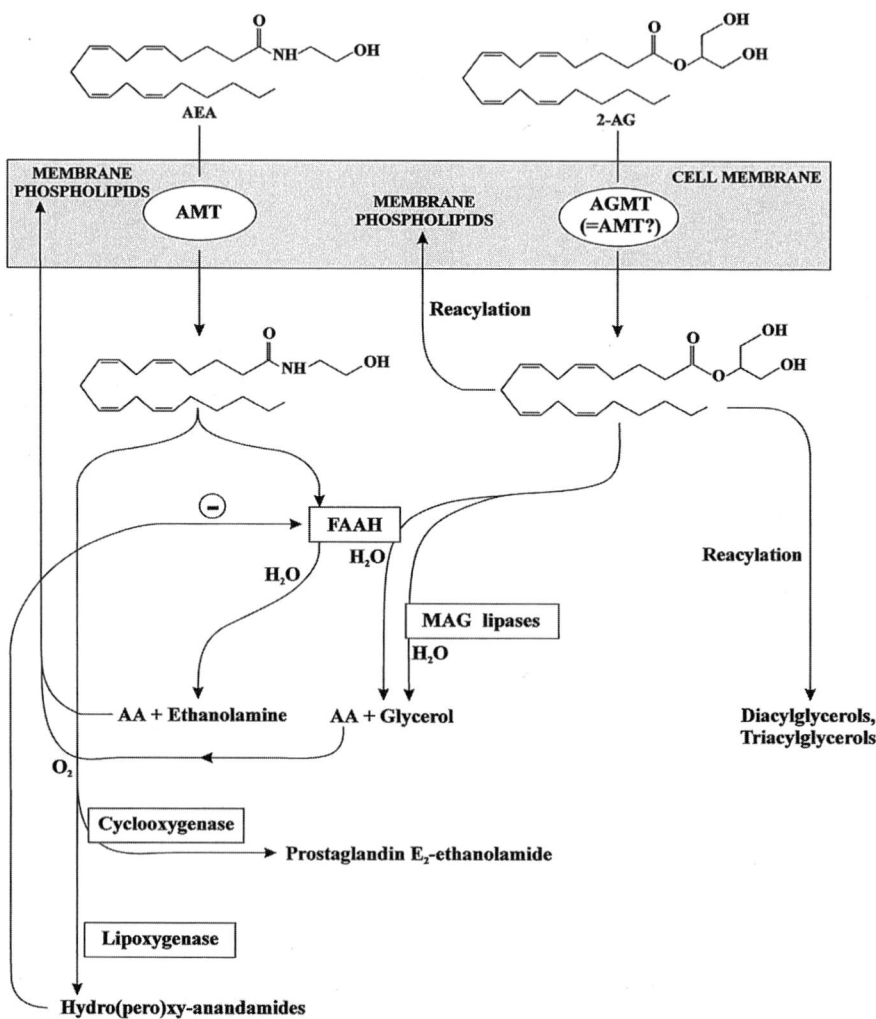

FIGURE 4 Degradation of AEA and 2-AG. AEA is transported into the cell by an AEA membrane transporter (AMT), which might or might not be identical to the 2-AG membrane transporter (AGMT). Once taken up, AEA is hydrolyzed by fatty acid amide hydrolase (FAAH), or can be oxidized by the enzymes of the "arachidonate cascade": cyclooxygenase, which generates a prostaglandin E_2-ethanolamide, or lipoxygenase, which produces hydro(pero)xy-anandamides able to back inhibit FAAH. 2-AG is hydrolyzed by FAAH or by monoacylglycerol (MAG) lipase. It can be reacylated to generate phosphoglycerides or neutral glycerolipids. Arachidonic acid (AA) released from AEA or 2-AG is immediately reincorporated into membrane lipids.

outside the cell (Hillard and Jarrahian, 2000). Indeed, it is increasingly evident that AEA can act backward, and AMT might play a role in regulating this retrograde signaling (MacDonald and Vaughan, 2001; Wilson and Nicoll, 2001; Baringa, 2001).

Once taken up by cells, both AEA and 2-AG are substrates for the enzyme fatty acid amide hydrolase (arachidonoylethanolamide amidohydrolase, EC 3.5.1.4;

FAAH), which breaks the amide or the ester bond and releases arachidonic acid and ethanolamine or glycerol, respectively (Fig. 4). FAAH is a membrane-bound enzyme found mainly in microsomal and mitochondrial fractions and has an optimum pH around 9 (reviewed by Ueda et al., 2000). However, in addition a distinct AEA hydrolase with optimum pH around 5 has been described in the mitochondrial-lysosomal fraction of a human megakaryoblastic (CMK) cell line, and found to exhibit a different substrate specificity compared to FAAH (Ueda et al., 1999). FAAH also shows an esterase activity for monoacylglycerols (Di Marzo et al., 1998b; Goparaju et al., 1998, 1999a) and methyl esters of fatty acids (Kurahashi et al., 1997; Goparaju et al., 1999a; Patricelli and Cravatt, 1999). In fact, 2-AG was hydrolyzed at a rate several fold higher than AEA by recombinant rat and porcine FAAH (Goparaju et al., 1998, 1999a). It has been proposed that FAAH controls the cellular uptake of AEA by creating and/or maintaining an inward concentration gradient that drives the facilitated diffusion of AEA through AMT (Deutsch et al., 2001). Although FAAH is not the only factor controlling AEA transport (Day et al., 2001), its pivotal role in AEA degradation may explain why FAAH (Maccarrone et al., 2000e), but not AMT (Maccarrone et al., 2001b), is modulated under pathophysiological conditions. In this context, the use of potent inhibitors of FAAH as therapeutic agents for sleep disorders or as analgesics has been proposed (Boger et al., 2000).

FAAH cDNA contains an "amidase signature" sequence (Cravatt et al., 1996). The sequence of human (Giang and Cravatt, 1997), porcine (Goparaju et al., 1999b), mouse (Giang and Cravatt, 1997) and rat FAAH (Cravatt et al., 1996) shows 73% overall identity and 90% identity in the "amidase signature" sequence. Although the mammalian FAAH mRNAs differ in size across species, the encoded proteins are all 579 amino acids in length (Ueda et al., 2000). Three domains have been identified in FAAH: a transmembrane domain at the N-terminus (Cravatt et al., 1996), which does not affect enzyme activity but instead directs protein oligomerization (Patricelli et al., 1998); a serine- and glycine-rich domain, which contains the amidase signature sequence (Kobayashi et al., 1997) and spans residues 215–257 in mammalian FAAH; and a proline-rich domain, which is homologous to the class II SH3-binding domain (Arreaza and Deutsch, 1999). In this context, it seems noteworthy that FAAH activity does not seem to be modulated by covalent modifications such as farnesylation, phosphorylation, glycosylation, or nitrosylation (Maccarrone et al., 2000d).

As for the cataytic mechanism, site-directed mutagenesis has demonstrated that a serine (S241) is the catalytic nucleophile in the active site of FAAH (Patricelli et al., 1999), though the enzyme does not use the typical serine–histidine–aspartic acid triad found in serine proteases (Patricelli and Cravatt, 1999). A lysine residue (K142) is also critical for the catalytic activity of FAAH, to such an extent that although wild-type FAAH hydrolyzes amides and esters with similar catalytic efficiencies, K142A FAAH hydrolyzes esters more than 500-fold faster than amides (Patricelli and Cravatt, 1999). These data suggest that lysine 142 is the general base involved in increased serine nucleophilicity. A strong preference of FAAH for acyl chains at least 9 carbons long has been shown, which is dictated by a single isoleucine residue (I491) (Patricelli and Cravatt, 2001). FAAH has been found and characterized in the same human cells bearing AMT, i.e., neuroblastoma (CHP100) and lymphoma (U937) cells (Maccarrone et al., 1998b), platelets (Maccarrone et al., 1999), endothelial cells (Maccarrone et al., 2000b), mastocytes (Maccarrone et al., 2000c), and peripheral lymphocytes (Maccarrone et al., 2001b). It has also been demonstrated in human brain, where it has an apparent Michaelis–Menten constant (K_m) of 2 μM, a maximum velocity (V_{max}) of 800 pmol/min/mg of protein, and an activation energy of 43.5 kJ/mol with AEA as substrate (Maccarrone et al., 1998b). The hydro(pero)xides generated from AEA by lipoxygenase activity were found to inhibit FAAH with apparent inhibition constants (K_i) in the low micromolar range (Maccarrone et al., 1998b). These hydro(pero)xy-anandamides are the most powerful natural inhibitors of FAAH yet discovered, and because they may be formed *in vivo* they might play a role in controlling AEA degradation (Fig. 4). On the other hand the products of AEA generated by cyclooxygenase-2 (Yu et al., 1997; Burstein et al., 2000) do not seem to affect FAAH activity (Maccarrone et al., 2000d). As reported above, 2-AG can be hydrolyzed by FAAH, as well as by an independent monoacylglycerol (MAG) lipase, into arachidonic acid and glycerol (Fig. 4). The latter enzyme has been found in porcine brain (Goparaju et al., 1999a) and in macrophages (Di Marzo et al., 1999), and is regulated by stimuli that do not modulate FAAH activity (Freysz et al., 1991; Farooqui et al., 1993). 2-AG can also be reacylated to form membrane phospholipids or neutral di- and triacylglycerols (Fig. 4), as shown in intact cells (Di Marzo et al., 1998b, 1999; Maccarrone et al., 2000b).

C. MOLECULAR TARGETS

The molecular targets of AEA and 2-AG are the receptors that together with these endogenous cannabinoids, the membrane transporters, and the hydrolases constitute the "endocannabinoid system." The receptor family includes the CB1 cannabinoid receptors, present mainly on central and peripheral neurons, the CB2 cannabinoid receptors expressed predominantly by immune cells, the non-CB1/non-CB2 cannabinoid receptors, the noncannabinoid receptors, and the vanilloid receptors (Reggio and Traore, 2000; Khanolkar et al., 2000; Howlett and Mukhopadhyay, 2000; Pertwee, 2000, 2001). Several biological actions of AEA, 2-AG, or both, occurring via activation of one or more of these targets, are summarized in Table II.

CB1 and CB2 receptors have been cloned and characterized by Matsuda et al., (1990) and by Munro et al. (1993), respectively, and have been shown to belong to the family of the "seven *trans*-membrane spanning receptors." Both are coupled to G-proteins, particularly those of the $G_{i/o}$ family (Howlett and Mukhopadhyay, 2000). Recently, the domains of the CB1 receptor that interact with different G-protein subtypes have been identified (Mukhopadhyay and Howlett, 2001). Signal transduction pathways regulated by CB receptor-coupled G-proteins include the inhibition of adenylyl cyclase, the regulation of ionic currents (inhibition of

TABLE II Effects of AEA, 2-AG, or Both at Different Molecular Targets

Molecular target	Biological actions
"Classic" cannabinoid (CB1 or CB2) receptors	Inhibition of adenylyl cyclase (i.e., of forskolin-induced cAMP formation)
	Inhibition of L-type, N-type, and P/Q-type Ca^{2+} channels
	Activation of inwardly rectifying K^+ channels
	Activation of the mitogen-activated protein kinase (MAPK) pathway
	Activation of cytosolic phospholipase A_2
	Activation of neuronal focal adhesion kinase (FAK)
	Activation of nitric oxide synthase (NOS)
	Suppression of prolactin receptors
	Decline in systemic and mesenteric vascular resistance
	Protection against apoptosis
"Nonclassic" (non-CB1/non-CB2) cannabinoid receptors (CBn)	Release of arachidonic acid
	Activation of the mitogen-activated protein kinase (MAPK) pathway
	Inhibition of gap junction activity
	Inhibition of gap junction-mediated and glutamate-triggered Ca^{2+} waves
	Activation of platelets
	Hypotension and mesenteric vasodilation
	Stimulation of cytokine-dependent proliferation of lymphoid cells
Non cannabinoid receptors and/or non-receptor-mediated actions	Inhibition of L-type Ca^{2+} channels
	Inhibition of Shaker-related voltage-gated K^+ channels
	Inhibition of serotonin 5-HT_3 receptor-mediated currents
	Activation of N-methyl-D-aspartate (NMDA)-mediated Ca^{2+} currents
	Release of arachidonic acid
	Increase in synaptosomal membrane lipid order
	Activation of protein kinase C
	Activation of platelets
Vanilloid (VR1) receptors	Vascular relaxation
	Release of the vasoactive calcitonin gene-related peptide (CGRP)
	Rise in intracellular Ca^{2+}
	Mitochondrial uncoupling
	Induction of apoptosis

voltage-gated L, N, and P/Q-type Ca^{2+} channels, activation of K^+ channels), the activation of focal adhesion kinase (FAK), of mitogen-activated protein kinase (MAPK), of cytosolic phospholipase A_2 and of nitric oxide synthase (NOS), and others (Table II). AEA and 2-AG show higher affinity for the CB1 than for the CB2 receptor and structure–activity relationship (SAR) studies have suggested either that only acyl chains that can assume a tightly folded (U-shaped) conformation can bind to CB1 receptors, or that ligand flexibility is very important for this binding (Reggio and Traore, 2000). The binding to CB1 reseptor requires that the endocannabinoid should have an aliphatic chain of 20 to 22 carbons, with at least three nonconjugated *cis* double bonds with a saturated tail of at least the last five carbons (Reggio and Traore, 2000). The head group can be either polar or nonpolar but should not be bulky. In the absence of X-ray crystallographic and nuclear magnetic resonance data, suitably tailored molecular probes have shed light on the structural requirements for ligand–receptor interactions (Khanolkar *et al.*, 2000). In general, after activation CB1 receptors undergo phosphorylation (Garcia *et al.*, 1998) and internalization (Hsieh *et al.*, 1999), which may be followed by recycling into the membrane if the time of treatment is short (Hsieh *et al.*, 1999). The carboxyl-terminus of the CB1 receptor is needed for internalization (Hsieh *et al.*, 1999), a process that has been recently elucidated in hippocampal neurons (Coutts *et al.*, 2001). Instead, the CB2 receptors seem to be constitutively active and phosphorylated (Bouaboula *et al.*, 1999), and it is not yet clear whether their dephosphorylation occurs through endocannabinoid-elicited signals, and what might be the nature of these endogenous signals (Bouaboula *et al.*, 1999).

Evidence has emerged that in addition to CB1 and CB2 receptors there are other molecular targets through which the endocannabinoids might induce a biological activity. "Nonclassic" CBn (i.e., non-CB1/non-CB2) cannabinoid receptors seem to mediate the effect of AEA and congeners on arachidonate release and other processes listed in Table II. In this context, it should be recalled that a peripheral "CB2-like" cannabinoid receptor has been postulated as the target on which PEA may act to prevent mast cell activation (Facci *et al.*, 1995) and relieve inflammatory pain (Calignano *et al.*, 1998), and that a new "endothelial receptor" for AEA, activated by abnormal cannabidiol (**IX**) and blocked by cannabidiol (**X**), induces hypotension and mesenteric vasodilation (Jarai *et al.*, 1999). Further noncannabinoid receptor-mediated activities have been demonstrated for the endocannabinoids, independent of CB1, CB2, or CBn receptors and involving in particular the regulation of ionic currents (Table II).

A new molecular target of AEA that is attracting great interest is the vanilloid receptor (VR1), a 6-membrane spanning protein with intracellular N- and C-terminals and a pore-loop between the fifth and sixth transmembrane helices (Caterina *et al.*, 1997, 1999). VR1 is activated by vanilloid ligands such as capsaicin (**XI**) as well as by noxious stimuli such as heat and acids, and thus it can be viewed as a molecular integrator of noxious stimuli in peripheral terminals of primary sensory neurons (Szallasi and Di Marzo, 2000). The observation that capsaicin and certain inhibitors of AEA transporter had similar structures (Di Marzo *et al.*, 1998c) led to the development of arvanil (**XII**), an anandamide–capsaicin hybrid molecule that shares the substituted phenyl polar head group of capsaicin, the hydrophobic *cis*-tetraene carbon chain of AEA and the polar carboxamido group of both AEA and capsaicin (Melck *et al.*, 1999). Arvanil was found to have higher affinity than AEA for CB1 receptors and the same affinity of capsaicin for VR1 receptors. Since then, a number of studies have pointed toward a physiological role for AEA as VR1 agonist (Zygmunt *et al.*, 1999; Smart *et al.*, 2000; Maccarrone *et al.*, 2000f), leading to the concept that AEA, besides being an endocannabinoid, is also a true "endovanilloid" (De Petrocellis *et al.*, 2001). Activation of VR1 by AEA occurs at an intracellular binding site (De Petrocellis *et al.*, 2001), exerting the effects summarized in Table II. Within them most interesting is the induction of apoptosis (programmed cell death) in human neuronal and immune cells (Maccarrone *et al.*, 2000f). The activation of CB1 receptors present in these cells counteracts this effect, suggesting an interplay between the different targets of AEA, which might have a broader regulatory meaning. Future research will answer to the possible implications (also in therapy) of the multitargeted actions of AEA and congeners, also in view of the fact that new hybrid molecules that bridge the properties of different neurotransmitters, such as *N*-acyl-dopamines (Bisogno *et al.*, 2000), will help to unravel the cross-talk between the different players in the game. The "endocannabinoid system," as we presently know it, is depicted in Fig. 5, where the interactions of AEA with cannabinoid and vanilloid receptors, with its membrane transporter and with FAAH are reported.

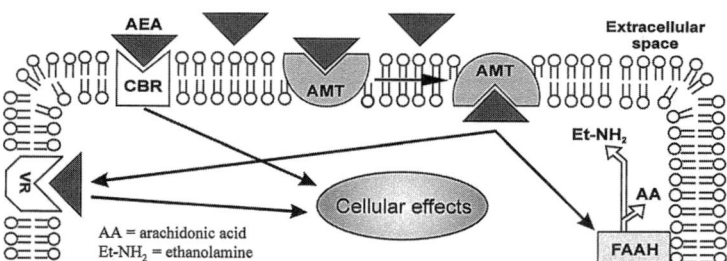

FIGURE 5 The endocannabinoid system in human cells. Binding of extracellular anandamide (AEA) to cannabinoid receptors (CBR) leads to different cellular effects (depending on the cell type), terminated by the AEA membrane transporter (AMT)-mediated uptake of AEA. Once taken up, AEA can be rapidly cleaved by membrane-bound fatty acid amide hydrolase (FAAH), releasing arachidonic acid (AA) and ethanolamine (Et-NH$_2$). Alternatively, AEA can activate vanilloid receptors (VR) by binding to an intracellular site, thus triggering cellular responses that sometimes may be opposite to those elicited by AEA through CBR.

III. ACTIONS IN THE CENTRAL NERVOUS SYSTEM

The levels of the endocannabinoids in brain and in neuronal cells (Table I) and the multiple actions of these compounds (Table II) suggest that they might play a role as modulators of the release and/or the activity of other neurotransmitters, or as true neurotransmitters themselves. However, probably due to their rapid degradation *in vivo* (Di Marzo and Deutsch, 1998), it has not always been possible to extend to the endocannabinoids the many pharmacological actions previously described for THC and other exogenous cannabinoids. Therefore, the physiological significance of the endocannabinoid system is still not fully understood, though important information is becoming available on the roles in the control of critical neuronal functions such as pain initiation (Calignano *et al.*, 1998, 2000; Walker *et al.*, 1999), psychomotor behavior (reviewed by Giuffrida and Piomelli, 2000), and vision (Yazulla *et al.*, 1999). Other actions of AEA, 2-AG, or both in the central nervous system have been nicely reviewed (Di Marzo, 1998; Howlett and Mukhopadhyay, 2000; Pertwee, 2000) and are summarized in Table III. An interesting aspect of the effects of (endo)cannabinoids relevant for neurodegenerative diseases is their use for the reversal, through appetite stimulation, of weight loss experienced by AIDS patients. The oral preparation of THC, dronabinol (Marinol), is already available in the United States for this purpose (Pertwee, 2000). The recent finding that hypothalamic endocannabinoids appear to be under negative control by leptin, the primary signal through which the hypothalamus senses nutritional state and modulates food intake and energy balance (Di Marzo *et al.*, 2001), suggests that indeed also endocannabinoids are natural orexigenic (appetite-stimulating) factors, which might have implications beyond the nutritional field (Mechoulam and Fride, 2001). More recently, evidence has accumulated showing that endocannabinoids are involved in two neuropathologies with high incidence in humans: multiple sclerosis and Parkinson's disease.

A. MULTIPLE SCLEROSIS

Multiple sclerosis (MS) is a chronic, demyelinating disease of the central nervous system, leading to neuronal damage and paresis. In addition, life-impairing symptoms often develop, such as pain, spasticity, and tremor, which are difficult to treat with conventional drugs. This prompted some patients to try alternative medicine such as the oral use of THC or smoking cannabis (Pertwee, 2000). Recently, it has been demonstrated that exogenous agonists of CB receptors, in particular of the type 1, can inhibit spasticity in the Biozzi ABH mouse chronic relapsing experimental allergic encephalomyelitis (CREAE) model (Baker *et al.*, 2000). This model shows demyelinization and axonal loss of nerve fibers, which result in a disorder that bears several similarities to MS, including relapsing-remitting paralytic episodes and tremor and spasticity of limb muscles during postrelapse remission (Di Marzo *et al.*, 2000a). In particular, in the CREAE model tremors

TABLE III Interference of AEA, 2-AG, or Both with Neuronal Functions and Neurotransmission in the Central Nervous System (CNS)

Biological actions	Localization in CNS
Control of pain initiation	Amygdala, thalamus, superior colliculus, periaqueductal gray, rostral ventromedial medulla
Control of psychomotor disorders	Basal ganglia
Control of the secretion of pituitary hormones Control of wake/sleep cycles Control of thermogenesis Control of appetite	Hypothalamus
Control of synaptic plasticity Impairment of working memory and of memory consolidation, possibly due to interference with acetylcholine release Inhibition of long-term potentiation Inhibition of glutamatergic transmission	Hippocampus
Control of tremor and spasticity	Whole brain, spinal cord
Inhibition of dopamine synthesis and/or release Inhibition of γ-aminobutyric acid (GABA)ergic transmission	Striatum
Interference with dopaminergic transmission	Basal ganglia
Potentiation of γ-aminobutyric acid (GABA)-mediated catalepsy Suppression of locomotion	Globus pallidus
Blockade of N-methyl-D-aspartate (NMDA) receptors	Cortex and cerebellum
Control of scotopic vision	Retina

occur during voluntary movements after sudden disturbance of the animal, and with higher frequency than in true MS. Cannabinoid receptor antagonism was demonstrated to transiently exacerbate spasticity that occurs during CREAE, suggesting that spasticity might be controlled by the equilibrium of the endocannabinoid system (Baker et al., 2000). Indeed, increased levels of the endocannabinoids AEA and 2-AG, and of the AEA congener PEA, were detected in areas of the central nervous system (such as the spinal cord) associated with nerve damage, in comparison to the levels of these compounds in normal and nonspastic CREAE mice (Baker et al., 2001). In this line, exogenously administered endocannabinoids and PEA were found to ameliorate spasticity, as did selective inhibitors of AEA uptake by AMT and hydrolysis by FAAH, which indirectly enhance its endogenous levels (Baker et al., 2001). These studies seem to suggest a protective role of endocannabinoids against neuronal damage associated with motor disorders, providing evidence for the tonic control of spasticity by the endocannabinoid system. They might also open new horizons for the therapy of MS, and possibly other neuromuscular diseases, pointing to agents that modulate endocannabinoid metabolism and actions. Moreover, they suggest that the equilibrium of the endocannabinoid system can respond to abnormal neuronal signaling and/or neurodegenerative effects in damaged nerves (as those observed during spastic events in CREAE), further supporting a link between the endocannabinoids and the networks of other neurotransmitters (Di Marzo et al., 2000a). Finally, they seem to give a solid ground to the still limited number of clinical studies indicating that THC analogues can ameliorate the symptoms of MS (Martyn et al., 1995).

B. PARKINSON'S DISEASE

The high density of CB1 cannabinoid receptors within the basal ganglia has suggested a potential role for endocannabinoids in the control of voluntary movement and in basal ganglia-related movement disorders such as Parkinson's disease. Stimulation of CB receptors in the globus pallidus by exogenous agonists reduces the uptake of the inhibitory neurotransmitter λ-aminobutyric acid (GABA) (Maneuf et al., 1996), and reduces voluntary movements, producing Parkinson-like symptoms (Pertwee and Wickens, 1991). Within the basal ganglia, CB receptors are particularly abundant on the terminals of GABAergic projections from the striatum to the globus pallidus or to the substantia nigra pars reticulata, i.e., the "indirect" or "direct" striatal output pathways, respectively (Herkenham et al., 1991). Stimulation of CB receptors also reduces the anti-parkinsonian actions of D2, but not D1, dopamine receptor agonists (Maneuf et al., 1997), suggesting a role for endocannabinoids in the modulation of signaling by the indirect striatal output. In reserpine-treated rats, an animal model of Parkinson's disease, suppression of locomotion has been recently associated with a remarkable increase of 2-AG (but not of AEA) in the globus pallidus, but not in other brain regions (Di Marzo et al., 2000b). Consistently, stimulation of locomotion in the reserpine-treated rats with selective agonists of D1 and D2 dopamine receptors was accompained by a reduction in the levels of both AEA and 2-AG in the globus pallidus, whereas full restoration of locomotion in the reserpine-treated rat was achieved by coadministration of a D2 dopamine receptor agonist and an antagonist of CB1 receptors (Di Marzo et al., 2000). These findings suggest that selection, initiation, and execution of movement by stimulation of the GABAergic indirect striatal output pathway, at least in the parkinsonian brain, may be under negative control by endocannabinoids. They also suggest that modulation of the endocannabinoid system might prove useful in the treatment of Parkinson's disease, and possibly of other basal ganglia-related movement disorders (Di Marzo et al., 2000b). Modulation of the expression of CB1 receptors seems a critical point in the development of Parkinson's disease, because the CB1 receptor mRNA falls in the striatum of reserpine-treated rats (Silverdale et al., 2001). This observation confirms the potential therapeutic use of CB receptor antagonists for the treatment of the symptoms of Parkinson's disease. The therapeutic potential of selective inhibitors of endocannabinoid degradation through uptake and hydrolysis is still under investigation.

IV. ACTIONS IN THE PERIPHERY

The endocannabinoids, by binding to both CB1 and CB2 receptors and thus inhibiting adenylyl cyclase or activating MAPK and cytosolic phospholipase A_2 (Table II), may act as signaling molecules between different immune cells or between sensory fibers and blood cells. Indeed, endogenous cannabinoids have been recognized as immune regulators (Berdyshev, 2000, and by Salzet *et al.*, 2000) and cardiovascular modulators (Kunos *et al.*, 2000), and have been shown to control inflammation and related disorders (De Petrocellis *et al.*, 2000). Also a role for the peripheral endocannabinoid system in the control of pain initiation (Calignano *et al.*, 2000) and in regulating the urinary function (Martin *et al.*, 2000) has been documented, along with a number of other activities (Table IV).

TABLE IV Peripheral Actions of AEA, 2-AG, or Both

Biological actions	Localization in the periphery
Profound decrease in blood pressure (hypotension) and heart rate (bradycardia)	Cardiovascular system
Reduction of sympathetic tone due to inhibition of norepinephrine release	
Induction of hypotension during hemorrhagic shock	
Induction of hypotension during endotoxic shock	
Vasodilation	
Platelet aggregation	
Repression of interleukin-2 (IL-2) transcription and secretion	Immune system
Stimulation of interleukin-6 (IL-6) synthesis	
Reduction of binding and promoter activity of the nuclear factor of activated T cells (NF-AT)	
Inhibition of tumor necrosis factor-α (TNF-α) production	
Inhibition of interferon-γ (IFN-γ) synthesis	
Down-regulation of rat mast cell activation	
Stimulation of hematopoietic cell growth	
Induction of lymphocyte apoptosis	
Inhibition of leukemia inhibitory factor (LIF) release	
Down-regulation of airway hyperreactivity	
Stimulation of serotonin release	
Inhibition of neutrophil recruitment	
Developmental arrest at the stage of two-cell embryos	Reproductive system
Inhibition of zona-hatching of blastocysts	
Inhibition of implantation	
Acceleration of trophoblast differentiation and outgrowth (low doses of endocannabinoids)	
Inhibition of trophoblast differentiation (high doses of endocannabinoids)	
Induction of blastocyst apoptosis	

AEA causes smooth muscle relaxation in various endothelium-intact and denuded arterial preparations, mimicking the hypotensive and bradycardic effects of THC (Wagner et al., 1998). In anesthetized rats, AEA changes blood pressure and heart rate in a sequential, triphasic manner, consisting of (1) a transient bradycardia and hypotension, followed by (2) a brief pressor increase, and then by (3) a relatively long-lasting depressor effect (Wagner et al., 1998). The initial vagal bradycardia and subsequent pressor changes are not mediated by CB1 receptors (Wagner et al., 1998), whereas the third phase, the prolonged hypotensive effect, is mediated by this type of receptors (Varga et al., 1995). The use of CB1 knockout mice has recently provided definitive evidence to support this concept, showing that AEA fails to elicit hypotension and bradycardia in these animals (Járai et al., 1999). More recently, CB1 receptors have been also shown to mediate at least in part the systemic and portal hemodynamic effects of AEA (Garcia et al., 2001). Endocannabinoids are also mediators of the hypotensive effects of hemorrhagic shock (Wagner et al., 1997) and endotoxic shock (Varga et al., 1998). In those conditions, both AEA generated by circulating macrophages and 2-AG produced by platelets are responsible for the hypotension, even though they act in different ways. 2-AG does not induce hypotension through CB1 or CB2 receptors since it is effective also in CB1 knockout mice, probably because it generates active arachidonate metabolite(s), which in turn cause hypotension via a noncannabinoid receptor mechanism (Járai et al., 2000). On the other hand, AEA seems to act through a non-CB1/non-CB2 "endothelial" cannabinoid receptor, which is fully activated by the neurobehaviorally inactive cannabinoid analogue "abnormal cannabidiol" (Járai et al., 1999). Interestingly, AEA can also induce vasodilation by activating VR1 receptors, thus extending to noncannabinoid receptors its activity in the cardiovascular system (Zygmunt et al., 1999). By the latter mechanism, AEA triggers VR1-mediated release of the calcitonin gene-related peptide (CGRP), which has vasodilatory effects (Zygmunt et al., 1999). Thus, the hypotensive effects of AEA have two components: one activity on the endothelium, where AEA binds to a unique nonclassic cannabinoid receptor, and the other one on sensory nerve terminals, which is endothelium independent and involves VR1 receptor activation and release of the vasoactive CGRP (Kunos et al., 2000).

The endocannabinoids are widely recognized as immunomodulatory molecules, able to regulate macrophage functions, lymphocyte response, and mast cell activation (Berdyshev, 2000; Salzet et al., 2000; De Petrocellis et al., 2000). The immunomodulatory potential of the endocannabinoids is largely based on their ability to regulate synthesis and/or release by immune cells of various cytokines that modulate the immune response (Table IV). Human peripheral lymphocytes have been shown to possess an active FAAH (Maccarrone et al., 2000e), as well as a specific AMT and CB1 and CB2 receptors (Maccarrone et al., 2001b). Also human mast cells (HMC-1) have been shown to have the biochemical tools to degrade AEA, i.e., a specific AMT and a FAAH, but unlike rat mast cells (Facci et al., 1995) they do not express functional cannabinoid receptors on their surface (Maccarrone et al., 2000c). The absence of cannabinoid receptors in HMC-1 cells is noteworthy, because CB2 receptors in rat mast cells (Facci et al., 1995) and peripheral CB1 receptors in rats (Richardson et al., 1998) have been proposed to down-regulate inflammation. AEA or PEA do not modulate either HMC-1 activation through noncannabinoid receptor-mediated mechanisms, indicating major differences between rats and humans in mast cell activation (Maccarrone et al., 2000c). Interestingly, these findings do not favor the hypothesis that an autacoid local inflammation antagonism (ALIA) mechanism might be operational in humans, as it seems to be the case in rats (Facci et al., 1995). Recently two additional roles of periperal endocannabinoids have been proposed, i.e., in the activation of human platelets and in the regulation of human fertility.

A. PLATELET ACTIVATION

AEA has been described as an endothelium-derived hyperpolarizing factor (EDHF), however this hypothesis was debated and recent data suggest that EDHF is instead a cytochrome P-450 metabolite (Fisslthaler et al., 1999). Whether or not it is an EDHF, AEA is likely to play an important role in the control of vascular tone, because both rat endothelial cells and macrophages can release it (Varga et al., 1998). AEA (≤ 1.2 mM) activates washed human platelets by a cannabinoid receptor-independent mechanism, which involves a rise in intracellular calcium and does not depend on the arachidonate cascade (Maccarrone et al., 1999). In the same study it was shown that human platelets have the biochemical machinery to degrade AEA, i.e., a high-affinity transporter (AMT) and a fatty acid amide hydrolase (FAAH). Afterward, these data have been largely confirmed by Braud et al. (2000), who found that AEA in rabbit platelets is active at physiological concentrations (≤ 10 μM) when used in combination with CaCl$_2$ and fibrinogen. Altogether these data suggest that AEA alone is an unlikely physiological agonist of platelets, but it can rather act in vivo as a coagonist in combination with other "classic" aggregating molecules such as arachidonic acid, fibrinogen, or thrombin. The role of endocannabinoids in the cross-talk between platelets and endothelium might be crucial in thrombosis. This would require an interplay between different blood cells in regulating the peripheral endocannabinoid system, and therefore the cardiovascular activities of these lipid mediators (Wagner et al., 1998, 1999; Járai et al., 1999, 2000). A further study has shown that 2-AG, unlike AEA, can activate human platelets also in platelet-rich plasma (Maccarrone et al., 2001c). The activation occurs at physiological concentrations of 2-AG and is inhibited by selective antagonists of both CB1 and CB2 receptors, which instead are ineffective on platelet activation by AEA (Maccarrone et al., 2000b). The activation by 2-AG is paralleled by an increase of intraplatelet calcium and inositol-1,4,5-trisphosphate and by a decrease of cyclic AMP. Moreover, activation by 2-AG of platelets preloaded with 2-AG induces an approximately 3-fold increase in 2-AG release, according to a CB receptor-dependent mechanism. On the other hand, ADP and collagen counteract the activation of platelets by 2-AG, whereas 5-hydroxytryptamine (serotonin) potentiates and extends its effects. Remarkably, ADP and collagen also reduce 2-AG release from 2-AG-activated platelets, whereas 5-hydroxytryptamine further increases it (Maccarrone et al., 2001c). Therefore, it can be suggested that 2-AG might elicit platelet activation by stimulating the release of further molecules of 2-AG from the target cells. The ability of physiological agonists to modulate the release of 2-AG from platelets might be relevant also for other actions of this endocannabinoid on cardiovascular cells. Taken together, these findings suggest a so far unnoticed interplay between the peripheral endocannabinoid system and physiological platelet agonists, which might open new avenues for the therapy of platelet-based cardiovascular diseases. They also identify in platelets new types of cannabinoid receptors that mediate platelet activation by 2-AG and are sensitive to both CB1 and CB2 receptor antagonists.

B. FERTILITY

Human reproduction is a remarkably inefficient process, with a loss of about half of all conceptuses before the expected term (Lockwood, 2000). However, the cellular and molecular mechanisms of spontaneous abortion remain largely unknown. Recently, we reported an association between decreased concentrations of FAAH in peripheral lymphocytes and early pregnancy failure in humans, showing that low FAAH predicts miscarriage (Maccarrone et al., 2000e). Progesterone (P), a hormone essential for the maintenance of pregnancy, is known to modulate immune function (Correale et al., 1998) and to elicit an immunological response critical for normal gestation (Szekeres-Bartho et al., 1996). Indeed, P has been shown to favor the development of human T lymphocytes producing type 2 T-helper (Th2) cytokines (interleukins 4 and 10), which inhibit Th1-type cytokines (interleukin-12 and interferon-γ) thus allowing the survival of fetal allograft and therefore a successful pregnancy (Piccinni et al., 1995; Piccinni and Romagnani, 1996). More recently, the P-induced Th2 bias has been found to stimulate the release of leukemia inhibitory factor (LIF) from T lymphocytes, mediated by interleukin-4 (Piccinni et al., 1998). Clinical data showing that women with unexplained recurrent abortions have a reduced LIF production suggest that the latter is indeed critical for implantation and maintenance of a fetus in humans (Piccinni et al., 1998; Sharkey, 1998; Taupin et al., 1999). In this framework, it was observed that P time and dose dependently enhance FAAH activity in peripheral lymphocytes, whereas AMT activity and AEA binding to CB receptors remain unchanged (Maccarrone et al., 2001b). Stimulation of FAAH occurs through up-regulation of gene expression at a transcriptional and translational level, and is partly mediated by the type 2 T-helper cytokines. In fact, lymphocyte treatment with interleukin-4 or with interleukin-10 has a stimulating effect on FAAH, whereas the type 1 T-helper cytokines interleukin-12 and interferon-γ reduce the activity and the protein expression of FAAH. Human chorionic gonadotropin, an important factor in pregnancy (Goldstein, 1994), widely used as a marker to monitor human gestation (Kovalevskaya et al., 1999), has no effect on FAAH activity. Interestingly, good

FAAH substrates such as AEA and 2-AG inhibit the release of LIF from human lymphocytes, whereas PEA, a poor substrate, is ineffective (Maccarrone et al., 2001b). These results represent the first evidence of a link between the hormone–cytokine network responsible for successful pregnancy and the peripheral endocannabinoid system, and suggest that FAAH, but not AMT or CB receptors, might be critical for this link. They might also represent a useful framework for the interpretation of a novel interaction between progesterone and exogenous cannabinoids, recently shown to regulate female sexual receptivity (Mani et al., 2001).

V. SUMMARY AND CONCLUSIONS

A role for the endogenous cannabinoid system in several aspects of human (patho)physiology has been proposed, through the activation of cannabinoid and vanilloid receptors, and via non-receptor-mediated actions. To place this evidence into perspective, we describe the major endocannabinoids and endocannabinoid-like compounds, the pathways of their formation and inactivation, and their molecular targets. We then review the biological actions of the endogenous cannabinoids in the central nervous system and in the periphery, with a focus on some recently discovered activities in multiple sclerosis, Parkinson's disease, platelet activation, and fertility. Taken together, the available evidence suggests that the endocannabinoids might play a role as modulators of neuronal networks by controlling the release and/or the activity of other neurotransmitters. Besides a role as neuromodulators, endogenous cannabinoids might act as true neurotransmitters, as supported by their concentrations in the brain and in neuronal cells. In line with this concept, the peripheral endocannabinoids interfere with hormone–cytokine networks, including those that regulate fertility. In perspective, investigations aimed at understanding how the endocannabinoid system is integrated within neuronal, hormonal, and cytokine networks will confirm the relevance of these newly discovered lipid mediators for human (patho)physiology.

REFERENCES

Arreaza, G., and Deutsch, D. G. (1999). Deletion of a proline-rich region and a transmembrane domain in fatty acid amide hydrolase. *FEBS Lett.* **454,** 57–60.

Baker, D., Pryce, G., Croxford, J. L., Brown, P., Pertwee, R. G., Huffman, J. W., and Layward, L. (2000). Cannabinoids control spasticity and tremor in a multiple sclerosis model. *Nature* **404,** 84–87.

Baker, D., Pryce, G., Croxford, J. L., Brown, P., Pertwee, R. G., Makriyannis, A., Khanolkar, A., Layward, L., Fezza, F., Bisogno, T., and Di Marzo, V. (2001). Endocannabinoids control spasticity in a multiple sclerosis model. *FASEB J.* **15,** 300–302.

Barinaga, M. (2001). How cannabinoids work in the brain. *Science* **291,** 2530–2531.

Beltramo, M., Rodriguez de Fonseca, F., Navarro, M., Calignano, A., Gorriti, M. A., Grammatikopoulos, G., Sadile, A. G., Giuffrida, A., and Piomelli, D. (2000). Reversal of dopamine D_2-receptor responses by an anandamide transport inhibitor. *J. Neurosci.* **20,** 3401–3407.

Berdyshev, E. V. (2000). Cannabinoid receptors and the regulation of immune response. *Chem. Phys. Lipids* **108,** 169–190.
Bisogno, T., Melck, D., Bobrov, M. Yu., Gretskaya, N. M., Bezuglov, V. V., De Petrocellis, L., and Di Marzo, V. (2000). *N*-Acyl-dopamines: Novel synthetic CB1 cannabinoid-receptor ligands and inhibitors of anandamide inactivation with cannabimimetic activity in vitro and in vivo. *Biochem. J.* **351,** 817–824.
Bisogno, T., Maccarrone, M., De Petrocellis, L., Jarrahian, A., Finazzi-Agrò, A., Hillard, C., and Di Marzo, V. (2001). The uptake by cells of 2-arachidonoylglycerol, an endogenous agonist of cannabinoid receptors. *Eur. J. Biochem.* **268,** 1982–1989.
Bloom, A. S., Edgemond, W. S., and Moldvan, J. C. (1997). Nonclassical and endogenous cannabinoids: Effects on the ordering of brain membranes. *Neurochem. Res.* **22,** 563–568.
Boger, D. L., Patterson, J. E., Guan, X., Cravatt, B. F., Lerner, R. A., and Gilula, N. B. (1998). Chemical requirements for inhibition of gap junction communication by the biologically active lipid oleamide. *Proc. Natl. Acad. Sci. USA* **95,** 4810–4815.
Boger, D. L., Sato, H., Lerner, A. E., Hedrick, M. P., Fecik, R. A., Miyauchi, H., Wilkie, G. D., Austin, B. J., Patricelli, M. P., and Cravatt, B. F. (2000). Exceptionally potent inhibitors of fatty acid amide hydrolase: The enzyme responsible for degradation of endogenous oleamide and anandamide. *Proc. Natl. Acad. Sci. USA* **97,** 5044–5049.
Bouaboula, M., Dussossoy, D., and Casellas, P. (1999). Regulation of peripheral cannabinoid receptor CB2 phosphorylation by the inverse agonist SR 144528. Implications for receptor biological responses. *J. Biol. Chem.* **274,** 20397–20405.
Braud, S., Bon, C., Touqui, L., and Mounier, C. (2000). Activation of rabbit blood platelets by anandamide through its cleavage into arachidonic acid. *FEBS Lett.* **471,** 12–16.
Burstein, S. H., Rossetti, R. G., Yagen, B., and Zurier, R. B. (2000). Oxidative metabolism of anandamide. *Prostaglandins Other Lipid Mediat.* **61,** 29–41.
Cadas, H., di Tomaso, E., and Piomelli, D. (1997). Occurrence and biosynthesis of endogenous cannabinoid precursor, N-arachidonoyl phosphatidylethanolamine, in rat brain. *J. Neurosci.* **17,** 1226–1242.
Calignano, A., La Rana, G., Giuffrida, A., and Piomelli, D. (1998). Control of pain initiation by endogenous cannabinoids. *Nature* **394,** 277–281.
Calignano, A., La Rana, Loubet-Lescoulié, P., and Piomelli, D. (2000). A role for the endogenous cannabinoid system in the peripheral control of pain initiation. *Prog. Brain Res.* **129,** 471–482.
Caterina, M. J., Schumacher, M. A., Tominaga, M., Rosen, T. A., Levine, J. D., and Julius, D. (1997). The capsaicin receptor: A heat-activated ion channel in the pain pathway. *Nature* **389,** 816–824.
Caterina, M. J., Rosen, T. A., Tominaga, M., Brake, A. J., and Julius, D. (1999). A capsaicin-receptor homologue with a high threshold for noxious heat. *Nature* **398,** 436–441.
Correale, J., Arias, M., and Gilmore, W. (1998). Steroid hormone regulation of cytokine secretion by proteolipid protein-specific CD+ T cell clones isolated from multiple sclerosis patients and normal control subjects. *J. Immunol.* **161,** 3365–3374.
Coutts, A. A., Anavi-Goffer, S., Ross, R. A., MacEwan, D. J., Mackie, K., Pertwee, R. G., and Irving, A. J. (2001). Agonist-induced internalization and trafficking of cannabinoid CB1 receptors in hippocampal neurons. *J. Neurosci.* **21,** 2425–2433.
Cravatt, B. F., Prospero-Garcia, O., Siuzdak, G., Gilula, N. B., Henriksen, S. J., Boger, D. L., and Lerner, R. A. (1995). Chemical characterization of a family of brain lipids that induce sleep. *Science* **268,** 1506–1509.
Cravatt, B. F., Giang, D. K., Mayfield, S. P., Boger, D. L., Lerner, R. A., and Gilula, N. B. (1996). Molecular characterization of an enzyme that degrades neuromodulatory fatty-acid amides. *Nature* **384,** 83–87.
Day, T. A., Rakhshan, F., Deutsch, D. G., and Barker, E. L. (2001). Role of fatty acid amide hydrolase in the transport of the endogenous cannabinoid anandamide. *Mol. Pharmacol.* **59,** 1369–1375.
De Petrocellis, L., Melck, D., Bisogno, T., and Di Marzo, V. (2000). Endocannabinoids and fatty acid amides in cancer, inflammation and related disorders. *Chem. Phys. Lipids* **108,** 191–209.

De Petrocellis, L., Bisogno, T., Maccarrone, M., Davis, J. B., Finazzi-Agrò, A., and Di Marzo, V. (2001). The activity of anandamide at vanilloid VR1 receptors requires facilitated transport across the cell membrane and is limited by intracellular metabolism. *J. Biol. Chem.* **276,** 12856–12863.

Deutsch, D. G., Goligorsky, M. S., Schmid, P. C., Krebsbach, R. J., Schmid, H. H. O., Das, S. K., Dey, S. K., Arreaza, G., Thorup, C., Stefano, G., and Moore, L. C. (1997). Production and physiological actions of anandamide in the vasculature of the rat kidney. *J. Clin. Invest.* **100,** 1538–1546.

Deutsch, D. G., Glaser, S. T., Howell, J. M., Kunz, J. S., Puffenbarger, R. A., Hillard, C. J., and Abumrad, N. (2001). The cellular uptake of anandamide is coupled to its breakdown by fatty acid amide hydrolase (FAAH). *J. Biol. Chem.* **276,** 6967–6973.

Devane, W. A., and Axelrod, J. (1994). Enzymatic synthesis of anandamide, and endogenous ligand for the cannabinoid receptor, by brain membranes. *Proc. Natl. Acad. Sci. USA* **91,** 6698–6701.

Devane, W. A., Hannus, L., Breuer, A., Pertwee, R. G., Stevenson, L. A., Griffin, G., Gibson, D., Mandelbaum, A., Etinger, A., and Mechoulam, R. (1992). Isolation and structure of a brain constituent that binds to the cannabinoid receptor. *Science* **258,** 1946–1949.

Di Marzo, V. (1998). "Endocannabinoids" and other fatty acid derivatives with cannabimimetic properties: Biochemistry and possible physiopathological relevance. *Biochim. Biophys. Acta* **1392,** 153–175.

Di Marzo, V., and Deutsch, D. G. (1998). Biochemistry of the endogenous ligands of cannabinoid receptors. *Neurobiol. Dis.* **5,** 386–404.

Di Marzo, V., and Fontana, A. (1995). Anandamide, an endogenous cannabinomimetic eicosanoid: "Killing two birds with one stone." *Prostaglandins Leuko. Essent. Fatty Acids* **53,** 1–11.

Di Marzo, V., Sepe, N., De Petrocellis, L., Berger, A., Crozier, G., Fride, E., and Mechoulam, R. (1998a). Trick or treat from food endocannabinoids? *Nature* **396,** 636.

Di Marzo, V., Bisogno, T., Sugiura, T., Melck, D., and De Petrocellis, L. (1998b). The novel endogenous cannabinoid 2-arachidonoylglycerol is inactivated by neuronal- and basophil-like cells: Connections with anandamide. *Biochem. J.* **331,** 15–19.

Di Marzo, V., Bisogno, T., Melck, D., Ross, R., Brockie, H., Stevenson, L., Pertwee, R. G., and De Petrocellis, L. (1998c). Interactions between synthetic vanilloids and the endogenous cannabinoid system. *FEBS Lett.* **436,** 449–454.

Di Marzo, V., Bisogno, T., De Petrocellis, L., Melck, D., Orlando, P., Wagner, J. A., and Kunos, G. (1999). Biosynthesis and inactivation of the endocannabinoid 2-arachidonoylglycerol in circulating and tumoral macrophages. *Eur. J. Biochem.* **264,** 258–267.

Di Marzo, V., Bifulco, M., and De Petrocellis, L. (2000a). Endocannabinoids and multiple sclerosis: A blessing from the "inner bliss"? *Trends Pharmacol.* **21,** 195–197.

Di Marzo, V., Hill, M. P., Bisogno, T., Crossman, A. R., and Brotchie, J. M. (2000b). Enhanced levels of endogenous cannabinoids in the globus pallidus are associated with a reduction in movement in an animal model of Parkinson's disease. *FASEB J.* **14,** 1432–1438.

Di Marzo, V., Goparaju, S. K., Wang, L., Liu, J., Batkai, S., Jarai, Z., Fezza, F., Miura, G. I., Palmiter, R. D., Sugiura, T., and Kunos, G. (2001). Leptin-regulated endocannabinoids are involved in maintaining food intake. *Nature* **410,** 822–825.

di Tomaso, E., Beltramo, M., and Piomelli, D. (1996). Brain cannabinoids in chocolate. *Nature* **382,** 677–678.

Facci, L., Dal Toso, R., Romanello, S., Buriani, A., Skaper, S. D., and Leon, A. (1995). Mast cells express a peripheral cannabinoid receptor with differential sensitivity to anandamide and palmitoylethanolamide. *Proc. Natl. Acad. Sci. USA* **92,** 3376–3380.

Farooqui, A. A., Anderson, D. K., and Horrocks, L. A. (1993). Effect of glutamate and its analogs on diacylglycerol and monoacylglycerol lipase activities of neuron-enriched cultures. *Brain Res.* **604,** 180–184.

Fisslthaler, B., Popp, R., Kiss, L., Potente, M., Harder, D. R., Fleming, I., and Busse, R. (1999). Cytochrome P450 2C is an EDHF synthase in coronary arteries. *Nature* **401,** 493–497.

Freysz, L., Farooqui, A. A., Horrocks, L. A., Massarelli, R., and Dreyfus, H. (1991). Stimulation of mono- and diacylglycerol lipase activities by gangliosides in chicken neuronal cultures. *Neurochem. Res.* **16,** 1241–1244.

Fride, E., and Mechoulam, R. (1993). Pharmacological activity of the cannabinoid receptor agonist, anandamide, a brain constituent. *Eur. J. Pharmacol.* **231,** 313–314.

Galiègue, S., Mary, S., Marchand, J., Dussossoy, D., Carrière, D., Carayon, P., Bouaboula, M., Shire, D., Le Fur, G., and Casellas, P. (1995). Expression of central and peripheral cannabinoid receptors in human immune tissues and leukocyte subpopulations. *Eur. J. Biochem.* **232,** 54–61.

Gaoni, Y., and Mechoulam, R. (1964). Isolation, structure and partial synthesis of an active constituent of hashish. *J. Am. Chem. Soc.* **86,** 1646–1647.

Garcia, D. E., Brown, S., Hille, B., and Mackie, K. (1998). Protein kinase C disrupts cannabinoid actions by phosphorylation of the CB1 cannabinoid receptor. *J. Neurosci.* **18,** 2834–2841.

Garcia, N., Jr., Járai, Z., Mirshahi, F., Kunos, G., and Sanyal, A. J. (2001). Systemic and portal hemodynamic effects of anandamide. *Am. J. Physiol. Gastrointest. Liver Physiol.* **280,** G14–G20.

Giang, D. K., and Cravatt, B. F. (1997). Molecular characterization of human and mouse fatty acid amide hydrolase. *Proc. Natl. Acad. Sci. USA* **94,** 2238–2242.

Giuffrida, A., and Piomelli, D. (2000). The endocannabinoid system: A physiological perspective on its role in psychomotor control. *Chem. Phys. Lipids* **108,** 151–158.

Giuffrida, A., Parsons, L. H., Kerr, T. M., Rodriguez de Fonseca, F., Navarro, M., and Piomelli, D. (1999). Dopamine activation of endogenous cannabinoid signaling in dorsal striatum. *Nat. Neurosci.* **2,** 358–363.

Goldstein, S. R. (1994). Embryonic death in early pregnancy: A new look at the first trimester. *Obstet. Gynecol.* **84,** 294–297.

Goparaju, S. K., Ueda, N., Yamaguchi, H., and Yamamoto, S. (1998). Anandamide amidohydrolase reacting with 2-arachidonoylglycerol, another cannabinoid receptor ligand. *FEBS Lett.* **422,** 69–73.

Goparaju, S. K., Ueda, N., Taniguchi, K., and Yamamoto, S. (1999a). Enzymes of porcine brain hydrolyzing 2-arachidonoylglycerol, an endogenous ligand of cannabinoid receptors. *Biochem. Pharmacol.* **57,** 417–423.

Goparaju, S. K., Kurahashi, Y., Suzuki, H., Ueda, N., and Yamamoto, S. (1999b). Anandamide amidohydrolase of porcine brain: cDNA cloning, functional expression and site-directed mutagenesis. *Biochim. Biophys. Acta* **1441,** 77–84.

Hampson, A. J., Grimaldi, M., Axelrod, J., and Wink, D. (1998). Cannabidiol and (−)delta9-tetrahydrocannabinol are neuroprotective antioxidants. *Proc. Natl. Acad. Sci. USA* **95,** 8268–8273.

Hansen, H. S., Moesgaard, B., Hansen, H. H., Schousboe, A., and Petersen, G. (1999). Formation of *N*-acyl-phosphatidylethanolamine and *N*-acylethanolamine (including anandamide) during glutamate-induced neurotoxicity. *Lipids* **34,** S327–S330.

Hansen, H. H., Hansen, S. H., Schousboe, A., and Hansen, H. S. (2000). Determination of the phospholipid precursor of anandamide and other *N*-acylethanolamine phospholipids before and after sodium azide-induced toxicity in cultured neocortical neurons. *J. Neurochem.* **75,** 861–871.

Hanus, L., Abu-Lafi, S., Fride, E., Breuer, A., Vogel, Z., Shalev, D. E., Kustanovich, I., and Mechoulam, R. (2001). 2-Arachidonyl glyceryl ether, an endogenous agonist of the cannabinoid CB1 receptor. *Proc. Natl. Acad. Sci. USA* **98,** 3662–3665.

Herkenham, M., Lynn, A. B., de Costa, B. R., and Richfield, E. K. (1991). Neuronal localization of cannabinoid receptors in the basal ganglia of the rat. *Brain Res.* **547,** 267–274.

Hillard, C. J., and Jarrahian, A. (2000). The movement of *N*-arachidonoylethanolamine (anandamide) across cellular membranes. *Chem. Phys. Lipids* **108,** 123–134.

Hillard, C. J., Edgemond, W. S., Jarrahian, A., and Campbell, W. B. (1997). Accumulation of *N*-arachidonoylethanolamide (anandamide) into cerebellar granule cells occurs via facilitated diffusion. *J. Neurochem.* **69,** 631–638.

Howlett, A. C., and Mukhopadhyay, S. (2000). Cellular signal transduction by anandamide and 2-arachidonoylglycerol. *Chem. Phys. Lipids* **108,** 53–70.

Hsieh, C., Brown, S., Derleth, C., and Mackie, K. (1999). Internalization and recycling of the CB1 cannabinoid receptor. *J. Neurochem.* **73,** 493–501.

Járai, Z., Wagner, J., Varga, K., Lake, K. D., Compton, D. R., Martin, B. R., Zimmer, A. M., Bonner, T. I., Buckley, N. E., Mezey, E., Razdan, R. K., Zimmer, A., and Kunos, G. (1999). Cannabinoid-induced mesenteric vasodilation through an endothelial site distinct from CB1 or CB2 receptors. *Proc. Natl. Acad. Sci. USA* **96,** 14136–14141.

Járai, Z., Wagner, J., Goparaju, S. K., Wang, L., Razdan, R. K., Sugiura, T., Zimmer, A. M., Bonner, T. I., Zimmer, A., and Kunos, G. (2000). Cardiovascular effects of 2-arachidonoyl glycerol in anesthetized mice. *Hypertension* **35,** 679–684.

Jarrahian, A., Manna, S., Edgemond, W. S., Campbell, W. B., and Hillard, C. J. (2000). Structure-activity relationships among *N*-arachidonylethanolamine (anandamide) head group analogues for the anandamide transporter. *J. Neurochem.* **74,** 2597–2606.

Khanolkar, A. D., Palmer, S. L., and Makriyannis, A. (2000). Molecular probes for the cannabinoid receptors. *Chem. Phys. Lipids* **108,** 37–52.

Kobayashi, M., Fujiwara, Y., Goda, M., Komeda, H., and Shimizu, S. (1997). Identification of active sites in amidase: Evolutionary relationship between amide bond- and peptide bond-cleaving enzymes. *Proc. Natl. Acad. Sci. USA* **94,** 11986–11891.

Kovalevskaya, G., Birken, S., Kakuma, T., Schlatterer, J., and O'Connor, J. F. (1999). Evaluation of nicked human chorionic gonadotropin content in clinical specimens by a specific immunometric assay. *Clin. Chem.* **45,** 68–77.

Kruszka, K. K., and Gross, R. W. (1994). The ATP- and CoA-independent synthesis of arachidonoylethanolamide. A novel mechanism underlying the synthesis of the endogenous ligand of the cannabinoid receptor. *J. Biol. Chem.* **269,** 14345–14348.

Kunos, G., Jarai, Z., Batkai, S., Goparaju, S. K., Ishac, E. J., Liu, J., Wang, L., and Wagner, J. A. (2000). Endocannabinoids as cardiovascular modulators. *Chem. Phys. Lipids* **108,** 159–168.

Kurahashi, Y., Ueda, N., Suzuki, H., Suzuki, M., and Yamamoto, S. (1997). Reversible hydrolysis and synthesis of anandamide demonstrated by recombinant rat fatty acid amide hydrolase. *Biochem. Biophys. Res. Commun.* **237,** 512–515.

Lambert, D. M., and Di Marzo, V. (1999). The palmitoylethanolamide and oleamide enigmas: Are these two fatty acid amides cannabimimetic? *Curr. Med. Chem.* **6,** 757–773.

Lockwood, C. J. (2000). Prediction of pregnancy loss. *Lancet* **355,** 1292–1293.

Maccarrone, M., Navarra, M., Corasaniti, M. T., Nisticò, G., and Finazzi-Agrò, A. (1998a). Cytotoxic effect of HIV-1 coat glycoprotein gp 120 on human neuroblastoma CHP 100 cells involves activation of the arachidonate cascade. *Biochem. J.* **333,** 45–49.

Maccarrone, M., van der Stelt, M., Rossi, A., Veldink, G. A., Vliegenthart, J. F. G., and Finazzi-Agrò, A. (1998b). Anandamide hydrolysis by human cells in culture and brain. *J. Biol. Chem.* **273,** 32332–32339.

Maccarrone, M., Bari, M., Menichelli, A., Del Principe, D., and Finazzi-Agrò, A. (1999). Anandamide activates human platelets through a pathway independent of the arachidonate cascade. *FEBS Lett.* **447,** 277–282.

Maccarrone, M., De Felici, M., Bari, M., Klinger, F., Siracusa, G., and Finazzi-Agrò, A. (2000a). Down-regulation of anandamide hydrolase in mouse uterus by sex hormones. *Eur. J. Biochem.* **267,** 2991–2997.

Maccarrone, M., Bari, M., Lorenzon, T., Bisogno, T., Di Marzo, V., and Finazzi-Agrò, A. (2000b). Anandamide uptake by human endothelial cells and its regulation by nitric oxide. *J. Biol. Chem.* **275,** 13484–13492.

Maccarrone, M., Fiorucci, L., Erba, F., Bari, M., Finazzi-Agrò, A., and Ascoli, F. (2000c). Human mast cells take up and hydrolyze anandamide under the control of 5-lipoxygenase, and do not express cannabinoid receptors. *FEBS Lett.* **468,** 176–180.

Maccarrone, M., Salvati, S., Bari, M., and Finazzi-Agrò, A. (2000d). Anandamide and 2-arachidonoylglycerol inhibit fatty acid amide hydrolase by activating the lipoxygenase pathway of the arachidonate cascade. *Biochem. Biophys. Res. Commun.* **278,** 576–583.

Maccarrone, M., Valensise, H., Bari, M., Lazzarin, N., Romanini, C., and Finazzi-Agrò, A. (2000e). Relation between decreased anandamide hydrolase concentrations in human lymphocytes and miscarriage. *Lancet* **355**, 1326–1329.

Maccarrone, M., Lorenzon, T., Bari, M., Melino, G., and Finazzi-Agrò, A. (2000f). Anandamide induces apoptosis in human cells via vanilloid receptors. Evidence for a protective role of cannabinoid receptors. *J. Biol. Chem.* **275**, 31938–31945.

Maccarrone, M., Attinà, M., Cartoni, A., Bari, M., and Finazzi-Agrò, A. (2001a). Gas chromatography-mass spectrometry analysis of endogenous cannabinoids in healthy and tumoral human brain and human cells in culture. *J. Neurochem.* **76**, 594–601.

Maccarrone, M., Valensise, H., Bari, M., Lazzarin, N., Romanini, C., and Finazzi-Agrò, A. (2001b). Progesterone up-regulates anandamide hydrolase in human lymphocytes. Role of cytokines and implications for fertility. *J. Immunol.* **166**, 7183–7189.

Maccarrone, M., Bari, M., Menichelli, A., Giuliani, E., Del Principe, D., and Finazzi-Agrò, A. (2001c). Human platelets bind and degrade 2-arachidonoylglycerol, which activates these cells through a cannabinoid receptor. *Eur. J. Biochem.* **268**, 819–825.

MacDonald, J. C., and Vaughan, C. W. (2001). Cannabinoids act backwards. *Nature* **410**, 527–530.

Maneuf, Y. P., Nash, J. E., Crossman, A. R., and Brotchie, J. M. (1996). Activation of the cannabinoid receptor by delta 9-tetrahydrocannabinol reduces gamma-aminobutyric acid uptake in the globus pallidus. *Eur. J. Pharmacol.* **308**, 161–164.

Maneuf, Y. P., Crossman, A. R., and Brotchie, J. M. (1997). The cannabinoid receptor agonist WIN 55, 212–2 reduces D2, but not D1, dopamine receptor-mediated alleviation of akinesia in the reserpine-treated rat model of Parkinson's disease. *Exp. Neurol.* **148**, 265–270.

Mani, S. K., Mitchell, A., and O'Malley, B. W. (2001). Progesterone receptor and dopamine receptors are required in Δ^9-tetrahydrocannabinol modulation of sexual receptivity in female rats. Proc. *Natl. Acad. Sci. USA* **98**, 1249–1254.

Martin, B. R., Compton, D. R., Thomas, B. F., Prescott, W. R., Little, P. J., Razdan, R. K., Johnson, M. R., Melvin, L. S., Mechoulam, R., and Ward, S. J. (1991). Behavioral, biochemical, and molecular modeling evaluations of cannabinoid analogs. *Pharmacol. Biochem. Behav.* **40**, 471–478.

Martin, R. S., Luong, L. A., Welsh, N. J., Eglen, R. M., Martin, G. R., and MacLennan, S. J. (2000). Effects of cannabinoid receptor agonists on neuronally-evoked contractions of urinary bladder tissues isolated from rat, mouse, pig, dog, monkey and human. *Br. J. Pharmacol.* **129**, 1707–1715.

Martyn, C. N., Illis, L. S., and Thom, J. (1995). Nabilone in the treatment of multiple sclerosis. *Lancet* **345**, 579.

Matsuda, L. A., Lolait, S. J., Brownstein, M. J., Young, A. C., and Bonner, T. I. (1990). Structure of a cannabinoid receptor and functional expression of the cloned cDNA. *Nature* **346**, 561–564.

Mechoulam, R. (1986). The pharmahistory of Cannabis sativa. *In* "Cannabinoids as Therapeutic Agents" (R. Mechoulam, Ed.), pp. 1–19. CRC Press, Boca Raton, FL.

Mechoulam, R., and Fride, E. (2001). A hunger for cannabinoids. *Nature* **410**, 763–764.

Mechoulam, R., and Hanus, L. (2000). A historical overview of chemical research on cannabinoids. *Chem. Phys. Lipids* **108**, 1–13.

Mechoulam, R., Ben-Shabat, S., Hanus, L., Ligumsky, M., Kaminski, N. E., Schatz, A. R., Gopher, A., Almog, S., Martin, B. R., Compton, D. R., Pertwee, R. G., Griffin, G., Bayewitch, M., Barg, J., and Vogel, Z. (1995). Identification of an endogenous 2-monoglyceride, present in canine gut, that binds to cannabinoid receptors. *Biochem. Pharmacol.* **50**, 83–90.

Mechoulam, R., Fride, E., and Di Marzo, V. (1998). Endocannabinoids. *Eur. J. Pharmacol.* **359**, 1–18.

Melck, D., Bisogno, T., De Petrocellis, L., Chuang, H.-h., Julius, D., Bifulco, M., and Di Marzo, V. (1999). Unsaturated long-chain N-acyl-vanillyl-amides (N-AVAMs): Vanilloid receptor ligands that inhibit anadamide-facilitated transport and bind to CB1 cannabinoid receptors. *Biochem. Biophys. Res. Commun.* **262**, 275–284.

Mukhopadhyay, S., and Howlett, A. C. (2001). CB1 receptor-G protein association. Subtype selectivity is determined by distinct intracellular domains. *Eur. J. Biochem.* **268**, 499–505.

Munro, S., Thomas, K. L., and Abu-Shaar, M. (1993). Molecular characterization of a peripheral receptor for cannabinoids. *Nature* **365**, 61–65.

Paria, B. C., and Dey, S. K. (2000). Ligand-receptor signaling with endocannabinoids in preimplantation embryo development and implantation. *Chem. Phys. Lipids* **108**, 211–220.

Paria, B. C., Deutsch, D. D., and Dey, S. K. (1996). The uterus is a potential site for anandamide synthesis and hydrolysis: Differential profiles of anandamide synthase and hydrolase activities in the mouse uterus during the periimplantation period. *Mol. Reprod. Dev.* **45**, 183–192.

Patricelli, M. P., and Cravatt, B. F. (1999). Fatty acid amide hydrolase competitively degrades bioactive amides and esters through a nonconventional catalytic mechanism. *Biochemistry* **38**, 14125–14130.

Patricelli, M. P., and Cravatt, B. F. (2001). Characterization and manipulation of the acyl chain selectivity of fatty acid amide hydrolase. *Biochemistry* **40**, 6107–6115.

Patricelli, M. P., Lashuel, H. A., Giang, D. K., Kelly, J. W., and Cravatt, B. F. (1998). Comparative characterization of a wild type and a transmembrane domain-deleted fatty acid amide hydrolase: Identification of the transmembrane domain as a site for oligomerization. *Biochemistry* **37**, 15177–15187.

Patricelli, M. P., Lovato, M. A., and Cravatt, B. F. (1999). Chemical and mutagenic investigations of fatty acid amide hydrolase: Evidence for a family of serine hydrolases with distinct catalytic properties. *Biochemistry* **38**, 9804–9812.

Perry, T. L., Berry, K., Hansen, S., Diamond, S., and Mok, C. (1971). Regional distribution of amino acids in human brain obtained at autopsy. *J. Neurochem.* **18**, 513–519.

Pertwee, R. G. (1997). Pharmacology of cannabinoid CB1 and CB2 receptors. *Pharmacol. Ther.* **74**, 129–180.

Pertwee, R. G. (2000). Cannabinoid receptor ligands: Clinical and neuropharmacological considerations, relevant to future drug discovery and development. *Exp. Opin. Invest. Drugs* **9**, 1553–1571.

Pertwee, R. G. (2001). Cannabinoid receptors and pain. *Prog. Neurobiol.* **63**, 569–611.

Pertwee, R. G., and Wickens, A. P. (1991). Enhancement by chlordiazepoxide of catalepsy induced in rats by intravenous or intrapallidal injections of enantiomeric cannabinoids. *Neuropharmacology* **30**, 237–244.

Piccinni, M. P., and Romagnani, S. (1996). Regulation of fetal allograft survival by hormone-controlled Th1- and Th2-type cytokines. *Immunol. Res.* **15**, 141–150.

Piccinni, M. P., Giudizi, M. G., Biagiotti, R., Beloni, L., Giannarini, L., Sampognaro, S., Parronchi, P., Manetti, R., Annunziato, F., Livi, C., Romagnani, S., and Maggi, E. (1995). Progesterone favors the development of human T helper cells producing Th2-type cytokines and promotes both IL-4 production and membrane CD30 expression in established Th1 cell clones. *J. Immunol.* **155**, 128–133.

Piccinni, M. P., Beloni, L., Livi, C., Maggi, E., Scarselli, G., and Romagnani, S. (1998). Defective production of both leukemia inhibitory factor and type 2 T-helper cytokines by decidual T cells in unexplained recurrent abortions. *Nat. Med.* **4**, 1020–1024.

Piomelli, D., Beltramo, M., Glasnapp, S., Lin, S. Y., Goutopoulos, A., Xie, X. Q., and Makriyannis, A. (1999). Structural determinants for recognition and translocation by the anandamide transporter. *Proc. Natl. Acad. Sci. USA* **96**, 5802–5807.

Pop, E. (1999). Cannabinoids, endogenous ligands and synthetic analogs. *Curr. Opin. Chem. Biol.* **3**, 418–425.

Randall, M. D., Alexander, S. P. H., Bennett, T., Boyd, E. A., Fry, J. R., Gardiner, S. M., Kemp, P. A., McCulloch, A. I., and Kendall, D. A. (1996). An endogenous cannabinoid as an endothelium derived vasorelaxant. *Biochem. Biophys. Res. Commun.* **229**, 114–120.

Reggio, P. H., and Traore, H. (2000). Conformational requirements for endocannabinoid interaction with the cannabinoid receptors, the anandamide transporter and fatty acid amidohydrolase. *Chem. Phys. Lipids* **108**, 15–35.

Richardson, J. D., Kilo, S., and Hargreaves, K. M. (1998). Cannabinoids reduce hyperalgesia and inflammation via interaction with peripheral CB1 receptors. *Pain* **75**, 111–119.

Salzet, M., Breton, C., Bisogno, T., and Di Marzo, V. (2000). Comparative biology of the endocannabinoid system: Possible role in the immune response. *Eur. J. Biochem.* **267,** 4917–4927.

Schmid, H. H. O. (2000). Pathways and mechanisms of *N*-acylethanolamine biosynthesis: Can anandamide be generated selectively? *Chem. Phys. Lipids* **108,** 71–87.

Self, D. W. (1999). Anandamide: A candidate neurotransmitter heads for the big leagues. *Nat. Neurosci.* **2,** 303–304.

Sharkey, A. (1998). Cytokines and implantation. *Rev. Reprod.* **3,** 52–57.

Silverdale, M. A., McGuire, S., McInnes, A., Crossman, A. R., and Brotchie, J. M. (2001). Striatal cannabinoid CB1 receptor mRNA expression is decreased in the reserpine-treated rat model of Parkinson's disease. *Exp. Neurol.* **169,** 400–406.

Skaper, S. D., Buriani, A., Dal Toso, R., Petrelli, L., Romanello, S., Facci, L., and Leon, A. (1996). The ALIAmide palmitoylethanolamide and cannabinoids, but not anandamide, are protective in a delayed postglutamate paradigm of excitotoxic death in cerebellar granule neurons. *Proc. Natl. Acad. Sci. USA* **93,** 3984–3989.

Smart, D., Gunthorpe, M. J., Jerman, J. C., Nasir, S., Gray, J., Muir, A. I., Chambers, J. K., Randall, A. D., and Davis, J. B. (2000). The endogenous lipid anandamide is a full agonist at the human vanilloid receptor (hVR1). *Br. J. Pharmacol.* **129,** 227–230.

Stella, N., Schweitzer, P., and Piomelli, D. (1997). A second endogenous cannabinoid that modulates long-term potentiation. *Nature* **388,** 773–778.

Sugiura, T., and Waku, K. (2000). 2-Arachidonoylglycerol and the cannabinoid receptors. *Chem. Phys. Lipids* **108,** 89–106.

Sugiura, T., Kondo, S., Sukagawa, A., Nakane, S., Shinoda, A., Itoh, K., Yamashita, A., and Waku, K. (1995). 2-Arachidonoylglycerol: A possible endogenous cannabinoid receptor ligand in brain. *Biochem. Biophys. Res. Commun.* **215,** 89–97.

Sugiura, T., Kodaka, T., Nakane, S., Kishimoto, S., Kondo, S., and Waku, K. (1998). Detection of an endogenous cannabimimetic molecule, 2-arachidonoylglycerol, and cannabinoid CB1 receptor mRNA in human vascular cells: Is 2-arachidonoylglycerol a possible vasomodulator? *Biochem. Biophys. Res. Commun.* **243,** 838–843.

Szallasi, A., and Di Marzo, V. (2000). New perspectives on enigmatic vanilloid receptors. *Trends Neurosci.* **23,** 491–497.

Szekeres-Bartho, J., Faust, Zs., Varga, P., Szereday, L., and Kelemen, K. (1996). The immunological pregnancy protective effect of progesterone is manifested via controlling cytokine production. *Am. J. Reprod. Immunol.* **35,** 348–351.

Taupin, J.-L., Minvielle, S., Thèze, J., Jacques, Y., and Moreau, J.-F. (1999). The interleukin-6 family of cytokines and their receptors. *In* "The Cytokine Network and Immune Functions" (J. Thèze, Ed.), pp. 31–44. Oxford University Press, New York.

Ueda, N., Kurahashi, Y., Yamamoto, S., and Tokunaga, T. J. (1995). Partial purification and characterization of the porcine brain enzyme hydrolyzing and synthesizing anandamide. *J. Biol. Chem.* **270,** 23823–23827.

Ueda, N., Yamanaka, K., Terasawa, Y., and Yamamoto, S. (1999). An acid amidase hydrolyzing anandamide as an endogenous ligand for cannabinoid receptors. *FEBS Lett.* **454,** 267–270.

Ueda, N., Puffenbarger, R. A., Yamamoto, S., and Deutsch, D. G. (2000). The fatty acid amide hydrolase (FAAH). *Chem. Phys. Lipids* **108,** 107–121.

Varga, K., Lake, K., Martin, B. R., and Kunos, G. (1995). Novel antagonist implicates the CB1 cannabinoid receptor in the hypotensive action of anandamide. *Eur. J. Pharmacol.* **278,** 279–283.

Varga, K., Wagner, J. A., Bridgen, D. T., and Kunos, G. (1998). Platelet- and macrophage-derived endogenous cannabinoids are involved in endotoxin-induced hypotension. *FASEB J.* **12,** 1035–1044.

Venance, L., Piomelli, D., Glowinski, J., and Giaume, C. (1995). Inhibition by anandamide of gap junctions and inercellular calcium signalling in striatal astrocytes. *Nature* **376,** 590–594.

Wagner, J. A., Varga, K., Ellis, E., Rzigalinski, B. A., Martin, B. R., and Kunos, G. (1997). Activation of peripheral CB1 cannabinoid receptors in haemorrhagic shock. *Nature* **390,** 518–521.

Wagner, J. A., Varga, K., and Kunos, G. (1998). Cardiovascular actions of cannabinoids and their generation during shock. *J. Mol. Med.* **76,** 824–836.

Wagner, J. A., Varga, C., Jàrai, Z., and Kunos, G. (1999). Mesenteric vasodilation mediated by endothelial anandamide receptors. *Hypertension* **33,** 429–434.

Walker, J. M., Huang, S. M., Strangman, N. M., Tsou, K., and Sañudo-Peña, M. C. (1999). Pain modulation by release of the endogenous cannabinoid anandamide. *Proc. Natl. Acad. Sci. USA* **96,** 12198–12203.

Wilson, R. I., and Nicoll, R. A. (2001). Endogenous cannabinoids mediate retograde signalling at hippocampal synapses. *Nature* **410,** 588–592.

Yazulla, S., Studholme, K. M., McIntosh, H. H., and Deutsch, D. G. (1999). Immunocytochemical localization of cannabinoid CB1 receptor and fatty acid amide hydrolase in rat retina. *J. Comp. Neurol.* **15,** 80–90.

Yu, M., Ives, D., and Ramesha, C. S. (1997). Synthesis of prostaglandin E_2 ethanolamide from anandamide by cyclooxygenase-2. *J. Biol. Chem.* **272,** 21181–21186.

Zygmunt, P. M., Petersson, J., Andersson, D. A., Chuang, H.-h., Sorgård, M., Di Marzo, V., Julius, D., and Högestätt, E. D. (1999). Vanilloid receptors on sensory nerves mediate the vasodilator action of anandamide. *Nature* **400,** 452–457.

Endomorphins and Related Opioid Peptides

Yoshio Okada,* Yuko Tsuda,*
Sharon D. Bryant,[†] and
Lawrence H. Lazarus[†]

*Faculty of Pharmaceutical Sciences and High Technology Research Center, Kobe Gakuin University, Kobe 651-2180, Japan, and
[†]Peptide Neurochemistry, LCBRA, National Institute of Environmental Health Sciences, Research Triangle Park, North Carolina 27709

I. Introduction
II. Discovery of Endomorphins
   A. *Localization of Endomorphins*
   B. *Action of Endomorphins*
III. Structure-Activity Relationship
IV. Conclusions and Future Directions
    References

Opioid peptides and their G-protein-coupled receptors (δ, κ, μ) are located in the central nervous system and peripheral tissues. The opioid system has been studied to determine the intrinsic mechanism of modulation of pain and to develop uniquely effective pain-control substances with minimal abuse potential and side effects. Two types of endogenous opioid peptides exist, one containing Try-Gly-Gly-Phe as the message domain (enkephalins, endorphins, dynorphins) and the other containing the Tyr-Pro-Phe/Trp sequence (endomorphins-1 and -2). Endomorphin-1 (Tyr-Pro-Trp-Phe-NH$_2$), which has high μ receptor affinity ($K_i = 0.36$ nM) and remarkable selectivity (4000- and

15,000-fold preference over the δ and κ receptors, respectively), was isolated from bovine and human brain. In addition, endomorphin-2 (Tyr-Pro-Phe-Phe-NH$_2$), isolated from the same sources, exhibited high μ receptor affinity ($K_i =$ 0.69 nM) and very high selectivity (13,000- and 7500-fold preference relative to δ and κ receptors, respectively). Both opioids bind to μ-opioid receptors, thereby activating G-proteins, resulting in regulation of gastrointestinal motility, manifestation of antinociception, and effects on the vascular systems and memory. To develop novel analgesics with less addictive properties, evaluation of the structure–activity relationships of the endomorphins led to the design of more potent and stable analgesics. Opioidmimetics and opioid peptides containing the amino acid sequence of the message domain of endomorphins, Tyr-Pro-Phe/Trp, could exhibit unique binding activity and lead to the development of new therapeutic drugs for controlling pain. © 2002, Elsevier Science (USA).

I. INTRODUCTION

Opioid peptides and receptors were investigated to study the intrinsic mechanisms involved in the modulation of pain and to develop novel and effective ligands to relieve pain without potential for abuse, while exhibiting minimal side effects. The G-protein coupled δ-, κ-, and μ-opioid receptors (Martin *et al.*, 1976; Lord *et al.*, 1977) are located throughout the central nervous system and peripheral tissues, such as the spinal cord, and reproductive and gastrointestinal tracts. These receptors and their endogenous ligands are primarily involved in the modulation and perception of pain, although other activities are attributed to them. Mammalian endogenous ligands, the endorphins, enkephalins, and dynorphins, bind with low to moderate specificity to three different receptors. Of these opioid peptides, enkephalins interact with the δ-opioid receptor (Hughes *et al.*, 1975), dynorphins are κ-opioid receptor ligands (Goldstein *et al.*, 1981), and β-endorphin binds to both μ- and δ-opioid receptors with comparable affinity (Ling *et al.*, 1976; Cox *et al.*, 1976). However, highly selective endogenous μ-opioid receptor ligands that produce potent and prolonged analgesia in test animals were reported by Zadina *et al.* (1997). These peptides, endomorphin-1 (H-Tyr-Pro-Trp-Phe-NH$_2$) and endomorphin-2 (H-Tyr-Pro-Phe-Phe-NH$_2$), were initially isolated from bovine brain and later from human brain cortex (Hackler *et al.*, 1997). The endomorphins and their analogues may provide a powerful molecular platform to examine the opioid pathways throughout the body, while paving the way for development of novel, short-acting analgesic drugs.

II. DISCOVERY OF ENDOMORPHINS

Opioid receptors belong to the large family of seven-transmembrane-spanning G-protein-coupled receptors. Each of the three subtypes has been cloned and their cDNA sequenced (Kieffer *et al.*, 1992, 1995; Evans *et al.*, 1992; Yasuda *et al.*, 1993; Chen *et al*, 1993; Stanasila *et al.*, 1999). Binding of opioid agonists to their respective membrane receptors transmits changes into the cell that causes a plethora of actions; for example, the closure of voltage-sensitive calcium channels (Porzig, 1990), activation of potassium channels (North, 1989), and a reduction of cyclic AMP formation (Childers, 1991; Nevo et al., 2000). These actions cause a reduction in the potential for excitatory neurotransmission, which appears to be the mechanism by which opioids produce their analgesic effect.

The weak selectivity of the enkephalins toward δ receptors (Hughes et al., 1975) and dynorphins for κ receptors (Goldstein et al., 1979) was demonstrated more than two decades ago. Morphine and their related alkaloids, and the enkephalin analogue DAMGO [d-Ala^2(N-Me)Phe^4,Gly^5-ol]-enkephalin, act primarily through the μ receptors (Matthes et al., 1996; Burford et al., 1998). Although morphine is clinically relevant, it is subject to abuse because of its well-known addictive properties. In 1992, the amidated tetrapeptide Tyr-W-MIF-1 [Tyr-Pro-Trp-Gly-NH$_2$] was isolated from human frontal cortex (Erchegyi et al., 1992) and the following year from bovine hypothalami (Hackler et al., 1993), and was shown to be selective for the μ receptor by factors 200- and 300-fold greater than for δ- and κ-opioid receptors, respectively. However, its affinity for the μ receptor ($K_i = 70$ nM) was lower than that of enkephalin, endorphin, or dynorphin (Zadina et al., 1994). Therefore to identify candidate peptides as potential μ receptor agonists, the amino acid in the fourth position of Tyr-W-MIF-1 was replaced with all the possible natural amino acids (Zadina et al., 1997). The results indicated that the replacement of Gly by Phe resulted in endomorphin-1 (Tyr-Pro-Trp-Phe-NH$_2$), which has very high μ receptor affinity ($K_i = 0.36$ nM) and remarkable selectivity; 4000- and 15,000-fold preference over the δ and κ receptors, respectively. In addition, endomorphin-1 was isolated from bovine (Zadina et al., 1997) and human brains (Hackler et al., 1997). Not only was this peptide more effective than the standard DAMGO in vitro, it also produced potent and prolonged analgesia in mice. A second peptide, endomorphin-2 (Tyr-Pro-Phe-Phe-NH$_2$), which contained Phe in place of Trp in the third position, was also isolated from the same sources (Zadina et al., 1997; Hackler et al., 1997). Endomorphin-2 exhibited high μ receptor affinity ($K_i = 0.69$ nM) and selectivity; 13,000- and 7500-fold preference over the δ and κ receptors, respectively. These peptides have the highest specificity for μ receptors than any of the endogenous substances heretofore described. They differ from all previously known endogenous opioid ligands due to their unique N-terminal sequence, their shorter length, and C-terminally amidation. These structural features are summarized in Tables I and II.

A. LOCALIZATION OF ENDOMORPHINS

Using radioimmunoassays, Zadina et al. (1997) reported that endomorphin-1 was found in several brain regions (thalamus, hypothalamus, cortex, and striatum) known to contain a high density of μ-opioid receptors (Mansour et al., 1995) (Table III).

Localization of endomorphin-2 and endomorphin-2-like immunoreactivity in the central nervous system (CNS) was also studied (Martin-Schild et al., 1997,

TABLE I Amino Acid Sequences of Mammalian Endogenous Opioid Peptides

Peptide	Sequence
Enkephalins	
Leu-enkephalin	YGGFL
Met-enkephalin	YGGFM
Endorphins	
β-Endorphin	YGGFMTSEKSQTPLVTLFKNAIIKNAYKKGQ
α-Endorphin	YGGFMTSEKSQTPLVT
γ-Endorphin	YGGFMTSEKSQTPLVTL
δ-Endorphin	YGGFMTSEKSQTPLVTLFKNAIIKNAY
Dynorphins	
α-Neoendorphin	YGGFLRKYPK
Dynorphin	YGGFLRRIRPKLKWENQ
Dynorphin 1–8	YGGFLRRI
Endomorphins	
Endomorphin-1	YPWF-NH$_2$
Endomorphin-2	YPFF-NH$_2$

1998; Pierce et al., 1998; Pierce and Wessendorf, 2000; Wu et al., 1999; Jiang et al., 2000).

In situ localization used polyclonal antisera raised against endomorphin-2 by immunocytochemical methods in the brain stem, spinal cord, and sensory ganglia of rats revealed a dense aggregation of endomorphin-2-like immunostaining that appeared as varicose fibers in the superficial laminae of the dorsal horn of the medulla and spinal cord. It was suggested that endomorphin-2 was synthesized

TABLE II Binding Activities of Opioid Peptides[a,b]

	Receptor binding K_1 (nM)			Binding selectivity	
	μ	δ	κ	δ/μ	κ/μ
DAMGO	0.34	190	1,300[c]	560	3,820
Endomorphin-1	0.36	1,510	5,430	4,180	15,100
Endomorphin-2	0.69	9,230	5,240	13,380	7,590
β-Endorphin	4.4			1.1	46
	2.1[d]	2.4[d]	96[d]		
Met-enkephalin	5.9			0.1	470
	9.5[d]	0.91[d]	4,440[d]		
Dynorphin	2.0			3.3	0.16
	0.73[d]	2.4[d]	0.12[d]		

[a] From Zadina et al. (1997).
[b] Binding to μ (^{3}H-DAMGO), δ(^{3}H-pCl-DPDPE), or κ (^{3}H-ethylketocyclazocine) (Zadina et al., 1994).
[c] Values taken from Zadina et al. (1994).
[d] Values taken from Corbett et al. (1993).

TABLE III Concentration of Endomorphin-1-Like Immunoreactivity in Different Regions of Bovine Brains[a]

Region	Immunoreactivity (pmol/g tissue)
Thalamus	16.1
Hypothalamus	12.4
Striatum	10.2
Frontal cortex	8.3

[a] From Zadina et al. (1997).

in ganglia and transported to the superficial dorsal horn, where it is released near neurons expressing μ-opioid receptors (Martin-Schild et al., 1997). Because endomorphin-2 was present in primary afferent fibers, it was further suggested that it could be an endogenous ligand for pre- and postsynaptic μ receptors and therefore could act as a potential modulator for pain perception (Martin-Schild et al., 1998). Endomorphin-2 also occurred in the primary afferent fibers in rodents and primates, and the release of neurotransmitters from these sources might be regulated by the release of endomorphin-2 (Pierce et al., 1998). Pierce and Wessendorf (2000) examined the distribution of endomorphin-2 immunoreactivity in the rat brain using an affinity-purified antiserum and measured the release of endomorphin-2-like immunoreactive material from the dorsal horn in isolated rat spinal cord (Williams et al., 1999). Those results provided the first evidence that endomorphin-2 immunoreactivity release was directly correlated with the dorsal horn laminae I and II since endomorphin-2 immunoreactivity containing fibers are concentrated in this region.

The first studies to demonstrate the existence of distinct endomorphin-1 and endomorphin-2 immunoreactivity in tissues outside the CNS used raidioimmunoassays in combination with high-performance liquid chromatography (HPLC) (Jessop et al., 2000). Endomorphins-1 and -2 were detectable in extracts of rat spleen and thymus, whereas endomorphin-2 immunoreactivity was observed in the human spleen.

B. ACTION OF ENDOMORPHINS

Opioid agonists exert their activity through changes in intracellular pathways elicited by G-proteins consisting of $G\alpha$-, $G\beta$-, and $G\gamma$-subunits coupled to their receptors (Chen et al., 1993; Liang et al., 1995). In this section, various methods that have led to identification of the endomorphins as μ agonists are explored.

1. Stimulation of [^{35}S]GTPγS Binding by Endomorphins-1 and -2

The interaction of endomorphin-1 or -2 with the μ receptor inhibits adenyl cyclase, activates an inwardly rectifying K^+ conductance, and inhibits Ca^{2+} conductance. The binding of an agonist causes an exchange mechanism to occur, whereby the guanosine diphosphate (GDP) bound to $G\alpha$ is exchanged for guanosine triphosphate (GTP). That in turn results in a conformational change in $G\alpha$, causing it to dissociate the receptor–G-protein complex and release $G\alpha$ and $G\beta\gamma$ to stimulate second messenger systems and alter ion-channel activities. The α subunit is inactivated by GTPase as it hydrolyzes GTP to GDP, which terminates the G-protein cycle. *In vitro,* this exchange process can be readily measured by replacing GTP with its radioactive nonhydrolyzable analogue guanosine-5'-O-(3-[^{35}S]thio)triphosphate ([^{35}S]GTPγS), which yields the permanent formation of a $G\alpha$–[^{35}S]GTPγS complex that accumulates in the cell membrane. Thus, the measurement of the binding of [^{35}S]GTPγS provides further evidence that the endomorphins interact with and function as agonists for μ-opioid receptors.

Narita *et al.* (1998, 2000) examined G-protein activation by the endomorphins in the mouse spinal cord by monitoring the binding of the [^{35}S]GTPγS. Endomorphin-1 increased [^{35}S]GTPγS binding in a concentration-dependent manner, and the stimulatory effect on [^{35}S]GTPγS binding was similar to that observed for endomorphin-2. Both endomorphins at 10 μM caused a maximal stimulation of 57–60%. The μ-agonist analogue DAMGO also stimulated [^{35}S]GTPγS binding, but a maximal stimulation of 103% at the same concentration, which was twice that observed with morphine.

To verify whether this effect was elicited by the endomorphins, the [^{35}S]GTPγS binding assays were conducted in the presence and absence of various antagonists: β-funaltrexamine (a noncompetitive μ-opioid receptor antagonist), naltrindole (antagonist selectivity for δ receptors), and nor-binaltorphimine (a κ-opioid receptor antagonist). The increase of [^{35}S]GTPγS binding by the endomorphins was completely and specifically blocked by β-funaltrexamine. Neither naltrindole nor nor-binaltorphimine (norBNI) had any effect on the stimulation of [^{35}S]GTPγS binding by the endomorphins. These results further suggested that the endomorphins stimulate G-proteins by the activation of μ-opioid receptors and are similar to morphine in regard to its efficacy for G-protein activation. Monory *et al.* (2000) reported that both endomorphins activated G-proteins and inhibited adenyl cyclase activity in membrane preparations expressing the μ-opioid receptor (Spetea *et al.,* 1998).

Stimulation of [^{35}S]GTPγS binding by the endomorphins in mouse brain (Kakizawa *et al.,* 1998; Mizoguchi *et al.,* 1999) and rat brain (Sim *et al.,* 1998) and with a cloned μ receptor (Alt *et al.,* 1998) was also reported. In support of these *in vitro* methods, mice lacking the μ-opioid receptor (knockout mice) also exhibited the absence of G-protein activation by endomorphins (Narita *et al.,* 1999; Connor *et al.,* 1999).

2. Antinociceptive Activity of Endomorphins-1 and -2

In vivo, the endomorphins are potent analgesics, both spinally and supraspinally. Endomorphin-1 had a significant 3-fold greater analgesic activity after both administrations (Zadina *et al.,* 1997). In fact, endomorphin-1 was also more potent spinally and supraspinally than endomorphin-2 (Goldberg *et al.,* 1998). The response by the endomorphins was readily reversed by naloxone (Goldberg *et al.,* 1998).

β-Funaltrexamine (β-FNA) effectively inhibited the action of both endomorphins. Naloxonazine, a μ_1 receptor antagonist (Paul et al., 1989; Hahn et al., 1982; Pick et at., 1991), significantly lowered the analgesia induced by both endomorphins. However, it was not as effective as β-FNA. Neither norBNI nor naltrindole was active against either endomorphin. The inactivity of the two endomorphins in CXBK mice, a strain that is insensitive to morphine (Moskowitz and Goodman, 1985; Reith et al., 1981; Baron et al., 1975; Pick et al., 1993), also verified their μ selectivity in vivo.

The endomorphins produced short-acting, naloxone-sensitive antinociception in the tail-flick test in mice (Stone et al., 1997), whereas endomorphin-1 exhibited antinociception using the rat tail pressure test (Hao et al., 2000). The analgesic properties of the endomorphins were further assessed in mice using the formalin test, which produced a dose-dependent analgesia that was shorter in duration than that obtained for morphine (Soignier et al., 2000).

The success in creating μ-opioid receptor knockout mice, which contain either zero, one, or two copies of the μ-opioid receptor gene, permitted evaluation of μ-opioid agonists on various pharmacological and physiological parameters. The intracerebroventricular (icv) injection of either endomorphin-1 or endomorphin-2 produced antinociception in both the hot plate and tail-flick tests in wild-type mice (Mizoguchi et al., 1999; Loh et al., 1998). However, antinociception was significantly reduced in heterozygous mice and abolished in the homozygous knockout mice.

Furthermore, the antinociceptive and antihyperalgesic effects on acute and inflammatory pain following intrathecal (it) injection of the endomorphins in rats caused dose-dependent, short-lasting antinociception; high doses caused motor impairment in the tail-flick test (Horvath et al., 1999). Przewlocka et al. (1999) compared DAMGO and morphine with endomorphins-1 and -2 on spinal analgesia following acute inflammatory and neuropathic pain in rats. Interestingly, the authors reported an antiallodynic effect by the endomorphins in rats subjected to sciatic nerve crushing, which was greater than with morphine and thus suggests the usefulness of endomorphins in the clinical therapy of neuropathic pain (Przewlocka et al., 1999; Jin et al., 1999). Endomorphin-1 produced antinociception without induction of immunomodulatory effects in rat. Therefore, it might be possible to develop therapeutic strategies for separating antinociception and immunomodulatory properties through the μ-opoid receptor (Carrigan et al., 2000). The differential antinociceptive effects of endomorphins-1 and -2 were also studied in mice (Sakurada et al., 1999; Tseng et al., 2000).

Considerable biological and pharmacological evidence supports the existence of μ-opioid receptor subtypes, μ_1 and μ_2 (Wolozin and Pasternak, 1981; Nishimura et al., 1984; Goodman and Pasternak, 1985; Pasternak and Wood, 1986). β-Funaltrexamine irreversibly binds to both μ_1 and μ_2 subtypes and inhibits both supraspinal and spinal antinociception, whereas naloxonazine selectively antagonizes the μ_1-opioid receptor. It appears that these receptor subtypes may have different antinociceptive pathways. For example, μ_1-opioid receptors appear to mediate supraspinal and μ_2-opioid receptors affect spinal antinociception. Using these antagonists, Sakurada *et al.* (1999) examined the role of the μ_1 and μ_2 subtypes on endomorphin antinociception in response to mechanical noxious stimuli. Both icv- and it-injected endomorphins produced potent and significant antinociception. However, the effect observed by endomorphin-1 was not reversed by naloxonadine, whereas that by endomorphin-2 was decreased. As anticipated from other studies, antinociception by endomorphins-1 and -2 was fully reversed by β-funaltrexamine. These results revealed that endomorphins-1 and -2 elicit antinociception through their differential activity at μ_1 and μ_2 subtypes following central and spinal administration (Sakurada *et al.*, 1999, 2000).

Earlier reports indicated that morphine or DAMGO antinociception via icv was less effective in diabetic than in nondiabetic mice (Kamei *et al.*, 1992a,b). Moreover, δ-opioid receptor-mediated antinociception was enhanced in diabetic relative to nondiabetic mice (Kamei *et al.*, 1992c). Thus, endomorphins-1 and -2 were examined using these diabetic and wild-type mice (Kamei *et al.*, 2000). Interestingly, both endomorphins inhibited the tail-flick response in both types of mice and no significant difference was observed between the antinociception by endomorphin-1 in either mouse strain. On the other hand, the effect by endomorphin-2 was greater in wild-type than in diabetic mice. In wild-type mice, antinociception by the endomorphins was reduced by β-funaltrexamine, but not by naltrindole or nor-binaltorphimine. However, in diabetic mice, β-funaltrexamine and naloxonazine reduced endomorphin-2-induced antinociception without affecting endomorphin-1 antinociception in diabetic mice. On the other hand, naltrindole and 7-benzilidenenaltrexone, a selective δ_1-opioid receptor antagonist, interestingly reduced antinociception by endomorphin-1 in diabetic mice. These results indicated that the antinociceptive effects of endomorphin-1 and endomorphin-2 in nondiabetic mice are mediated through the activation of the μ_1-opioid receptor, whereas in diabetic mice, endomorphin-1 and endomorphin-2 may produce antinociception through different opioid receptors, such as δ_1 and μ_1 (Kamei *et al.*, 2000).

Subcutaneous injection of formalin into the rat hind-paw produces a biphasic nociceptive response. Phase 1 is basically an acute pain response, whereas phase 2 is responsible for injury-induced sensitization and hyperalgesia (Coderre *et al.*, 1990). Lidocaine, a common local analgesic, and μ-opioid receptor agonists inhibit nociceptive C-fiber activity that results in antinociception (Fraser *et al.*, 1992). Spinal morphine and lidocaine injections depress the behavioral responses in the formalin test (Hao and Ogawa, 1998; Yamamoto and Yaksh, 1992). Combining two drugs produced a synergistic antinociceptive effect, reduced the dose required, and thereby minimized the severity of the side effects. Intrathecal lidocaine produced dose-dependent inhibition of phase-2 behavioral response and based on isobolographic analysis confirmed the combination of intrathecal endomorphin-1 and lidocaine produced a synergistic suppression of the phase-2 behavioral response (Hao *et al.*, 1999).

3. Various Activities of Endomorphins -1 and -2

In addition to antinociception, opioid peptides seem to be involved in numerous physiological activities. These include behavioral and locomotor activity, affects on the neuroendocrine system (Mansour *et al.*, 1995), gastrointestinal motility (Kromer, 1988), the production and release of neuroendocrine hormones and immune system function (Herz, 1993), eating and drinking (Holzman, 1975), some aspects of sexual behavior (Pfaus and Gorzalka, 1987; Gessa *et al.*, 1979), and alcohol abuse (Gianoulakis and de Waele, 1994). It is well known that endomorphin-1 and -2 immunoreactivities are found both throughout the CNS and outside the CNS (Jessop *et al.*, 2000). Furthermore, in the gastrointestinal tract (Zadina *et al.*, 1997; Hahn and Allescher, 1998) and guinea-pig small intestine (Tonini *et al.*, 1998), both endomorphins interact with μ receptor sites and they inhibited electrical field stimulation-induced tachykinin-mediated contractions of the guinea pig bronchus (Fischer and Undem, 1999).

a. Effects on Vascular Systems

The rostral ventrolateral medulla brain region (RVLM), which is the nucleus in the CNS that regulates peripheral cardiovascular functions (Spyer, 1994; Sun, 1996), contains an abundance of μ receptors (Arvidsson *et al.*, 1995; Mansour *et al.*, 1995, 1988). It was suggested that this high concentration of μ receptors might implicate endomorphin-1 and -2 in the regulation of cardiovascular activity through RVLM neurons (Holaday, 1983). In fact, endomorphin-1 and, to a lesser extent, endomorphin-2 exerted an inhibitory modulation of the electrical activity of RVLM neurons in rat brain slices *in vitro* (Chu *et al.*, 1999).

Endomorphin-1 and -2 decreased systemic arterial pressure in the rabbit (Champion *et al.*, 1997a). This vasodepressor response to the endomorphins was inhibited by naloxone, demonstrating the involvement of μ receptors. It was also shown that the endomorphins were naloxone sensitive in their vasodilator activity in the rat (Champion *et al.*, 1997b,c) where the roles of nitric oxide, prostaglandins, and the K^+-ATP channels on mediating the response to endomorphin-1 and other opioid peptides were studied (Champion *et al.*, 1998a; Champion and Kadowitz, 1999). Vasodilator responses to endomorphin-1 were attenuated by the nitric oxide synthetase inhibitor N^ω-nitro-1-arginine methyl ester. Endomorphin-1 and -2 induced decreases in arterial pressure when injected intravenously (iv) in both rats (Czapla *et al.*, 1998) and mice (Champion *et al.*, 1998b). The biphasic mechanism was characterized by an initial increase followed by a decrease in systemic arterial pressure (Champion *et al.*, 1998c). The iv administered endomorphins in urethane-anesthetized rats lowered heart rate and mean arterial pressure through an activation of vagal afferent nerves (Kwock and Dun, 1998). Similarly, the endomorphins apparently relaxed the tonus of vascular smooth muscle, as investigated *in vitro* on isolated rings from rat aorta (Hugghins *et al.*, 2000).

b. Effects on Memory

It has been reported that the opioid neuronal systems are involved in the memory process. For example, β-endorphin and enkephalin impair the memory process, although it is facilitated by naloxone (Rigter *et al.*, 1979; Izquirdo, 1980; Castellano and Pavone, 1985; Izquierdo *et al.*, 1985; Izquierdo and Netto, 1985). Drugs that have amnesic properties, such as scopolamine, morphine, and dizocipline (Sarter *et al.*, 1988; Parada-Turska and Turski, 1990; Stone *et al.*, 1991), as well as [d-Ala^2,NMePhe^4,Gly-ol]enkephalin (DAMGO) and Tyr-d-Arg-Phe-β-Ala-NH$_2$ (TAPA), μ-opioid selective receptor agonists, impair spontaneous alteration performance (Ukai *et al.*, 1993; Itoh *et al.*, 1994).

The endomorphins produced a decrease in the spontaneous alteration performance test in mice without affecting total working memory, where β-funaltrexamine almost completely reversed this effect, and both naltrindole and norbinaltorphimine were ineffective (Ukai *et al.*, 2000). Thus, it was concluded that endomorphins impair spatial working memory through μ-opioid receptors (Ukai *et al.*, 2000).

Long-term memory for passive-avoidance learning in the day-old chick has two periods of protein synthesis. The first phase occurs up to 90 min posttraining and the other between 4 and 5 h after training; the second wave of protein synthesis occurs in the lobus parolfactorius (Freeman *et al.*, 1995). The involvement of the μ-opioid receptor in passive-avoidance was studied by Freeman and Young (2000) who found that bilateral intracranical injections of β-funaltrexamine caused amnesia for the learned task in chicks. This suggested involvement of μ-opioid receptors in long-term memory formation for the passive-avoidance learning task during the second wave of neuronal activity. Endomorphin-2 did not disrupt this process.

c. Inhibitory Effects on Acetylcholine Release

The inhibition of acetylcholine (ACh) release primarily through μ receptors in isolated preparations of longitudinal muscle with myenteric plexus (LMMP) from guinea-pig ileum by endogenous opioid peptides appeared to be associated with muscarinic autoinhibition (Nishiwaki *et al.*, 1998a,b). The endomorphins inhibited ACh release from LMMP in the guinea-pig ileum and its effect had a component in common with muscarinic autoinhibition (Nishiwaki *et al.*, 1998c). The endomorphins also inhibited the release of ACh from the rat stomach (Yokotani and Osumi, 1998). The inhibitory effect of these peptides was naloxone sensitive on the electrically evoked Ach release from the cholinergic nerves in isolated guinea pig and human trachea (Patel *et al.*, 1999).

d. Other Activities

The most common side effect of morphine administered epidurally or intrathecally is pruritus, which is generalized or localized to the face, neck, and upper thorax. Opioid antagonists block pruritus induced by morphine (Saiah *et al.*, 1994). Facial scratching is induced by icv or intramedullary injections of DAMGO, but not by δ- or κ-opioid receptor agonists (Tohda *et al.*, 1997; Thomas and Hammond, 1995). Endomorphins induced facial scratching, which was characterized by bell-shaped dose–response curves and inhibited by subcutaneous pretreatment with naloxone (Yamaguchi *et al.*, 1998).

Opiates and endogenous opioid peptides have various immunomodulatory properties (Einstein *et al.*, 1996). For example, morphine inhibited the chemotactic (Chao *et al.*, 1997) and phagocytic (Sowa *et al.*, 1997) functions of microglia. These brain cells support the productive infection by HIV-1 (Gendelman *et al.*, 1997), and, interestingly, the κ-opioid receptor ligand U50,488 (*trans*-($\pm$)-3,4-dichloro-*N*-methyl-[2-(1-pyrrolidinyl)-cyclohexyl]benzenacetamide) was found to suppress its replication (Chao *et al.*, 1996). Morphine, on the other hand, increased HIV-1 expression in chronically infected promonocytes when they were cocultured with a mixture of glial cells and neurons (Peterson *et al.*, 1994). Endomorphin-1 also potentiated HIV-1 expression, but in a bell-shaped dose–response curve, and was blocked by pretreatment with either β-funaltrexamine or the G-protein inhibitor pertussis toxin (Peterson *et al.*, 1999). It was also demonstrated that endomorphin-1 and -2 might modulate the production of superoxide anions in neutrophils via μ-opioid receptors (Azuma *et al.*, 2000).

III. STRUCTURE–ACTIVITY RELATIONSHIP

In addition to the endomorphin tetrapeptides, morphiceptin (Tyr-Pro-Phe-Pro-NH$_2$) (Chang *et al.*, 1981) and hemorphin-4 (Tyr-Pro-Trp-Thr) (Brantl *et al.*, 1986) are formed by enzymatic degradation of the milk protein casin and hemoglobin, respectively. Tyr-MIF-1 (Tyr-Pro-Leu-Gly-NH$_2$), where MIF is melanocyte-stimulating hormone-release-inhibiting factor (Pro-Leu-Gly-NH$_2$) (Horvath and Kastin, 1989,1990), and Tyr-W-MIF-1 (Tyr-Pro-Trp-Gly-NH$_2$) (Erchegyi *et al.*, 1992; Hackler *et al.*, 1993) were isolated from bovine hypothalamus and human brain cortex. As in the endomorphins, Pro at the second position confers μ-opioid receptor selectivity; however, the affinity of these opioid peptides depends on the nature of the amino acid in the fourth position. The absence of Trp or Phe in position 3 or 4 in Tyr-MIF-1 may account for the low potency as seen in Table IV. The replacement of Leu by Trp in position 3 of Tyr-MIF-1 improved its ability to hyperpolarize neuronal membrane potentials (Yang *et al.*, 1999). Endomorphin-1 (Tyr-Pro-Trp-Phe-NH$_2$) and endomorphin-2 (Tyr-Pro-Phe-Phe-NH$_2$) are almost equally potent. Therefore, the replacement of Trp by Phe in position 3 does not affect the ability to bind within μ-opioid receptors or reduce neuronal excitation (Yang *et al.*, 1999). The potency of opioid tetrapeptides containing the Tyr-Pro-Trp-X-NH$_2$ or Tyr-Pro-Phe-X-NH$_2$ sequence was determined by the nature of the fourth residue and the order of effectiveness was as follows: Phe^4 (endomorphin-1 or endomorphin-2) > Pro^4 (morphiceptin) > Gly^4 (Tyr-W-MIF-1).

To study the structure–activity relationships of endomorphins, 16 stereoisomeric analogues of endomorphin-2 and des-Phe^4-endomorphin-2 were synthesized

TABLE IV Inhibition of Locus Coeruleus Neuronal Activities by Opioid Tetrapeptides[a]

Tetrapeptide	Sequence	IC$_{50}$(nM)	HC$_{5mv}$[b] (nM)
Endomorphin-1	Tyr-Pro-Trp-Phe-NH$_2$	8.8	22.1
Endomorphin-2	Tyr-Pro-Phe-Phe-NH$_2$	5.3	16.1
Morphiceptin	Tyr-Pro-Phe-Pro-NH$_2$	65	335
Hemorphin-4	Tyr-Pro-Trp-Thr	6.7×10^3	36.9×10^3
Tyr-MIF-1	Tyr-Pro-Leu-Gly-NH$_2$	37.5×10^3	76.2×10^3
Tyr-W-MIF-1	Tyr-Pro-Trp-Gly-NH$_2$	3.8×10^3	6.7×10^3

[a] From Yang et al. (1999).
[b] HC$_{5mv}$, 5-mV hyperpolarization.

and their opioid receptor binding activities were examined (Table V) (Okada et al., 2000).

The change in conformation by the introduction of d-amino acid(s) lowered their binding activity for μ-opioid receptors, indicating that the three-dimensional structure of endomorphin-2 with the natural l-configuration was the most preferentially recognized. These data demonstrated an important role of Pro at position 2

TABLE V Receptor Binding of Stereoisomeric Endomorphin-2 Derivatives[a]

Compounds	Tyr	Pro	Phe	Phe-NH$_2$	K_i values (nM) δ	μ	Binding selectivity δ/μ
Endomorphin-2							
[1]	l	l	l	l	7,250	1.33	5,450
[2]	d	d	d	d	16,580	1,040	16
[3]	l	l	d	d	1,250	24.3	51
[4]	d	d	l	l	19,460	2,760	7
[5]	d	l	l	l	4,120	32.1	128
[6]	l	d	d	d	13,280	2,010	6.6
[7]	l	l	l	d	8,160	45.9	177
[8]	d	d	d	l	7,200	108	67
[9]	l	l	d	l	4,230	203	21
[10]	d	d	l	d	18,620	7,050	2.6
[11]	l	d	l	l	30,640	512	60
[12]	d	l	d	d	21,260	364	58
[13]	d	l	l	d	4,190	557	7.5
[14]	l	d	d	l	16,660	4,710	3.5
[15]	d	l	d	l	14,580	652	22
[16]	l	d	l	d	26,210	1,310	20
[17]	l	l	l	—	15,900	46.3	343

[a] From Okada et al. (2000).

TABLE VI Bioactivity of Endomorphin-1 Analogues Based on the Guinea Pig Ileum Assay[a]

Peptide	IC_{50} (nM)	Agonist potency ratio[b]
Endomorphin-1 (EM-1)	6.7	10.9
[d-Tyr^1]EM-1	303	0.24
[d-Pro^2]EM-1	28%[c]	—
[d-Trp^3]EM-1	553	0.14
[d-Phe^4]EM-1	127	0.99

[a] From Paterlini et al. (2000).
[b] The IC_{50} of the agonist divided by the IC_{50} of morphine in the same preparation.
[c] Maximum response at 1 μM peptide.

of endomorphin-2 and were supported by the data of Paterlini et al. (2000) who synthesized stereoisomeric analogues of endomorphin-1 as shown in Table VI.

The ability of [d-Tyr^1] and [d-Trp^3] to activate the μ-opioid receptor with greatly reduced potency may be due to the partial similarity with the bioactive conformation of endomorphin-1. The severe reduction in potency of the d-Trp analogue suggests that this residue is more critical for receptor recognition and a drastic loss of activity occurred in [d-Pro^2]-endomorphin-1. Molecular simulations of trans-[d-Pro^2]-endomorphin-1 using NOE-derived distance constraints afforded well-defined structures in which Tyr and Trp side chains stacked against the proline ring. The inactivity of [d-Pro^2]-endomorphin-1 was explained by comparison with endomorphin-1 (Podlogar et al., 1998), in which these peptides showed an orientation opposite to the Trp^3 residue with respect to Tyr^1. [des-Phe^4]-Endomorphin-2 exhibited binding activity with μ-opioid receptors suggesting that Tyr-Pro-Phe-derivatives could provide unique opioid mimetics (Okada et al., 2000).

In several cases, the substitution of an α-amino acid for their β-isomer in biologically active peptides resulted in increased activity and enzymatic stability. Some residues containing a β-isomer in Tyr-Pro-Trp-Phe-NH$_2$ were studied and this is summarized in Table VII (Cardillo et al., 2000).

The substitution of a Tyr, Trp, or Phe residue with a β-isomer in endomorphin-1 prevented specific binding interactions. On the other hand, the tetrapeptide containing β-Pro [Tyr-β-(R)-Pro-Trp-Phe-NH$_2$] displayed a higher affinity than endomorphin-1.

Several peptidases metabolize opioid peptides. Dipeptidyl peptidase IV (EC.3.4.14.5), a membrane-bound serine proteinase, removes dipeptides from the amino-terminus of peptides containing proline as the penultimate amino acid (Kato et al., 1978; Mentlein, 1999; Sugimoto-Watanabe et al., 1999; Ronai et al., 1999) as found in endomorphin-1 and -2. The metabolic fate of endomorphins was also investigated (Peter et al., 1999). Because dipeptidyl peptidase IV has an absolute requirement for l-Pro, a more metabolically stable [d-Pro^2]endomorphin-2

TABLE VII Bioactivity of β-Amino Acid Containing Endomorphin-1 Derivatives[a]

Compound	K_i (nM)	IC_{50}(nM)	n_H
DAMGO	1.64	9.89	0.88
Tyr-Pro-Trp-PheNH$_2$	11.1	56	0.78
Tyr-Pro-Trp-βPheNH$_2$	908	680	nd[b]
Tyr-Pro-βTrp-PheNH$_2$	>10^3	nd	nd
βTyr-Pro-Trp-PheNH$_2$	>10^3	nd	nd
Tyr-D-Pro-Trp-PheNH$_2$	54	470	0.66
Tyr-β-(S)-Pro-Trp-PheNH$_2$	10.4	72.0	0.81
Tyr-β-(R)-Pro-Trp-PheNH$_2$	0.33	1.80	0.77

[a] From Cardillo *et al.* (2000).
[b] nd, not determined.

analogue should produce longer analgesic activity at a lower dose. Endomorphin-2 was degraded by dipeptidyl peptidase IV, whereas the d-Pro^2 analogue was totally resistant (Shane *et al.,* 1999) and was more potent than endomorphin-2 in increasing tail-flick latencies with longer duration of action. However, both peptides were equipotent in increasing jump thresholds. Ala-Pyrrolidonyl-2-nitrile, a potent, stable, and specific inhibitor of dipeptidyl peptidase IV (Li *et al.,* 1995), produced a dose-dependent analgesia and potentiated the analgesic action of endomorphin-2 (Shane *et al.,* 1999). The data suggest that dipetidyl peptidase IV plays a role in the inactivation of endomorphin-2 *in vivo* and could modulate its central analgesic actions.

IV. CONCLUSIONS AND FUTURE DIRECTIONS

G-protein-coupled receptors constitute the largest known family of cell-surface receptors and sequenced genes encoding many medically important G$_i$-coupled receptors, such as the opioid, serotonin, and dopamine receptors. Drugs that target these receptors have found application in the treatment of pain, depression, various psychoses, and Parkinson's disease. The μ receptor-selective alkaloid morphine is still an important drug with major clinical and therapeutic applications for analgesia. Based on the potent and selective activity of the endomorphins, it has been suggested that they could provide new molecular tools for examining opiate pathways in the brain and spinal cord, leading to the development of potent analgesics with less addictive properties than the traditional opioid receptor agonists.

To develop novel analgesics with fewer side effects, it is important that peptide mimetics or peptide analogues are more stable, potent, and selective than endomorphins. Because Tyr-Pro-Phe-NH$_2$ exhibited μ-opioid receptor binding, new opioid

TABLE VIII Binding Activity and Biological Potencies of Dmt Containing Endomorphin-2 Analogues[a]

Compound	Receptor binding K_i (nM)		Binding selectivity δ/μ	IC_{50} (nM)		pA_2 value vs. DADLE
	δ	μ		GPI assay	MVD assay	
Dmt-Pro-Phe-NH-1-naphthalene	19.9	0.30	68	0.494	5.47	—
Dmt-Pro-Phe-NH-3-quinoline	190	0.33	580	9.14	>10,000	6.01
Dmt-Pro-Phe-NH-8-quinoline	33.1	0.49	68	44.5	2,980	5.87
Dmt-Pro-Phe-NH-5-isoquinoline	98.3	0.19	517	0.939	>10,000	6.14

[a] From Takahashi et al. (2001).

analogues were synthesized based on this sequence, such as Tyr-Pro-Phe-NH-1-naphthalene (Okada et al., 1999), 2′,6′-dimethyl-l-tyrosine (Dmt) analogues, Dmt-Pro-Phe-NH-1-naphthalene, and Dmt-Pro-Phe-NH-5-isoquinoline (Table VIII). These molecules exhibited extraordinary binding activity with K_i μ values of 0.30 and 0.19 nM, respectively, higher than any previously identified μ-opioid peptides. In addition, these derivatives exhibited IC_{50} values of 0.49 and 5.5 nM in guinea pig ileum (GPI) and mouse vas deferens (MVD) assays, respectively, whereas others had IC_{50} values of 0.94 nM in GPI assay and >10,000 nM in MVD assays. The marked increase in binding affinity and bioactivity can be attributed solely to involvement of Dmt in the ligand-binding domain as discussed in detail elsewhere (Salvadori et al., 1995; Bryant et al., 1998; Lazarus et al., 1998). Thus, these compounds act as agonists toward μ-opioid receptors and antagonists against δ-opioid receptor and have antinociceptive activity by icv administration in mice (Takahashi et al., 2000). Clearly further studies on endomorphin derivatives will contribute to the molecular and chemical design of more potent μ-selective compounds, the determination of the function of each opioid receptor subtype, and the development of new therapeutic drugs.

REFERENCES

Alt, A., Mansour, A., Akil, H., Medizihradsky, F., Traynor, J. R., and Woods, J. H. (1998). Stimulation of guanosine-5′-O-(3-[35S]thio)triphosphate binding by endogenous opioids acting at a cloned mu receptor. J. Pharmacol. Exp. Ther. **286,** 282–288.

Arvidsson, U., Riedl, M., Chakarabarti, S., Lee, J. H., Nakano, A., Dado, H., Loh, H., Law, P. Y., Wessendorf, M. W., and Elde, R. (1995). Distribution and targeting of a μ-opioid receptor (MOR1) in brain and spinal cord. J. Neurosci. **15,** 3328–3341.

Azuma, Y., Wang, P. -L., Shinohara, M., and Ohura, K. (2000). Immunomodulation of the neutrophil respiratory burst by endomorphin 1 and 2. Immunol. Lett. **75,** 55–59.

Baron, A., Shuster, L., Eleftheriou, B. E., and Bailey, D. W. (1975). Opiate receptors in mice: Genetic differences. Life Sci. **17,** 633–640.

Brantl, V., Gramsch, C., Lottspeich, F., Mertz, R., Jaeger, K.-H., and Herz, A. (1986). Novel opioid peptides derived from hemoglobin: Hemorphins. Eur. J. Pharmacol. **125,** 309–310.

Bryant, S. D., Salvadori, S., Cooper, P. S., and Lazarus, L. H. (1998). New δ-opioid antagonists as pharmacological probes. *Trends Pharmcol. Sci.* **19,** 42–46.

Burford, N. T., Tolbert, L. M., and Sadee, W. (1998). Specific G protein activation and μ-opioid receptor internalization caused by morphine, DAMGO and endomorphin I. *Eur. J. Pharmacol.* **342,** 123–126.

Cardillo, G., Gentilucci, L., Melchiorre, P., and Spampinato, S. (2000). Synthesis and biological activity of endomorphin-1 analogues containing β-amino acids. *Bioorg. Med. Chem. Lett.* **10,** 2755–2758.

Carrigan, K. A., Nelson, C. J., and Lysle, D. T. (2000). Endomorphin-1 induces antinociception without immunomodulatory effects in the rat. *Psychopharmacology* **151,** 299–305.

Castellano, C., and Pavone, F. (1985). Dose- and strain-dependent effects of dermorphin and [d-Ala^2, d-Leu^5]enkephalin on passive avoidance behavior in mice. *Behav. Neurosci.* **99,** 1120–1127.

Champion, H. C., and Kadowitz, P. J. (1999). Vasodepressor response to [d-Ala^2]-endomorphin 2 (TAPP) are mediated an L-NAME-sensitive mechanism in the rat. *J. Cardiovasc. Pharmacol.* **33,** 280–284.

Champion, H. C., Zadina, J. E., Kastin, A. J., Hackler, L., Ge, L. J., and Kadowitz, P. J. (1997a). The endogenous mu-opioid receptor agonists, endomorphin 1 and 2 have novel hypotensive activity in the rabbit. *Biochem. Biophys. Res. Commun.* **235,** 567–570.

Champion, H. C., Zadina, J. E., Kastin, A. J., and Kadowitz, P. J. (1997b). The endogenous μ-opioid agonists, endomorphin 1 and 2, have vasodilator activity in the hindquarters vascular bed of the rat. *Life Sci.* **61,** 409–415.

Champion, H. C., Zadina, J. E., Kastin, A. J., Hackler, L., Ge, L.-J., and Kadowitz, P. J. (1997c). Endomorphin 1 and 2, endogenous ligands for the μ-opioid receptor, decrease cardiac output and total peripheral resistance in the rat. *Peptides* **18,** 1393–1397.

Champion, H. C., Bivalacqua, T. J., Friedman, D. E., Zadina, J. E., Kastin, A. J., and Kadowitz, P. J. (1998a). Nitric oxide release mediates vasodilator response to endomorphin 1 but not nociceptin/OFQ in the hindquarters vascular bed of the rat. *Peptides* **19,** 1595–1602.

Champion, H. C., Zadina, J. E., Kastin, A. J., and Kadowitz, P. J. (1998b). Endomorphin 1 and 2 have vasodepressor activity in the anesthetized mouse. *Peptides* **19,** 925–929.

Champion, H. C., Bivalacqua, T. J., Lambert, D. G., McWilliams, S. M., Zadina, J. E., Kastin, A. J., and Kadowitz, P. J. (1998c). Endomorphin 1 and 2, the endogenous μ-opioid agonists, produce biphasic changes in systemic arterial pressure in the cat. *Life Sci.* **63,** 131–136.

Chang, K.-J., Killian, A., Hazum, E., and Chang, J.-K. (1981). Morphiceptin (NH4-Tyr-Pro-Phe-Pro-CONH2): A potent and specific agonist for morphine (μ) receptors. *Science* **212,** 75–77.

Chao, C. C., Gekker, G., Hu, S., Sheng, W. S., Shark, K. B., Bu, D. F., Archer, S., Bidlak, T. H., and Peterson, P. K. (1996). Kappa opioid receptors in human microglia downregulate human immunodeficiency virus 1 expression. *Proc. Natl. Acad. Sci. USA* **93,** 8051–8056.

Chao, C. C., Hu, S., Shark, K. B., Sheng, W. S., Gekker, G., and Peterson, P. K. (1997). Activation of mu-opioid receptors inhibits microglial cell chemotaxis. *J. Pharmacol. Exp. Ther.* **281,** 998–1004.

Chen, Y., Mestek, A., Liu, J., Hurley, J. A., and Yu, L. (1993). Molecular cloning and functional expression of a μ-opioid receptor from the rat brain. *Mol. Pharmacol.* **44,** 8–12.

Childers, S. R. (1991). Opioid receptor coupled second messenger system. *Life Sci.* **48,** 1991–2003.

Chu, X. P., Xu, N. S., Li, P., and Wang, J. Q. (1999). Endomorphin-1 and endomorphin-2, endogenous ligand for the μ-opioid receptor, inhibit electrical activity of rat rostral ventrolateral medulla neurons *in vitro*. *Neuroscience* **93,** 681–686.

Coderre, T. J., Vaccarino, A. L., and Melzack, R. (1990). Central nervous system plasticity in the tonic pain response to subcutaneous formalin injection. *Brain Res.* **535,** 155–158.

Connor, M., Schuller, A., Pintar, J. E., and Christie, MacD. J. (1999). μ-Opioid receptor modulation of calcium channel current in periaqueductal grey neurons from C57B16/J mice and mutant mice lacking MOR-1. *Br. J. Pharmacol.* **126,** 1553–1558.

Corbett, A. D., Pareson, A. I., and Kosteritz, H. W. (1993). *In* Opioids I (Handbook of Experimental Pharmacology), Vol. 104/1 (A. Herz, Ed.), pp. 645–675. Springer-Verlag, New York.

Cox, B. M., Goldstein, A., and Li, C. H. (1976). Opioid activity of a peptide, β-lipotropin-(61–91), derived from β-lipotropin. *Proc. Natl. Acad. Sci. USA* **73,** 1821–1823.

Czapla, M. A., Champion, H. C., Zadina, J. E., Kastin, A. J., Hackler, L., Ge, L.-J., and Kadowitz, P. J. (1998). Endomorphin 1 and 2, endogenous μ-opioid agonists, decrease systemic arterial pressure in the rat. *Life Sci.* **62,** 175–179.

Einstein, T. K., Hilburger, M. D., and Lawrence, D. M. P. (1996). Immunomodulation by morphine and other opioids. *In* "Drugs of Abuse, Immunity and Infections" (H. Friedman, T. W. Klein, and S. Specter, Eds.), pp. 103–120. CRC press, Boca Raton, FL.

Erchegyi, J., Kastin, A. J., and Zadina, J. E. (1992). Isolation of a novel tetrapeptide with opiate, and antiopiate activity from human brain cortex: Tyr-Pro-Try-Gly-NH$_2$ (Tyr-W-MIF-1). *Peptides* **13,** 623–631.

Evans, C. J., Keith, D. E. , Jr., Morrison, H., Magendzo, K., and Edwards, R. H. (1992). Cloning of a delta opioid receptor by functional expression. *Science* **258,** 1952–1955.

Fischer, A., and Undem, B. J. (1999). Naloxone blocks endomorphn-1 but not endomorphin-2 induced inhibition of tachykinergic contractions of guinea pig isolated bronchus. *Br. J. Pharmacol.* **127,** 605–608.

Fraser, H. M., Chapman, V., and Dickenson, A. H. (1992). Spinal local anaesthetic actions on afferent evoked responses and wind-up of nociceptive neurones in the rat spinal cord: Combination with morphine produces marked potentiation of antinociception. *Pain* **49,** 33–41.

Freeman, F. M., and Young, I. G. (2000). Identification of the opioid receptors involved in passive-avoidance learning in the day-old chick during the second wave of neuronal activity. *Brain Res.* **854,** 230–239.

Freeman, F. M., Rose, S. P. R., and Scholey, A. B. (1995). Two time windows of anisomycin-induced amnesia for passive-avoidance training in the day-old chick. *Neurobiol. Learn. Mem.* **63,** 291–295.

Gendelman, H. E., Ghorpade, A., and Persidsky, Y. (1997). The neuropathogenesis of HIV-1-associated dementia. *In* "Defense of the Brain: Current Concepts in the Immunopathogenesis and Clinical Aspects of CNS Infections" (P. K. Peterson and J. S. Remington, Eds.), pp. 290–304. Blackwell, Malden, MA.

Gessa, G. L., Paglietti, E., and Quarantotti, B. P. (1979). Induction of copulatory behavior in sexually inactive rat by naloxone. *Science* **204,** 203–205.

Gianoulakis, C., and de Waele, J.-P. (1994). Genetics of alcoholism: Role of the endogenous opioid system. *Metab. Brain Dis.* **9,** 105–125.

Goldberg, I. E., Rossi, G. C., Letchworth, S. R., Mathis, J. P., Ryan-Moro, J., Leventhal, L., Su, W., Emmel., D., Bolan, E. A., and Pasternak, G. A. (1998). Pharmacological characterization of endomorphin- and endomorphin-2 in mouse brain. *J. Pharmacol. Exp. Ther.* **286,** 1007–1013.

Goldstein, A., Tachibana, S., Lowney, L. I., Hunkapiller, M., and Hood, L. (1979). Dynorphin (1–13), an extraordinarily potent opioid peptide. *Proc. Natl. Acad. Sci. USA* **76,** 6666–6670.

Goldstein, A. G., Fischli, W., Lowney, L. I., Hunkapilier, M., and Hood, L. (1981). Porcine pituitary dynorphin: Complete amino acid sequence of biologically active heptapeptide. *Proc. Natl. Acad. Sci. USA* **78,** 7219–7223.

Goodman, R. R., and Pasternak, G. W. (1985). Visualization of mu-1 opiate receptors in rat brain by using a computerized autoradiographic subtraction technique. *Proc. Natl. Acad. Sci. USA* **82,** 6667–6671.

Hackler, L., Kastin, A. J., Erchegyi, J., and Zadina, J. E. (1993). Isolation of Tyr-W-MIF-1 from bovine hypothalami. *Neuropeptides* **24,** 159–164.

Hackler, L., Zadina, J. E., Ge, L. J., and Kastin, A. J. (1997). Isolation of relatively large amounts of endomorphin-1 and endomorphin-2 from human brain cortex. *Peptides* **18,** 1635–1639.

Hahn, A., and Allescher, D. (1998). Effect of endomorphin-1 and endomorphin-2 on the ascending and descending reflex pathway in rat intestine. *Gastroenterology* **114,** A795.

Hahn, E. F., Carroll-Buatti, M., and Pasternak, G. W. (1982). Irreversible opiate agonists and antagonists: The 14-hydroxydihydromorphinone azines. *J. Neurosci.* **2,** 572–576.

Hao, S., and Ogawa, H. (1998). Sevoflurane suppresses behavioral response in the rat formalin test: Combination with intrathecal lidocaine produced profound suppression of the responses. *Neurosci. Lett.* **248,** 124–126.

Hao, S., Takahata, O., and Iwasaki, H. (1999). Isobolographic analysis of interaction between spinal endomorphin-1 a newly isolated endogenous opioid peptide, and lidocaine in the rat formalin test. *Neurosci. Lett.* **276,** 177–180.

Hao, S., Takahata, O., and Iwasaki, H. (2000). Intrathecal endomorphin-1 produces antinociceptive activities modulated by alpha 2-adrenoceptors in the rat tail flick, tail pressure and formalin test. *Life Sci.* **66,** 195–204.

Herz, A. (Ed.) (1993). "Opioids II." Springer-Verlag, Berlin.

Holaday, J. W. (1983). Cardiovascular effects of endogenous opiate system. *Annu. Rev. Pharmacol. Toxicol.* **23,** 541–594.

Holtzman, S. G. (1975). Effects of narcotic antagonists on fluid intake in the rat. *Life Sci.* **16,** 1465–1470.

Horvath, A., and Kastin, A. J. (1989). Isolation of tyrosine-melanocyte-stimulating hormone release-inhibiting factor 1 from bovine brain tissue. *J. Biol. Chem.* **264,** 2175–2179.

Horvath, A., and Kastin, A. J. (1990). Evidence for presence of Tyr-MIF-1 (Tyr-Pro-Leu-Gly-NH$_2$) in human brain cortex. *Int. J. Peptide. Protein Res.* **36,** 281–284.

Horvath, G., Szikszay, Tomboly, C., and Benedek, G. (1999). Antinociceptive effects of intrathecal endomorphin-1 and -2 in rats. *Life Sci.* **65,** 2635–2641.

Hosohata, K., Burkey, T. H., Alfaro-Lopez, J., Varga, E., Hruby, V. J., Roeske, W. R., and Yamamura, H. I. (1998). Endomorphin-1 and endomorphin-2 are partial agonists at the human μ-opioid receptor. *Eur. J. Pharmacol.* **346,** 111–114.

Hugghins, S. Y., Champion, H. C., Cheng, G., Kadowitz, P. J., and Jeter, J. R., Jr. (2000). Vasorelaxant responses to endomorphins, nociception, albuterol, and adrenomedullin in isolated rat aorta. *Life Sci.* **67,** 471–476.

Hughes, J., Smith, T. W., Kosterlitz, H. W., Forthergill, L. A., Morgan, B. A., and Morris, H. R. (1975). Identification of two related pentapeptides from the brain with potent opiate agonist activity. *Nature* **258,** 577–580.

Itoh, J., Ukai, M., and Kameyama, T. (1994). Dynorphin A-(1–13) potently improves the impairment of spontaneous alteration performance induced by the μ-selective opioid receptor agonist DAMGO in mice. *J. Pharmcol. Exp. Ther.* **269,** 15–21.

Izquierdo, I. (1980). Effect of β-endorphin and naloxone on acquisition, memory, and retrieval of shuttle avoidance and habituation learning in rat. *Psychopharmacology* **69,** 111–115.

Izquierdo, I., and Netto, C. A. (1985). Role of β-endorphin in behavioral regulation. *Ann. N. Y. Acad. Sci.* **444,** 162–177.

Izquierdo, I., De Almeida, M. A. M. R., and Emiliano, V. R. (1985). Unlike β-endorphin, dynorphin 1–13 does not cause retrograde amnesia for shuttle avoidance or inhibitory avidance learning in rat. *Psychopharmacology* **87,** 216–218.

Jessop, D. S., Major, G. N., Coventry, T. L., Kaye, S. J., Fulford, A. J., Harbuz, M. S., and DeBree, F. M. (2000). Novel opioid peptides endomorphin-1 and endomorphin-2 are present in mammalian immune tissues. *J. Neuroimmunol.* **106,** 53–59.

Jiang, Y., Klodesky, C. M., and Chang, S. L. (2000). Endomorphin-1 and endomorphin-2 induce the expression of c-FOS immunoreactivity in the rat brain. *Brain Res.* **873,** 291–296.

Jin, S. X., Lei, L. G., Wang, Y., Da, D. F., and Zhao, Z. Q. (1999). Endomorphin-1 reduces carrageenan-induced Fos expression in the rat spinal horn. *Neuropeptides* **33,** 281–284.

Kakizawa, K., Shimohira, I., Sakurada, S., Fujimura, T., Murayama, K., and Ueda, H. (1998). Parallel stimulations of in vitro and in situ [^{35}S]GTPγS binding by endomorphin 1 and DAMGO in mouse brains. *Peptides* **19,** 755–758.

Kamei, J., Kawashima, N., and Kasuya, Y. (1992a). Role of spleen or spleen products in the deficiency in morphine-induced analgesia in diabetic mice. *Brain Res.* **576,** 139–142.

Kamei, J., Ohhashi, Y., Aoki, T., Kawashima, N., and Kasuya, Y. (1992b). Streptozotocin-induced diabetes selectively alters the potency of analgesia produced by δ opioid agonists, but not by μ and κ opioid agonists. *Brain Res.* **571,** 199–203.

Kamei, J., Kawashima, N., and Kasuya, Y. (1992c). Paradoxical analgesia produced by naloxone in diabetic mice is attributable to supersensitivity of δ opioid receptor. *Brain Res.* **592,** 101–105.

Kamei, J., Zushida, K., Ohsawa, M., and Nagase, H. (2000). The antinociceptive effects of endomorphin-1 and endomorphin-2 in diabetic mice. *Eur. J. Pharmcol.* **391,** 91–96.

Kato, T., Nagatsu, T., Kimura, T., and Sakakibara, S. (1978). Studies on substrate specificity of X-prolyl dipeptidyl-aminopeptidase using new chromogenic substrate X-Y-*p*-nitroanilides. *Experientia* **15,** 319–320.

Kieffer, B. L. (1995). Recent advances in molecular recognition and signal transduction of active peptides: Receptors for opioid peptides. *Cell. Mol. Neurobiol.* **15,** 615–635.

Kieffer, B. L., Befort, K., Gaveriaux-Ruff, C., and Hirth, C. G. (1992). The δ-opioid receptor: Isolation of a cDNA by expression cloning and pharmacological characterization. *Proc. Natl. Acad. Sci. USA* **89,** 12048–12052.

Kromer, W. (1988). Endogenous and exogenous opioids in the control of gastrointestinal motility and secretion. *Pharmacol. Rev.* **40,** 121–162.

Kwock, E. H., and Dun, N. J. (1998). Endomorphins decrease heart rate and blood pressure possibly by activating vagal afferents in anesthetized rats. *Brain Res.* **803,** 204–207.

Lazarus, L. H., Bryant, S. D., Cooper, P. S., Guerrini, R., Balboni, G., and Salvadori, S. (1998). Design of δ-opioid peptide antagonists for emerging drug applications. *Drug Discov. Today* **3,** 284–294.

Li, J., Wilk, E., and Wilk, S. (1995). Aminoacylpyrrolidine-2-nitriles: Potent and stable inhibitors of dipeptidyl peptidase IV (CD 26). *Arch. Biochem. Biophys.* **323,** 148–154.

Liang, Y., Mestek, A., Yu, L., and Carr, L. G. (1995). Cloning and characterization of the promoter region of the mouse μ-opioid receptor gene. *Brain Res.* **679,** 82–88.

Ling, N., Burgus, R., and Guillemin, R. (1976). Isolation, primary structure and synthesis of α-endorphin and γ-endorphin, two peptides of hypothalamic-hypophysial origin with morphinomimetic activity. *Proc. Natl. Acad. Sci. USA* **73,** 3042–3046.

Loh, H. H., Liu, H. C., Cavalli, A., Yang, W., Chen, Y. F., and Wei, L. N. (1998). Mu opioid receptor knockout in mice: Effects on ligand-induced analgesia and morphine lethality. *Mol. Brain Res.* **54,** 321–326.

Lord, J. A. H., Waterfield, A. A., Hughes, J., and Kosterlitz, H. W. (1977). Endogenous opioid peptides: Multiple agonists and receptors. *Nature* **267,** 495–499.

Mansour, A., Khachaturian, H., Lewis, M. E., Akil, H., and Watson, S. J. (1988). Anatomy of CNS opioid receptors. *Trends Neurosci.* **11,** 308–314.

Mansour, A., Fox, C. A., Akil, H., and Watson, S. J. (1995). Opioid-receptor mRNA expression in the rat CNS: Anatomical and functional implications. *Trends Neurosci.* **18,** 22–29.

Martin, W. R., Eades, C. G., Thompson, J. A., Huppler, R. E., and Gilbert, P. E. (1976). The effect of morphine- and morphine-like drugs in the nondependent and morphine-dependent chronic spinal dog. *J. Pharmacol. Exp. Ther.* **197,** 517–532.

Martin-Schild, S., Gerall, A. A., Kastin, A. J., and Zadina, J. E. (1998). Endomorphin-2 is an endogenous opioid in primary sensory afferent fibers. *Peptides* **19,** 1783–1789.

Martin-Schild, S., Zadina, J. E., Gerall, A. A., Vigh, S., and Kastin, A. J. (1997). Localization of endomorphin-2-like immunoreactivity in the rat medulla and spinal cord. *Peptides* **18,** 1641–1649.

Matthes, H. W., Maldonado, R., Simonin, F., Valverde, O., Slowe, S., Kitchen, I., Befort, K., Dierich, A., Le Meur, M., Dolle, P., Tzavara, E., Hanoune, J., Roques, B. P., and Kieffer, B. L. (1996). Loss of morphine-induced analgesia, reward effect and withdrawal symptoms in mice lacking the μ-opioid-receptor gene. *Nature* **383,** 819–823.

Mentlein, R. (1999). Dipeptidyl-peptidase IV (CD26)—role in the inactivation of regulatory peptides. *Regul. Peptides* **85,** 9–24.

Mizoguchi, H., Narita, M., Oji, D. E., Suganuma, C., Nagase, H., Sora, I., Uhl, G. R., Cheng, E. Y., and Tseng, L. F. (1999). The μ-opioid receptor gene—dose dependent reductions in G-protein activation in the pons/medulla and antinociception induced by endomorphins in μ-opioid receptor knockout mice. *Neuroscience* **94,** 203–207.

Monory, K., Bourin, M. C., Spetea, M., Tomboly, C., Toth, G., Matthes, H. W., Kieffer, B. L., Hanoune, J., and Borsodi, A. (2000). Specific activation of the μ opioid receptor (MOR) by endomorphin 1 and endomorphin 2. *Eur. J. Neurosci.* **12,** 577–584.

Moskowitz, A. S., and Goodman, R. R. (1985). Autoradiographic analysis of mμ_1, mμ_2 and delta opioid binding in the central nervous system of C57BL/6BY and CXBK (opioid receptor deficient) mice. *Brain Res.* **360**, 108–116.

Narita, M., Mizoguchi, H., Oji, G. S., Tseng, E. L., Suganuma, C., Nagase, H., and Tseng, L. F. (1998). Characterization of endomorphin-1 and -2 on [^{35}S]GTPγS binding in mouse spinal cord. *Eur. J. Pharmacol.* **351**, 383–387.

Narita, M., Mizoguchi, H., Narita, Mi., Sora, I., Uhl, G. R., and Tseng, L. F. (1999). Absence of G-protein activation by μ-opioid receptor agonists in the spinal cord of μ-opioid receptor knockout mice. *Br. J. Pharmacol.* **126**, 451–456.

Narita, M., Mizoguchi, H., Narita, M., Dun, N. J., Hwang, B. H., Endoh, T., Suzuki, T., Nagase, H., Suzuki, T., and Tseng, L. F. (2000). G-Protein activation by endomorphins in the mouse priaqueductal gray matter. *J. Biomed. Sci.* **7**, 221–225.

Nevo, I., Avidor-Reiss, T., Levy, R., Bayewitch, M., and Vogel, Z. (2000). Acute and chronic activation of the μ-opioid receptor with endogenous ligand endomorphin differentially regulates adenyl cyclase isozymes. *Neurol. Pharmacol.* **39**, 364–371.

Nishimura, S. L., Rech, L. D., and Pasternak, G. W. (1984). Biochemical characterization of high affinity ^{3}H-opioid binding: Further evidence for mu sites. *Mol. Pharmacol.* **25**, 29–37.

Nishiwaki, H., Saitoh, N., Nishio, H., Takeuchi, T., and Hata, F. (1998a). Relationship between muscarinic autoinhibition and the inhibitory effect of morphine on acetylcholine release from myenteric plexus of guinea pig ileum. *Jpn. J. Pharmacol.* **77**, 271–278.

Nishiwaki, H., Saitoh, N., Nishio, H., Takeuchi, T., and Hata, F. (1998b). Relationship between inhibitory effect of endogenous opioid via mu-receptors and muscarinic autoinhibition in acetylcholine release from myenteric plexus of guinea pig ileum. *Jpn. J. Pharmacol.* **77**, 279–286.

Nishiwaki, H., Satoh, N., Nishio, H., Takeuchi, T., and Hata, F. (1998c). Inhibitory effect of endomorphin-1 and -2 on acetylcholine release from myenteric plexus of guinea pig ileum. *Jpn. J. Pharmacol.* **78**, 83–86.

North, R. A. (1989). Drug receptors and the inhibition of nerve cells. *Br. J. Pharmacol.* **98**, 13–28.

Okada, Y., Shimizu, Y., Takahashi, M., Fukumizu, A., Tsuda, Y., Bryant, S. D., and Lazarus, L. H. (1999). Design and synthesis of μ-selective opioid mimetics with endomorphin sequences. Abstract. 16th American Peptide Symposium, Minneapolis, MN, 1999, p. 273.

Okada, Y., Fukumizu, A., Takahashi, M., Shimizu, Y., Tsuda, Y., Bryant, S. D., and Lazarus, L. H. (2000). Synthesis of stereoisomeric analogues of endomorphin-2, H-Tyr-Pro-Phe-Phe-NH$_2$, and examination of their opioid receptor binding activities and solution conformation. *Biochem. Biophys. Res. Commun.* **276**, 7–11.

Parada-Turska, J., and Turski, W. A. (1990). Excitatory amino acid antagonists and memory: Effects of drugs acting at *N*-methyl-d-aspartate receptors in learning and memory task. *Neuropharmacology* **29**, 1111–1116.

Pasternak, G. W., and Wood, P. L. (1986). Multiple mu opiate receptors. *Life Sci.* **38**, 1888–1898.

Patel, H. J., Venkatesan, P., Halfpenny, J., Yacoub, M. H., Fox, A., Barnes, P. J., and Belvisi, M. G. (1999). Modulation of acetylcholine release from parasympathetic nerves innervating guinea-pig and human trachea by endomorphin-1 and -2. *Eur. J. Pharmacol.* **374**, 21–24.

Paterlini, M. G., Avitabile, F., Ostrowski, B. G., Ferguson, D. M., and Portoghese, P. S. (2000). Stereochemical requirements for receptor recognition of the μ-opioid peptide endomorphin-1. *Biophys. J.*, **78**, 590–599.

Paul, D., Standifer, K. M., Inturrisi, C. E., and Pasternak, G. W. (1989). Pharmacological characterization of morphine-6β-glucuronide, a very potent morphine metabolite. *J. Pharmacol. Exp. Ther.,* **251**, 477–483.

Peter, A., Toth, G., Tomboly, C., Laus, G., and Tourwe. (1999). Liquid chromatographic study of the enzymatic degradation of endomorphins, with identification by electrospray ionization mass spectrometry. *J. Chromatogr.* **846**, 39–48.

Peterson, P. K., Gekker, G., and Hu, S. (1994). Morphine amplifies HIV-1 expression in chronically infected promonocytes cocultured with human brain cells. *J. Neuroimmunol.* **50**, 167–175.

Peterson, P. K., Gekker, G., Hu, S., Lokensgard, J., Protoghese, P. S., and Chao, C. C. (1999). Endomorphin-1 potentiates HIV-1 expression in human brain cell cultures: Implication of an atypical μ-opioid receptor. *Neuropharmacology* **38,** 273–278.
Pfaus, J. G., and Gorzalka, B. B. (1987). Opioids and sexual behavior. *Neurosci. Biobehav. Rev.* **11,** 1–34.
Pick, C. G., Paul, D., and Pasternak, G. W. (1991). Comparison of naloxonazine and β-funaltrexamine antagonism of μ_1 and μ_2 opioid actions. *Life Sci.* **48,** 2005–2011.
Pick, C. G., Nejat, R., and Pasternak, G. W. (1993). Independent expression of two pharmacologically distinct supraspinal mu analgesic systems in genetically different mouse strains. *J. Pharmacol. Exp. Ther.* **265,** 166–171.
Pierce, T. L., and Wessendorf, M. W. (2000). Immunocytochemical mapping of endomorphin-2-imunoreactivity in rat brain. *J. Chem. Neuroanat.* **18,** 181–207.
Pierce, T. L., Grahek, M. D., and Wessendorf, M. W. (1998). Immunoreactivity for endomorphin-2 occurs in primary afferents in rat and monkey. *Neuroreport* **9,** 385–389.
Podlogar, B. L., Paterlini, M. G., Ferguson, D. M., Leo, G. C., Demeter, D. A., Brown, F. K., and Reitz, A. B. (1998). Conformational analysis of the endogenous μ-opioid agonist endomorphin-1 using NMR spectroscopy and molecular modeling. *FEBS Lett.* **439,** 13–20.
Porzig, H. (1990). Pharmacological modulation of voltage dependent calcium channels in intact cells. *Rev. Physiol. Biochem. Pharmacol.* **114,** 209–262.
Przewlocka, B., Mika, J., Labuz, D., Toth, G., and Przewlocki, R. (1999). Spinal analgesic action of endomorphins in acute, inflammatory and neuropathic pain in rats. *Eur. J. Pharmacol.* **367,** 189–196.
Reith, M. E. A., Sershen, H., Vadasz, C., and Lajtha, A. (1981). Strain differences in opiate receptors in mouse brain. *Eur. J. Pharmacol.* **74,** 377–380.
Rigter, H., Hannan, T. J., Messing, R. B., Martinez, J. L., Jr., Vasquez, B. J., Jensen, R. A., Veliquette, J., and McGaugh, J. L. (1979). Enkephalins interfere with acquisition of an active avoidance response. *Life Sci.* **26,** 337–345.
Ronai, A. Z., Timar, J., Mako, E., Erdo, F., Gyarmati, Z., Toth, G., Orosz, G., Furst, S., and Szekely, J. I. (1999). Diprotin A, an inhibitor of dipeptidyl aminopeptidase IV (EC 3. 4. 14. 5) produces naloxone-reversible analgesia in rat. *Life Sci.* **64,** 145–152.
Saiah, M., Borgeat, A., Wilder-Smith O. H. G., Rifat, K., and Suter, P. M. (1994). Epidual-morphine-induced pruritus: Propofol versus naloxone. *Anesth. Analg* **78,** 1110–1113.
Sakurada, S., Zadina, J. E., Kastin, A. J., Katsuyama, S., Fujimura, T., Murayama, K., Yuki, M., Ueda, H., and Sakurada, T. (1999). Differential involvement of μ-opioid receptor subtypes in endomorphin-1 and -2-induced antinociception. *Eur. J. Pharmacol.* **372,** 25–30.
Sakurada, S., Hayashi, T., Yuhki, M., Fujimura, T., Murayama, K., Yonezawa, A., Sakurada, C., Takeshita, M., Zadina, J. E., Kastin, A. J., and Sakurada, T. (2000). Differential antagonism of endomorphin-1 and endomorphin-2 spinal antinociception by naloxonazine and 3-methylnaltrexone. *Brain Res.* **881,** 1–8.
Salvadori, S., Attila, M., Balboni, G., Bianchi, C., Bryant, S. D., Crescenzi, O., Guerrini, R., Picone, D., Tancredi, T., Temussi, P. A., and Lazarus, L. H. (1995). δ Opioidmimetic antagonists: Prototypes for designing a new generation of ultraselective opioid peptides. *Mol. Med.* **1,** 678–689.
Shane, R., Wilk, S., and Bodnar, R. J. (1999). Modulation of endomorphin-2-induced analgesia by dipeptidyl peptidase IV. *Brain Res.* **815,** 278–286.
Sim, L. J., Liu, Q., Childer, S. R., and Selley, D. E. (1998). Endomorphin-stimulated [^{35}S]GTPγS binding in rat brain: Evidence for partial agonist activity at μ-opioid receptor. *J. Neurochem.* **70,** 1567–1576.
Soignier, R. D., Vaccarino, A. L., Brennan, A. M., Kastin, A. J., and Zadina, J. E. (2000). Analgesic effects of endomorphin-1 and endomorphin-2 in the formalin test in mice. *Life Sci.* **67,** 907–912.
Sowa, G., Gekker, G., Lipovsky, M. M., Hu, S., Chao, C. C., Molitor, T. W., and Peterson, P. K. (1997). Inhibition of swine microglial cell phagocytosis of *Cryptococcus neoformans* by femtomolar concentrations of morphine. *Biochem. Pharmacol.* **53,** 823–828.

Spetea, M., Monory, K., Tomboly, C., Toth, G., Tzavara, E., Benyhe, S., Hanoune, J., and Borsodi, A. (1998). In vitro binding and signaling profile of the novel μ opioid receptor agonist endomorphin 2 in rat brain membranes. *Biochem. Biophys. Res. Commun.* **250,** 720–725.

Spyer, R. M. (1994). Central nervous mechanisms contributing to cardiovascular control. *J. Physiol.* **474,** 1–19.

Stanasila, L., Massote, D., Kieffer, B. L., and Pattus, F. (1999). Expression of δ, κ and μ human opioid receptors in *Escherichia coli* and reconstitution of the high-affinity state for agonist with heterotrimeric G protein. *Eur. J. Biochem.* **260,** 430–438.

Sarter, M., Bodewitz, G., and Stephens, D. N. (1988). Attenuation of scopolamine-induced impairment of spontaneous alteration behaviour by antagonist but not inverse agonist and agonist β-carbolines. *Psychopharmacology* **94,** 491–495.

Stone, L. S., Fairbanks, C. A., Lauphlin, T. M., Nguyen, H. O., Bushy, T. M., Wessendorf M. W., and Wikox, G. L. (1997). Spinal analgesic actions of the new endogenous opioid peptides endomorphin-1 and -2. *Neuroreport* **8,** 3131–3135.

Stone, W. S., Walser, B., Gold, S. D., and Gold, P. E. (1991). Scopolamine- and morphine-induced impairment of spontaneous alteration performance in mice: Reversal with glucose and with cholinergic and adrenergic agonists. *Behav. Neurosci.* **105,** 264–271.

Sugimoto-Watanabe, A., Kubota, K., Fujibayashi, K., and Saito, K. (1999). Antinociceptive effect and enzymatic degradation of endomorphin-1 in newborn rat spinal cord. *Int. J. Pharmacol.* **81,** 264–270.

Sun, M. K. (1996). Pharmacology of reticular spinal vasomotor neurons in cardiovascular regulation. *Pharmacol. Res.* **48,** 465–494.

Takahashi, M., Fukumizu, A., Shimizu, Y., Tsuda, Y., Yokoi, T., Bryant, S. D., Lazarus, L. H., Ambo, A., Sasaki, Y., Kita, A., Oka, M., and Okada, Y. (2000). Development of μ-receptor selective opioid mimetics derived from endomorphin sequences. In "Peptide Science 1999: Proceedings of the 36th Japanese Peptide Symposium" Kyoto, Japan (N. Fujii Ed.), pp. 437–440. Japanese Peptide Society, Japan.

Takahashi, M., Fukumizu, A., Shimizu, Y., Tsuda, Y., Bryant, S. D., Lazarus, L. H., Anbo, A., Sasaki, Y., Kita, A., Oka, M., and Okada, Y. (2001). Development of bifunctional opioid ligands containing dimethyltyrosine. In "Peptides 2000: Proceedings of the 26th European Peptide Symposium," Montepellier, France (J. Martinez and J.-A. Fehrentz Eds.), pp. 753–754. Editions EDK, Paris.

Thomas, D. A., and Hammond, D. L. (1995). Microinjection of morphine into the rat medullary dorsal horn produces a dose-dependent increase in facial scratching. *Brain Res.* **695,** 267–270.

Tohda, C., Yamagichi, T., and Kuraishi, Y. (1997). Intracisternal injection of opioid induces itch-associated response through μ-opioid receptors in mice. *Jpn. J. Pharmacol.* **74,** 77–82.

Tonini, M., Fiori, E., Balestra, B., Spelta, V., D'Agostino, G., Di Nucci, A., Brecha, N. C., and Sternini, C. (1998). Endomorphin-1 and endomorphin-2 activate μ-opioid receptors in myenteric neurons of the guinea-pig small intestine. *Naunyn-Schmiedebergs Arch. Pharmacol.* **358,** 686–689.

Tseng, L. T., Narita, M., Suganuma, C., Mizogichi, H., Ohsawa, M., Nagase, H., and Kampine J. P. (2000). Differential antinociceptive effects of endomorphin-1 and endomorphin-2 in mouse. *J. Pharmacol. Exp. Ther.* **292,** 576–583.

Ukai, M., Mori, K., Hashimoto, S., Kobayashi, T., Sasaki, Y., and Kameyama, T. (1993). Tyr-d-Arg-Phe-β-Ala-NH$_2$, a novel dermorphin analog, impairs memory consolidation in mice. *Eur. J. Pharmacology* **239,** 237–240.

Ukai, M., Watanabe, Y., and Kameyama, T. (2000). Effects of endomorphin-1 and -2, endogenous μ-opioid receptor agonists, on spontaneous alteration performance in mice. *Eur. J. Pharmacol.* **395,** 211–215.

Williams, C. A., Wu, S. Y., Dun, S. L., Kwok, E. H., and Dun, N. J. (1999). Release of endomorphin-2 like substances from the rat spinal cord. *Neuosci. Lett.* **273,** 25–28.

Wolozin, B. L., and Pasternak, G. W. (1981). A classification of multiple morphine and enkephalin binding sites in the central nervous system. *Proc. Natl. Acad. Sci. USA* **78,** 6181–6185.

Wu, S. Y., Dun, S. L., Wright, M. T., Chang, J. K., and Dun, N. J. (1999). Endomorphin-like immunoreactivity in the rat dorsal horn and inhibition of substantia gelatinosa neurons *in vitro*. *Neuroscience* **89,** 317–321.

Yamaguchi, T., Kitagawa, K., and Kuraishi, Y. (1998). Itch-associated response and antinociception induced by intracisternal endomorphins in mice. *Jpn. J. Pharmacol.* **78,** 337–343.

Yamamoto, T., and Yaksh, T. L. (1992). Comparison of the antinociceptive effects of pre- and posttreatment with intrathecal morphine and MK801, and NMDA antagonist, on the formalin test in the rat. *Anesthesiology* **77,** 756–763.

Yang, Y. -R., Chiu, T. -H., and Chen, C. -L. (1999). Structure-activity relationship of naturally occurring and synthetic opioid tetrapeptides acting on locus coeruleus neurons. *Eur. J. Pharmacol.* **372,** 229–236.

Yasuda, K., Raynor, K., Kong, H., Breder, C. D., Takeda, J., Reisine, T., and Bell, G. I. (1993). Cloning and functional comparison of κ and δ opioid receptors from mouse brain. *Proc. Natl. Acad. Sci. USA* **90,** 6736–6740.

Yokotani, K., and Osumi, Y. (1998). Involvement of μ-receptor in endogenous opioid peptide-mediated inhibition of acetylcholine release from the rat stomach. *Jpn. J. Pharmacol.* **78,** 93–95.

Zadina, J. E., Kastin, A. J., Ge, L. J., and Hackler, L. (1994). Mu, delta and kappa opiate receptor binding of Tyr-MIF-1 and of Tyr-W-MIF-1, its active fragments and two potent analogs. *Life Sci.* **55,** PL461–466.

Zadina, J. E., Hackler, L., Ge, L. J., and Kastin, A. B. (1997). A potent and selective endogenous agonist for the μ-opiate receptor. *Nature* **386,** 499–502.

Leptin and Melanocortin Signaling in the Hypothalamus

CHRISTIAN BJØRBÆK AND
ANTHONY N. HOLLENBERG

Division of Endocrinology, Beth Israel Deaconess Medical Center and Harvard Medical School, Boston, Massachusetts 02215

  I. Introduction
 II. The Leptin System
III. Discovery of Leptin Receptors
 IV. Hypothalamic Signal Transduction by the Leptin Receptor
  V. Hypothalamic Gene Regulation by Leptin
 VI. Human Mutations in the Leptin Signaling System
VII. The Melanocortin System
VIII. The Melanocortin Receptors
 IX. Proopiomelanocortin
  X. Agouti-Related Peptide
 XI. Melanocortin Coreceptors
XII. Targets of Melanocortin Signaling
XIII. Human Mutations in the Melanocortin System
XIV. Future Directions
     References

The regulation of body weight in humans is coordinated by the interplay between food intake and energy expenditure. The identification of the adipocyte-secreted hormone leptin as a key regulator on both of these processes has shed new light on the pathways involved in their regulation. Indeed, mutations in the gene's encoding leptin and its cognate receptor cause severe obesity in humans. Leptin's actions are mediated principally by target neurons in the hypothalamus where it acts to alter food intake, energy expenditure, and neuroendocrine-function. Recently, it has become clear that a number of critical neuropeptides are regulated by leptin in the hypothalamus. Among these is the proopiomelanocortin (POMC)-derived peptide, α-melanocyte-stimulating hormone (α-MSH), which is produced in the arcuate nucleus and is a potent negative regulator of food intake. Like leptin, mutations in POMC or in central melanocortin receptors lead to obesity in humans. Thus, an understanding of the mechanisms by which the leptin and melanocortin pathways signal in the hypothalamus is critical in order to begin to clarify the pathways involved in regulating body weight in humans. © 2002, Elsevier Science (USA).

I. INTRODUCTION

Body weight is maintained within a narrow range, despite the intake of large quantities of food during long periods of time and under conditions of variable nutritional supply. This relative stability of body weight strongly argues for the existence of a regulatory system controlling energy balance. The notion that the brain is important for the regulation of appetite is well established and is consistent with early results from brain lesions in rats showing that the hypothalamus is critical in the regulation of energy homeostasis. It has therefore been proposed that energy stores would have the ability to transmit signals about nutritional status to centers in the central nervous system, which regulate energy intake and expenditure. The factor(s) mediating these signals to the brain and the mechanisms of action have remained elusive until recently.

Results from studies of several genetic models of obesity in mice have had an unprecedented impact on our understanding of molecular signals that regulate energy homeostasis and body weight. Indeed these mouse models have identified two intertwined hypothalamic signaling pathways, namely leptin and melanocortin. The hormone leptin is expressed in adipose tissue in proportion to fat mass and acts on the central nervous system to inform key regulatory centers about energy stores. Leptin receptors (ObR) are present within the hypothalamus and are coexpressed with neupeptides involved in regulation of body weight. Administration of recombinant leptin to rodents results in a profound decrease in food intake and loss of fat mass. Furthermore, lack of functional leptin or its receptor results in massive obesity and in neuroendocrine defects in both rodents and humans. Thus a significant body of evidence demonstrates that leptin is a critical controller of appetite, energy expenditure, neuroendocrine function, and metabolism.

A key mediator of leptin action is the central melanocortin signaling system, which has received tremendous attention in the past few years. This complex system involves peptides derived from the proopiomelanocortin (POMC) precursor,

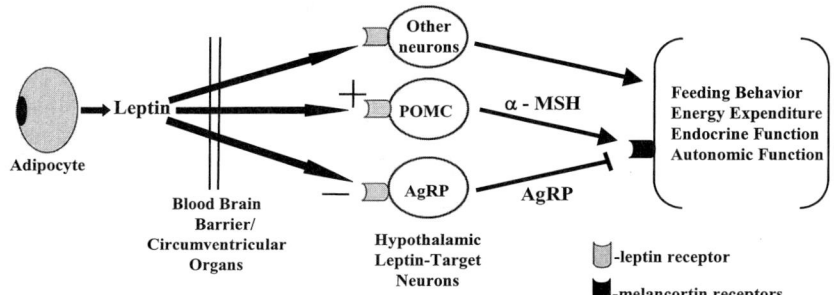

FIGURE 1 Schematic model of leptin/melanocortin pathways in the hypothalamus. Leptin is secreted from adipose tissue and acts on its receptor in the hypothalamus. Leptin positively regulates anorexigenic neuropeptides including the POMC-derived α-MSH, and negatively regulates the potent orexigenic neuropeptide, AgRP. Via neuronal projections, α-MSH acts as an agonist on cells that express melanocortin receptors. AgRP is delivered to the same target neurons and is a melanocortin receptor antagonist. Leptin acts in concert with a complex neuronal circuitry to inhibit food intake, increase energy expenditure, and regulate endocrine and autonomic function.

primarily α-melanocyte-stimulating hormone (α-MSH), their receptors, principally the melanocortin 3 and 4 receptors (MC3/4-R), and the melanocortin antagonist, agouti-related peptide (AgRP). α-MSH has strong anorexigenic properties while AgRP is a potent stimulator of food intake. Hypothalamic POMC and AgRP neurons are primary targets of leptin, and consistent with leptin's satiety effects, POMC expression is stimulated and AgRP expression is inhibited by leptin, suggesting that the melanocortin system is a conduit of leptin action. Genetic and pharmacological data in rodents and the identification of mutations in both POMC and the MC4 receptor leading to human obesity have firmly established this system as critical regulator of body weight. This article will focus on past and recent developments in the unraveling of the central leptin–melanocortin signaling system (Fig. 1).

II. THE LEPTIN SYSTEM

The genetically obese *ob/ob* and *db/db* mice were first described more than 40 years ago and have subsequently been the subject of study in more than a thousand papers. Recessive single mutations in these mice cause a near identical phenotype that resembles morbid obesity in humans (Coleman, 1978). Early parabiosis studies suggested that a humoral factor, which was missing in the *ob/ob* mice, was causing the severe phenotype in those mice (Coleman, 1973). Furthermore, these cross-circulation experiments also provided clues implying that the *db* gene encoded a receptor for the circulating factor lacking in the *ob/ob* mouse. These two strains of mice have within the past 7 years played a dramatic role in the identification of the molecular basis for a regulatory system controlling energy balance.

By the use of positional cloning, the identity of the *ob* gene was reported in 1994 (Zhang *et al.*, 1994). It was shown that the gene encoded a 16-kDa polypeptide (leptin) that was truncated at codon 105 in the *ob/ob* mice, resulting in a defective protein. Tissue-expression studies showed that leptin mRNA was almost exclusively expressed in adipose tissue. Furthermore, leptin was predicted to be a secreted protein due to the presence of a signal sequence. Subsequent studies demonstrated that leptin was indeed present in the circulation of mice and humans, that the levels were highly correlated with the amount of fat tissue, and that serum levels fall after weight loss (Maffei *et al.*, 1995). Generating considerable excitement and consistent with the predictions of Coleman *et al.*, it was shown that administration of recombinant leptin to *ob/ob* and wild-type, but not to *db/db*, mice resulted in marked weight loss (Pelleymounter *et al.*, 1995; Campfield *et al.*, 1995; Halaas *et al.*, 1995; Stephens *et al.*, 1995). The weight-reducing effect was restricted to adipose tissue without significantly affecting lean body mass. Early studies also showed that low doses of leptin were effective in reducing food intake and body weight when administered directly into the brain ventricles of rodents (Campfield *et al.*, 1995; Stephens *et al.*, 1995). Combined, these results therefore argue that leptin is an adiposity signal that acts directly on receptors expressed in the central nervous system.

III. DISCOVERY OF LEPTIN RECEPTORS

Approximately a year following the cloning of the leptin gene, the first identification of leptin receptors was reported. By using an expression-cloning strategy, a novel gene highly expressed in the choroid plexus of the mouse was identified. The cDNA was predicted to encode a 894-amino acid-long protein that had strong similarity to cytokine receptors (Tartaglia *et al.*, 1995). A highly homologous human receptor with a longer intercellular domain was also identified. Shortly following this finding, the identity of the *db* gene was reported independently using a positional cloning approach and a candidate gene approach (Lee *et al.*, 1996; Chen *et al.*, 1996). It was shown that the *db* gene encoded the leptin receptor that was discovered by Tartaglia *et al.* (1995). Of several splice variants described by Lee *et al.* (1996), a long isoform was abnormally spliced in the *db/db* mice generating a mutant protein lacking most of the cytoplasmic domain. The predictions made by Coleman *et al.* more than 20 years earlier were essentially proven to be correct.

The crystal structure of leptin was resolved in 1997 and showed that leptin is a member of the cytokine family (Zhang *et al.*, 1997), consistent with its receptor (ObR) belonging to the cytokine receptor class I superfamily. ObR is most closely related to gp130, the signal-transducing membrane protein of the interleukin-6 signaling complex, the leukemia inhibitory factor (LIF) receptor, and the granulocyte colony-stimulating factor (G-CSF) receptor (Tartaglia *et al.*, 1995). Five alternatively spliced isoforms of ObR have been identified in the mouse (Lee *et al.*, 1996). Four (ObR a, b, c, d) have identical sequences until amino acid 889, which includes the extracellular domain, a transmembrane domain, and a short 29-residue intracellular sequence. Three of these (ObR a, c, d) have short additional intracellular extensions consisting of 5, 3, and 11 amino acids, respectively. Several distinct short isoforms have subsequently been found in other species including the rat and humans (Lee et al., 1996; Chua et al., 1996; Bennett et al., 1996; Cioffi et al., 1996; Wang et al., 1996). The long ObRb form has an additional 273 intracellular residues (total 1162 aa). The fifth form (ObRe) consists of 796 residues and is a soluble form that does not contain the transmembrane domain.

Due in part to the multiple forms of ObR, tissue-distribution data must be evaluated with care. A further complication is that different results have been reported depending on the study, the assay used, or the species analyzed. Finally, mRNA studies do not appear to completely reflect the existence of functional leptin receptors. It can, however, be concluded that the short ObRa form is expressed almost ubiquitously with high levels in lung, kidney, liver, and muscle (Tartaglia et al., 1995; Lee et al., 1996; Fei et al., 1997). Surprisingly, the function of ObRa at these sites still remains largely unknown. In the brain, the highest expression of this isoform is found in the choroid plexus (Tartaglia et al., 1995) and in microvessels (Bjorbaek et al., 1998a). The expression at the choroid plexus may be important for leptin uptake or efflux from the cerebrospinal fluid. Expression in brain microvessels constituting the blood–brain barrier is consistent with studies suggesting a role of ObRa in receptor-mediated transport of leptin into the brain (Kastin et al., 1999; Hileman et al., 2000; Maresh et al., 2001). This function of ObRa is, however, not supported by all studies and needs to be further examined (Kowalski et al., 2001). Although present in the brain (Guan et al., 1997; Hileman et al., 2002) less is known about the other short membrane-bound isoforms and specific functions of these receptors have not been reported. The soluble ObRe isoform binds leptin in the circulation, and the significance of this is also unclear (Sinha et al., 1996; Gavrilova et al., 1997; Li et al., 1998).

The long form, ObRb, has been reported to be present in many tissues, although the levels are mostly much lower than those of ObRa. An increasing body of evidence supports direct actions of leptin via ObRb in peripheral tissues, but the exact biological function at these sites requires further studies. More importantly, ObRb mRNA is highly expressed in selected nuclear groups in the rodent and human brain. Within the hypothalamus, dense expression is detected in the arcuate, dorsomedial (DMH), ventromedial (VMH), and ventral premamillary nuclei (PMV), and moderate expression is found in the periventricular hypothalamic nucleus and yet lower levels in the paraventricular nucleus (PVH) (Mercer et al., 1996a; Elmquist et al., 1998). Outside the hypothalamus, high mRNA levels have been found in the thalamus and in the Purkinje and granular cell layers of the cerebellum in rodents and in humans (Guan et al., 1997; Elmquist et al., 1998; Savioz et al., 1997; Mercer et al., 1998). Surprisingly, no clear evidence of functional ObRb proteins at these latter sites has been reported, suggesting that the protein is not expressed, or that leptin does not reach these regions and/or that ObRb serves a novel function that has yet to be identified. In contrast, leptin administration into the bloodstream does affect the function of neurons located in the hypothalamus, and, as described below, these actions of leptin are critical for its function.

IV. HYPOTHALAMIC SIGNAL TRANSDUCTION BY THE LEPTIN RECEPTOR

ObRb mainly exists as a homodimer at the cell surface and does not readily form heterodimers with the short receptor isoforms (White *et al.,* 1997; Nakashima *et al.,* 1997; White and Tartaglia, 1999). Analysis of ligand–receptor complexes shows that the extracellular domain interacts with leptin in a 1:1 ratio (Devos *et al.,* 1997). Leptin receptors lack intrinsic catalytic activity, and analogous with other cytokine receptor systems, leptin binding results in activation of intracellular Janus tyrosine kinases (JAKs), specifically JAK2, that are associated with conserved JAK-binding motifs present in the membrane proximal region of leptin receptors (Ghilardi and Skoda, 1997). Activated JAK2 then phosphorylate phosphotyrosine residues in the intracellular domain of ObRb and of proteins that are associated with ObRb (Bjorbaek *et al.,* 1997). The short forms of leptin receptors have been reported to have limited signaling capabilities but additional studies are needed to confirm this observation *in vivo* (Bjorbaek *et al.,* 1997; Murakami *et al.,* 1997).

The murine ObRb receptor contains three intracellular tyrosine residues, located at positions 985, 1077, and 1138. These amino acids are conserved among known species of long form leptin receptors. Tyrosine phosphorylation sites provide binding motifs for src homology 2 (SH2) domain-containing proteins, such as STATs (signal transducer and activator of transcription). Tyrosine 1077 is likely to be buried in a hydrophobic pocket and does not become readily phosphorylated in *in vitro* systems (Li and Friedman, 1999; Banks *et al.,* 2000). Tyrosine 1138 is part of a YXXQ motif, generating a consensus STAT3 binding site (White *et al.,* 1997; Baumann *et al.,* 1996; Bjorbaek *et al.,* 2000). STAT3 proteins bound to Y1138 become tyrosine phosphorylated in response to JAK activation, then form dimers and finally translocate to the nucleus to regulate gene transcription. Whereas several different STAT isoforms are activated by leptin in cell systems (Baumann *et al.,* 1996; Ghilardi *et al.,* 1996; Rosenblum *et al.,* 1996), only STAT3 is activated in the hypothalamus (Vaisse *et al.,* 1996). The severely obese *db/db* mice lack the intracellular domain of ObRb, and therefore cannot mediate activation of STAT3 in the hypothalamus (Ghilardi *et al.,* 1996). STAT3 activation is a crucial component in leptin's pathway to regulate body weight as recently demonstrated by specific knockout of the Y1138 residue in mice, resulting in a severe obese phenotype that is similar to that of the *db/db* mouse (Bates *et al.,* 2001).

In addition to STAT3 activation, studies show that ObRb can stimulate other signaling pathways. Rapid activation of the p42 and p44 mitogen-activated protein kinases (MAPK or ERK1/2) by leptin has been reported in tissue culture systems (Bjorbaek *et al.,* 1997; Takahashi *et al.,* 1997). Tyrosine 985 of ObRb is required for full activation of this pathway as demonstrated by transfection of 293 cells with mutant receptors (Banks *et al.,* 2001). Maximal MAPK activation depends on binding of the SH2-containing protein tyrosine phosphatase, SHP-2, to Y985 (Bjorbaek *et al.,* 2001a). However, MAPK can also be activated via a pathway that is independent of Y985 and is likely to emerge directly from JAK2 (Bjorbaek *et al.*, 2001a). Rapid activation of MAPK by leptin *in vivo* has been reported in fat, liver, and hypothalamus, probably via direct effects on ObRb expressed at these sites (Bjorbaek *et al.*, 2001a; Y. B. Kim *et al.*, 2000). Y985 and the MAPK pathway have been shown to be required for stimulation of c-fos mRNA expression in transfection systems (Banks *et al.*, 2000). Several *in vitro* and *in vivo* studies have reported conflicting results regarding regulation of additional intracellular signaling proteins, including insulin receptor substrate (IRS) 1 and 2 phosphorylation and activation of IRS-associated phosphoinositol-3 kinase (PI3-K) (Bjorbaek *et al.*, 1997; Y. B. Kim *et al.*, 2000; Cohen *et al.*, 1996; Berti *et al.*, 1997; Kellerer *et al.*, 1997). Further studies are needed to clarify these latter findings and to determine whether these signaling events are regulated by leptin in the hypothalamus.

In addition to regulation of MAPK, Y985 of ObRb also mediates negative regulation of leptin receptor signaling. Recent studies show that ObRb induces expression of suppressor of cytokine signaling (SOCS)-3 in the hypothalamus (Bjorbaek *et al.*, 1998b). Consistent with this, fasting of rats, a state of low leptin levels, leads to a fall in hypothalamic SOCS-3 mRNA expression (Baskin *et al.*, 2000). Furthermore, a murine model of hyperleptinemia is associated with elevated hypothalamic SOCS-3 levels (Bjorbaek *et al.*, 1998b). Activation of SOCS-3 expression depends of Y1138 and binding of activated STAT3 to sequences in the proximal promoter of the *socs-3* gene (Banks *et al.*, 2000; Auernhammer *et al.*, 1998). SOCS-3 is an inhibitor of proximal ObRb signaling and acts by binding to Y985, possibly also to JAK2, eventually leading to inhibition of JAK2 kinase activity (Bjorbaek *et al.*, 2000; Eyckerman *et al.*, 2000). Another potential negative regulator of proximal leptin signaling is the protein tyrosine phosphatase 1B (PTP1B), which may act by directly dephosphorylating JAK2, thereby inhibiting its kinase activity (Zabolonty *et al.*, 2002, Cheng *et al.*, 2002). Although not proven *in vivo*, SOCS-3 is likely to play a critical role in negative feedback regulation of the hypothalamic leptin receptor signaling system (Bjorbaek *et al.*, 1999). Further studies are required to determine whether overreactivity of these negative regulatory systems may play a role in the development of leptin-resistant obesity.

Although some intracellular signaling events in leptin-responsive neurons lead to changes in gene expression, very rapid actions resulting in modulation of synaptic transmission in the hypothalamus have also been reported (Glaum *et al.*, 1996). Electrophysiological recordings of POMC neurons in the arcuate nucleus show that leptin increases the frequency of action potentials. Furthermore, neuronal depolarization occurs via regulation of a nonspecific cation channel and can be measured within minutes following leptin treatment (Cowley *et al.*, 2001). Studies of direct membrane effects on neurons in coronal slices from the PVH also show that leptin elicited dose-related depolarization by increasing conductance of a cation channel (Powis *et al.*, 1998). In addition, leptin has been reported to affect the activity of ATP-sensitive K^+ channels of glucose-sensitive cells in the hypothalamus (Spanswick *et al.*, 1997), leading to neuronal inhibition (hyperpolarization). PI3-K may also be activated by leptin in the hypothalamus (Niswender *et al.*, 2001) and it is conceivable that this signaling event plays a role in the

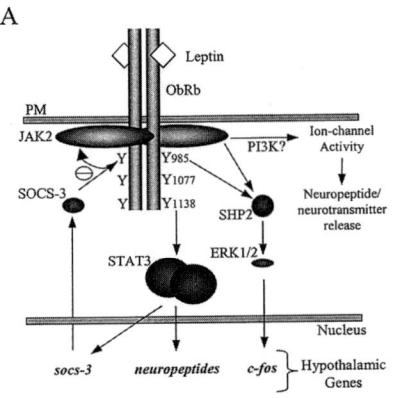 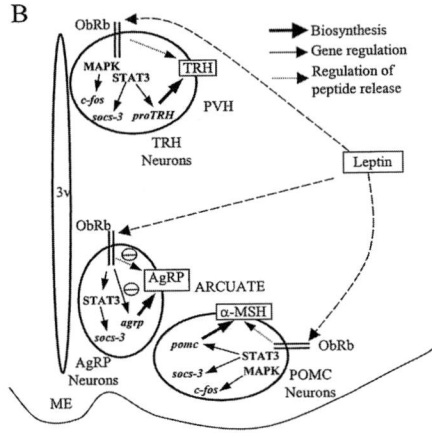

FIGURE 2 A model of leptin receptor signaling. (A) Activation of ObRb by leptin activates JAK2 kinases resulting in tyrosine phosphorylation of the receptor and of proteins bound to the receptor, including STAT3 and SHP2. STAT3 regulates a number of genes, including hypothalamic neuropeptides and SOCS-3. SOCS-3 proteins act in a negative feedback loop to inhibit JAK activity. ObRB binds SHP2, which plays a positive role in activation of the MAPK/ERK pathway leading to activation of c-*fos* expression. Leptin receptors also have rapid, nongenomic effects on neuronal firing rates via regulation of ion channels, likely leading to modulation of neuropeptide and neurotransmitter release. (B) ObRb is highly expressed in two distinct cell populations in the arcuate nucleus that express AgRP and POMC. Leptin directly acts on these neurons leading to inhibition of AgRP expression and release, and to stimulation of POMC expression and α-MSH release. Leptin also acts on TRH neurons located in the PVH.

regulation of ion channels, but further studies are clearly required to determine this. Taken together these data show that leptin can produce rapid stimulatory and inhibitory effects on neuronal activity in specific regions of the hypothalamus and suggest that these effects are important for the physiological function of leptin. Specifically, such signaling events might serve to regulate secretion of neuropeptides and neurotransmitters (Fig. 2A). In summary, although much progress has been made, further studies are clearly needed to fully understand the intercellular signaling pathways and the genes that are regulated by leptin in the hypothalamus. In addition, the relative importance of individual leptin-responsive neuronal groups in leptin biology remains to be elucidated.

V. HYPOTHALAMIC GENE REGULATION BY LEPTIN

It has long been acknowledged that neuropeptides that influence body weight are expressed within the hypothalamus. One member of this group of peptides is neuropeptide Y (NPY). NPY is a potent stimulator of food intake and its expression increases during starvation, under conditions in which circulating leptin levels fall (Woods *et al.*, 1998). A potential link between leptin and NPY was thus suggested.

Subsequent colocalization studies have clearly demonstrated that ObRb mRNA is expressed in the majority of NPY neurons present in the arcuate nucleus (Mercer *et al.,* 1996b). Furthermore, arcuate NPY mRNA levels are elevated in *ob/ob* and *db/db* mice, and injection of leptin reduces NPY mRNA in normal and *ob/ob* mice (Stephens *et al.,* 1998; Schwartz *et al.,* 1996; Ahima *et al.,* 1996). This has led to the conclusion that leptin acts directly on NPY neurons resulting in inhibition of NPY expression. Some studies, but not all (Beck *et al.,* 1998; King *et al.,* 2000), suggest that leptin also inhibits release of NPY peptides from hypothalamic tissue (Stephens *et al.,* 1998; Jang *et al.,* 2000). NPY neurons also express agouti-related peptide (AgRP), another potent stimulator of food intake (Ollmann *et al.,* 1997; Broberger *et al.,* 1998; Hahn *et al.,* 1998). The expression of AgRP is strongly up-regulated by fasting, is elevated in *ob/ob* mice, and is inhibited by leptin *in vivo* (Stephens *et al.,* 1998; Mizuno and Mobbs, 1999). Furthermore, leptin inhibits release of AgRP peptides from hypothalamic tissue *in vitro* (Li *et al.,* 2000). The intracellular signaling pathways by which ObRb negatively regulates NPY and AgRP expression and secretion are presently unknown. Although activation of STAT3 by leptin has been demonstrated in neurons located in the arcuate nucleus that contain the AgRP/NPY cells (Hubschle *et al.,* 2001), further studies are required to determine whether this transcription factor plays a role in inhibition of AgRP and NPY mRNA expression. Surprisingly, mice lacking NPY do not have an obese phenotype and leptin-deficient mice cross-bred with NPY knockout mice still have marked obesity, although attenuated as compared to mice lacking only leptin (Erickson *et al.,* 1996). These results imply that NPY is not the major conduit of leptin function. Furthermore, studies of mice lacking functional AgRP suggest that neither this peptide is strictly required for normal regulation of feeding behavior or body weight (Qian *et al.,* 2002). The lack of clear phenotypes of the NPY and AgRP knockout mice, together with the strong pharmacological data, implies that other compensatory systems exist, that has yet to be identified.

Starvation causes a rapid reduction in thyroid hormone levels in rodents (Blake *et al.,* 1991). This adaptive response, which reduces energy expenditure, is associated with a reduction in hypothalamic thyrotropin-releasing hormone (TRH) expression that can be reversed by the administration of leptin (Ahima *et al.,* 1996; Legradi *et al.,* 1997), suggesting that leptin plays an important role in neuroendocrine regulation of the thyroid axis. This effect of leptin is believed to occur via both direct actions on TRH neurons in the PVH and via indirect signals from leptin-responsive neurons that project to the hypophysiotrophic TRH neurons. In support of a direct effect is the fact that leptin receptor mRNA is found in the PVH (Elmquist *et al.,* 1998), that SOCS-3 mRNA expression is induced in TRH neurons following leptin injection in rats (Harris *et al.,* 2001), and that leptin rapidly regulates neuronal polarization of neurons in brain slices of the PVH (Powis *et al.,* 1998). Furthermore, leptin directly regulates the TRH promoter in a heterologous transfection system via a novel STAT3-response element located in the proximal promoter (Harris *et al.,* 2001). In addition, leptin dose dependently stimulates pro-TRH biosynthesis in primary hypothalamic neuronal cultures (Nillni *et al.,* 2000). Solid data also support an indirect positive regulation of the hypophysiotrophic

TRH neurons by leptin, via synaptic input from leptin-responsive neurons located in the arcuate (Legradi *et al.,* 1998; Fekete *et al.,* 2000). Leptin has also been reported to stimulate TRH peptide release from hypothalamic cultures and from hypothalamic tissue via both direct and indirect pathways (Nillni *et al.,* 2000; M. S. Kim *et al.,* 2000).

A promising target of leptin is the POMC-producing neurons, a distinct group of cells located in the arcuate nucleus of the hypothalamus. As described below, a large body of genetic and pharmacological evidence suggests that the POMC-derived neuropeptide, α-MSH, induces an anorexigenic response in rodents (Shimizu *et al.,* 1989; Tsujii and Bray, 1989). Double *in situ* studies show at least half of the POMC neurons express ObRb (Cheung *et al.,* 1997), strongly suggesting that these cells are direct targets of leptin signaling. In addition, leptin rapidly induces SOCS-3 mRNA in at least half of POMC cells (Elias *et al.,* 1999) and POMC mRNA is decreased in *ob/ob* and *db/db* mice (Thornton *et al.,* 1997; Schwartz *et al.,* 1997; Mizuno *et al.,* 1998). Recent data also show that leptin can directly induce the proximal POMC promoter via ObRb and STAT3 in a heterologous cell system (Bjorbaek *et al.,* 2001b). Consistent with this finding, POMC gene expression is induced in the pituitary by cytokine signaling pathways, possibly through STAT3 binding sites in the POMC promoter (Bousquet *et al.,* 2000). The significance of theses sites in the arcuate nucleus has yet to be determined. The majority of POMC neurons also express cocaine and amphetamine-regulated transcript (CART) (Elias *et al.,* 1998), a peptide with anorexigenic properties (Kristensen *et al.,* 1998). CART is widely expressed in the hypothalamus including the DMH, PMV, and in the TRH neurons localized in the PVH (Elias *et al.,* 2001). Like POMC, CART expression in the arcuate is stimulated by leptin, consistent with the satiety effects of leptin. Combined, these data show that the POMC neurons, in addition to the AgRP-producing cells, form a direct link between the leptin and the melanocortin system.

An anatomical picture of the initial circuits in hypothalamic leptin action is beginning to emerge (Elias *et al.,* 1999). Both AgRP and POMC neurons are regulated by leptin. When leptin levels are low, POMC expression is reduced and AgRP expression is activated. When leptin levels are high or when leptin is administered to rodents, POMC expression is increased while AgRP expression is decreased. As discussed previously, leptin is likely to regulate gene expression through STAT3, activation of which overlaps very well with ObRb expression in the hypothalamus (Hubschle *et al.,* 2001). SOCS-3 expression is also stimulated by leptin in hypothalamic cells that express ObRb, and is therefore an outstanding marker for cells that respond directly to leptin (Bjorbaek *et al.,* 1998b). c-Fos expression is increased in the hypothalamus in response to leptin (Woods and Stock, 1996), but only in a subpopulation of leptin-responsive cells (Elias *et al.,* 1999). c-Fos is increased in POMC neurons in response to leptin administration, possibly via the MAPK pathway. Furthermore, SOCS-3 mRNA and c-Fos proteins are stimulated in the PVH, consistent with at least some direct effects of leptin on cells in this region, specifically on the hypophysiotrophic TRH neurons. Fos is, however, not induced in AgRP neurons consistent with these cells being inhibited by leptin, implying that leptin differentially activates downstream signaling pathways, i.e., the MAPK pathway, in these different target neurons. Leptin also has very rapid effects on neuronal activity that do not involve regulation of gene transcription. Both stimulatory and inhibitory effects on action potentials on different neurons have been reported, consistent with results suggesting differential effects on peptide secretion. In summary, intracellular signaling pathways by which leptin regulates hypothalamic neuropeptide genes and peptide release are becoming more clear. However, further studies in this area are critical for a better understanding of effectors of leptin action and of the mechanisms leading to leptin resistance that appears to characterize human obesity (Maffei *et al.*, 1995) (Fig. 2B).

VI. HUMAN MUTATIONS IN THE LEPTIN SIGNALING SYSTEM

Despite considerable evidence suggesting that genetic defects play an important role in development of human obesity, no mutations causing human obesity had been identified until the late 1990s. Furthermore, although multiple studies in rodents point to profound effects of leptin on appetite and fat mass, no data regarding the importance of leptin in human physiology were available until 1997 when two extremely obese siblings with homozygous mutations in the leptin gene were reported (Montague *et al.*, 1997). The children both had the same point mutation, resulting in a truncated protein that negatively affects expression and secretion from adipocytes, and most likely ablates its biological function (Rau *et al.*, 1999). Serum leptin levels were very low and close to the detection limits. As in *ob/ob* mice, the humans with leptin deficiency have normal birthweight, followed by rapid weight gain (above the 99th centile) associated with hyperphagia and impaired satiety, strongly supporting a critical role for leptin in influencing appetite in humans. In addition, human and murine leptin deficiency are both characterized by hyperinsulinemia, although less severe in leptin-deficient humans. Furthermore, leptin is required for normal pubertal development both in mice and in humans, as a subsequent study showed hypogonadotrophic hypogonadism in two adults with congenital leptin deficiency (Strobel *et al.*, 1998). However, several differences from mice lacking functional leptin were noted. Humans with leptin deficiency do not exhibit impaired growth, nor do they have hypercortisolnemia or hyperglycemia. Furthermore, there is no evidence of low energy expenditure, body temperature, or oxygen consumption as is the case in the *ob/ob* mice (Farooqi *et al.*, 1999). With the availability of recombinant leptin, leptin therapy was subsequently initiated in one of the two children. As in mice, leptin administration led to a dramatic suppression of food intake and sustained loss of body weight over a period of 12 months. The weight loss was almost entirely due to loss of fat mass (95%). Combined, the data confirm a critical role of leptin in control of energy balance in humans, an effect that is mainly due to negative regulation of appetite (Farooqi *et al.*, 1999).

In 1998, less than a year following the finding of the first humans with inherited leptin deficiency, humans with mutations in the leptin receptor gene were reported. As in *db/db* mice, homozygous mutations resulting in premature truncation of leptin receptors lead to early-onset morbid obesity (Clement *et al.,* 1998). Consistent with the adult human leptin-deficient counterparts, the leptin receptor-deficient patients are characterized by hypogonadism of central origin, supporting a role for ObR signaling in sexual maturation. In addition, humans with receptor mutations are characterized by normal glucose concentrations. However, in contrast to the patients lacking leptin, mutation of the leptin receptor results in mild delay in linear growth, likely caused by inadequate secretion of growth hormone. This finding is consistent with studies in rodents showing a central stimulatory effect of leptin on the growth hormone axis, supporting a role of the leptin receptor in regulation of growth. Furthermore, indications of hypothalamic thyroid dysfunction were noted in human ObR mutants, consistent with effects of leptin on this system in normal mice (Ahima *et al.,* 1996), however in contrast to lack of clear defects in *ob/ob* and *db/db* mice and in leptin-deficient humans. The reasons for these differences in the growth hormone axis and possibly the thyroid axis between the leptin and leptin receptor-deficient humans are unknown, and further studies are clearly needed to determine the role of leptin in these systems in humans.

The incidence of genetic defects in the leptin and the leptin receptor genes is, however, extremely low and are these mutations not found in the vast majority of obese humans. Common human obesity is characterized by increased expression and serum levels of leptin (Maffei *et al.,* 1995; Lonnquist *et al.,* 1995; Considine *et al.,* 1996a) and of normal coding regions of both genes (Considine *et al.,* 1995, 1996b,c; Niki *et al.,* 1996; Maffei *et al.,* 1996), leading to the notion of leptin resistance. The molecular basis for resistance to leptin in human obesity is unknown. However, as described below, additional mutations in the leptin–melanocortin pathway have recently been described that may account for up to 5% of cases with severe obesity.

VII. THE MELANOCORTIN SYSTEM

A relationship between melanocortin action and obesity can be linked to the beginning of the twentieth century with the description of the *lethal yellow* (A^y) *agouti* mouse (Leibel *et al.,* 1997). A^y animals are phenotypically characterized by both a yellow coat color and the development of late-onset obesity associated with hyperphagia, increased linear growth, and hyperinsulinemia (Yen *et al.,* 1994). In addition to the A^y allele, there are several other autosomal dominant yellow mutations that also affect energy balance. These mutations were subsequently mapped to the *agouti* locus, which along with *extension,* had been determined to be distinct loci responsible for the regulation of coat color (Bultman *et al.,* 1992; Robbins *et al.,* 1993). The A^y animal models suggested that melanocortin signaling could be related to body weight and energy expenditure. Indeed, both the *agouti* and *extension* genes were cloned in the early 1990s, which quickly led to the development of a model for melanocortin signaling in the periphery, with the realization that signaling elsewhere may influence body weight. The *extension* gene was found to encode the melanocortin-1 receptor (MC1-R) (Robbins *et al.*, 1993), which has high affinity for α-MSH, whereas the *agouti* gene encoded a 131 amino acid protein produced by cells near the hair follicle. Binding of α-MSH to the MC1-R in melanocytes results in the production of eumelanin (black) pigment. However, Agouti antagonizes the action of α-MSH at the MC1-R and in its presence the pheomelanin pigment (yellow) is produced (Lu *et al.*, 1994). Thus, functional overexpression of Agouti in skin would lead to a yellow coat color. The dominant yellow mutations in the *agouti* gene also result, through rearrangements in the *agouti* promoter, in the ubiquitous expression of the Agouti peptide, including the brain where Agouti is not normally expressed (Michaud *et al.*, 1994; Klebig *et al.*, 1995). It is now clear that antagonism of melanocortin action in the brain results in the obese phenotype of the A^y mouse. We will review below the components of the central melanocortin system, particularly in the hypothalamus, including mutations in rodents and humans that definitively demonstrate that this pathway is critical for body weight regulation.

VIII. THE MELANOCORTIN RECEPTORS

The significance of the melanocortin system in the biology of energy balance was further established by the identification of novel melanocortin receptors that were found to be expressed in the central nervous system (CNS) only. It had long been known that melanocortin signaling in both melanocytes and adrenal cortical cells was mediated by G-protein-coupled receptors (GPCR) coupled to cAMP production (Mountjoy, 2000). Thus, the first melanocortin receptors (MC1-R and MC2-R) were cloned by the use of degenerate polymerase chain reaction (PCR) primers designed against conserved transmembrane domains in GPCRs (Chhajlani *et al.*, 1992; Mountjoy *et al.*, 1992). The human and murine MC1-R were found to be highly conserved and to be expressed in melanocytes and melanoma. RT-PCR data suggest that the MC1-R is also expressed in macrophages and periaqueductal gray matter. The human MC1-R is highly specific for α-MSH and ACTH, both of which activate the receptor. Mutations in the MC1-R affect coat color in mice and sequence variations occur in this gene in 80% of humans with red hair and/or fair skin that tans poorly (Valverde *et al.*, 1995). The hMC2-R shares some amino acid homology with the hMC1-R. The hMC2-R is expressed in all three layers of the adrenal cortex and is specifically activated by ACTH but not by any of the MSH peptides (Mountjoy, 2000).

After the identification of the MC1- and 2-Rs, degenerate PCR was used against homologous regions of these receptors to identify three subsequent MC-Rs. The MC3-R and MC4-R were found to be expressed primarily in brain whereas the MC5-R is mainly expressed in exocrine glands and in skeletal muscle. Functional deletion of the MC5-R in mice leads to exocrine gland dysfunction (Chen et al., 1997). The hMC3-R shares homology with the hMC1- and MC2-Rs and is conserved across species (Gantz et al., 1993a; Desarnaud et al., 1994). It is predominantly expressed in brain and in the hypothalamus in particular. In situ hybridization studies have demonstrated its expression in the ventromedial and arcuate nucleus of the hypothalamus, which are areas known to be important in the regulation of body weight. Specifically, the MC3-R has been shown to be expressed on both POMC and AgRP neurons (Bagnol et al., 1999). The MC3-R is unique in that it does not appear to exert specificity in context of melanocortins and thus can be activated by a variety of MSH peptides, including γ-MSH and α-MSH (Hruby et al., 1995; Li et al., 1996). Recently, genetic knockout experiments in mice of the MC3-R demonstrate that it contributes significantly to body weight regulation. Mice lacking the receptor have increased fat mass but are not hyperphagic (Chen et al., 2000; Butler et al., 2000). They are not profoundly overweight but appear to exert themselves less as compared to wild-type mice and may possess a metabolic defect.

In contrast to the MC3-R, the MC4-R is more widely expressed in brain and includes every major division of the CNS suggesting that it is involved in a number of different actions. These include processing of visual and auditory stimuli and autonomic function. Its expression in the PVH, the lateral hypothalamus, and the DMH suggest a role in regulating the function of neurons that control endocrine function and body weight (Mountjoy et al., 1994; Cone, 2000). Structurally, it is most related to the MC3-R and the MC5-R (Mountjoy et al., 1994; Gantz et al., 1993b) and shares similar ligand specificity with the MC1-R such that it prefers α-MSH and ACTH. The MC4-R can also bind to the AgRP and Agouti peptides. In contrast, the MC3-R prefers AgRP whereas the MC1-R prefers Agouti (Fong et al., 1997; Ollmann et al., 1998). Like all of the melanocortin receptors, the MC4-R couples to $G\alpha_s$ and the activation of adenyl cyclase, which in turn increases intracellular cAMP and activates protein kinase A in heterologous cell lines. It is presumed, though it has not been demonstrated, that the MC4-R signals this way in vivo (Fig. 3A).

After the MC4-R was identified, it was speculated that it played a role in body weight regulation because of its expression in areas of the hypothalamus and because of its ability to be antagonized by Agouti in vitro (Lu et al., 1994). This role was confirmed through the development and study of the MC4-R knockout mouse (Huszar et al., 1997). These mice were found to recapitulate the phenotype found in A^y mice and in transgenic mice that have had the Agouti peptide overexpressed in the brain (Klebig et al., 1995). The phenotype of the MC4-R knockout mice includes increased body weight, increased linear growth, hyperphagia, hyperinsulinemia, diabetes in males, and hyperleptinemia. Although these mice developed more profound obesity than that seen in A^y mice, this discrepancy is most likely due to the dosage of Agouti required to antagonize MC4-R rather than an alternate mechanism at work (Huszar et al., 1997; Perry et al., 1995). This is supported by the finding that mice with one deleted MC4-R allele (heterozygotes) are affected in

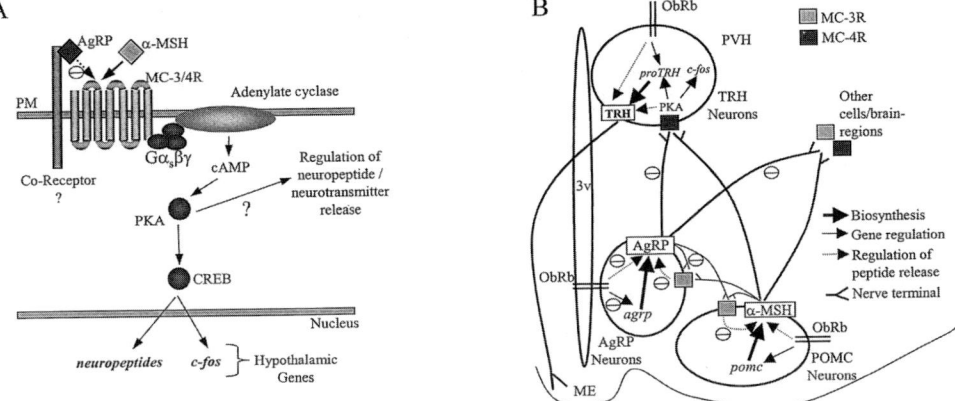

FIGURE 3 A model of melanocortin receptor signaling in the hypothalamus. (A) The G-protein-coupled receptors MC3-R and MC4-R are activated by POMC-derived melanocortin peptides, including α-MSH. AgRP is a melanocortin receptor antagonist and may require binding to coreceptors for full activity. The MC3-R and MC4-R are coupled to adenylate cyclase via Gα$_s$ proteins leading to generation of intracellular cAMP and activation of protein kinase A (PKA). This results in phosphorylation of transcription factors including the cAMP-responsive element-binding protein (CREB). Activation of the melanocortin receptors also has nongenomic effects that may mediate rapid effects on neuropeptide release. (B) Leptin stimulates α-MSH biosynthesis in POMC neurons. These cells also express MC3-R, which may play an autoinhibitory role. Cells containing α-MSH project widely throughout the brain, and include TRH neurons in the PVH that express the MC4-R. AgRP neurons project to targets that strongly overlap with those of the POMC neurons. The MC3-R is also expressed on the AgRP cells where it may play a role in autoregulation of AgRP. TRH neurons in the PVH are strongly innervated by POMC and AgRP neurons, and α-MSH and AgRP peptide levels at this site will determine MC4-R signaling to stimulate proTRH expression and TRH release at the median eminence (ME).

an intermediate fashion as compared to homozygous knockout animals, implying that gene dosage is important in context of the MC4-R. Although the knockout experiment provided evidence for the loss of function of MC4-R signaling resulting in the development of obesity, the role of the MC4-R in mediating a satiety signal was not yet shown. This was subsequently accomplished by using the MC3-R and MC4-R synthetic agonist MTII (Hruby *et al.*, 1995) and the synthetic antagonist SHU9119 (Al-Obeidi *et al.*, 1989). Intracerebroventricular (icv) injection of MTII was initially shown to suppress feeding in food-deprived mice as compared to untreated controls and its actions were subsequently confirmed in three other models of hyperphagia (*ob* mice, A^y mice, and coinjected NPY mice) (Fan *et al.*, 1997). The MTII effect in food-deprived mice was blocked by the coadministration of SHU9119. The specific effect of MTII to inhibit food intake via the MC4-R has been confirmed in the MC3-R knockout mice (Chen *et al.*, 2000). Thus, the melanocortin system and in particular the MC4-R is a key regulator of appetite.

To discriminate between the role of the MC3-R and MC4-R, double knockouts have been generated. These mice are heavier than the MC4-R knockout mice alone suggesting that the MC3-R affects body weight in a mechanism separate from the

MC4-R (Chen *et al.*, 2000). The reason for these differences is not clear but likely relates to tissue-specific expression, ligand recognition, and perhaps mechanisms of signal transduction as the MC3-R has been reported to couple to both $G\alpha_s$ and $G\alpha_q$ (Konda *et al.*, 1994). Based on the above, it has been speculated that the MC4-R $-/-$ obese phenotype is caused by hyperphagia whereas the MC3-R $-/-$ mice possess a metabolic defect that allows for greater fat accumulation. However, more recent studies in MC4-R $-/-$ mice suggest that they also have greater caloric efficiency than wild-type mice and evidence of decreased locomotor activity, suggesting that MC4-R $-/-$ mice also have a metabolic defect (Ste Marie *et al.*, 2000). In fact, this study and others suggest that the MC4-R, through the sympathetic nervous system, regulates heat production via the up-regulation of uncoupling protein-1 (UCP-1) expression in brown fat (Satoh *et al.*, 1998; Commins *et al.*, 1999, 2000). In addition, the MC4-R $-/-$ animals do not increase locomotor activity in response to a high fat diet as compared to wild-type controls (Butler *et al.*, 2001), suggesting that the MC4-R regulates energy expenditure as well as food intake.

IX. PROOPIOMELANOCORTIN

Proopiomelanocortin (POMC), long known to be expressed in the pituitary and to be essential for normal adrenal function, was cloned in 1979. Soon after it was also determined to be expressed in the brain, specifically in the arcuate nucleus and the nucleus of the solitary tract in the brainstem (Gee *et al.*, 1983). POMC is posttranslationally processed in the arcuate nucleus and the pituitary by prohormone convertases, including PC1 and PC2 (Benjannet *et al.*, 1991). However, processing is cell specific and leads to the production of primarily α-MSH (from ACTH) and β-endorphin (from β-lipotropin) in the arcuate nucleus (Smith and Funder, 1988). In contrast, the corticotroph in the pituitary produces primarily ACTH. α-MSH consists of the 13 N-terminal amino acids of ACTH and is remarkably conserved across species. It is through the actions of α-MSH that the POMC precursor neuron mediates its effects on energy balance, though γ- and β-MSH may also have activity in the CNS. Indeed, the cloning of the MC3-R and MC4-R and their localization to the brain established a role for the hypothalamic expression of POMC (Mountjoy *et al.*, 1992, 1994; Cone *et al.*, 1993) and suggested that regulation of the POMC gene and its peptides would play a role in body weight regulation. More recently, it has been demonstrated that POMC knockout mice develop hyperphagia and obesity (as well as defective adrenal function and a yellowish pigmentation) that can be reversed by treatment with the melanocortin receptor agonist MTII (Yaswen *et al.*, 1999), conclusively demonstrating a role for POMC in energy homeostasis.

The identification of leptin and its ability to regulate POMC expression, as described previously, suggests that the actions of POMC are downstream of leptin. Indeed, the inhibition of feeding by leptin can be reversed by central administration of SHU9119 (Satoh *et al.*, 1998). However, it is also clear that POMC neurons are regulated by other signaling pathways and have actions independent of leptin in regulating energy homeostasis. This was demonstrated by Boston *et al.* (1997) who crossed A^y and *ob* mice to generate mice without a functional leptin gene and with Agouti-mediated inhibition of α-MSH signaling. Indeed in this model, both leptin deficiency and the inhibited melanocortin pathway contributed independently to the development of obesity. This is supported by the fact that leptin does have some anorexigenic effects in MC4-R $-/-$ mice (Marsh *et al.*, 1999a), suggesting that leptin also acts via separate pathways from the melanocortin system. Furthermore, other pathways that target POMC neurons are likely to be important in regulation of body weight. For example, POMC neurons also express MC3-R (Bagnol *et al.*, 1999), suggesting that melanocortins autoregulate their own expression. Recently, POMC neurons have been targeted with green fluorescent protein (GFP) *in vivo* using regulatory elements in the *pomc* gene (Cowley *et al.*, 2001). Using an electrophysiological approach it has been discerned that the POMC neurons receive input from AgRP neurons, which also synthesize γ-aminobutyric acid (GABA). Leptin depolarizes POMC neurons and hyperpolarizes AgRP/GABA neurons decreasing these latter inhibitory signals to the POMC neuron. The leptin signal may be further modulated through the MC3-R and melanocortin peptides. Thus, POMC expression and release of its peptides are controlled by multiple inputs from a variety of nearby neurons and peripheral signals.

X. AGOUTI-RELATED PEPTIDE

Based on the function and structure of Agouti and given that it is not normally expressed in brain, Ollman *et al.* (1997) and Shutter *et al.* (1997) searched an expressed sequence tag (EST) database for an Agouti homologue. They identified a novel gene (*agrp*) from mice and humans that encoded a protein identical in size to Agouti and containing a characteristic pattern of cysteines, also found in Agouti. Agouti-related peptide (AgRP) was found to be expressed primarily in the adrenal gland and in the hypothalamus with low levels of expression in the testis and lung, in both mice and humans (Shutter *et al.*, 1997). Murine and human AgRP share high homology, with human AgRP containing an extra amino acid (132 vs. 131 residues). Agouti and AgRP share the most homology at the C-terminus, which contains a domain with 10 cysteine residues that are essential for function (Dinulescu and Cone, 2000). Based on its structural similarities to Agouti, AgRP is likely to function in a similar fashion. However, the mechanism of competition between AgRP and α-MSH for the MC3-R and MC4-R is not entirely clear. Two models have been proposed: in the first it is proposed that an octapeptide loop present in the C-terminus of both Agouti and AgRP mimics the structure of α-MSH (Tota *et al.*, 1999); in the second it is proposed that the antagonist binds to a separate site and blocks α-MSH binding through an allosteric mechanism (Yang *et al.*, 1999). More recent data suggest that similar residues in the melanocortin receptors recognize both agonist and antagonist (Yang *et al.*, 1999; Haskell-Luevano *et al.*, 2001). Finally, AgRP may act independently of α-MSH as it has recently been shown to have the capacity to act as an inverse agonist on the MC4-R in the absence of α-MSH due to constitutive activity of the MC4-R (Nijenhuis *et al.*, 2001). It has also been speculated that AgRP has activity independent of melanocortin receptors. Consistent with this, AgRP does have activity in MC4-R knockout mice but these could be mediated by the MC3-R (Marsh *et al.*, 1999b).

Hypothalamic expression of AgRP is confined to the arcuate nucleus in neurons that also express NPY (Broberger *et al.*, 1998; Wilson *et al.*, 1999). These cells are known to project to neurons, such as those found in the PVH and in the lateral hypothalamus (Elmquist *et al.*, 1999; Cowley *et al.*, 1999). These target neurons also receive separate input from POMC neurons and express MC3-R or MC4-R, indicating that AgRP is poised to act as a functional antagonist of melanocortin signaling *in vivo* (Bagnol *et al.*, 1999). Indeed, this role has been substantiated by studies in which centrally administered AgRP functions to stimulate feeding (Rossi *et al.*, 1998). Further confirmation of its role *in vivo* has come from experiments in which transgenic expression of AgRP recapitulates the MC4-R −/− and A^y phenotype (Ollmann *et al.*, 1997; Graham *et al.*, 1997). Thus, AgRP functions as an antagonist of α-MSH and is likely downstream of leptin action (Fig. 3B).

XI. MELANOCORTIN CORECEPTORS

Further insight into melanocortin signaling has been gained from the study of additional mouse mutant strains one termed *mahogony*. This strain has since the 1960s been known to suppress Agouti action and to block yellow pigment synthesis (Lane and Green, 1960). In addition, it became clear in the last few years that the *mahogony* mutation also blocked the obesity phenotype present in A^y mice suggesting that Agouti function required Mahogony both peripherally in skin at the MC1-R and centrally in the hypothalamus at the MC4-R (Miller *et al.*, 1997; Dinulescu *et al.*, 1998). In 1999, two groups used positional cloning to identify the *mahogany* gene, which was found to encode a transmembrane protein of 200 kDa termed attractin (Atrn) (Gunn *et al.*, 1999; Nagle *et al.*, 1999). Atrn mRNA is widely expressed including many areas of the brain required for the regulation of feeding behavior. Hypotheses regarding the action of Atrn suggested that it may function as an accessory receptor for Agouti in the skin and potentially play a role in the function of neuronal networks in the brain.

Recent work by He *et al.* (2001) demonstrates that mice carrying a more severe Atrn mutation (*mg-3J*) than the original *mahogany* mouse have spongy degeneration of the central nervous system. Furthermore, the phenotype of the *Zitter* rat, a well-described model of spongiform encephelopathy, has a loss-of-function mutation in Atrn (Kuramoto *et al.*, 2001), confirming its role in normal brain development. This suggests that some of the physiological effects seen in cross-bred *mahogany*/A^y mice are independent of Agouti function and are due to neuronal degeneration. Biochemical studies have demonstrated that the N-terminal region of Agouti (that lacks homology to AgRP) interacts with Atrn, suggesting that Agouti, but not AgRP, action may require Atrn. Indeed, crossing of the Atrn *mg-3j* mutant mice with A^y mice was found to suppress the effects of Agouti but did not influence the development of obesity in mice overexpressing AgRP (He *et al.*, 2001; Gunn *et al.*, 2001). Taken together, these data suggest that Agouti action requires Atrn, in contrast, Atrn does not interact with AgRP. Thus, mice carrying mutant Atrn alleles, which causes reduced body weight, are likely affected by a central process that is independent of AgRP signaling in the brain.

Despite the lack of effect of Atrn on AgRP signaling the potential for an Atrn-like molecule that interacts with AgRP must be considered. Recently Reizes *et al.* (2001) overexpressed syndecan-1, a heparin sulfate proteoglycan known to be a modulator of receptor–ligand interactions, in mice using a viral promoter. Surprisingly, these mice became hyperphagic and developed obesity at week 8. Syndecan-1, normally expressed only in the periphery, was found to be expressed in the PVH, arcuate, DMH, and lateral hypothalamic areas in these transgenic mice. Furthermore, syndecan-1 was found to interact with both AgRP and Agouti *in vitro* and enhance the actions of Agouti *in vivo*. This identification of syndecan-1 as a potential *in vitro* modulator of AgRP action led the authors to examine the expression of syndecan-3, a related isoform known to be normally expressed in the brain. Indeed, syndecan-3 binds AgRP *in vitro*, is expressed in the hypothalamus, and is up-regulated by food deprivation, consistent with a potential role as an endogenous modulator of AgRP action. Despite these recent insights, much work needs to be done in understanding the role of these potential coreceptors.

XII. TARGETS OF MELANOCORTIN SIGNALING

Despite the extensive work in discerning the role of melanocortin signaling in the regulation of food intake and energy expenditure little is known about the targets regulated the MC3-R and MC4-R. Clearly the melanocortin receptor antagonists and agonists have significant effects *in vivo* on feeding behavior and energy expenditure, but the gene targets that mediate these effects remain poorly described. In addition, more immediate effects of melanocortin signaling may occur independent of transcription.

In 1996, Ahima *et al.* first described the hypothyroidism that results from leptin deficiency in the face of starvation. This was subsequently shown to be centrally mediated by a fall in TRH mRNA in the PVH (Legradi *et al.*, 1997), which could be prevented by either leptin or centrally administered α-MSH (Fekete *et al.*, 2000). It was also demonstrated that α-MSH nerve terminals innervate TRH neurons in the PVH (Fekete *et al.*, 2000). Furthermore, α-MSH stimulated TRH release in hypothalamic cultures and AgRP blocks the effect of leptin on the release of TRH in this same culture system (Kim *et al.*, 2000). Given these data, it appears that TRH neurons are direct targets of melanocortin signaling. Consistent with this most TRH neurons in the PVH coexpress MC4-R (Harris *et al.*, 2001). Because the MC4-R activates the cAMP pathway it was shown that the TRH promoter was upregulated by α-MSH in a heterologous system via binding of the cAMP response element binding protein (CREB) to a site in the proximal promoter conserved across species. Thus, the TRH gene becomes one of the first targets separate from immediate early genes such as c-*fos,* known to be regulated at the transcriptional level by melanocortin signaling.

XIII. HUMAN MUTATIONS IN THE MELANOCORTIN SYSTEM

Proof of the importance of the melanocortin system in energy balance in humans has come from the identification of mutations that lead to obesity (Wardlaw *et al.*, 2001). Krude *et al.* (1998) identified the first two patients with POMC mutations who presented with adrenal insufficiency, red hair pigmentation, and early-onset obesity. The two individuals had separate genetic mutations. The first patient was a compound heterozygote with two mutations in exon 3 that resulted in premature termination of the POMC message, leading to an absence of functional ACTH and α-MSH. Heterozygous mutations were confirmed in the parents who were phenotypically normal. The second patient was homozygous for a mutation in exon 2 leading to aberrant translation from an out-of-frame start codon that resulted in the lack of POMC production. Once the diagnosis of adrenal insufficiency was made and replacement steroids begun, both children were observed to be markedly hyperphagic leading to the development of severe obesity. More recently, other pediatric patients have been described with novel POMC mutations (Krude and Gruters, 2000) leading to obesity. In addition, POMC processing defects, such as those found in a patient with PC1 mutations, also leads to significant obesity (Jackson *et al.*, 1997). The potential importance of POMC mutations in human obesity is further underscored by population studies in Mexican-Americans and African-Americans linking serum leptin levels to a locus on chromosome 2, which includes the *pomc* gene (Rotimi *et al.*, 1999; Comuzzie *et al.*, 1997). However, although a French kindred displayed a similar linkage to the *pomc* locus, no mutations were found within the *pomc* coding sequence suggesting that mutations in the *pomc* regulatory sequences or other nearby genes may exist (Delplanque *et al.*, 2000).

Although mutations in the *pomc* gene are extremely rare causes of obesity, mutations within the MC4-R are not. There have been a number of studies documenting the prevalence of MC4-R mutations since the original reports of heterozygous mutations causing obesity (Vaisse *et al.*, 1998; Yeo *et al.*, 1998). Vaisse *et al.* (2000) found a 4% frequency of MC4-R mutations in a large population of severely obese patients. Interestingly, functional analysis of the mutant receptors in heterologous cells revealed that they ranged from loss of function to constitutive activation and that there was variable obesity among families that did not relate to the functional severity of the mutation. Farooqi *et al.* studied 243 patients with severe early-onset obesity and found that 3.3% had MC4-R mutations (Farooqi and O'Rahilly, 2000; Farooqi *et al.*, 2000). One of the mutations, N62S, was found in a homozygous form in five children whereas heterozygous carriers were not obese. The N62S was more responsive to α-MSH than the missense or nonsense mutations described in functional assays. Those with mutations in the MC4-R also had hyperphagia, a tendency toward tall stature, and hyperinsulinemia, a similar phenotype to that found in the MC4-R −/− mouse. Interestingly, these children also had a marked increase in bone mineral density. To date, mutations have not been found in other components of the melanocortin system including the *agrp* and *mc3-r* genes. Still, MC4-R mutations remain the most common known cause of monogenic obesity.

XIV. FUTURE DIRECTIONS

The recent insight gained from unraveling the signaling pathways that control feeding behavior and energy expenditure in the hypothalamus has created tremendous interest in the biology of obesity in humans. Certainly, the leptin and melanocortin pathways are key regulators of body weight in humans, however many of the mechanisms by which they act remain to be elucidated. In the next few years much attention will be paid to the targets of both leptin and melanocortin signaling and further identification of downstream mediators controling feeding behavior and energy expenditure. In addition, the leptin and melanocortin pathways are ideal candidates for drug discovery; indeed leptin has already been used in humans and selective melanocortin receptor agonists are not far behind. Thus, the discovery of the relevance of central leptin and melanocortin action to human obesity has opened up a whole new field in which effective treatments for obesity will soon be identified.

ACKNOWLEDGMENTS

CB was supported by a grant from NIH (DK-60673) and ANH was supported by grants DK-57658 and DK-56123. We apologize if certain authors and papers have not been appropriately acknowledged in this review, but due to space limitations, some references have been excluded.

REFERENCES

Ahima, R. S., Prabakaran, D., Mantzoros, C., Qu, D., Lowell, B., Maratos-Flier, E., and Flier, J. S. (1996). Role of leptin in the neuroendocrine response to fasting. *Nature* **382**(6588), 250–252.

Al-Obeidi, F., Castrucci, A. M., Hadley, M. E., and Hruby, V. J. (1989). Potent and prolonged acting cyclic lactam analogues of alpha-melanotropin: design based on molecular dynamics. *J. Med. Chem.* **32**(12), 2555–2561.

Auernhammer, C. J., Chesnokova, V., Bousquet, C., and Melmed, S. (1998). Pituitary corticotroph SOCS-3: Novel intracellular regulation of leukemia-inhibitory factor-mediated proopiomelanocortin gene expression and adrenocorticotropin secretion. *Mol. Endocrinol.* **12**(7), 954–961.

Bagnol, D., Lu, X. Y., Kaelin, C. B., Day, H. E., Ollmann, M., Gantz, I., Akil, H., Barsh, G. S., and Watson, S. J. (1999). Anatomy of an endogenous antagonist: Relationship between Agouti-related protein and proopiomelanocortin in brain. *J. Neurosci.* **19**(18), RC26.

Banks, A. S., Davis, S. M., Bates, S. H., and Myers, M. G., Jr. (2000). Activation of downstream signals by the long form of the leptin receptor. *J. Biol. Chem.* **275**(19), 14563–14572.

Baskin, D. G., Breininger, J. F., and Schwartz, M. W. (2000). SOCS-3 expression in leptin-sensitive neurons of the hypothalamus of fed and fasted rats. *Regul. Pept.* **92**(1–3), 9–15.

Bates, S. H., Stearns, W. H., Schubert, M., Banks, A. S., Hak, A. K., Schwartz, M. W., and Myers, J. M. G. (2001). Leptin receptor-STAT3 signaling integrates energy balance and metabolic homeostasis. Endocrine Society, Denver 2001.

Baumann, H., Morella, K. K., White, D. W., Dembski, M., Bailon, P. S., Kim, H., Lai, C. F., and Tartaglia, L. A. (1996). The full-length leptin receptor has signaling capabilities of interleukin 6-type cytokine receptors. *Proc. Natl. Acad. Sci. USA* **93**(16), 8374–8378.

Beck, B., Kozak, R., Stricker-Krongrad, A., and Burlet, C. (1998). Neuropeptide Y release in the paraventricular nucleus of Long-Evans rats treated with leptin. *Biochem. Biophys. Res. Commun.* **242**(3), 636–639.

Benjannet, S., Rondeau, N., Day, R., Chretien, M., and Seidah, N. G. (1991). PC1 and PC2 are proprotein convertases capable of cleaving proopiomelanocortin at distinct pairs of basic residues. *Proc. Natl. Acad. Sci. USA* **88**(9), 3564–3568.

Bennett, B. D., Solar, G. P., Yuan, J. Q., Mathias, J., Thomas, G. R., and Matthews, W. (1996). A role for leptin and its cognate receptor in hematopoiesis. *Curr. Biol.* **6**(9), 1170–1180.

Berti, L., Kellerer, M., Capp, E., and Haring, H. U. (1997). Leptin stimulates glucose transport and glycogen synthesis in C2C12 myotubes: Evidence for a P13-kinase mediated effect.. *Diabetologia* **40**(5), 606–609.

Bjorbaek, C., Uotani, S., da Silva, B., and Flier, J. S. (1997). Divergent signaling capacities of the long and short isoforms of the leptin receptor. *J. Biol. Chem.* **272**(51), 32686–32695.

Bjorbaek, C., Elmquist, J. K., Michl, P., Ahima, R. S., van Bueren, A., McCall, A. L., and Flier, J. S. (1998a). Expression of leptin receptor isoforms in rat brain microvessels. *Endocrinology* **139**(8), 3485–3491.

Bjorbaek, C., Elmquist, J. K., Frantz, J. D., Shoelson, S. E., and Flier, J. S. (1998b). Identification of SOCS-3 as a potential mediator of central leptin resistance. *Mol. Cell* **1**(4), 619–625.

Bjorbaek, C., El-Haschimi, K., Frantz, J. D., and Flier, J. S. (1999). The role of SOCS-3 in leptin signaling and leptin resistance. *J. Biol. Chem.* **274**(42), 30059–30065.

Bjorbaek, C., Lavery, H. J., Bates, S. H., Olson, R. K., Davis, S. M., Flier, J. S., and Myers, M. G., Jr. (2000). SOCS3 mediates feedback inhibition of the leptin receptor via Tyr985. *J. Biol. Chem.* **275**(51), 40649–40657.

Bjorbaek, C., Buchholz, R. M., Davis, S. M., Bates, S. H., Pierroz, D. D., Gu, H., Neel, B. G., Myers, M. G., Jr., and Flier, J. S. (2001a). Divergent roles of SHP-2 in ERK activation by leptin receptors. *J. Biol. Chem.* **276**(7), 4747–4755.

Bjorbaek, C., Tzameli, I., Fox, G., Flier, J. S., and Nillni, E. A. (2001b). Regulation of POMC expression by leptin: Roles of STAT3 and cAMP. Endocrine Society, Denver 2001.

Blake, N. G., Eckland, D. J., Foster, O. J, and Lightman, S. L. (1991). Inhibition of hypothalamic thyrotropin-releasing hormone messenger ribonucleic acid during food deprivation. *Endocrinology* **129**(5), 2714–2718.

Boston, B. A., Blaydon, K. M., Varnerin, J., and Cone, R. D. (1997). Independent and additive effects of central POMC and leptin pathways on murine obesity. *Science* **278**(5343), 1641–1644.

Bousquet, C., Zatelli, M. C., and Melmed, S. (2000). Direct regulation of pituitary proopiomelanocortin by STAT3 provides a novel mechanism for immuno-neuroendocrine interfacing. *J. Clin. Invest.* **106**(11), 1417–1425.

Broberger, C., Johansen, J., Johansson, C., Schalling, M., and Hokfelt, T. (1998). The neuropeptide Y/agouti gene-related protein (AGRP) brain circuitry in normal, anorectic, and monosodium glutamate-treated mice. *Proc. Natl. Acad. Sci. USA* **95**(25), 15043–15048.

Bultman, S. J., Michaud, E. J., and Woychik, R. P. (1992). Molecular characterization of the mouse agouti locus. *Cell* **71**(7), 1195–1204.

Butler, A. A., Kesterson, R. A., Khong, K., Cullen, M. J., Pelleymounter, M. A., Dekoning, J., Baetscher, M., and Cone, R. D. (2000). A unique metabolic syndrome causes obesity in the melanocortin-3 receptor-deficient mouse. *Endocrinology* **141**(9), 3518–3521.

Butler, A. A., Marks, D. L., Fan, W., Kuhn, C. M., Bartolome, M., and Cone, R. D. (2001). Melanocortin-4 receptor is required for acute homeostatic responses to increased dietary fat. *Nat. Neurosci.* **4**(6), 605–611.

Campfield, L. A., Smith, F. J., Guisez, Y., Devos, R., and Burn, P. (1995). Recombinant mouse OB protein: Evidence for a peripheral signal linking adiposity and central neural networks. *Science* **269**(5223), 546–549.

Chen, A. S., Marsh, D. J., Trumbauer, M. E., Frazier, E. G., Guan, X. M., Yu, H., Rosenblum, C. I., Vongs, A., Feng, Y., Cao, L., Metzger, J. M., Strack, A. M., Camacho, R. E., Mellin, T. N., Nunes, C. N., Min, W., Fisher, J., Gopal-Truter, S., MacIntyre, D. E., Chen, H. Y., and Van der Ploeg, L. H. (2000). Inactivation of the mouse melanocortin-3 receptor results in increased fat mass and reduced lean body mass. *Nat. Genet.* **26**(1), 97–102.

Chen, H., Charlat, O., Tartaglia, L. A., Woolf, E. A., Weng, X., Ellis, S. J., Lakey, N. D., Culpepper, J., Moore, K. J., Breitbart, R. E., Duyk, G. M., Tepper, R. I., and Morgenstern, J. P. (1996). Evidence that the diabetes gene encodes the leptin receptor: Identification of a mutation in the leptin receptor gene in db/db mice. *Cell* **84**(3), 491–495.

Chen, W., Kelly, M. A., Opitz-Araya, X., Thomas, R. E., Low, M. J., and Cone, R. D. (1997). Exocrine gland dysfunction in MC5-R-deficient mice: Evidence for coordinated regulation of exocrine gland function by melanocortin peptides. *Cell* **91**(6), 789–798.

Cheng, A., Uetani, N., Simoncic, P. D., Chaubey, V. P., Lee-Loy, A., McGlade, C. J., Kennedy, B. P., and Tremblay, M. L. (2002). Attenuation of leptin action and regulation of obesity by protein tyrosine phosphatase 1B. *Dev. Cell* **2**(4), 497–503.

Cheung, C. C., Clifton, D. K., and Steiner, R. A. (1997). Proopiomelanocortin neurons are direct targets for leptin in the hypothalamus. *Endocrinology* **138**(10), 4489–4492.

Chhajlani, V., and Wikberg, J. E. (1992). Molecular cloning and expression of the human melanocyte stimulating hormone receptor cDNA. *FEBS Lett.* **309**(3), 417–420.

Chua, S. C., Jr., Chung, W. K., Wu-Peng, X. S., Zhang, Y., Liu, S. M., Tartaglia, L., and Leibel, R. L. (1996). Phenotypes of mouse *diabetes* and rat *fatty* due to mutations in the OB (leptin) receptor. *Science* **271**(5251), 994–996.

Cioffi, J. A., Shafer, A. W., Zupancic, T. J., Smith-Gbur, J., Mikhail, A., Platika, D., and Snodgrass, H. R. (1996). Novel B219/OB receptor isoforms: Possible role of leptin in hematopoiesis and reproduction. *Nat. Med.* **2**(5), 585–589.

Clement, K., Vaisse, C., Lahlou, N., Cabrol, S., Pelloux, V., Cassuto, D., Gourmelen, M., Dina, C., Chambaz, J., Lacorte, J. M., Basdevant, A., Bougneres, P., Lebouc, Y., Froguel, P., and Guy-Grand, B. (1998). A mutation in the human leptin receptor gene causes obesity and pituitary dysfunction [see comments]. *Nature* **392**(6674), 398–401.

Cohen, B., Novick, D., and Rubinstein, M. (1996). Modulation of insulin activities by leptin. *Science* **274**(5290), 1185–1188.

Coleman, D. L. (1973). Effects of parabiosis of obese with diabetes and normal mice. *Diabetologia* **9**(4), 294–298.

Coleman, D. L. (1978). Obese and diabetes: Two mutant genes causing diabetes-obesity syndromes in mice. *Diabetologia* **14**(3), 141–148.

Commins, S. P., Watson, P. M., Padgett, M. A., Dudley, A., Argyropoulos, G., and Gettys, T. W. (1999). Induction of uncoupling protein expression in brown and white adipose tissue by leptin. *Endocrinology* **140**(1), 292–300.

Commins, S. P., Watson, P. M., Levin, N., Beiler, R. J., and Gettys, T. W. (2000). Central leptin regulates the UCP1 and ob genes in brown and white adipose tissue via different beta-adrenoceptor subtypes. *J. Biol. Chem.* **275**(42), 33059–33067.

Comuzzie, A. G., Hixson, J. E., Almasy, L., Mitchell, B. D., Mahaney, M. C., Dyer, T. D., Stern, M. P., MacCluer, J. W., and Blangero, J. (1997). A major quantitative trait locus determining serum leptin levels and fat mass is located on human chromosome 2. *Nat. Genet.* **15**(3), 273–276.

Cone, R. D. (2000). The melanocortin-4 receptor *In* "The Melanocortin Receptors". (R. D. Cone, Ed.), pp. 405–449. Humana Press, Totowa, NJ.

Cone, R. D., Mountjoy, K. G., Robbins, L. S., Nadeau, J. H., Johnson, K. R., Roselli-Rehfuss, L., and Mortrud, M. T. (1993). Cloning and functional characterization of a family of receptors for the melanotropic peptides. *Ann. N.Y. Acad. Sci.* **680**, 342–363.

Considine, R. V., Considine, E. L., Williams, C. J., Nyce, M. R., Magosin, S. A., Bauer, T. L., Rosato, E. L., Colberg, J., and Caro, J. F. (1995). Evidence against either a premature stop codon or the absence of obese gene mRNA in human obesity. *J. Clin. Invest.* **95**(6), 2986–2988.

Considine, R. V., Sinha, M. K., Heiman, M. L., Kriauciunas, A., Stephens, T. W., Nyce, M. R., Ohannesian, J. P., Marco, C. C., McKee, L. J., Bauer, T. L., *et al.* (1996a). Serum immunoreactive-leptin concentrations in normal-weight and obese humans. *N. Engl. J. Med.* **334**(5), 292–295.

Considine, R. V., Considine, E. L., Williams, C. J., Nyce, M. R., Zhang, P., Opentanova, I., Ohannesian, J. P., Kolaczynski, J. W., Bauer, T. L., Moore, J. H., and Caro, J. F. (1996b). Mutation screening and identification of a sequence variation in the human ob gene coding region. *Biochem. Biophys. Res. Commun.* **220**(3), 735–739.

Considine, R. V., Considine, E. L., Williams, C. J., Hyde, T. M., and Caro, J. F. (1996c). The hypothalamic leptin receptor in humans: Identification of incidental sequence polymorphisms and absence of the db/db mouse and fa/fa rat mutations. *Diabetes* **45**(7), 992–994.

Cowley, M. A., Pronchuk, N., Fan, W., Dinulescu, D. M., Colmers, W. F., and Cone, R. D. (1999). Integration of NPY, AGRP, and melanocortin signals in the hypothalamic paraventricular nucleus: Evidence of a cellular basis for the adipostat. *Neuron* **24**(1), 155–163.

Cowley, M. A., Smart, J. L., Rubinstein, M., Cerdan, M. G., Diano, S., Horvath, T. L., Cone, R. D., and Low, M. J. (2001). Leptin activates anorexigenic POMC neurons through a neural network in the arcuate nucleus. *Nature* **411**(6836), 480–484.

Delplanque, J., Barat-Houari, M., Dina, C., Gallina, P., Clement, K., Guy-Grand, B., Vasseur, F., Boutin, P., and Froguel, P. (2000). Linkage and association studies between the proopiomelanocortin (POMC) gene and obesity in caucasian families. *Diabetologia* **43**(12), 1554–1557.

Desarnaud, F., Labbe, O., Eggerickx, D., Vassart, G., and Parmentier, M. (1994). Molecular cloning, functional expression and pharmacological characterization of a mouse melanocortin receptor gene. *Biochem. J.* **299**(Pt. 2), 367–373.

Devos, R., Guisez, Y., Van der Heyden, J., White, D. W., Kalai, M., Fountoulakis, M., and Plaetinck, G. (1997). Ligand-independent dimerization of the extracellular domain of the leptin receptor and determination of the stoichiometry of leptin binding. *J. Biol. Chem.* **272**(29), 18304–18310.

Dinulescu, D. M., and Cone, R. D. (2000). Agouti and agouti-related protein: Analogies and contrasts. *J. Biol. Chem.* **275**(10), 6695–6698.

Dinulescu, D. M., Fan, W., Boston, B. A., McCall, K., Lamoreux, M. L., Moore, K. J., Montagno, J., and Cone, R. D. (1998). Mahogany (mg) stimulates feeding and increases basal metabolic rate independent of its suppression of agouti. *Proc. Natl. Acad. Sci.USA* **95**(21), 12707–12712.

Elias, C. F., Aschkenasi, C., Lee, C., Kelly, J., Ahima, R. S., Bjorbaek, C., Flier, J. S., Saper, C. B., and Elmquist, J. K. (1999). Leptin differentially regulates NPY and POMC neurons projecting to the lateral hypothalamic area. *Neuron* **23**(4), 775–786.

Elias, C. F., Lee, C., Kelly, J., Aschkenasi, C., Ahima, R. S., Couceyro, P. R., Kuhar, M. J., Saper, C. B., and Elmquist, J. K. (1998). Leptin activates hypothalamic CART neurons projecting to the spinal cord. *Neuron* **21**(6), 1375–1385.

Elias, C. F., Lee, C. E., Kelly, J. F., Ahima, R. S., Kuhar, M., Saper, C. B., and Elmquist, J. K. (2001). Characterization of CART neurons in the rat and human hypothalamus. *J. Comp. Neurol.* **432**(1), 1–19.

Elmquist, J. K., Bjorbaek, C., Ahima, R. S., Flier, J. S., and Saper, C. B. (1998). Distributions of leptin receptor mRNA isoforms in the rat brain. *J. Comp. Neurol.* **395**(4), 535–547.

Elmquist, J. K., Elias, C. F., and Saper, C. B. (1999). From lesions to leptin: Hypothalamic control of food intake and body weight. *Neuron* **22**(2), 221–232.

Erickson, J. C., Hollopeter, G., and Palmiter, R. D. (1996). Attenuation of the obesity syndrome of ob/ob mice by the loss of neuropeptide Y [see comments]. *Science* **274**(5293), 1704–1707.

Eyckerman, S., Broekaert, D., Verhee, A., Vandekerckhove, J., and Tavernier, J. (2000). Identification of the Y985 and Y1077 motifs as SOCS3 recruitment sites in the murine leptin receptor. *FEBS Lett.* **486**(1), 33–37.

Fan, W., Boston, B. A., Kesterson, R. A., Hruby, V. J., and Cone, R. D. (1997). Role of melanocortinergic neurons in feeding and the agouti obesity syndrome [see comments]. *Nature* **385**(6612), 165–168.

Farooqi, I. S., and O'Rahilly, S. (2000). Recent advances in the genetics of severe childhood obesity. *Arch. Dis. Child.* **83**(1), 31–34.

Farooqi, I. S., Jebb, S. A., Langmack, G., Lawrence, E., Cheetham, C. H., Prentice, A. M., Hughes, I. A., McCamish, M. A., and O'Rahilly, S. (1999). Effects of recombinant leptin therapy in a child with congenital leptin deficiency. *N. Engl. J. Med.* **341**(12), 879–884.

Farooqi, I. S., Yeo, G. S., Keogh, J. M., Aminian, S., Jebb, S. A., Butler, G., Cheetham, T., and O'Rahilly, S. (2000). Dominant and recessive inheritance of morbid obesity associated with melanocortin 4 receptor deficiency. *J. Clin. Invest.* **106**(2), 271–279.

Fei, H., Okano, H. J., Li, C., Lee, G. H., Zhao, C., Darnell, R., and Friedman, J. M. (1997). Anatomic localization of alternatively spliced leptin receptors (Ob-R) in mouse brain and other tissues. *Proc. Natl. Acad. Sci. USA* **94**(13), 7001–7005.

Fekete, C., Legradi, G., Mihaly, E., Huang, Q.-H., Tatro, J. B., Rand, W. M., Emerson, C. H., and Lechan, R. M. (2000). (alpha)-Melanocyte-stimulating-hormone is contained in nerve terminals innervating thyrotropin-releasing hormone synthesizing neurons in the hypothalamic paraventricular nucleus and prevents fasting induced suppression of prothyrotropin-releasing hormone gene expression. *J. Neurosci.* **20**, 1550–1558.

Fong, T. M., Mao, C., MacNeil, T., Kalyani, R., Smith, T., Weinberg, D., Tota, M. R., and Van der Ploeg, L. H. (1997). ART (protein product of agouti-related transcript) as an antagonist of MC-3 and MC-4 receptors. *Biochem. Biophys. Res. Commun.* **237**(3), 629–631.

Gantz, I., Konda, Y., Tashiro, T., Shimoto, Y., Miwa, H., Munzert, G., Watson, S. J., DelValle, J., and Yamada, T. (1993a). Molecular cloning of a novel melanocortin receptor. *J. Biol. Chem.* **268**(11), 8246–8250.

Gantz, I., Miwa, H., Konda, Y., Shimoto, Y., Tashiro, T., Watson, S. J., DelValle, J., and Yamada, T. (1993b). Molecular cloning, expression, and gene localization of a fourth melanocortin receptor. *J. Biol. Chem.* **268**(20), 15174–15179.

Gavrilova, O., Barr, V., Marcus-Samuels, B., and Reitman, M. (1997). Hyperleptinemia of pregnancy associated with the appearance of a circulating form of the leptin receptor. *J. Biol. Chem.* **272**(48), 30546–30551.

Gee, C. E., Chen, C. L., Roberts, J. L., Thompson, R., and Watson, S. J. (1983). Identification of proopiomelanocortin neurones in rat hypothalamus by in situ cDNA-mRNA hybridization. *Nature* **306**(5941), 374–376.

Ghilardi, N., and Skoda, R. C. (1997). The leptin receptor activates janus kinase 2 and signals for proliferation in a factor-dependent cell line. *Mol. Endocrinol.* **11**(4), 393–399.

Ghilardi, N., Ziegler, S., Wiestner, A., Stoffel, R., Heim, M. H., and Skoda, R. C. (1996). Defective STAT signaling by the leptin receptor in diabetic mice. *Proc. Natl. Acad. Sci. USA* **93**(13), 6231–6235.

Glaum, S. R., Hara, M., Bindokas, V. P., Lee, C. C., Polonsky, K. S., Bell, G. I., and Miller, R. J. (1996). Leptin, the obese gene product, rapidly modulates synaptic transmission in the hypothalamus. *Mol. Pharmacol.* **50**(2), 230–235.

Graham, M., Shutter, J. R., Sarmiento, U., Sarosi, I., and Stark, K. L. (1997). Overexpression of Agrt leads to obesity in transgenic mice. *Nat. Genet.* **17**(3), 273–274.

Guan, X. M., Hess, J. F., Yu, H., Hey, P. J., and van der Ploeg, L. H. (1997). Differential expression of mRNA for leptin receptor isoforms in the rat brain. *Mol. Cell. Endocrinol.* **133**(1), 1–7.

Gunn, T. M., Miller, K. A., He, L., Hyman, R. W., Davis, R. W., Azarani, A., Schlossman, S. F., Duke-Cohan, J. S., and Barsh, G. S. (1999). The mouse mahogany locus encodes a transmembrane form of human attractin. *Nature* **398**(6723), 152–156.

Gunn, T. M., Inui, T., Kitada, K., Ito, S., Wakamatsu, K., He, L., Bouley, D. M., Serikawa, T., and Barsh, G. S. (2001). Molecular and phenotypic analysis of attractin mutant mice. *Genetics* **158**(4), 1683–1695.

Hahn, T. M., Breininger, J. F., Baskin, D. G., and Schwartz, M. W. (1998). Coexpression of Agrp and NPY in fasting-activated hypothalamic neurons. *Nat. Neurosci.* **1**(4), 271–272.

Halaas, J. L., Gajiwala, K. S., Maffei, M., Cohen, S. L., Chait, B. T., Rabinowitz, D., Lallone, R. L., Burley, S. K., and Friedman, J. M. (1995). Weight-reducing effects of the plasma protein encoded by the obese gene. *Science* **269**(5223), 543–546.

Harris, M., Aschkenasi, C., Elias, C. F., Chandrankunnel, A., Nillni, E. A., Bjorbaek, C., Elmquist, J. K., Flier, J. S., and Hollenberg, A. N. (2001). Transcriptional regulation of the thyrotropin-releasing hormone gene by leptin and melanocortin signaling. *J. Clin. Invest.* **107**, 1–11.

Haskell-Luevano, C., Cone, R. D., Monck, E. K., and Wan, Y. P. (2001). Structure activity studies of the melanocortin-4 receptor by in vitro mutagenesis: Identification of agouti-related protein (AGRP), melanocortin agonist and synthetic peptide antagonist interaction determinants. *Biochemistry* **40**(20), 6164–6179.

He, L., Gunn, T. M., Bouley, D. M., Lu, X. Y., Watson, S. J., Schlossman, S. F., Duke-Cohan, J. S., and Barsh, G. S. (2001). A biochemical function for attractin in agouti-induced pigmentation and obesity. *Nat. Genet.* **27**(1), 40–47.

Hileman, S. M., Pierroz, D. D., Masuzaki, H., Bjorbaek, C., El-Haschimi, K., Banks, W. A., and Flier, J. S. (2002). Characterization of short isoforms of the leptin receptor in rat cerebral microvessels and of brain uptake of leptin in mouse models of obesity. *Endocrinology* **143**(3), 775–783.

Hileman, S. M., Tornoe, J., Flier, J. S., and Bjorbaek, C. (2000). Transcellular transport of leptin by the short leptin receptor isoform ObRa in Madin-Darby canine kidney cells. *Endocrinology* **141**(6), 1955–1961.

Hruby, V. J., Lu, D., Sharma, S. D., Castrucci, A. L., Kesterson, R. A., al-Obeidi, F. A., Hadley, M. E., and Cone, R. D. (1995). Cyclic lactam alpha-melanotropin analogues of Ac-Nle4-cyclo[Asp5, D-Phe7,Lys10] alpha-melanocyte-stimulating hormone-(4-10)-NH2 with bulky aromatic amino acids at position 7 show high antagonist potency and selectivity at specific melanocortin receptors. *J. Med. Chem.* **38**(18), 3454–3461.

Hubschle, T., Thom, E., Watson, A., Roth, J., Klaus, S., and Meyerhof, W. (2001). Leptin-induced nuclear translocation of STAT3 immunoreactivity in hypothalamic nuclei involved in body weight regulation. *J. Neurosci.* **21**(7), 2413–2424.

Huszar, D., Lynch, C. A., Fairchild-Huntress, V., Dunmore, J. H., Fang, Q., Berkemeier, L. R., Gu, W., Kesterson, R. A., Boston, B. A., Cone, R. D., Smith, F. J., Campfield, L. A., Burn, P., and Lee, F. (1997). Targeted disruption of the melanocortin-4 receptor results in obesity in mice. *Cell* **88**(1), 131–141.

Jackson, R. S., Creemers, J. W., Ohagi, S., Raffin-Sanson, M. L., Sanders, L., Montague, C. T., Hutton, J. C., and O'Rahilly, S. (1997). Obesity and impaired prohormone processing associated with mutations in the human prohormone convertase 1 gene. *Nat. Genet.* **16**(3), 303–306.

Jang, M., Mistry, A., Swick, A. G., and Romsos, D. R. (2000). Leptin rapidly inhibits hypothalamic neuropeptide Y secretion and stimulates corticotropin-releasing hormone secretion in adrenalectomized mice. *J. Nutr.* **130**(11), 2813–2820.

Kastin, A. J., Pan, W., Maness, L. M., Koletsky, R. J., and Ernsberger, P. (1999). Decreased transport of leptin across the blood-brain barrier in rats lacking the short form of the leptin receptor. *Peptides* **20**(12), 1449–1453.

Kellerer, M., Koch, M., Metzinger, E., Mushack, J., Capp, E., and Haring, H. U. (1997). Leptin activates PI-3 kinase in C2C12 myotubes via janus kinase-2 (JAK-2) and insulin receptor substrate-2 (IRS-2) dependent pathways. *Diabetologia* **40**(11), 1358–1362.

Kim, M. S., Small, C. J., Stanley, S. A., Morgan, D. G. A., Seal, L. J., Kong, W. M., Edwards, C. M. B., Abusana, S., Sunter, D., Ghatei, M. A., and Bloom, S. R. (2000). The central melanocortin system affects the hypothalamo-pituitary thyroid axis and may mediate the effect of leptin. *J. Clin. Invest.* **105,** 1005–1011.

Kim, Y. B., Uotani, S., Pierroz, D. D., Flier, J. S., and Kahn, B. B. (2000). In vivo administration of leptin activates signal transduction directly in insulin-sensitive tissues: Overlapping but distinct pathways from insulin. *Endocrinology* **141**(7), 2328–2339.

King, P. J., Widdowson, P. S., Doods, H., and Williams, G. (2000). Regulation of neuropeptide Y release from hypothalamic slices by melanocortin-4 agonists and leptin. *Peptides* **21**(1), 45–48.

Klebig, M. L., Wilkinson, J. E., Geisler, J. G., and Woychik, R. P. (1995). Ectopic expression of the agouti gene in transgenic mice causes obesity, features of type II diabetes, and yellow fur. *Proc. Natl. Acad. Sci. USA* **92**(11), 4728–4732.

Konda, Y., Gantz, I., DelValle, J., Shimoto, Y., Miwa, H., and Yamada, T. (1994). Interaction of dual intracellular signaling pathways activated by the melanocortin-3 receptor. *J. Biol. Chem.* **269**(18), 13162–13166.

Kowalski, T. J., Liu, S. M., Leibel, R. L., and Chua, S. C., Jr. (2001). Transgenic complementation of leptin-receptor deficiency. I. Rescue of the obesity/diabetes phenotype of LEPR-null mice expressing a LEPR-B transgene. *Diabetes* **50**(2), 425–435.

Kristensen, P., Judge, M. E., Thim, L., Ribel, U., Christjansen, K. N., Wulff, B. S., Clausen, J. T., Jensen, P. B., Madsen, O. D., Vrang, N., Larsen, P. J., and Hastrup, S. (1998). Hypothalamic CART is a new anorectic peptide regulated by leptin. *Nature* **393**(6680), 72–76.

Krude, H., and Gruters, A. (2000). Implications of proopiomelanocortin (POMC) mutations in humans: The POMC deficiency syndrome. *Trends Endocrinol. Metab.* **11**(1), 15–22.

Krude, H., Biebermann, H., Luck, W., Horn, R., Brabant, G., and Gruters, A. (1998). Severe early-onset obesity, adrenal insufficiency and red hair pigmentation caused by POMC mutations in humans. *Nat. Genet.* **19**(2), 155–157.

Kuramoto, T., Kitada, K., Inui, T., Sasaki, Y., Ito, K., Hase, T., Kawagachi, S., Ogawa, Y., Nakao, K., Barsh, G. S., Nagao, M., Ushijima, T., and Serikawa, T. (2001). Attractin/mahogany/zitter plays a critical role in myelination of the central nervous system. *Proc. Natl. Acad. Sci. USA* **98**(2), 559–564.

Lane, P. W., and Green, M. C. (1960). Mahogany, a recessive color mutation in linkage group V of the mouse. *J. Hered.* **51,** 228–230.

Lee, G. H., Proenca, R., Montez, J. M., Carroll, K. M., Darvishzadeh, J. G., Lee, J. I., and Friedman, J. M. (1996). Abnormal splicing of the leptin receptor in diabetic mice. *Nature* **379**(6566), 632–635.

Legradi, G., Emerson, C. H., Ahima, R. S., Flier, J. S., and Lechan, R. M. (1997). Leptin prevents fasting-induced suppression of prothyrotropin-releasing hormone messenger ribonucleic acid in neurons of the hypothalamic paraventricular nucleus. *Endocrinology* **138**(6), 2569–2576.

Legradi, G., Emerson, C. H., Ahima, R. S., Rand, W. M., Flier, J. S., and Lechan, R. M. (1998). Arcuate nucleus ablation prevents fasting-induced suppression of ProTRH mRNA in the hypothalamic paraventricular nucleus. *Neuroendocrinology* **68**(2), 89–97.

Leibel, R. L., Chung, W. K., and Chua, S. C., Jr. (1997). The molecular genetics of rodent single gene obesities. *J. Biol. Chem.* **272**(51), 31937–31940.

Li, C., and Friedman, J. M. (1999). Leptin receptor activation of SH2 domain containing protein tyrosine phosphatase 2 modulates Ob receptor signal transduction. *Proc. Natl. Acad. Sci. USA* **96**(17), 9677–9682.

Li, C., Ioffe, E., Fidahusein, N., Connolly, E., and Friedman, J. M. (1998). Absence of soluble leptin receptor in plasma from dbPas/dbPas and other db/db mice. *J. Biol. Chem.* **273**(16), 10078–10082.

Li, J. Y., Finniss, S., Yang, Y. K., Zeng, Q., Qu, S. Y., Barsh, G., Dickinson, C., and Gantz, I. (2000). Agouti-related protein-like immunoreactivity: Characterization of release from hypothalamic tissue and presence in serum. *Endocrinology* **141**(6), 1942–1950.

Li, S. J., Varga, K., Archer, P., Hruby, V. J., Sharma, S. D., Kesterson, R. A., Cone, R. D., and Kunos, G. (1996). Melanocortin antagonists define two distinct pathways of cardiovascular control by alpha- and gamma-melanocyte-stimulating hormones. *J. Neurosci.* **16**(16), 5182–5188.

Lonnqvist, F., Arner, P., Nordfors, L., and Schalling, M. (1995). Overexpression of the obese (ob) gene in adipose tissue of human obese subjects. *Nat. Med.* **1**(9), 950–953.

Lu, D., Willard, D., Patel, I. R., Kadwell, S., Overton, L., Kost, T., Luther, M., Chen, W., Woychik, R. P., Wilkison, W. O., *et al.* (1994). Agouti protein is an antagonist of the melanocyte-stimulating-hormone receptor. *Nature* **371**(6500), 799–802.

Maffei, M., Halaas, J., Ravussin, E., Pratley, R. E., Lee, G. H., Zhang, Y., Fei, H., Kim, S., Lallone, R., Ranganathan, S., *et al.* (1995). Leptin levels in human and rodent: measurement of plasma leptin and ob RNA in obese and weight-reduced subjects. *Nat. Med.* **1**(11), 1155–1161.

Maffei, M., Stoffel, M., Barone, M., Moon, B., Dammerman, M., Ravussin, E., Bogardus, C., Ludwig, D. S., Flier, J. S., Talley, M., *et al.* (1996). Absence of mutations in the human OB gene in obese/diabetic subjects. *Diabetes* **45**(5), 679–682.

Maresh, G. A., Maness, L. M., Zadina, J. E., and Kastin, A. J. (2001). In vitro demonstration of a saturable transport system for leptin across the blood-brain barrier. *Life Sci.* **69**(1), 67–73.

Marsh, D. J., Hollopeter, G., Huszar, D., Laufer, R., Yagaloff, K. A., Fisher, S. L., Burn, P., and Palmiter, R. D. (1999a). Response of melanocortin-4 receptor-deficient mice to anorectic and orexigenic peptides. *Nat. Genet.* **21**(1), 119–122.

Marsh, D. J., Miura, G. I., Yagaloff, K. A., Schwartz, M. W., Barsh, G. S., and Palmiter, R. D. (1999b). Effects of neuropeptide Y deficiency on hypothalamic agouti-related protein expression and responsiveness to melanocortin analogues. *Brain Res.* **848**(1–2), 66–77.

Mercer, J. G., Hoggard, N., Williams, L. M., Lawrence, C. B., Hannah, L. T., and Trayhurn, P. (1996a). Localization of leptin receptor mRNA and the long form splice variant (Ob-Rb) in mouse hypothalamus and adjacent brain regions by in situ hybridization. *FEBS Lett.* **387**(2–3), 113–116.

Mercer, J. G., Hoggard, N., Williams, L. M., Lawrence, C. B., Hannah, L. T., Morgan, P. J., and Trayhurn, P. (1996b). Coexpression of leptin receptor and preproneuropeptide Y mRNA in arcuate nucleus of mouse hypothalamus. *J. Neuroendocrinol.* **8**(10), 733–735.

Mercer, J. G., Moar, K. M., and Hoggard, N. (1998). Localization of leptin receptor (Ob-R) messenger ribonucleic acid in the rodent hindbrain. *Endocrinology* **139**(1), 29–34.

Michaud, E. J., Bultman, S. J., Klebig, M. L., van Vugt, M. J., Stubbs, L. J., Russell, L. B., and Woychik, R. P. (1994). A molecular model for the genetic and phenotypic characteristics of the mouse lethal yellow (Ay) mutation. *Proc. Natl. Acad. Sci. USA* **91**(7), 2562–2566.

Miller, K. A., Gunn, T. M., Carrasquillo, M. M., Lamoreux, M. L., Galbraith, D. B., and Barsh, G. S. (1997). Genetic studies of the mouse mutations mahogany and mahoganoid. *Genetics* **146**(4), 1407–1415.

Mizuno, T. M., and Mobbs, C. V. (1999). Hypothalamic agouti-related protein messenger ribonucleic acid is inhibited by leptin and stimulated by fasting. *Endocrinology* **140**(2), 814–817.

Mizuno, T. M., Kleopoulos, S. P., Bergen, H. T., Roberts, J. L., Priest, C. A., and Mobbs, C. V. (1998). Hypothalamic pro-opiomelanocortin mRNA is reduced by fasting and [corrected] in ob/ob and db/db mice, but is stimulated by leptin. *Diabetes* **47**(2), 294–297. [Published erratum appears in *Diabetes* (1998). 47(4), 696.]

Montague, C. T., Farooqi, I. S., Whitehead, J. P., Soos, M. A., Rau, H., Wareham, N. J., Sewter, C. P., Digby, J. E., Mohammed, S. N., Hurst, J .A., Cheetham, C. H., Earley, A. R., Barnett, A. H., Prins, J. B., and O'Rahilly, S. (1997). Congenital leptin deficiency is associated with severe early-onset obesity in humans. *Nature* **387**(6636), 903–908.

Mountjoy, K. (2000). Cloning of the melanocortin receptors. *In* "The Melanocortin Receptors" (R. D. Cone, Ed.), pp. 209–239. Humana Press, Totowa, NJ.

Mountjoy, K. G., Robbins, L. S., Mortrud, M. T., and Cone, R. D. (1992). The cloning of a family of genes that encode the melanocortin receptors. *Science* **257**(5074), 1248–1251.

Mountjoy, K. G., Mortrud, M. T., Low, M. J., Simerly, R. B., and Cone, R. D. (1994). Localization of the melanocortin-4 receptor (MC4-R) in neuroendocrine and autonomic control circuits in the brain. *Mol. Endocrinol.* **8**(10), 1298–1308.

Murakami, T., Yamashita, T., Iida, M., Kuwajima, M., and Shima, K. (1997). A short form of leptin receptor performs signal transduction. *Biochem. Biophys. Res. Commun.* **231**(1), 26–29.

Nagle, D. L., McGrail, S. H., Vitale, J., Woolf, E. A., Dussault, B. J., Jr., DiRocco, L., Holmgren, L., Montagno, J., Bork, P., Huszar, D., Fairchild-Huntress, V., Ge, P., Keilty, J., Ebeling, C., Baldini, L., Gilchrist, J., Burn, P., Carlson, G. A., and Moore, K. J. (1999). The mahogany protein is a receptor involved in suppression of obesity. *Nature* **398**(6723), 148–152.

Nakashima, K., Narazaki, M., and Taga, T. (1997). Leptin receptor (OB-R) oligomerizes with itself but not with its closely related cytokine signal transducer gp130. *FEBS Lett.* **403**(1), 79–82.

Nijenhuis, W. A., Oosterom, J., and Adan, R. A. (2001). AgRP(83-132) acts as an inverse agonist on the human-melanocortin-4 receptor. *Mol. Endocrinol.* **15**(1), 164–171.

Niki, T., Mori, H., Tamori, Y., Kishimoto-Hashirmoto, M., Ueno, H., Araki, S., Masugi, J., Sawant, N., Majithia, H. R., Rais, N., et al. (1996). Human obese gene: Molecular screening in Japanese and Asian Indian NIDDM patients associated with obesity. *Diabetes* **45**(5), 675–678.

Nillni, E. A., Vaslet, C., Harris, M., Hollenberg, A., Bjorbak, C., and Flier, J. S. (2000). Leptin regulates prothyrotropin-releasing hormone biosynthesis. Evidence for direct and indirect pathways. *J. Biol. Chem.* **275**(46), 36124–36133.

Niswender, K. D., Norton, G. J., Stearns, W. H., Rhodes, C. J., Myers, M. G., Jr., and Schwartz, M. W. (2001). Intracellular signaling. Key enzymes in leptin-induced anorexia. *Nature* **413**, 794–795.

Ollmann, M. M., Wilson, B. D., Yang, Y. K., Kerns, J. A., Chen, Y., Gantz, I., and Barsh, G. S. (1997). Antagonism of central melanocortin receptors in vitro and in vivo by agouti-related protein. *Science* **278**(5335), 135–138. [Published erratum appears in *Science* **281**(5383), 1615.]

Ollmann, M. M., Lamoreux, M. L., Wilson, B. D., and Barsh, G. S. (1998). Interaction of Agouti protein with the melanocortin 1 receptor in vitro and in vivo. *Genes Dev.* **12**(3), 316–330.

Pelleymounter, M. A., Cullen, M. J., Baker, M. B., Hecht, R., Winters, D., Boone, T., and Collins, F. (1995). Effects of the obese gene product on body weight regulation in ob/ob mice. *Science* **269**(5223), 540–543.

Perry, W. L., Hustad, C. M., Swing, D. A., Jenkins, N. A., and Copeland, N. G. (1995). A transgenic mouse assay for agouti protein activity. *Genetics* **140**(1), 267–274.

Powis, J. E., Bains, J. S., and Ferguson, A. V. (1998). Leptin depolarizes rat hypothalamic paraventricular nucleus neurons. *Am. J. Physiol.* **274**(5 Pt. 2), R1468–R1472.

Qian, S., Chen, H., Weingarth, D., Trumbauer, M. E., Novi, D. E., Guan, X., Yu, H., Shen, Z., Feng, Y., Frazier, E., Chen, A., Camacho, R. E., Shearman, L. P., Gopal-Truter, S., MacNeil, D. J., Van der Ploeg, L. H., and Marsh, D. J. (2002). Neither agouti-related protein nor neuropeptide Y is critically required for the regulation of energy homeostasis in mice. *Mol. Cell. Biol.* **22**(14), 5027–5035.

Rau, H., Reaves, B. J., O'Rahilly, S., and Whitehead, J. P. (1999). Truncated human leptin (delta133) associated with extreme obesity undergoes proteasomal degradation after defective intracellular transport. *Endocrinology* **140**(4), 1718–1723.

Reizes, O., Lincecum, J., Wang, Z., Goldberger, O., Huang, L., Kaksonen, M., Ahima, R., Hinkes, M. T., Barsh, G. S., Rauvala, H., and Bernfield, M. (2001). Transgenic expression of syndecan-1 uncovers a physiological control of feeding behavior by syndecan-3. *Cell* **106**(1), 105–116.

Robbins, L. S., Nadeau, J. H., Johnson, K. R., Kelly, M. A., Roselli-Rehfuss, L., Baack, E., Mountjoy, K. G., and Cone, R. D. (1993). Pigmentation phenotypes of variant extension locus alleles result from point mutations that alter MSH receptor function. *Cell* **72**(6), 827–834.

Rosenblum, C. I., Tota, M., Cully, D., Smith, T., Collum, R., Qureshi, S., Hess, J. F., Phillips, M. S., Hey, P. J., Vongs, A., Fong, T. M., Xu, L., Chen, H. Y., Smith, R. G., Schindler, C., and Van der Ploeg, L. H. (1996). Functional STAT 1 and 3 signaling by the leptin receptor (OB-R): reduced expression of the rat fatty leptin receptor in transfected cells. *Endocrinology* **137**(11), 5178–5181.

Rossi, M., Kim, M. S., Morgan, D. G., Small, C. J., Edwards, C. M., Sunter, D., Abusnana, S., Goldstone, A. P., Russell, S. H., Stanley, S. A., Smith, D. M., Yagaloff, K., Ghatei, M. A., and Bloom, S. R. (1998). A C-terminal fragment of Agouti-related protein increases feeding and antagonizes the effect of alpha-melanocyte stimulating hormone in vivo. *Endocrinology* **139**(10), 4428–4431.

Rotimi, C. N., Comuzzie, A. G., Lowe, W. L., Luke, A., Blangero, J., and Cooper, R. S. (1999). The quantitative trait locus on chromosome 2 for serum leptin levels is confirmed in African-Americans. *Diabetes* **48**(3), 643–644.

Satoh, N., Ogawa, Y., Katsuura, G., Numata, Y., Masuzaki, H., Yoshimasa, Y., and Nakao, K. (1998). Satiety effect and sympathetic activation of leptin are mediated by hypothalamic melanocortin system. *Neurosci. Lett.* **249**(2–3), 107–110.

Savioz, A., Charnay, Y., Huguenin, C., Graviou, C., Greggio, B., and Bouras, C. (1997). Expression of leptin receptor mRNA (long form splice variant) in the human cerebellum. *Neuroreport* **8**(14), 3123–3126.

Schwartz, M. W., Seeley, R. J., Campfield, L. A., Burn, P., and Baskin, D. G. (1996). Identification of targets of leptin action in rat hypothalamus. *J. Clin. Invest.* **98**(5), 1101–1106.

Schwartz, M. W., Seeley, R. J., Woods, S. C., Weigle, D. S., Campfield, L. A., Burn, P., and Baskin, D. G. (1997). Leptin increases hypothalamic pro-opiomelanocortin mRNA expression in the rostral arcuate nucleus. *Diabetes* **46**(12), 2119–2123.

Shimizu, H., Shargill, N. S., Bray, G. A., Yen, T. T., and Gesellchen, P. D. (1989). Effects of MSH on food intake, body weight and coat color of the yellow obese mouse. *Life Sci.* **45**(6), 543–552.

Shutter, J. R., Graham, M., Kinsey, A. C., Scully, S., Luthy, R., and Stark, K. L. (1997). Hypothalamic expression of ART, a novel gene related to agouti, is up-regulated in obese and diabetic mutant mice. *Genes Dev.* **11**(5), 593–602.

Sinha, M. K., Opentanova, I., Ohannesian, J. P., Kolaczynski, J. W., Heiman, M. L., Hale, J., Becker, G. W., Bowsher, R. R., Stephens, T. W., and Caro, J. F. (1996). Evidence of free and bound leptin in human circulation. Studies in lean and obese subjects and during short-term fasting. *J. Clin. Invest.* **98**(6), 1277–1282.

Smith, A. I., and Funder, J. W. (1988). Proopiomelanocortin processing in the pituitary, central nervous system, and peripheral tissues. *Endocr. Rev.* **9**(1), 159–179.

Spanswick, D., Smith, M. A., Groppi, V. E., Logan, S. D., and Ashford, M. L. (1997). Leptin inhibits hypothalamic neurons by activation of ATP-sensitive potassium channels. *Nature* **390**(6659), 521–525.

Ste Marie, L., Miura, G. I., Marsh, D. J., Yagaloff, K., and Palmiter, R. D. (2000). A metabolic defect promotes obesity in mice lacking melanocortin-4 receptors. *Proc. Natl. Acad. Sci. USA* **97**(22), 12339–12344.

Stephens, T. W., Basinski, M., Bristow, P. K., Bue-Valleskey, J. M., Burgett, S. G., Craft, L., Hale, J., Hoffmann, J., Hsiung, H. M., Kriauciunas, A., *et al.* (1995). The role of neuropeptide Y in the antiobesity action of the obese gene product. *Nature* **377**(6549), 530–532.

Strobel, A., Issad, T., Camoin, L., Ozata, M., and Strosberg, A. D. (1998). A leptin missense mutation associated with hypogonadism and morbid obesity. *Nat. Genet.* **18**(3), 213–215.

Takahashi, Y., Okimura, Y., Mizuno, I., Iida, K., Takahashi, T., Kaji, H., Abe, H., and Chihara, K. (1997). Leptin induces mitogen-activated protein kinase-dependent proliferation of C3H10T1/2 cells. *J. Biol. Chem.* **272**(20), 12897–12900.

Tartaglia, L. A., Dembski, M., Weng, X., Deng, N., Culpepper, J., Devos, R., Richards, G. J., Campfield, L. A., Clark, F. T., Deeds, J., *et al.* (1995). Identification and expression cloning of a leptin receptor, OB-R. *Cell* **83**(7), 1263–1271.

Thornton, J. E., Cheung, C. C., Clifton, D. K., and Steiner, R. A. (1997). Regulation of hypothalamic proopiomelanocortin mRNA by leptin in ob/ob mice. *Endocrinology* **138**(11), 5063–5066.

Tota, M. R., Smith, T. S., Mao, C., MacNeil, T., Mosley, R. T., Van der Ploeg, L. H., and Fong, T. M. (1999). Molecular interaction of Agouti protein and Agouti-related protein with human melanocortin receptors. *Biochemistry* **38**(3), 897–904.

Tsujii, S., and Bray, G. A. (1989). Acetylation alters the feeding response to MSH and beta-endorphin. *Brain Res. Bull.* **23**(3), 165–169.

Vaisse, C., Halaas, J. L., Horvath, C. M., Darnell, J. E., Jr., Stoffel, M., and Friedman, J. M. (1996). Leptin activation of Stat3 in the hypothalamus of wild-type and ob/ob mice but not db/db mice. *Nat. Genet.* **14**(1), 95–97.

Vaisse, C., Clement, K., Guy-Grand, B., and Froguel, P. (1998). A frameshift mutation in human MC4R is associated with a dominant form of obesity [letter]. *Nat. Genet.* **20**(2), 113–114.

Vaisse, C., Clement, K., Durand, E., Hercberg, S., Guy-Grand, B., and Froguel, P. (2000). Melanocortin-4 receptor mutations are a frequent and heterogeneous cause of morbid obesity. *J. Clin. Invest.* **106**(2), 253–262.

Valverde, P., Healy, E., Jackson, I., Rees, J. L., and Thody, A. J. (1995). Variants of the melanocyte-stimulating hormone receptor gene are associated with red hair and fair skin in humans. *Nat. Genet.* **11**(3), 328–330.

Wang, M. Y., Zhou, Y. T., Newgard, C. B., and Unger, R. H. (1996). A novel leptin receptor isoform in rat. *FEBS Lett.* **392**(2), 87–90.

Wardlaw, S. L. (2001). Obesity as a neuroendocrine disease: Lessons to be learned from proopiomelanocortin and melanocortin receptor mutations in mice and men. *J. Clin. Endocrinol. Metab.* **86**(4), 1442–1446.

White, D. W., and Tartaglia, L. A. (1999). Evidence for ligand-independent homo-oligomerization of leptin receptor (OB-R) isoforms: A proposed mechanism permitting productive long-form signaling in the presence of excess short-form expression. *J. Cell. Biochem.* **73**(2), 278–288.

White, D. W., Kuropatwinski, K. K., Devos, R., Baumann, H., and Tartaglia, L. A. (1997). Leptin receptor (OB-R) signaling. Cytoplasmic domain mutational analysis and evidence for receptor homo-oligomerization. *J. Biol. Chem.* **272**(7), 4065–4071. [Published erratum appears in *J. Biol. Chem.* **272**(18), 12248].

Wilson, B. D., Bagnol, D., Kaelin, C. B., Ollmann, M. M., Gantz, I., Watson, S. J., and Barsh, G. S. (1999). Physiological and anatomical circuitry between Agouti-related protein and leptin signaling. *Endocrinology* **140**(5), 2387–2397.

Woods, A. J., and Stock, M. J. (1996). Leptin activation in hypothalamus [letter]. *Nature* **381**(6585), 745.

Woods, S. C., Figlewicz, D. P., Madden, L., Porte, D., Jr., Sipols, A. J., and Seeley, R. J. (1998). NPY and food intake: Discrepancies in the model. *Regul. Pept.* **75–76**, 403–408.

Yang, Y. K., Dickinson, C. J., Zeng, Q., Li, J. Y., Thompson, D. A., and Gantz, I. (1999). Contribution of melanocortin receptor exoloops to Agouti-related protein binding. *J. Biol. Chem.* **274**(20), 14100–14106.

Yang, Y. K., Fong, T. M., Dickinson, C. J., Mao, C., Li, J. Y., Tota, M. R., Mosley, R., Van Der Ploeg, L. H., and Gantz, I. (2000). Molecular determinants of ligand binding to the human melanocortin-4 receptor. *Biochemistry* **39**(48), 14900–14911.

Yaswen, L., Diehl, N., Brennan, M. B., and Hochgeschwender, U. (1999). Obesity in the mouse model of pro-opiomelanocortin deficiency responds to peripheral melanocortin. *Nat. Med.* **5**(9), 1066–1070.

Yen, T. T., Gill, A. M., Frigeri, L. G., Barsh, G. S., and Wolff, G. L. (1994). Obesity, diabetes, and neoplasia in yellow A(vy)/- mice: Ectopic expression of the agouti gene. *FASEB J.* **8**(8), 479–488.

Yeo, G. S., Farooqi, I. S., Aminian, S., Halsall, D. J., Stanhope, R. G., and O'Rahilly, S. (1998). A frameshift mutation in MC4R associated with dominantly inherited human obesity [letter]. *Nat. Genet.* **20**(2), 111–112.

Zabolotny, J. M., Bence-Hanulec, K. K., Stricker-Krongrad, A., Haj, F., Wang, Y., Minokoshi, Y., Kim, Y. B., Elmquist, J. K., Tartaglia, L. A., Kahn, B. B., and Neel, B. G. (2002). PTP1B regulates leptin signal transduction *in vivo*. *Dev. Cell* **2**(4), 489–495.

Zhang, F., Basinski, M. B., Beals, J. M., Briggs, S. L., Churgay, L. M., Clawson, D. K., DiMarchi, R. D., Furman, T. C., Hale, J. E., Hsiung, H. M., Schoner, B. E., Smith, D. P., Zhang, X. Y., Wery, J. P., and Schevitz, R. W. (1997). Crystal structure of the obese protein leptin-E100. *Nature* **387**(6629), 206–209.

Zhang, Y., Proenca, R., Maffei, M., Barone, M., Leopold, L., and Friedman, J. M. (1994). Positional cloning of the mouse obese gene and its human homologue [see comments]. *Nature* **372**(6505), 425–432. [Published erratum appears in *Nature* (1995). **374**(6521), 479.]

Function and Regulation of Cytosolic Molecular Chaperone CCT

Hiroshi Kubota

Department of Molecular and Cellular Biology and CREST/JST, Institute for Frontier Medical Sciences, Kyoto University, Kyoto 606-8397, Japan

    I. Introduction
    II. Function of CCT
    III. Structure of the CCT Complex
    IV. Evolution of CCT and Its Eight Different Subunits
    V. Expression of CCT and Its Subunits
    VI. Possible Roles of the Different Subunits
    VII. Conclusions
        References

Molecular chaperones are a group of proteins that assists in the folding of newly synthesized proteins or in the refolding of denatured proteins. The cytosolic chaperonin-containing t-complex polypeptide 1 (CCT) is a molecular chaperone that plays an important role in the folding of proteins in the eukaryotic cytosol. Actin, tubulin, and several other proteins are known to be folded by CCT, and an estimated 15% of newly translated proteins in mammalian cells are folded with the assistance of CCT. CCT differs from other chaperonin family proteins in its subunit composition, which consists of eight subunit species comprising the CCT 16-mer double-ring-like complex.

CCT preferentially recognizes quasinative (or partially folded) intermediates, whereas its *Escherichia coli* homologue GroEL recognizes more unfolded intermediates, especially those displaying hydrophobic surfaces. Molecular evolutionary analyses have suggested that each subunit species has a specific function in addition to contributing to a common ATPase activity. Consistent with this view, it has been suggested that each subunit recognizes specific substrate proteins (or their parts) and that they collectively modulate the ATPase activity of the complex. The overall expression of CCT in mammalian cells is primarily dependent on cell growth, but each subunit exhibits an individual pattern of expression. Recent progress in CCT research is reviewed, focusing particularly on CCT function and expression. From these observations, the possible roles of the distinct subunits in CCT-assisted folding in the eukaryotic cytosol are discussed. © 2002, Elsevier Science (USA).

I. INTRODUCTION

Many newly synthesized cellular proteins require assistance from molecular chaperones to adopt a functional conformation (Hartl, 1996; Bukau and Horwich, 1998). Proteins denatured by heat or other stress also require help from molecular chaperones to recover their native conformations. Inhibition of the activities of molecular chaperones can cause severe damage to cells, because they lack functional proteins and accumulate aggregates of unfolded proteins.

The chaperonin-containing *t*-complex polypeptide 1 (CCT), also called TRiC, is a molecular chaperone essential for cell growth, and CCT assists in the folding of proteins in the eukaryotic cytosol (Willison and Kubota, 1994; Kubota *et al.*, 1995a; Lewis *et al.*, 1996; Stoldt *et al.*, 1996; Willison and Horwich, 1996; Cowan, 1998; Willison, 1999; Leroux and Hartl, 2000; Carrascosa *et al.*, 2001; Willison and Grantham, 2001). As the name indicates, CCT contains the subunit *t*-complex polypeptide 1, which was first found as a protein abundantly expressed in testicular germ cells and early stage preimplantation embryos (Sanchez and Erickson, 1985; Willison *et al.*, 1986; Silver *et al.*, 1979, 1987). As yeast mutants lacking any of the CCT subunits are not viable (Ursic and Culbertson, 1991; Chen *et al.*, 1994; Li *et al.*, 1994; Miklos *et al.*, 1994; Vinh and Drubin, 1994), CCT is considered to be an essential protein in eukaryotes. Since it was discovered that the heterooligomeric complex CCT is the eukaryotic homologue of the *Escherichia coli* chaperonin GroEL (Gupta, 1990; Lewis *et al.*, 1992; Rommelaere *et al.*, 1993) and it assists in the folding of actin and tubulin (Frydman *et al.*, 1992; Gao *et al.*, 1992; Yaffe *et al.*, 1992), the functions of the CCT complex have been investigated at biochemical, molecular biological, and structural biological levels. It is now evident that CCT plays essential roles in the folding of newly synthesized actin and tubulin (Tian *et al.*, 1995; Frydman and Hartl, 1996; Farr *et al.*, 1997). CCT appears to be also important for the folding of many other proteins, including cyclin E (Won *et al.*, 1998) and the VHL tumor suppressor (Feldman *et al.*, 1999). CCT uses energy produced by adenosine triphosphate (ATP) hydrolysis to facilitate the correct folding of target proteins in a manner similar to that of GroEL, but with significant differences (Melki and Cowan, 1994; Tian et al., 1995; Farr et al., 1997). Immunoprecipitation experiments suggests that approximately 15% of newly synthesized proteins require assistance from CCT for folding (Thulasiraman et al., 1999). Thus, the range of unfolded proteins recognized by CCT may be more narrow and specialized than that recognized by the E. coli chaperonin GroEL, because GroEL exhibits affinity for a wide range of unfolded proteins [30–50% of total proteins (Viitanen et al., 1992; Horwich et al., 1993)] by recognizing exposed hydrophobic surfaces whereas CCT preferentially recognizes partially folded (quasinative) intermediates (Melki and Cowan, 1994; Tian et al., 1995).

The structure of CCT resembles a double doughnut with 16 subunits (Marco et al., 1994; Llorca et al., 1999b), and consists of eight different subunit species (Rommelaere et al., 1993; Kubota et al., 1994). This heterooligomeric composition is distinct from other chaperonins: bacterial GroEL and mitochondrial heat shock protein 60 (Hsp60) are homooligomers that contain 14 subunits, and the plastid Rubisco subunit-binding protein and many archeal chaperonins are composed of two subunit species. Judging from the fact that most molecular chaperones consist of one or two subunit species, CCT can be considered one of the most complex chaperones. The eight distinct subunits of CCT share approximately 30% amino acid sequence identity with each other (Kubota et al., 1995a), and the expression patterns of each of these subunits are similar but not identical (Kubota et al., 1999b; Yokota et al., 2000b). Although phylogenetic analysis of the eight subunits suggests that each has a specific function other than contributing to the common ATPase activity (Kim et al., 1994; Kubota et al., 1994), the role of each subunit in CCT-assisted folding has not been sufficiently clarified.

Here, I review recent progress in CCT research, with particular emphasis on the function and expression of CCT and its eight different subunits. From these observations, I discuss possible roles of the different subunits in CCT-assisted protein folding.

II. FUNCTION OF CCT

It is now widely accepted that many newly synthesized polypeptides require assistance from molecular chaperones to adopt a correctly folded conformation (Hartl, 1996; Bukau and Horwich, 1998). One of the most important functions of chaperones is to inhibit the formation of aggregates by trapping the exposed hydrophobic surfaces of misfolded proteins. Another is to stimulate the correct folding of target proteins by partially denaturing misfolded proteins. Thus, most molecular chaperones bind unfolded proteins but do not bind correctly folded proteins.

CCT assists in folding newly synthesized polypeptides in the eukaryotic cytosol. The chaperone activity of CCT was discovered using actin and tubulin as substrates about 10 years ago (Frydman et al., 1992; Gao et al., 1992; Yaffe et al., 1992), and these two proteins have been confirmed as important targets of CCT-assisted folding. However, accumulating evidence indicates that CCT is also important for folding a number of proteins unrelated to actin and tubulin (Table I). First, five other proteins—myosin (Srikakulam and Winkelmann, 1999), cyclin E (Won et al., 1998), the VHL tumor suppresser (Feldman et al., 1999), transducin (Farr et al., 1997), and luciferase (Frydman et al., 1994; Gebauer et al., 1998)—have been shown to be folded by CCT. Second, the CCT complex can bind in vitro at least nine other unfolded/disassembled proteins: the Mason–Pfizer monkey virus (MPMV) Gag protein (Hong et al., 2001), hepatitis B virus (HBV) capsid protein (Lingappa et al., 1994), 22-kDa peroxisome membrane protein (Pmp22p) (Pause et al., 1997), $p21^{ras}$, cyclin B, cap-binding protein, cofilin/actin-depolymerizing factor (ADF) (Melki et al., 1997), enolase (McCallum et al., 2000), and prepro-α factor (Plath and Rapoport, 2000). In addition, the Epstein–Barr virus-encoded nuclear protein EBNA-3 (Kashuba et al., 1999) and mouse testicular haploid expressed gene product THEG (Yanaka et al., 2000) were found to interact with the CCT ε-subunit in a yeast two-hybrid system; binding has been confirmed by in vitro analysis. It is also known that the MPMV Gag protein can be immunoprecipitated

TABLE I Proteins Folded and/or Bound by CCT[a]

Target protein	CCT-assisted folding	Binding subunit	Experiment
Actin (and actin-RVP)	+	WC, δ	CL, EM, IP, GF, NE, SG, YM
Tubulin (α, β, and γ)	+	WC	CL, IP, GF, NE, SG
Myosin	+	WC	GF, IP, PS
Cyclin E	+	WC	GF, IP, YM
VHL tumor suppressor	+	WC	GF, IP, PS
Transducin	+	WC	GF, IP, PS
Luciferase	+	WC	CL
MPMV Gag	ND	WC, γ	IP, GF, YT
HBV capsid protein	ND	WC	IP, SG
Pmp22p	ND	WC	IP, GF, CL
Phosducin-like protein	ND	WC	IP, NE
$p21^{ras}$	ND	WC	NE
Cyclin B	ND	WC	NE
Cap-binding protein	ND	WC	NE
Cofilin/ADF	ND	WC	NE
Enolase	ND	WC	CL
Prepro-α factor	ND	WC	IP, CL
EBV EBNA-3	ND	ε	GP, YT
THEG	ND	ε	IP, YT

[a] ADF, actin-depolymerizing factor; THEG, testicular haploid expressed gene; ND, not determined; WC, whole complex (subunit species not specified); CL, crosslinking; EM, electron microscopy; GP, GST-fusion protein pull down assay; IP, immunoprecipitation; GF, gel filtration column chromatography; NE, native gel electrophoresis; PS, protease sensitivity; SG, sucrose gradient centrifugation; YM, yeast mutant analysis; YT, yeast two-hybrid analysis.

with the CCT γ-subunit (Hong et al., 2001). Third, Thulasiraman et al. (1999) reported that more than 70 different newly translated polypeptides were distinguished by two-dimensional gel electrophoresis of CCT-bound polypeptides that have been immunoprecipitated with anti-CCT antibody. They estimated that proteins folded by the CCT-assisted pathway constitute approximately 15% of all newly synthesized proteins, and these proteins tended to fall in the 30–60 kDa range, suggesting that CCT may preferentially recognize proteins of this size. Apart from CCT substrates, phosducin-like protein in its native state was found to bind CCT very recently (McLaughlin et al., 2002).

On the other hand, some proteins have been reported to be unable to bind CCT even when unfolded. Melki et al. (1997) found that α- and β-globin and cofactor A do not form a binary complex with CCT, judged by native gel electrophoresis analysis. In conclusion, these observations indicate that a broad range of proteins, but clearly not all, is recognized by CCT. Elucidation of the characters common to proteins recognized by CCT would be important for understanding its role in cellular protein folding.

The regions of actin and tubulin recognized by CCT have been investigated using deletion mutants and synthetic polypeptides (Rommelaere et al., 1999; Dobrzynski et al., 1996, 2000; Hynes and Willison, 2000; Ritco-Vonsovici and Willison, 2000). The sequences recognized by CCT are dispersed throughout the primary structures, although several regions bind CCT more strongly than others. These studies suggested that hydrophobic residues play a role in binding CCT. However, the hydrophobic interaction between CCT and its substrates may be weaker than those between GroEL and its target proteins. CCT is considered to have a high affinity for late-forming quasinative folding intermediates, whereas GroEL can recognize early-forming intermediates with exposed hydrophobic surfaces. Because GroEL and Hsp60 cannot assist in correctly folding actin or tubulin, the distinct affinity for specific folding intermediates may play an important role in CCT function. Consistent with this view, CCT and GroEL show distinct affinities for different regions of actin (Rommelaere et al., 1999). On the other hand, electron microscopic observation of CCT–substrate complexes indicate that individual subunits may specifically recognize target proteins and their parts: actin was found to bind δ and another subunit (β or ε) (Llorca et al., 1999a), and a subunit-specific interaction with tubulin was also proposed very recently (Llorca et al., 2000). However, sequences or secondary structures specifically recognized by each subunit species have not yet been determined. The nature of the interactions between CCT and its substrate proteins needs to be investigated in further detail.

CCT uses energy produced from ATP hydrolysis to assist in the folding of target proteins (Frydman et al., 1992; Gao et al., 1992), as does the E. coli chaperonin GroEL (Fenton and Horwich, 1997). CCT binds unfolded substrates in the absence of ATP and releases them when they have undergone conformational changes that are driven by ATP binding and hydrolysis. CCT preferentially binds partially folded intermediates of substrate proteins (Melki and Cowan, 1994; Tian et al., 1995) probably by recognizing hydrophilic residues using hydrophilic surface of apical domains (Pappenberger *et al.*, 2002). However, the exact mechanism of CCT-assisted folding as mediated by ATP binding and hydrolysis is still controversial. Based on studies using a chaperonin trap that binds but does not release substrates, Farr *et al.* (1997) reported that multiple rounds of binding and release of substrates by CCT are important for folding. In contrast, Llorca *et al.* (2001) created three-dimensional reconstructions of electron microscopic images of CCT molecules bound to substrate proteins in the presence of AMP-PNP (a nonhydrolyzable ATP analogue) and reported that substrate proteins are enclosed in the CCT cavity. However, Szpikowska *et al.* (1998) previously reported that CCT is more resistant to the tryptic cleavage at the apical domain in the presence of ATP than in the presence of AMP-PNP or the absence of nucleotide, and that the AMP-PNP bound form and the nucleotide-free form exhibit similar susceptibility to the cleavage. This appears to conflict with the model of Llorca *et al.* (2001), and suggests that effects on CCT structure may not be identical between ATP and AMP-PNP. More information is required for resolving the discrepancy and for determining the exact mechanism of CCT-assisted folding.

Most CCT subunit proteins obtained from mammalian cells can be detected as large complexes of approximately 900 kDa by size fractionation techniques (Yokota *et al.*, 2000b, 2001b), although a minor proportion of CCT subunits can be found as monomers or small oligomers (Liou and Willison, 1997; Roobol and Carden, 1999; Roobol *et al.*, 1999a,b). Dynamic exchange of subunits between the large complex and monomers (Liou *et al.*, 1998) may occur *in vivo* and this may allow a damaged subunit to be replaced with a newly synthesized subunit for the regeneration of a functional CCT complex. Although it has been suggested that the monomer form of particular subunits may have special functions (Roobol *et al.*, 1999b), this requires further confirmation.

HSC70 (a cognate of HSP70) and HSP40 (a DnaJ homologue) are known to cooperate with CCT to assist in the folding of luciferase *in vitro* (Frydman *et al.*, 1994; Frydman and Hartl, 1996; Gebauer *et al.*, 1998). BAG-1 indirectly inhibits the CCT-dependent folding of luciferase by interacting with HSC70 (Gebauer *et al.*, 1998). Hop/p60 directly interacts with CCT and can stimulate the exchange of CCT-bound adenosine diphosphate (ADP) by ATP (Gebauer *et al.*, 1998). Recently, a heterooligomeric complex called prefoldin or GimC, which is composed of six subunit species of 14–23 kDa each, was shown to be able to transfer unfolded actin to CCT *in vitro* (Vainberg *et al.*, 1998; Hansen *et al.*, 1999). Mutants bearing deletions of the genes encoding prefoldin/GimC subunits are viable but have defects in microtubule formation (Geissler *et al.*, 1998). These results indicate that actin and tubulin are the targets commonly recognized by CCT and prefoldin/GimC, and that these chaperones cooperate to fold actin and tubulin. However, the lethality conferred by the deletion of genes encoding CCT subunits (Ursic and Culbertson, 1991; Chen *et al.*, 1994; Li *et al.*, 1994; Miklos *et al.*, 1994; Vinh and Drubin, 1994) indicates that in yeast, protein folding is more dependent on CCT than prefoldin/GimC. Thus, the respective roles of CCT and prefoldin/GimC in the folding of newly synthesized proteins may be partly distinct. As it has been reported that CCT can bind ribosome-bound nascent polypeptide chains *in vitro* (Frydman *et al.*,

1994; Frydman and Hartl, 1996; McCallum *et al.,* 2000) and that the newly synthesized proteins associated with CCT *in vivo* are sequestered in a protected environment (Siegers *et al.,* 1999; Thulasiraman *et al.,* 1999), it is possible that these molecular chaperones and cochaperones cooperate with CCT in a restricted portion of the cytosol in which the translation and folding machinery is concentrated.

III. STRUCTURE OF THE CCT COMPLEX

CCT has a cylindrical structure similar to that of GroEL and other chaperonins, consisting of two stacked rings of assembled subunits (Fig. 1; see color insert.) (Marco *et al.,* 1994; Llorca *et al.,* 1998, 1999a,b, 2000, 2001). Each of the two rings of the CCT complex are considered to contain eight different subunits (Table II) (Rommelaere *et al.,* 1993; Kubota *et al.,* 1994, 1995b; Liou and Willison, 1997), and therefore 16 subunits comprise a CCT molecule. An order for the eight different subunits in a single ring proposed by Liou and Willison (1997) is as follows: $\alpha-\varepsilon-\zeta-\beta-\gamma-\vartheta-\delta-\eta-(\alpha)$. In the absence of ATP, the two CCT rings are equivalent in size and shape (Llorca *et al.,* 1999b). In the presence of ATP, one ring becomes more closed and the other become more open at the opening of the cylinder (Llorca *et al.,* 1999b), although CCT exhibits symmetrical structure with closed entrance in the presence of AMP-PNP (Llorca *et al.,* 2001). The asymmetrical change in ring shape might take place when CCT-bound ATP is hydrolyzed. The asymmetrical double-ring structure is similar to that of GroEL and is suggestive of interring communication. Substrate proteins bind the apical domain located at the entrance to the cavity of the cylinder (Llorca *et al.,* 1999a, 2000). ATP induces a structural change in the apical domain, causing it to rotate toward the axis of the cylinder (Llorca *et al.,* 1998), and this may cause substrate proteins to be released. In contrast, AMP-PNP induced the entrance to close, trapping substrates inside

TABLE II Charcteristics of the CCT Subunit Species

Subunit	Number of amino acid	Molecular mass	Mobility on 2D gel		Amino acid identity to $CCT\alpha$ (%)
			Relative MW (kDa)	pI	
$CCT\alpha$	556	60,513	60	6.50	—
$CCT\beta$	535	57,456	53	6.65	35.4
$CCT\gamma$	545	60,636	65	6.85	32.5
$CCT\delta$	539	58,073	56	>7.20	32.1
$CCT\varepsilon$	541	59,631	61	6.35	31.9
$CCT\zeta - 1$	531	58,011	62	6.90	27.1
$CCT\zeta - 2^a$	531	58,191	57	7.10	26.3
$CCT\eta$	544	59,658	57	>7.20	34.6
$CCT\vartheta$	548	59,562	62	6.25	27.5

[a] $CCT\zeta-2$ is the testis-specific subtype of $CCT\zeta$.

the cavity (Llorca *et al.*, 2001). Although the structural characteristics of CCT are appreciably similar to those of GroEL, CCT differs from GroEL in several aspects, in addition to the effect of ATP. One distinctive feature of CCT is its heterooligomeric character and another is the number of subunits comprising the molecule (CCT is a 16-mer whereas GroEL is a 14-mer). Furthermore, GroES-like cochaperonin protein, which covers the chaperonin cylinder like a lid, does not have a counterpart in the eukaryotic CCT-assisted folding system. Autonomous closing of the entrance of the CCT cavity, as observed with AMP-PNP, may be able to replace the function of GroES (Llorca *et al.*, 2001).

IV. EVOLUTION OF CCT AND ITS EIGHT DIFFERENT SUBUNITS

CCT and other members of the chaperonin family (also called the HSP60 family) (Kubota *et al.*, 1995a; Gupta, 1996) can be divided into two groups, based on amino acid sequence identity and on structural similarity: GroEL, Hsp60, and Rubisco subunit binding protein belong to group I, whereas CCT and archeal chaperonins belong to group II (Willison and Kubota, 1994). The eight subunits of CCT share 30–40% amino acid sequence identity with archeal group II chaperonins and 15–20% identity with group I chaperonins. CCT subunits share approximately 30% amino acid sequence identity with each other (Table I). Phylogenetic analyses of amino acid sequences suggest that CCT and other chaperonins have evolved from a common ancestor (Gupta, 1990) and that gave rise to the group I and II chaperonin ancestors (Willison and Kubota, 1994). Hsp60, GroEL, and Rubisco subunit binding protein evolved from a group I ancestor, whereas archeal chaperonins and CCT evolved from a group II ancestor. The eight species of subunit are estimated to have diverged from a eukaryotic homooligomeric progenitor of CCT at nearly the same time approximately 2×10^9 years ago (Kubota *et al.*, 1994), although the order of evolution of the subunits can be estimated by careful phylogenetic analysis. The genes encoding the δ- and ε-subunits as well as the α-, β-, and η-subunits are suggested to be products of the most recent duplications (Archibald *et al.*, 2000), although functional and structural similarities of the gene products within these sets may have been lost over the course of evolution. Because the eight CCT subunits arose such a long time ago and have been maintained in all eukaryotes thus far investigated, they likely have essential roles in the survival of eukaryotic organisms. This is consistent with the fact that a null mutation of any of the genes encoding a CCT subunit confers lethality.

Although CCT subunits are approximately 30% identical to each other at the amino acid level, the greatest extent of identity concerns putative ATP-binding domains (equatorial domains) (Kim *et al.*, 1994; Kubota *et al.*, 1995a). The sequence of the apical domain, the region that binds substrate proteins, is less conserved among the eight subunits, but is highly conserved for a given subunit in a wide variety of eukaryotic organisms (Kim *et al.*, 1994; Archibald *et al.*, 2000). In the apical domains of each subunit species, more than several amino acid residues,

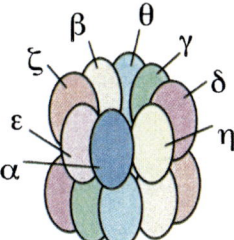

FIGURE 1   Structural model of CCT. The orientation of the eight subunits is as proposed by Liou and Willison (1997).

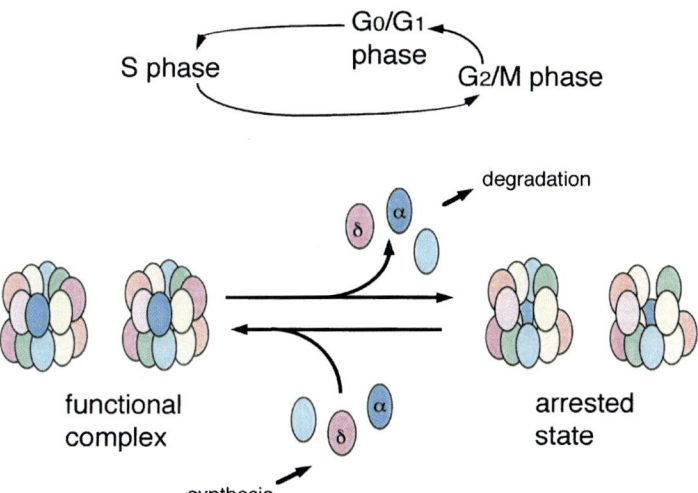

FIGURE 2   A model for the changes in CCT submit contents and function during the cell cycle. This model is based on the results of Yokota *et al.* (2001b).

including charged residues, are conserved from the mouse to amitochondriate parasitic eukaryotes such as *Giardia* or *Trichomonas* (Archibald et al., 2000). Thus, apical domains may impart specific functions to each subunit. These observations support the idea that specific substrates (or their parts) are recognized by the same CCT subunit in all eukaryotes.

V. EXPRESSION OF CCT AND ITS SUBUNITS

CCT is ubiquitously expressed in all eukaryotic organisms and in all mammalian cell types thus far investigated, consistent with the fact that CCT is essential for cellular survival. However, the levels of CCT expression differ appreciably among cell types. The first CCT subunit identified, *t*-complex polypeptide 1 (CCTα), was first found as a protein highly expressed in mouse testicular germ cells (Silver *et al.*, 1979) and preimplantation embryos (Sanchez and Erickson, 1985). The abundant expression of CCTα in testis germ cells is probably required for folding of an important substrate—tubulin, and in preimplantation embryos CCT may be required for the rapid protein synthesis in dividing blastomeres. The main factors inducing CCT expression are considered to be two-fold: CCT is up-regulated concomitant with the accelerated protein synthesis in rapidly proliferating cells, and it is further up-regulated in conjunction with the expression of particular substrates in specific cell types. One of the best examples of CCT expression coupled to the synthesis of specific substrate is provided by the testis. CCT expression is strongly up-regulated in testicular germ cells (Silver *et al.*, 1987; Kubota *et al.*, 1992, 1999b), in which tubulin is synthesized very rapidly to produce the main component of sperm tails. The expression of CCT in neuronal cells may also be related to tubulin synthesis (Roobol and Carden, 1993; Roobol *et al.*, 1995), as tubulin is important for neurite development. It is also known that the expression of CCT in a ciliate, *Tetrahymena pyriformis,* is coordinately up-regulated with the induction of tubulin synthesis during cilia formation (Soares *et al.*, 1994; Cyrne *et al.*, 1996), and that it is also altered by treatment with the microtuble-depolymerizing agent colchicine (Domingues *et al.*, 1999) and microtuble-polymerizing agent taxol (Casalou *et al.*, 2001). In addition, the high level expression of CCT in the cranial neural crest of *Xenopus* embryos (Dunn and Mercola, 1996) might be related to the synthesis of actin and tubulin, which are likely to be required for the motility of neural crest cells. Similarly, the abundantly expressed CCT in the brain, spinal cord, and somites of axolotl embryos (Sun *et al.*, 1995), which is consistent with the expression patterns of a reporter driven by the *CCT-1* gene (encoding α-subunit) promoter in the nematode *Caenorhabditis elegans* (Leroux and Candido, 1997), may play a role in the synthesis of actin, tubulin and myosin. In plants, the expression of CCT is also correlated with that of tubulin and actin (Nong *et al.*, 2002).

Recently, in an analysis of many types of cultured cells we found that the expression of each of the CCT subunits is highly dependent on the cellular growth rate (Yokota *et al.*, 1999). The abundant expression of CCT subunits in rapidly growing cells is reduced during growth arrest caused by starvation for growth factors or is associated with chemically induced differentiation. After the readdition of growth factors, the expression of CCT recovers rapidly in a cell cycle-dependent manner. The highest level of CCT expression is induced at the G_1/S transition through early S phase. Many of the CCT expression patterns reported previously can be explained by growth-dependent expression: CCT expression levels are high in embryos at early developmental stages (Sanchez and Erickson, 1985; Kubota et al., 1992; Sevigny et al., 1995) and in tissues containing rapidly proliferating cells (Kubota et al., 1999b). In contrast, CCT expression levels decrease during neuronal differentiation of mouse embryonal carcinoma P19 cells (Roobol and Carden, 1999). It is also known that the expression of CCT in the slime mold *Dictyostelium discoideum* rapidly decreases concomitant with growth arrest during asexual development (Iijima et al., 1998). Increased expression levels of CCT subunits in simian virus 40 (SV40)-transformed keratinocytes relative to normal keratinocytes (Hynes et al., 1996) is probably growth dependent. Furthermore, we found very recently that CCT expression is up-regulated in tumor tissues containing rapidly proliferating cancer cells, and that the levels of CCT subunits are highly correlated with the level of proliferating cell nuclear antigen (Yokota et al., 2001a).

Interestingly, CCT isolated from S phase-arrested cells is more active for substrate binding and folding when compared to CCT isolated from asynchronously dividing cells or M phase-arrested cells (Yokota et al., 2001b). In addition, CCT prepared from M phase-arrested cells has significantly lower activity than CCT prepared from asynchronous cells. Thus, CCT activity was high in S phase cells, moderate in asynchronous cells, and low in M phase cells. As asynchronous cells are at various phases of the cell cycle, CCT purified from asynchronous cells is likely to be a mixture of active (S phase type) and inactive (M phase type) CCT complexes. Differences in the activity of CCT can be explained by changes in particular subunit species. The levels of α- and δ-subunits were much lower in the M phase-arrested cells than in the S phase-arrested cells, and the levels of these subunits in asynchronous cells were intermediate between those of S phase-arrested cells and M phase-arrested cells (Yokota et al., 2001b). These results suggest that the α- and δ-subunits play critical roles in the substrate binding and folding activities of CCT. One of the mechanisms that influences the levels of the α- and δ-subunits seems to be their rapid turnover (synthesis and degradation) rates: the rates of synthesis of these two subunits change more extensively than those of the other subunits during cell cycle arrest, and they are significantly high in S phase-arrested cells (Table III).

The rates of degradation are also important in accounting for the overall level of each CCT subunit. The half-life of the CCT complex in mammalian cultured cells has been estimated to be 6–7.5 h, regardless of cellular growth status (Yokota et al., 1999). This is consistent with the fact that the level of CCT in cultured cells decreases quickly when its synthesis is blocked by growth factor starvation. Given that the half-life of CCT is 6 h and that CCT synthesis is completely stopped, the level of CCT in the cell decreases to 6% by 24 h and to 0.4% by 48 h. A proteasome-dependent system is involved in the breakdown of CCT because the degradation

TABLE III Turnover of CCT Subunits in Mammalian Cell during the Cell Cycle

Subunit	Synthesis rate (AS = 1)			Degradation rate ($t^{1/2}$, h)		
	SA	MA	AS	SA	MA	AS
CCTα	1.97	0.63	1.00	5.3	5.0	5.5
CCTβ	1.09	1.07	1.00	7.8	7.9	8.0
CCTγ	1.21	1.09	1.00	7.3	7.6	7.4
CCTδ	1.79	0.80	1.00	4.3	4.3	4.0
CCTε	0.92	0.88	1.00	5.3	5.3	5.3
CCTζ−1	1.12	0.65	1.00	4.7	5.3	5.0
CCTη	0.88	0.82	1.00	9.1	9.0	8.5
CCTϑ	2.50	1.24	1.00	7.2	5.7	6.9

SA, S phase-arrested cells; MA, M phase-arrested cells; AS, asynchronized cells. Data are taken from Yokota et al. (2001b).

rate of CCT is significantly reduced by treating cells with the proteasome-specific inhibitor lactacystin and by inactivation of the ubiquitin-activating enzyme E1 (Yokota et al., 2000a). Interestingly, the ε-subunit of CCT has been reported to interact with the 26 S proteasome (Tokumoto et al., 2000), suggesting a possibility of direct degradation of CCT by proteasome. The half-life of each of the subunit species varies from 4.0 to 9.1 h, where as the degradation rate of each subunit is not affected by cell growth (Yokota et al., 2001b) (Table III). Therefore, the relative frequency of each subunit changes when the overall synthesis of CCT subunits is inhibited.

Despite the fact that the expression of many members of the chaperonin family is induced by heat shock, accumulating evidence indicates that CCT is not heat inducible (Lewis et al., 1992; Ursic and Culbertson, 1992; Soares et al., 1994; Leroux and Candido, 1995; Palmedo and Ammermann, 1997; Campos and Hamdan, 2000). However, we recently found that chemical stress such as treatment with arsenite or proline analogue can up-regulate CCT expression (Yokota et al., 2000b). It is also reported that cadmium treatment can stimulate CCT expression in the hypotrichous soil ciliate *Oxytricha granullifera* (Palmedo and Ammermann, 1997). It is reasonable that CCT is induced by chemical stress, because the correct folding of newly synthesized proteins can be inhibited in the presence of chemical stressors. As all of the CCT genes contain heat shock element (HSE)-like *cis*-acting elements that are recognizable by heat shock transcription factors (HSFs) *in vitro*, and overexpression of these factors can stimulate CCT expression in assays with reporter genes (Kubota et al., 1999a), these *cis*-acting elements may play a role in the induction of CCT by chemical stress. It may also be possible that these HSE sequences play a role in CCT expression in testis because the transcription factor HSF2 is highly expressed in this tissue. CCT may play roles in cold shock response (Somer et al., 2002) and apoptosis (Zilkha-Falb et al., 2000) as upregulation during these responses have been observed.

The independent genes encoding the eight CCT subunits have been cloned from mouse genomic DNA libraries, and their nucleotide sequences have been determined (Kubota *et al.*, 1999b). All of them contain CpG-rich islands in the 5′-flanking region and first intron, and most of them contain possible Sp1 binding sites in these regions. The Sp1 sites may contribute to the constitutive expression of CCT. In addition, HSE sequences have been found in the CpG-rich regions, as described above. Reporter gene assays of all of the mouse CCT genes (Kubota *et al.*, 1999b) suggest that the transcriptional activities of the *Ccta, Cctd,* and *Cctq* genes (encoding α-, δ- and ϑ-subunits, respectively) are stronger than that of the combination of SV40 early promoter and enhancer in HeLa cells. In this assay, *Cctb* and *Cctg* (encoding β- and γ-subunits, respectively) have a transcriptional activity comparable to that of the SV40 promoter/enhancer. Thus, these five genes are considered to be transcribed strongly in rapidly growing cells. As reporter constructs containing the other CCT subunit genes were significantly less active for transcription, these genes may have enhancers outside the regions used for reporter construction. Only one of the mouse CCT genes, *Ccta*, has been investigated in detail for promoter/enhancer activity and *cis*-acting DNA elements. Two 18-bp *cis*-acting elements that strongly enhance the transcriptional activity of the *Ccta* gene have been found at around -70 and -20 bp with respect to the transcription initiation site, and these two elements have very similar nucleotide sequences (Kubota *et al.*, 2000). Both the elements are recognized by the zinc finger transcription factors ZNF143 and ZNF76, which are the mammalian homologues of *Xenopus* selenocystein tRNA gene transcription factor (Staf) (Schaub *et al.*, 1997). The Staf family of proteins is known to stimulate the transcription of small nuclear RNAs (snRNAs) such as U1 and U6, as well as selenocysteine tRNA. As snRNAs including U1 and U6 are important for the splicing of mRNAs, the chaperonin subunit gene *Ccta* may be coregulated with the tRNA and snRNA genes to promote the efficient production of proteins. A detailed transcriptional analysis of other *Cct* genes is necessary for the elucidation of the mechanisms that coordinate the expression of CCT subunit genes.

VI. POSSIBLE ROLES OF THE DIFFERENT SUBUNITS

One of the most important roles of the different subunits may be subunit-specific recognition of substrate proteins and their parts, as has been suggested by electron microscopic analyses (see Section III) and by molecular evolutionary analysis (see Section IV). This idea is consistent with the observation that CCT recognizes quasinative late-forming folding intermediates rather than early-forming intermediates with exposed hydrophobic surfaces, because hydrophilic interactions are considered to be more specific than are hydrophobic interactions. However, the specificity of each subunit for target polypeptides has not been determined and remains to be clarified.

Different CCT subunits may have distinct roles in modulating the subunit–subunit cooperativity in ATP-binding/hydrolysis activities and/or in the control of the hierarchy of subunits in these activities, as proposed by Lin et al. (1997). In mutational analysis of CCT subunit genes in budding yeast, Lin et al. (1997) found that mutations in the gene encoding the ζ-subunit, *CCT6*, are well tolerated by the cell, including those in putative ATPase domains, and the *CCT6* (L19S) mutation can suppress the growth defects conferred by mutations in *CCT1*, *CCT2*, and *CCT3* genes (encoding the α-, β-, and γ-subunits, respectively). The three mutations suppressed by the *CCT6* (L19S) allele all occurred in putative ATPase domains. These observations indicate that the function of each CCT subunit can be affected by the other subunits and suggest that each subunit may have distinct ATP binding/hydrolysis activities. It is possible that interactions among subunits may create allosteric effects with respect to ATPase function, which in turn may play a role in CCT-assisted protein folding.

The differential levels of expression of the CCT subunits may play a role in the regulation of CCT function *in vivo*. Cell cycle arrest causes more rapid changes in the levels of the α- and δ-subunits than in the levels of the other subunits (Table IV), leading to the production of a less functional CCT complex in M phase than in S phase (see Section V). Figure 2 (see color insert) shows a model for changes in the contents of particular subunit species concomitant with CCT function, proposed from our recent study (Yokota et al., 2001b). Assuming that the S phase CCT is the complete 16-mer complex, containing two each of the eight subunit species, the M phase CCT is considered to be an incomplete 15-mer or smaller complex containing reduced numbers of the α- and δ-subunits. The frequent turnover of these two subunits allows for a more rapid suppression of CCT activity than would

TABLE IV Expression Levels Different Between CCT Subunit Species[a]

Subunit species	Up-regulation by S phase arrest[a]	Down-regulation by M phase arrest[b]	Down-/up-regulation at growth factor starvation/readdition[c]	Down-regulation at neuronal differentiation[d]	Up-regulation by chemical stress[e]
CCTα	+++	+++	++	+	+++
CCTβ	+	+	+	+++	++
CCTγ	+	+	+	++	++
CCTδ	+++	+++	+++	++	++
CCTε	+	+	++	+++	+++
CCTζ–1	+	++	++	+	+++
CCTη	+	+	+	ND	++
CCTϑ	++	+	+	+++	+

[a] Maximam change in each experiment was set as +++.
[b] Yokota et al. (2001b).
[c] Yokota et al. (1999).
[d] Roobol and Carden (1999).
[e] Yokota et al. (2000b).

be provided by a complete turnover of the entire complex, at times when the CCT function is not required, such as at M phase. At the G_1/S transition, CCT activity is quickly recovered by the rapid synthesis of these subunits. Consistent with these observations, the level of the δ-subunit was also found to change most rapidly when growth factors are added to starved cultured cells (Yokota *et al.*, 1999) (Table IV). Different levels of expression of each subunit species were also observed during the neuronal differentiation of mouse P19 cells (Roobol and Carden, 1999). However, the levels of the β-, ε-, and θ-subunits decreased more rapidly than other subunits during differentiation in this case. When CCT levels rise in response to chemical stress, the α-, ε-, and ζ-subunits increase most rapidly (Yokota *et al.*, 2000b). In addition, differential levels of expression between the α- and ε-subunits have been found in a particular portion of developing maize seedlings (Himmelspach *et al.*, 1997). Thus, the levels of expression of each subunit species can change rapidly in response to different circumstances. The roles of differential expression remain to be further investigated.

VII. CONCLUSIONS

Here, I have reviewed recent progress in studies of CCT function and expression, with a main focus on the role of different subunits. Accumulating evidence suggests that subunit-specific functions contribute to the CCT-assisted folding of proteins. In addition, the expression of each of the subunits seems to be regulated differentially, although with a significant level of shared similarities. However, the mechanisms controlling subunit-specific functions and expression remain to be determined. Elucidation of these mechanisms will help in understanding how the most complex molecular chaperone works, and therefore these mechanisms represent one of the most important issues in CCT research.

ACKNOWLEDGMENTS

I thank Professors Kazuhiro Nagata and Takashi Yura for helpful discussions and Dr. Shin-ich Yokota for drawing the figures and for suggestions.

REFERENCES

Archibald, J. M., Logsdon, J. M., Jr., and Doolittle, W. F. (2000). Origin and evolution of eukaryotic chaperonins: Phylogenetic evidence for ancient duplications in CCT genes. *Mol. Biol. Evol.* **17,** 1456–1466.

Bukau, A., and Horwich, A. L. (1998). The Hsp70 and Hsp60 chaperone machines. *Cell* **92,** 351–366.

Campos, E. G., and Hamdan, F. F. (2000). Cloning of the chaperonin t-complex polypeptide 1 gene from Schistosoma mansoni and studies of its expression levels under heat shock and oxidative stress. *Parasitol. Res.* **86,** 253–258.

Carrascosa, J. L., Llorca, O., and Valpuesta, J. M. (2001). Structural comparison of prokaryotic and eukaryotic chaperonins. *Micron* **32,** 43–50.

Casalou, C., Cyrne, L., Rosa, M. R., and Soares, H. (2001). Microtubule cytoskeleton perturbation induced by taxol and colchicine affects chaperonin containining TCP-1 (CCT) subunit gene expression in Tetrahymena cells. *Biochim. Biophys. Acta* **1522,** 9–21.

Chen, X., Sullivan, D. S., and Huffaker, T. C. (1994). Two yeast genes with similarity to TCP-1 are required for microtubule and actin function in vivo. *Proc. Natl. Acad. Sci. USA* **91,** 9111–9115.

Cowan, N. J. (1998). Mammalian cytosolic chaperonin. *Methods Enzymol.* **290,** 230–241.

Cyrne, L., Guerreiro, P., Cardoso, A. C., Rodrigues-Pousada, C., and Soares, H. (1996). The *Tetrahymena* chaperonin subunit CCTη gene is coexpressed with CCTγ gene during cilia biogenesis and cell sexual reproduction. *FEBS Lett.* **383,** 277–283.

Dobrzynski, J. K., Sternlicht, M. L., Farr, G. W., and Sternlicht, H. (1996). Newly-synthesized β-tubulin demonstrates domain-specific interactions with the cytosolic chaperonin. *Biochemistry* **35,** 15870–15882.

Dobrzynski, J. K., Sternlicht, M. L., Peng, I., Farr, G. W., and Sternlicht, H. (2000). Evidence that beta-tubulin induces a conformation change in the cytosolic chaperonin which stabilizes binding: Implications for the mechanism of action. *Biochemistry* **39,** 3988–4003.

Domingues, S., Soares, H., Rodrigues-Pousada, C., and Cyrne, L. (1999). Structure of *Tetrahymena* CCTϑ and its expression under colchicine treatment. *Biochim. Biophys. Acta* **1446,** 443–449.

Dunn, M. K., and Mercola, M. (1996). Cloning and expression of Xenopus CCTγ, a chaperonin subunit developmentally regulated in neural-derived and myogenic lineages. *Dev. Dyn.* **205,** 387–394.

Farr, G. W., Scharl, E. C., Schumacher, R. J., Sondek, S., and Horwich, A. L. (1997). Chaperonin-mediated folding in the eukaryotic cytosol proceeds through rounds of release of native and nonnative forms. *Cell* **89,** 927–937.

Feldman, D. E., Thulasiraman, V., Ferreyra, R. G., and Frydman, J. (1999). Formation of the VHL-elongin BC tumor suppressor complex is mediated by the chaperonin TRiC. *Mol. Cell* **4,** 1051–1061.

Fenton, W. A., and Horwich, A. L. (1997). GroEL-mediated protein folding. *Protein Sci.* **1997,** 743–767.

Frydman, J., and Hartl, F. U. (1996). Principles of chaperone-assisted protein folding: Difference between in vitro and in vivo mechanisms. *Science* **272,** 1497–1502.

Frydman, J., Nimmesgern, E., Erdjument-Bromage, H., Wall, J. S., Tempst, P., and Hartl, F.-U. (1992). Function in protein folding of TRiC, a cytosolic ring complex containing TCP-1 and structurally related subunits. *EMBO J.* **11,** 4767–4778.

Frydman, J., Nimmesgern, E., Ohtuka, K., and Hartl, F.-U. (1994). Folding of nascent polypeptide chains in a high molecular mass assembly with molecular chaperones. *Nature* **370,** 111–117.

Gao, Y., Thomas, J. O., Chow, R. L., Lee, G.-H., and Cowan, N. J. (1992). A cytoplasmic chaperonin that catalyze β-actin folding. *Cell* **69,** 1043–1050.

Gebauer, M., Melki, R., and Gehring, U. (1998). The chaperone cofactor Hop/p60 interacts with the cytosolic chaperonin-containing TCP-1 and affects its nucleotide exchange and protein folding activities. *J. Biol. Chem.* **273,** 29475–29480.

Geissler, S., Siegers, K., and Schiebel, E. (1998). A novel protein complex promoting fomation of functional α- and γ-tubulin. *EMBO J.* **17,** 952–966.

Gupta, R. S. (1990). Sequence and structural homology between a mouse *t*-complex protein TCP-1 and the 'chaperonin' family of bacterial (GroEL, 60-65 kDa heat shock antigen) and eukaryotic proteins. *Biochem. Int.* **20,** 833–841.

Gupta, R. Y. (1996). Evolutionaly relationships of chaperonins. *In* "The Chaperonins" (R. J. Ellis, Ed.). pp. 27–64. Academic Press, San Diego.

Hansen, W. J., Cowan, N. J., and Welch, W. J. (1999). Prefoldin-nascent chain complexes in the folding of cytoskeletal proteins. *J. Cell Biol.* **145,** 265–277.

Hartl, F. U. (1996). Molecular chaperones in cellular protein folding. *Nature* **381,** 571–580.

Himmelspach, R., Nick, P., Schäfer, E., and Ehmann, B. (1997). Developmental and light-dependent changes of the cytosolic chaperonin containing TCP-1 (CCT) subunits in maize seedlings, and the localization in coleoptiles. *Plant J.* **12,** 1299–1310.

Hong, S., Choi, G., Park, S., Chung, A.-S., Hunter, E., and Rhee, S. S. (2001). Type D retrovirus Gag protein interacts with the cytosolic chaperonin TRiC. *J. Virol.* **75,** 2526–2534.

Horwich, A. L., Low, K. B., Fenton, W. A., Hirshfield, I. N., and Furtak, K. (1993). Folding in vivo of bacterial cytoplasmic proteins: Role of groEL. *Cell* **74,** 909–917.

Hynes, G. M., and Willison, K. R. (2000). Individual subunits of the eukaryotic cytosolic chaperonin mediate interactions with binding sites located on subdomains of beta-actin. *J. Biol. Chem.* **275,** 18985–18994.

Hynes, G., Celis, J. E., Lewis, V. A., Carne, A., U, S., Lauridsen, J. B., and Willison, K. R. (1996). Analysis of chaperonin-containing TCP-1 subunits in the human keratinocyte two-dimensional protein database: Further characterisation of antibodies to individual subunits. *Electrophoresis* **17,** 1720–1727.

Iijima, M., Shimizu, H., Tanaka, Y., and Urushihara, H. (1998). A *Dictyostelium discoideum* homologue to *Tcp-1* is essential for growth and development. *Gene* **213,** 101–106.

Kashuba, E., Pokrovskaja, K., Klein, G., and Szekely, L. (1999). Epstein-Barr virus encoded nuclear protein EBNA-3 interacts with the ε-subunit of the t-complex protein 1 chaperonin complex. *J. Hum. Virol.* **2,** 33–37.

Kim, S., Willison, K. R., and Horwich, A. L. (1994). Cytosolic chaperonin subunits have a conserved ATPase domain but diverged polypeptide-binding domains. *Trends Biochem. Sci.* **19,** 543–548.

Kubota, H., Willison, K., Ashworth, A., Nozaki, M., Miyamoto, H., Yamamoto, H., Matsushiro, A., and Morita, T. (1992). Structure and expression of the gene encoding mouse *t*-complex polypeptide (*Tcp-1*). *Gene* **120,** 207–215.

Kubota, H., Hynes, G., Carne, A., Ashworth, A., and Willison, K. (1994). Identification of six *Tcp-1*-related genes encoding divergent subunits of the TCP-1-containing chaperonin. *Curr. Biol.* **4,** 89–99.

Kubota, H., Hynes, G., and Willison, K. (1995a). The chaperonin containing *t*-complex polypeptide 1 (TCP-1): Multisubunit machinery assisting in protein folding and assembly in the eukaryotic cytosol. *Eur. J. Biochem.* **230,** 3–16.

Kubota, H., Hynes, G., and Willison, K. (1995b). The eighth *Cct* gene, *Cctq*, encoding the theta subunit of the cytosolic chaperonin containing TCP-1. *Gene* **154,** 231–236.

Kubota, H., Matsumoto, S., Yokota, S., Yanagi, Y., and Yura, T. (1999a). Transcriptional activation of mouse cytosolic chaperonin CCT subunit genes by heat shock transcription factors HSF1 and HSF2. *FEBS Lett.* **461,** 125–129.

Kubota, H., Yokota, S., Yanagi, H., and Yura, T. (1999b). Structures and coregulated expression of the genes encoding mouse cytosolic chaperonin CCT subunits. *Eur. J. Biochem.* **262,** 492–500.

Kubota, H., Yokota, S., Yanagi, H., and Yura, T. (2000). Transcriptional regulation of the mouse cytosolic chaperonin subunit gene Ccta/t-complex polypeptide 1 by selenocysteine tRNA gene transcription activating factor family zinc finger proteins. *J. Biol. Chem.* **275,** 28641–28648.

Leroux, M. R., and Candido, E. P. M. (1995). Molecualr analysis of Caenorhabditis elegans tcp-1, a gene encoding a chaperonin protein. *Gene* **156,** 241–246.

Leroux, M. R., and Candido, E. P. M. (1997). Subunit characterization of the Caenorhabditis elegans chaperonin containing TCP-1 and expression pattern of the gene encoding CCT-1. *Biochem. Biophys. Res. Commun.* **241,** 687–692.

Leroux, M. R., and Hartl, F. U. (2000). Protein folding: Versatility of the cytosolic chaperonin TRiC/CCT. *Curr. Biol.* **10,** R260–264.

Lewis, S. A., Tian, G., Vainberg, I. E., and Cowan, N. J. (1996). Chaperonin-mediated folding of actin and tubulin. *J. Cell Biol.* **132,** 1–4.

Lewis, V. A., Hynes, G. M., Zheng, D., Saibil, H., and Willison, K. (1992). T-complex polypeptide-1 is a subunit of a heteromeric particle in the eukaryotic cytosol. *Nature* **358,** 249–252.

Li, W.-Z., Lin, P., Frydman, J., Boal, T. R., Cardillo, T. S., Richard, L. M., Toth, D., Lichtman, M. A., Hartle, F.-U., Sherman, F., and Segel, G. B. (1994). Tcp20, a subunit of the eukaryotic TRiC chaperonin from humans and yeast. *J. Biol. Chem.* **269,** 18616–18622.

Lin, P., Cardillo, T. S., Richard, L. M., Segel, G. B., and Sherman, F. (1997). Analysis of mutationally altered forms of the Cct6 subunit of the chaperonin from Saccharomyces cerevisiae. *Genetics* **147,** 1609–1633.

Lingappa, J. R., Martin, R. L., Wong, M. L., Ganem, D., Welch, W. J., and Lingappa, V. R. (1994). A eukaryotic cytosolic chaperonin is associated with a high molecular weight intermediate in the assembly of hepatitis B virus capsid, a multimeric particle. *J. Cell Biol.* **125,** 99–111.

Liou, A. K. F., and Willison, K. R. (1997). Elucidation of the subunit orientation in CCT (chaperonin containing TCP1) from the subunit composition of CCT micro-complexes. *EMBO J.* **16,** 4311–4316.

Liou, A. K. F., McCormack, E. A., and Willison, K. R. (1998). The chaperonin containing TCP-1 (CCT) displays a single-ring mediated disassembly and reassembly cycle. *Biol. Chem.* **379,** 311–319.

Llorca, O., Smyth, M. G., Marco, S., Carrascosa, J. L., Willison, K. R., and Valpuesta, J. M. (1998). ATP binding induces large conformational changes in the apical and equatorial domains of the eukaryotic chaperonin containing TCP-1 complex. *J. Biol. Chem.* **273,** 10091–10094.

Llorca, O., McCormack, E. A., Hynes, G., Grantham, J., Cordell, J., Carrascosa, J. L., Willison, K. R., Fernandez, J. J., and Valpuesta, J. M. (1999a). Eukaryotic type II chaperonin CCT interacts with actin through specific subunits. *Nature* **402,** 693–696.

Llorca, O., Smyth, M. G., Carrascosa, J. L., Willison, K. R., Radermacher, M., Stenbacher, S., and Valpuesta, J. M. (1999b). 3D reconstruction of the ATP-bound form of CCT reveals the asymmetric folding conformation of a type II chaperonin. *Nat. Struct. Biol.* **6,** 639–642.

Llorca, O., Martin-Benito, J., Ritco-Vonsovici, M., Grantham, J., Hynes, G. M., Willison, K. R., Carrascosa, J. L., and Valpuesta, J. M. (2000). Eukaryotic chaperonin CCT stabilizes actin and tubulin folding intermediates in open quasi-native conformations. *EMBO J.* **19,** 5971–5979.

Llorca, O., Martin-Benito, J., Grantham, J., Ritoco-Vonsovici, M., Willison, K. R., Carrascosa, J., and Valpuesta, J. M. (2001). The 'sequential allosteric ring' mechanism in the eukaryotic chaperonin-assisted folding of actin and tubulin. *EMBO J.* **20,** 4065–4075.

Marco, S., Carrascosa, J. L., and Valpuesta, J. M. (1994). Reversible interaction of β-actin along the channel of the TCP-1 cytoplasmic chaperonin. *Biophys. J.* **67,** 364–368.

McCallum, C. D., Do, H., Johnson, A. E., and Frydman, J. (2000). The interaction of the chaperonin tailless complex polypeptide 1 (TCP1) ring complex (TRiC) with ribosome-bound nascent chains examined using photo-cross-linking. *J. Cell Biol.* **149,** 591–602.

McLaughlin, J. N., Thulin, C. D., Hart, S. J., Resing, K. A., Ahn, N. G., and Willardson, B. M. (2002). Regulatory interaction of phosducin-like protein with the cytosolic chaperonin complex. *Proc. Natl. Acad. Sci. USA* **99,** 7962–7967.

Melki, R., and Cowan, N. J. (1994). Facilitated folding of actines and tubulins occurs via a nucleotide-dependent interaction between cytoplasmic chaperonin and distinctive folding intermediates. *Mol. Cell. Biol.* **14,** 2895–2904.

Melki, R., Batelier, G., Soulie, S., and Williams, R. C. J. (1997). Cytoplasmic chaperonin containing TCP-1: Structural and functional characterization. *Biochemistry* **36,** 5817–5826.

Miklos, D., Caplan, S., Martens, D., Hynes, G., Pitluk, Z., Brown, C., Barrell, B., Horwich, A. L., and Willison, K. (1994). Primary structure and function of a second essential member of heterooligomeric TCP1 chaperonin complex of yeast, TCP1b. *Proc. Natl. Acad. Sci. USA* **91,** 2743–2747.

Nong, V. H., Arahira, M., Phan, V. C., Kim, C. S., Zhang, D., Udaka, K., and Fukazawa, C. (2002). Molecular Cloning and Characterization of a Group II Chaperonin delta-Subunit from Soybean. *J. Biochem.* (Tokyo) **132,** 291–300.

Palmedo, G., and Ammermann, D. (1997). Cloning and characterization of the *Oxytrica granulifera* chaperonin containing tailless complex polypeptide 1 γ gene. *Eur. J. Biochem.* **247,** 877–883.

Pappenberger, G., Wilsher, J. A., Roe, S. M., Counsell, D. J., Willison, K. R., and Pearl, L. H. (2002). Crystal structure of the CCTgamma apical domain: implications for substrate binding to the eukaryotic cytosolic chaperonin. *J. Mol. Biol.* **318,** 1367–1379.

Pause, B., Diestelkotter, P., Heid, H., and Just, W. W. (1997). Cytosolic factors mediate protein insertion into the peroxisomal membrane. *FEBS Lett.* **414,** 95–98.

Plath, K., and Rapoport, T. A. (2000). Spontaneous release of cytosolic proteins from posttranslational substrates before their transport into the endoplasmic reticulum. *J. Cell Biol.* **151,** 167–178.

Ritco-Vonsovici, M., and Willison, K. R. (2000). Defining the eukaryotic cytosolic chaperonin-binding sites in human tubulins. *J. Mol. Biol.* **304**, 81–98.

Rommelaere, H., vanTroys, M., Gao, Y., Melki, R., Cowan, N. J., Vandekerckhove, J., and Ampe, C. (1993). Eukaryotic cytosolic chaperonin contains *t*-complex polypeptide 1 and seven related subunits. *Proc. Natl. Acad. Sci. USA* **90**, 11975–11979.

Rommelaere, H., De Neve, M., Melki, R., Vandekerckhove, J., and Ampe, C. (1999). The cytosolic class II chaperonin CCT recognizes delinated hydrophobic sequences in its target proteins. *Biochemistry* **38**, 3246–3257.

Roobol, A., and Carden, M. J. (1993). Identification of chaperonin particles in mammalian brain cytosol and t-complex polypeptide 1 as one of their components. *J. Neurochem.* **60**, 2327–2330.

Roobol, A., and Carden, M. J. (1999). Subunits of the eukaryotic cytosolic chaperonin CCT do not always behave as components of a uniform hetero-oligomeric particle. *Eur. J. Cell. Biol.* **78**, 21–32.

Roobol, A., Holmes, F. E., Hayes, N. V. L., Baines, A. J., and Carden, A. J. (1995). Cytoplasmic chaperonin complexes enter neurites developing in vitro and differ in subunit composition within single cells. *J. Cell Sci.* **108**, 1477–1488.

Roobol, A., Grantham, J., Whitaker, H. C., and Carden, M. J. (1999a). Disassembly of the cytosolic chaperonin in mammalian cell extracts at intracellular levels of K+ and ATP. *J. Biol. Chem.* **274**, 19220–19227.

Roobol, A., Sahyoun, Z. P., and Carden, M. J. (1999b). Selected subunits of the cytosolic chaperonin associate with microtubules assembled in vitro. *J. Biol. Chem.* **274**, 2408–2415.

Sanchez, E. R., and Erickson, R. P. (1985). Expression of the *Tcp-1* locus of the mouse during early embryogenesis. *J. Embryol. Exp. Morphol.* **89**, 113–122.

Schaub, M., Myslinski, E., Schuster, C., Krol, A., and Carbon, P. (1997). Staf, a promiscuous activator for enhanced transcription by RNA polymerases II and III. *EMBO J.* **16**, 173–181.

Sevigny, G., Kothary, R., Tremblay, E., DeRepentigny, Y., Joly, E. C., and Bibor-Hardy, V. (1995). The cytosolic chaperonin subunit TRiC-P5 begins to be expressed at the two-cell stage in mouse embryos. *Biochem. Biophys. Res. Commun.* **216**, 279–283.

Siegers, K., Waldmann, T., Leroux, M. R., Grein, K., Shevchenko, A., Schiebel, E., and Hartl, F. U. (1999). Compartmentation of protein folding *in vivo:* Sequestration of non-native polypeptide by the chaperonin-GimC system. *EMBO J.* **18**, 75–84.

Silver, L. M., Artzt, K., and Bennett, D. (1979). A major testicular cell protein specified by a mouse T/t complex gene. *Cell* **17**, 275–284.

Silver, L. M., Kleen, K. C., Distel, R. J., and Hecht, N. B. (1987). Synthesis of mouse t complex proteins during haploid stages of spermatogenesis. *Dev. Biol.* **119**, 605–608.

Soares, H., Penque, D., Mouta, C., and Rodrigues-Pousada, C. (1994). A *Tetrahymena* orthologue of the mouse chaperonin subunit CCTγ and its coexpression with tubulin during cilia recovery. *J. Biol. Chem.* **269**, 29299–29307.

Somer, L., Shmulman, O., Dror, T., Hashmueli, S., and Kashi, Y. (2002). The eukaryote chaperonin CCT is a cold shock protein in Saccharomyces cerevisiae. *Cell Stress Chaperones* **7**, 47–54.

Srikakulam, R., and Winkelmann, D. A. (1999). Myosin II folding is mediated by a molecular chaperonin. *J. Biol. Chem.* **274**, 27265–27273.

Stoldt, V., Rademacher, F., Kehren, V., Ernst, J. F., Pearce, D. A., and Sherman, F. (1996). The Cct eukaryotic chaperonin subunits of *Saccharomyces cerevisiae* and other yeasts. *Yeast* **12**, 523–529.

Sun, H. B., Neff, A. W., Mescher, A. L., and Malacinski, G. M. (1995). Expression of the axolotl homologue of mouse chaperonin t-complex protein 1 during early development. *Biochim. Biophys. Acta* **1260**, 157–166.

Szpikowska, B. K., Swiderek, K. M., Sherman, M. A., and Mas, M. T. (1998). MgATP binding to the nucleotide-binding domains of the eukaryotic cytoplasmic chaperonin induces conformational changes in the putative substrate-binding domains. *Protein Sci.* **7**, 1524–1530.

Thulasiraman, V., Yang, C.-F., and Frydman, J. (1999). *In vivo* newly translated polypeptides are sequestered in a protected folding environment. *EMBO J.* **18**, 85–95.

Tian, G., Vainberg, I. E., Tap, W. D., Lewis, S. A., and Cowan, N. J. (1995). Specificity in chaperonin-mediated protein folding. *Nature* **375**, 250–253.

Tokumoto, M., Horiguchi, R., Nagahama, Y., Ishikawa, K., and Tokumoto, T. (2000). Two proteins, a goldfish 20S proteasome subunit and the protein interacting with 26S proteasome, change in the meiotic cell cycle. *Eur. J. Biochem.* **267,** 97–103.

Ursic, D., and Culbertson, M. R. (1991). The yeast homolog to mouse Tcp-1 affects microtubule-mediated processes. *Mol. Cell. Biol.* **11,** 2629–2640.

Ursic, D., and Culbertson, M. R. (1992). Is yeast TCP1 a chaperonin? *Nature* **356,** 392.

Vainberg, I. E., Lewis, S. A., Rommelaere, H., Ampe, C., Vandekerckhove, J., Klein, H. L., and Cowan, N. J. (1998). Prefoldin, a chaperone that delivers unfolded proteins to cytosolic chaperonin. *Cell* **93,** 863–873.

Viitanen, P. V., Gatenby, A. A., and Lorimer, G. H. (1992). Purified chaperonin 60 (groEL) interacts with the nonnative states of a multitude of Escherichia coli proteins. *Protein Sci.* **1,** 363–369.

Vinh, D. B.-N., and Drubin, D. G. (1994). A yeast TCP-1 like protein is required for actin function in vivo. *Proc. Natl. Acad. Sci. USA* **91,** 9116–9120.

Willison, K. R. (1999). Composition and function of the eukaryotic cytosolic chaperonin-containing TCP-1. *In* "Molecular Chaperones and Folding Catalysts: Regulation, Cellular Function and Mechanisms" (B. Bukau, Ed.), pp. 555–571. Harwood Academic Publishers, Amsterdam.

Willison, K. R., and Horwich, A. L. (1996). Structure and function of chaperonins in archaebacteria and eukaryotic cytosol. *In* "The Chaperonins" (R. J. Ellis, Ed.), pp. 107–136. Academic Press, San Diego.

Willison, K. R., and Kubota, H. (1994). The structure, function, and genetics of the chaperonin-containing TCP-1 (CCT) in eukaryotic cytosol. *In* "The Biology of Heatshock Proteins and Molecular Chaperones" (R. I. Morimoto, A. Tissieres, and C. Georgopoulos, Eds.), pp. 299–312. Cold Spring Harbor Laboratory Press, New York.

Willison, K. R., Dudley, K., and Potter, J. (1986). Molecular cloning and sequence analysis of a haploid expressed gene encoding t complex polypeptide 1. *Cell* **44,** 727–738.

Willison, K. R., and Grantham, J. (2001). The roles of cytosolic chaperonin CCT in normal eukaryotic cell growth. *In* "Molecular Chaperones: Frontiers in Molecular Biology" (P. Lund, Ed.), pp. 90–118. Oxford University Press, Oxford, UK.

Won, K.-A., Schumacher, R. J., Farr, G. W., Horwich, A. L., and Reed, S. I. (1998). Maturation of human cyclin E requires the function of eukaryotic chaperonin CCT. *Mol. Cell. Biol.* **18,** 7584–7589.

Yaffe, M. B., Farr, G. W., Miklos, D., Horwich, A. L., Sternlicht, M. L., and Sternlicht, H. (1992). TCP1 complex is a molecular chaperone in tubulin biogenesis. *Nature* **358,** 245–248.

Yanaka, N., Kobayashi, K., Wakimoto, N., Yamada, E., Imahie, H., Imai, Y., and Mori, C. (2000). Insertional mutation of the murine kisimo locus caused a defect in spermatogenesis. *J. Biol. Chem.* **275,** 14791–14794.

Yokota, S., Yanagi, H., Yura, T., and Kubota, H. (1999). Cytosolic chaperonin is up-regulated during cell growth. Preferential expression and binding to tubulin at G(1)/S transition through early S phase. *J. Biol. Chem.* **274,** 37070–37078.

Yokota, S., Kayano, T., Ohta, T., Kurimoto, M., Yanagi, H., Yura, T., and Kubota, H. (2000a). Proteasome-dependent degradation of cytosolic chaperonin CCT. *Biochem. Biophys. Res. Commun.* **279,** 712–717.

Yokota, S., Yanagi, H., Yura, T., and Kubota, H. (2000b). Upregulation of cytosolic chaperonin CCT subunits during recovery from chemical stress that causes accumulation of unfolded proteins. *Eur. J. Biochem.* **267,** 1658–1664.

Yokota, S., Yamamoto, Y., Shimizu, K., Momoi, H., Kamikawa, T., Yamaoka, Y., Yanagi, H., Yura, T., and Kubota, H. (2001a). Increased expression of cytosolic chaperonin CCT in human hepatocellular and colonic carcinoma. *Cell Stress Chaperones* **6,** 345–350.

Yokota, S., Yanagi, H., Yura, T., and Kubota, H. (2001b). Cytosolic chaperonin CCT changes contents of particular subunit species concomitant with substrate binding and folding activities during the cell cycle. *Eur. J. Biochem.* **268,** 4664–4673.

Zilkha-Falb, R., Barzilai, A., Djaldeti, R., Ziv, I., Melamed, E., and Shirvan, A. (2000). Involvement of T-complex protein-1delta in dopamine triggered apoptosis in chick embryo sympathetic neurons. *J. Biol. Chem.* **275,** 36380–36387.

HERBAL FACTORS IN THE TREATMENT OF AUTOIMMUNITY-RELATED HABITUAL ABORTION

Tomoyuki Fujii

Department of Obstetrics and Gynecology, Faculty of Medicine, University of Tokyo, Tokyo 113-8655, Japan

I. Introduction
II. Sairei-to and Tokishakuyaku-san and Their Constituents
III. Habitual Abortion and Its Causes
IV. Clinical Effects of Sairei-to and Tokishakuyaku-san on Habitual Abortion
V. Effects of Sairei-to and Tokishakuyaku-san on the Immune Reaction of Mothers
VI. Conclusions
References

Deterioration in the balance of T helper-1 (Th1)/T helper-2 (Th2) during pregnancy may cause complications such as habitual abortion. Two types of immunity-related abortion are, at present, recognized. One is caused by autoimmune disorders as exemplified by the antiphospholipid antibody syndrome, in which the Th1/Th2 balance is excessively shifted to polarization of Th2 . The other is caused by alloimmune fetal–maternal disorder, a condition associated with increased serum activity of Th1 cytokines. In Japan, herbal medicines such as Sairei-to (Sai) and Tokishakuyaku-san (Toki), which are prepared in granule forms and are manufactured, have been used in the treatment of these immunity-related habitual abortions and were reported to be clinically effective for these patients. The clinical effect of these herbal medicines can be explained by how they function in the maternal immune system. Sai and Toki enhance Th1 cytokine release from peripheral blood mononuclear cells (PBMCs) and might suppress the production of autoantibodies from B cells. However, Sai and Toki do not affect cytokine release from decidual mononuclear cells (DMCs), which are directly in contact with fetal trophoblasts. These herbal medicines might not enhance the killer activity of DMCs. Thus, the differential effects of Toki and Sai on the release of Th1/Th2 cytokines from PBMCs and DMCs may reveal the rationale for the use of these medicines in the treatment of autoimmunity-related habitual abortion. © 2002, Elsevier Science (USA).

I. INTRODUCTION

Japanese traditional herbal medicines have a history of use dating back more than 2000 years, and extensive knowledge and cumulative experience regarding their use have been acquired. These herbal medicines have been manufactured and used as formulations, usually in granule or powder form, and have generally been administered orally for treatment. The Ministry of Health and Welfare of Japan has approved 146 formulations of herbal medicines as extract preparations covered by the national health insurance system in Japan. Which herbal medicines have traditionally been used depends on the condition of the patients, i.e., whether they show signs of Yin–Yang or Kyo–Jitsu (in Japanese). Yin–Yang is a concept referring to blood, hormones, the autonomic nervous system, and other regulatory functions of the body's internal environment. Kyo–Jitsu is a concept referring to vital energy in the viscera or bowels. Yin and Kyo mean deficiency (or hypo) and Yang and Jitsu mean excess (or hyper). For patients showing signs of Yin and Kyo, nourishing herbal medicines such as Tokishakuyaku-san or Unkei-to are used; for patients showing signs of Yang and Jitsu, suppressing herbal medicines such as Keishibukuryo-gan or Tokakujoki-to are used. In the field of obstetrics and gynecology, herbal medicines have been used in Japan to treat, in particular, hormonal disorders or infertility (Table I).

II. SAIREI-TO AND TOKISHAKUYAKU-SAN AND THEIR CONSTITUENTS

Sairei-to (Sai) and Tokishakuyaku-san (Toki) are two widely used herbal medicines in Japan. Sai consists of 12 crude ingredients extracted from herbs, i.e., Atractylodis Lanceae rhizoma, Alismatis rhizoma, Bupleuri radix, Cinnamomi cortex, Ginseng radix, Glycyrrhizae radix, Hoelen, Pinelliae tuber, Polyporus, Scutellariae radix, Zingiberis rhizoma, and Zizyphi fructus. Toki consists of 6 crude ingredients extracted from herbs, i.e., Angelicae radix, Alismatis rhizoma, Atractylodis Lanceae rhizoma, Cnidii rhizoma, Hoelen, and Paeoniae radix.

TABLE I Japanese Herbal Medicines Commonly Used in the Field of Obstetrics and Gynecology.

Herbal medicine	Ying–Yang	Jitsu–Kyo	Clinical indication
Kamishoyo-san	Ying	Kyo	Ovarian dysfunction, dysmenorrhea, climacteric syndrome
Shimotsu-to	Ying	Kyo	Puerperium, ovarian dysfunction
Tokikenchu-to	Ying	Kyo	Dysmenorrhea
Tokishakuyaku-san	Ying	Kyo	Ovarian dysfunction, sterility, dysmenorrhea, habitual abortion, premature labor
Unkei-to	Ying	Kyo	Ovarian dysfunction, dysmenorrhea, sterility, climacteric syndrome
Nyoshin-san	Middle	Middle	Perinatal neurosis, ovarian dysfunction
Sairei-to	Middle	Middle	Habitual abortion, preeclampsia
Daiobotanpi-to	Yang	Jitsu	Ovarian dysfunction, dysmenorrhea
Keishibukuryo-gan	Yang	Jitsu	Ovarian dysfunction, dysmenorrhea, climacteric syndrome
Tsudo-san	Yang	Jitsu	Ovarian dysfunction, dysmenorrhea, climacteric syndrome
TokakuJoki-to	Yang	Jitsu	Ovarian dysfunction, dysmenorrhea

Sai is a medicine used to treat patients with signs that fall into an intermediate group between Yin and Yang and between Kyo and Jitsu. This medicine has been empirically used and is known to be effective in treating nausea, loss of appetite, feelings of thirst, oligouria, diarrhea, stomach colitis, and edema. The diuretic activity of Sai is almost the same as that of thiazide, furosemide, and digitalis. As for antiinflammatory activity, Sai suppresses the growth of human fibroblasts and enhances the activity of steroid hormones such as dexamethasone and increases the secretion of corticosterone from the adrenal cortex.

Toki is a medicine used to treat patients showing signs of Yin and Kyo. This herbal medicine has been empirically used and is known to be effective in treating anemia, malaise, climacteric syndrome, irregular menstruation, dysmenorrhea, infertility, palpitations, chronic nephritis, edema associated with pregnancy, and beriberi. Toki induces ovulation cycles in anovulatory mice, suppresses uterine contractions during pregnancy, reduces the heart rate of ritodrin-induced tachycardia in rabbits, and suppresses the growth of thrombosis in the blood vessels. Sai and Toki have been empirically used for the treatment of habitual abortion in Japan.

III. HABITUAL ABORTION AND ITS CAUSES

Habitual abortion, defined as three or more successive spontaneous abortions, has a multifaceted pathophysiology. The known causes of habitual abortion are chromosomal anomalies of patients and/or their husbands, endocrine disorders, anatomical disorders such as uterine anomalies including septate uterus, infections such as ureaplasma, and autoimmune disorders such as antiphospholipid antibody syndrome and alloimmune disorder, i.e. a disordered immune reaction for maintaining pregnancy. Most of the chromosomal anomalies found in patients (or their husbands) involve balanced translocation. One-third of the pregnancies will end uneventfully (half of them will have the same balanced translocation as that of their parents) and two-thirds will be aborted because of imbalanced translocation. For these patients, genetic counseling is required. The hormonal disorders include hyper- or hypothyroidism and the patients need to receive hormonal therapy. For patients with septate uterus, an operative procedure, in particular the resection of the septum using a resectoscope, is effective. Antimicrobiol therapy such as antibiotics should be used for patients with infections.

Patients with immunological disorders are thought to constitute the largest group of habitual aborters. As for autoimmune disorders, recent studies have postulated an association between autoantibodies including antiphospholipid antibodies and some disorders of pregnancies including the failure of *in vitro* fertilization (IVF) and embryo transfer (Geva *et al.,* 1995), habitual abortions, intrauterine growth restriction (El-Roeiy *et al.,* 1991), and preeclampsia (Branch *et al.,* 1989), thus suggesting an emerging clinical entity termed the "reproductive autoimmune failure syndrome" (RAFS) (Gleicher and El-Roeiy, 1988). Antiphospholipid antibodies are autoantibodies that react against phospholipids, among which cardiolipin (CL) has been the most commonly employed to detect the circulating antibodies against phospholipids. The autoimmune anticardiolipin antibody (aCL) is reported to react with CL only in the presence of a 50-kDa serum cofactor (Matsuura *et al.,* 1990) that was identified as human β_2-glycoprotein I (β_2-GPI) (McNeil *et al.,* 1990). β_2-GPI enhances CL binding with autoimmune aCL but inhibits binding by aCL, which is associated with syphilis. Autoimmune aCL does not recognize β_2-GPI or CL alone. It recognizes either the complex of CL and β_2-GPI or the novel epitope on the β_2-GPI, which appears after binding to CL (Koike and Matsuura, 1991). β_2-GPI acts as an inhibitor of the intrinsic blood coagulation pathway (Schousboe, 1985), adenosine diphosphate (ADP)-dependent aggregation (Nimpf *et al.,* 1987), and prothrombinase activation of activated platelets (Nimpf *et al.,* 1986). It is assumed that β_2-GPI-dependent aCL may interfere with the β_2-GPI function when β_2-GPI recognizes the phospholipid on the activated platelets. As a result, intravascular coagulation may be induced, presumably in the placenta, which, in turn, undergoes functional deterioration, leading to intrauterine fetal death or spontaneous abortion (Christiansen, 1996; Lim *et al.,* 1996).

It has been reported that although the risk of pregnancy disorders including habitual abortion in patients with β_2-GPI-related aCL is very high, the incidence of the presence of aCL in habitual abortion is very low (Maejima *et al.,* 1997). In addition, the cell membrane of platelets contains little CL but rather contains noncharged phospholipids such as phosphatidylethanolamine. A new combination of antiphospholipid antibody and its serum cofactor protein has been reported. It is a combination of antiphosphatidylethanolamine antibody (aPE) and kininogen (Sugi and McIntyre, 1995). This combination, which acts in a manner similar to the combination of aCL and β_2-GPI, induces intravascular coagulation. The presence of kininogen-dependent aPE in habitual aborters is reported to be much higher than that of β_2-GPI-dependent aCL (Sugi *et al.*, 1999). For patients with antiphospholipid antibodies, prednisolone and low-dose aspirin or heparin is administered. Prednisolone is used to decrease the titer of antiphospholipid antibodies and low-dose aspirin or heparin is used to suppress intravascular coagulation in the placenta.

As for alloimmune disorders, most are considered to be included in the category of patients with unknown causes. Recent multiple lines of evidence suggest that perturbation of the exquisite cytokine network needed to maintain pregnancy immunologically may result in the malconstruction of the placenta and bring on complications of pregnancy including habitual abortion. For habitual aborters in whom the cause is unknown, which might include patients with alloimmune disorders, immunotherapy such as injection of the husband's lymphocytes or immunoglobulin is performed in many hospitals.

IV. CLINICAL EFFECTS OF SAIREI-TO AND TOKISHAKUYAKU-SAN ON HABITUAL ABORTION

In Japan, Sai and Toki have been used in the treatment of habitual abortion, especially in cases involving autoimmune disorders or unknown causes containing alloimmune disorders. The clinical effects of Sai and Toki have been reported by several hospitals in Japan. In 1992, Asano reported treating habitual aborters with lupus anticoagulant using Sai. In 1996, Takakuwa *et al.* reported that they prescribed 9 g/day of Sai for 12 habitual aborters without genetic impairment, müllerian anomalies, hormonal deficiencies, infections, metabolic disorders, and/or manifested autoimmune diseases such as systemic lupus erythematosus. All the patients revealed positive antinuclear antibody and one or more antiphospholipid antibodies such as aCL or antiphosphatidylinositol antibody. Nine grams/day of Sai was prescribed before and during pregnancy without the prescription of corticosteroid hormones or low-dose aspirin. After administration of Sai, the titer of antiphospholipid antibodies was decreased in all patients and 10 of the patients delivered healthy babies. No adverse effects were observed.

Toki reduces the titer of aCL and is reported to be effective in treating habitual abortion (Shiraishi *et al.*, 1996). In 1999, Chishima *et al.* reported that the administration of Toki to habitual-abortion-model mice, CBA/J × DBA/2, increased the litter number and tended to reduce the abortion rate of these mice. Toki is reported to suppress the aggregation of platelets in the presence of placental chorioepithelial brush border membrane (Iioka *et al.*, 1990). This might be one explanation of the mechanism of its clinical effect in preventing abortion.

V. EFFECTS OF SAIREI-TO AND TOKISHAKUYAKU-SAN ON THE IMMUNE REACTION OF MOTHERS

In recent years, persuasive evidence has accumulated to suggest the existence of functionally polarized responses of T cells as determined by the cytokines they produce (Abbas et al., 1996; Romagnani, 1996). More specifically, T helper 1 (Th1) cells predominantly produce interferon (IFN)-γ and tumor necrosis factor (TNF)-β, both of which are involved in cell-mediated immunity. By contrast, T helper 2 (Th2) cells mostly produce interleukin (IL)-4, IL-5, IL-10, and IL-13, which are responsible for antibody responses. Pregnancy is thought to be a phenomenon in which the Th1/Th2 balance is shifted to Th2 polarization with the Th1 reaction being suppressed (Wegmann et al., 1993). Deterioration of the Th1/Th2 balance during pregnancy may cause complications such as habitual abortion. As for immunity-related habitual abortion, two types are, at present, recognized. One is caused by autoimmune disorders as exemplified by the antiphospholipid antibody syndrome, in which the Th1/Th2 balance is excessively shifted to Th2 polarization. The other is caused by an alloimmune fetal–maternal disorder, a condition associated with increased serum activity of Th1 cytokines (Hamai et al., 1998a,b; Hara et al., 1996; Hill et al., 1995; Lim et al., 1996).

We investigated whether Toki and Sai modulate Th1 and Th2 cytokine release from peripheral blood mononuclear cells (PBMCs) (Fujii et al., 2000). The effects of Sai and Toki were examined in relation to human leukocyte antigen (HLA)-G, a nonclassic HLA class I antigen expressed on trophoblasts and a putative crucial player involved in fetomaternal immune interplay. Toki and Sai increased the release of Th1 group cytokines, TNF-α (Fig. 1A) and IFN-γ (Fig. 2A), while preserving the inhibitory effect of HLA-G on the release of these cytokines. As for Th2 group cytokine release, Toki had no effect in modulating IL-4 release (Fig. 3A) regardless of the presence of HLA-G, whereas Sai nullified the effect of the presence of HLA-G in stimulating the release of IL-4 without affecting its release in the absence of HLA-G. Thus, Toki and Sai (Table II) shift the Th1/Th2 balance to Th1 polarization and may have therapeutic potential particularly in autoimmunity-related habitual abortion in which the Th2 response is pathologically enhanced, but not in habitual abortion involving alloimmune fetomaternal derangement, a condition, rather, of an enhanced Th1 response.

However, an unanswered question is whether Toki and Sai might enhance Th1 cytokine release in decidual tissues and thereby stimulate killer activity, thus working counterproductively by accelerating maternal alloimmune reactions toward fetal tissues. To address this, we compared the effects of these medicines as related to HLA-G on the release of cytokines from decidual mononuclear cells (DMCs) and PBMCs on the assumption that they might act differently on these different cell types (Fujii et al., 2001). Regarding Th1 cytokines, Toki marginally increased the release of TNF-α, but not INF-γ from DMCs whereas Sai did not

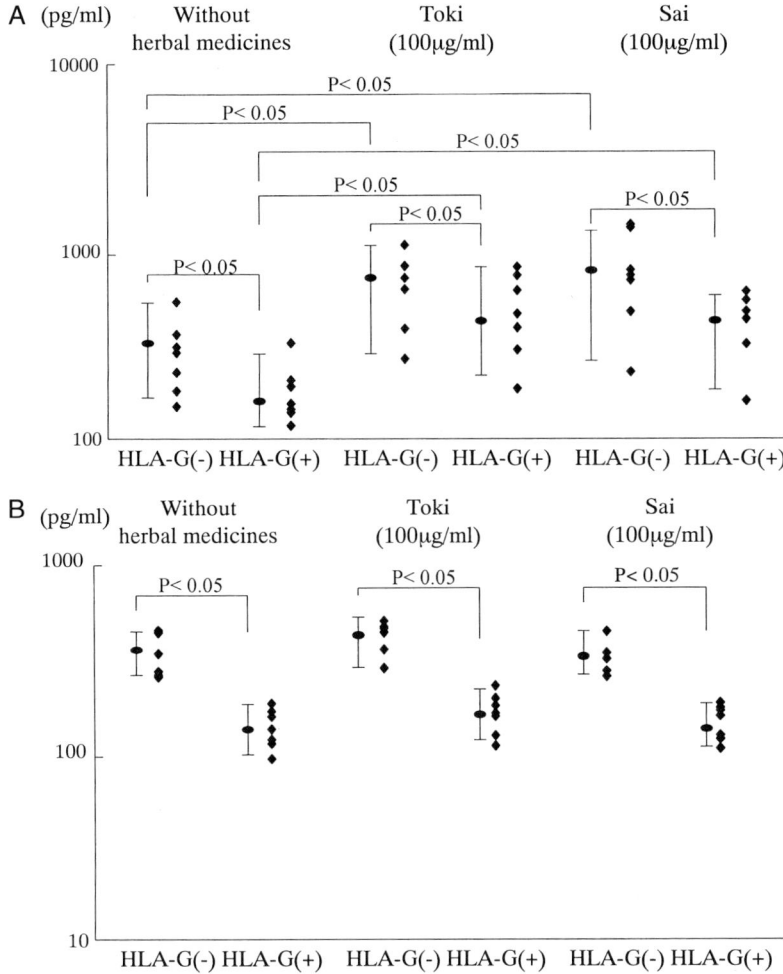

FIGURE 1 Tumor necrosis factor (TNF)-α concentrations in the culture media. Peripheral blood mononuclear cells (PBMCs) from seven healthy nulligravidous women (A) or decidual mononuclear cells (DMCs) from seven women who underwent legal first trimester abortion (B) were cultured with either .221 cells or .221-G1 cells for 48h in the medium supplemented with either Toki or Sai (100 μg/ml). As controls, mononuclear cells were cultured with either .221 cells or .221-G1 cells without herbal medicines. The concentrations of TNF-α in the culture media were measured and plotted in a log scale. Each bar represents the median concentration and the 5 to 95 percentile range. $p < 0.05$: significantly different by Wilcoxon's test.

affect the release of either (Figs. 1B and 2B). Both Toki and Sai had no effect in modulating the release of IL-4 (Fig. 3B), a member of the Th2 cytokines. Interestingly, the presence of HLA-G reduced the release of Th1 cytokines from DMCs regardless of the addition of Toki, Sai, or neither. These findings are in sharp contrast to the findings from PBMCs in which these medicines seem to act

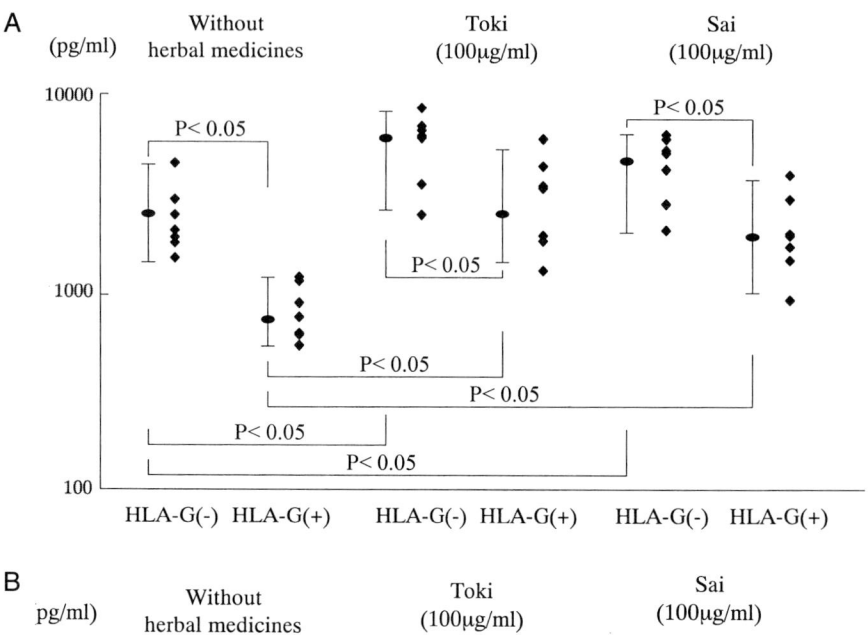
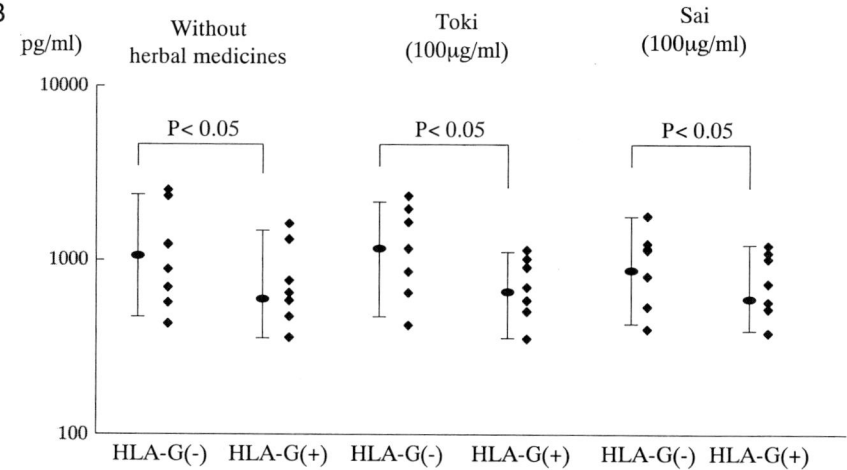

FIGURE 2 Interferon (IFN)-γ concentrations in the culture media. PBMCs (A) or DMCs (B) were cultured with either .221 cells or .221-G1 cells for 48 h in the medium supplemented with either Toki or Sai. As controls, mononuclear cells were cultured with either .221 cells or .221-G1 cells without herbal medicines. The concentrations of IFN-γ in the culture media were measured and plotted in a log scale. Each bar represents the median concentration and the 5 to 95 percentile range. $p < 0.05$: significantly different by Wilcoxon's test.

to enhance Th1 polarization and attenuate Th2 polarization. We can speculate that Sai and Toki enhance Th1 cytokine release from the peripheral blood lymphocytes and suppress the production of autoantibodies from B cells. On the other hand, Sai and Toki do not affect cytokine release from lymphocytes located in the decidua where maternal lymphocytes directly recognize the fetal trophoblasts and

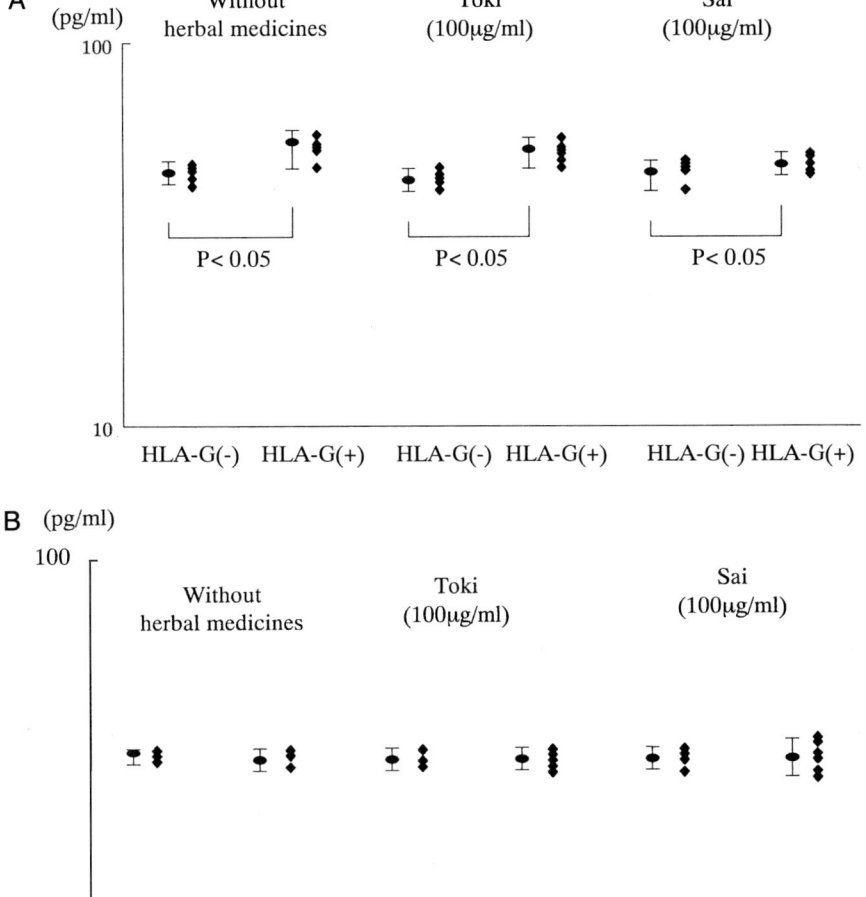

FIGURE 3 Interleukin (IL)-4 concentrations in the culture media. PBMCs (A) or DMCs (B) were cultured with either .221 cells or .221-G1 cells for 48 h in the medium supplemented with either Toki or Sai. As controls, mononuclear cells were cultured with either .221 cells or .221-G1 cells without herbal medicines. The concentrations of IL-4 in the culture media were measured and plotted in a log scale. Each bar represents the median concentration and the 5 to 95 percentile range. $p < 0.05$: significantly different by Wilcoxon's test.

react to maintain pregnancy (Table II) and do not enhance the killer activity of these decidual lymphocytes. Thus, the differential effect of Toki and Sai on the release of Th1/Th2 cytokines between DMCs and PBMCs may reveal the mechanism by which these medicines work in treating of autoimmunity-related habitual abortion.

TABLE II Cytokine Release from Peripheral and Decidual Mononuclear Cells When Toki and Sai Were Supplied.

	TNF-α	IFN-γ	IL-4
Toki			
PBMCs	↑	↑	→
DMCs	↑	→	→
Sai			
PBMCs	↑	↑	→
DMCs	→	→	→

↑: increase, ↓: decrease, →: unchanged

VI. CONCLUSIONS

Herbal medicines such as Sai and Toki are useful in treating habitual abortion. These medicines actually affect the release of cytokines from lymphocytes and Toki was reported to inhibit the aggregation of platelets. However, it still needs to be determined whether treatment with herbal medicines is truly clinically effective. No high-level clinical evidence exists. Further clinical research such as a randomized controlled trial is needed.

ACKNOWLEDGMENTS

We thank Dr. Daniel E. Geraghty (Fred Hutchinson Cancer Research Center, Seattle, WA) for providing the HLA-G cDNA-transfected cells (.221-G1 cells). The authors also thank Tsumura & Co. (Tokyo, Japan) for providing the Toki and Sai.

REFERENCES

Abbas, A. K., Murphy, K. M., and Sher, A. (1996). Functional diversity of helper T lymphocytes. *Nature* **383**, 787–793.

Asano, K. (1992). Effect of Sireitou therapy for habitual aborters with antiphospholipid antibodies. *Proc. 6th Jpn. Soc. Reprod. Immunol.* 217–219.

Branch, D. W., Andres, R., Digre, K. B., Rote, N. S., and Scott, J. R. (1989). The association of antiphospholipid antibodies with severe preeclampsia. *Obstet. Gynecol.* **73**, 541–545.

Chishima, F., Shiraishi, N., Hayakawa, S., and Sato, K. (1999). Habitual pregnancy loss and herbal medicine [Japanese]. *Recent Prog. Kampo Med. Obstet. Gynecol.* **16**, 30–34.

Christiansen, O. B. (1996). A fresh look at the causes and treatments of habitual miscarriage, especially its immunological aspects. *Hum. Reprod. Update* **2**, 271–293.

El-Roeiy, A., Myers, S. A., and Gleicher, N. (1991). The relationship between autoantibodies and intrauterine growth retardation in hypertensive disorders of pregnancy. *Am. J. Obstet. Gynecol.* **164**, 1253–1261.

Fujii, T., Kanai, T., Kozuma, S., Hamai, Y., Hyodo, H., Yamashita, T., Miki, A., Unno, N., and Taketani, Y. (2000). Theoretical basis for herbal medicines, Tokishakuyaku-san and Sairei-to, in the treatment of autoimmunity-related habitual abortion by correcting T helper-1/T helper-2 balance. *Am. J. Reprod. Immunol.* **44**, 342–346.

Fujii, T., Kanai, T., Kozuma, S., Baba, K., Miki, A., Hyodo, H., Yamashita, T., Unno, N., and Taketani, Y. (2001). Herbal medicines, Sairei-to and Tokishakuyaku-san, differently modulate the release of cytokines from decidual versus peripheral blood mononuclear cells. *Am. J. Reprod. Immunol.* **46**, 369–372.

Geva, E., Amit, A., Lerner-Geva, L., Azema, F., Yovel, I., and Lessing, J. B. (1995). Autoimmune disorders: Another possible cause for in-vitro fertilization and embryo transfer failure. *Hum. Reprod.* **10**, 2560–2563.

Gleicher, N., and El-Roeiy, A. (1988). The reproductive autoimmune failure syndrome. *Am. J. Obstet. Gynecol.* **159**, 223–227.

Hamai, Y., Fujii, T., Kozuma, S., Shibata, Y., and Taketani, Y. (1998a). Secretion of interleukin-2 from unstimulated peripheral blood mononuclear cells is a possible pathogenic mechanism in habitual abortion. *Am. J. Reprod. Immunol.* **40**, 63–64.

Hamai, Y., Fujii, T., Yamashita, T., Miki, A., Kozuma, S., Geraghty, D. E., and Taketani, Y. (1998b). Peripheral blood mononuclear cells from women with habitual abortion exhibit an aberrant reaction to release cytokines upon the direct contact of human leukocyte antigen (HLA)-G expressing cells. *Am. J. Reprod. Immunol.* **40**, 408–413.

Hara, N., Fujii, T., Yamashita, T., Kozuma, S., Okai, T., and Taketani, Y. (1996). Altered expression of human leukocyte antigen G (HLA-G) on extravillous trophoblasts in preeclampsia: Immunohistochemical demonstration with anti-HLA-G specific antibody '87G' and anti-cytokeratin antibody 'CAM5.2'. *Am. J. Reprod. Immunol.* **36**, 349–358.

Hill, J. A., Polger, K., and Anderson, D. J. (1995). T-helper 1-type immunity to trophoblast in women with habitual spontaneous abortion. *JAMA* **273**, 1933–1936.

Iioka, H., Moriyama, I., Hisanaga, H., Morimoto, K., Akada, S., and Ichijo, M. (1990). Tokishakuyaku-san inhibits the aggregation of platelets [Japanese]. *Kampo Med.* **14**, 200–203.

Koike, T., and Matsuura, E. (1991). What is the 'true' antigen for anticardiolipin antibodies. *Lancet* **337**(8742), 671–672.

Lim, K. J. H., Odukoya, O. A., Li, T. C., and Cooke, I. D. (1996). Cytokines and immuno-endocrine factors in habitual miscarriage. *Hum. Reprod. Update* **2**, 469–481.

Maejima, M., Fujii, T., Okai, T., Kozuma, S., Shibata, Y., and Taketani, Y. (1997). β_2–Glycoprotein I-dependent anticardiolipin antibody in early habitual spontaneous abortion. *Hum. Reprod.* **12**, 2140–2142.

Matsuura, E., Igarashi, Y., Fujimoto, M., Ichikawa, K., and Koike, T. (1990). Anticardiolipin cofactor(s) and differential diagnosis of autoimmune disease. *Lancet* **336**(8708), 177–178.

McNeil, H. P., Simpson, R. J., Chesterman, C. N., and Krilis, A. (1990). Antiphospholipid antibodies are directed against a complex antigen that includes a lipid binding inhibitor of coagulation: b2-Glycoprotein I (antiprotein H). *Proc. Natl. Acad. Sci. USA* **87**, 4120–4124.

Nimpf, J., Bovers, E. M., Bomans, P. H., Till, U., Wurm, H., Kostner, G. M., and Zwaal, R. F. (1986). Prothrombinase activity of human platelets is inhibited by glycoprotein I. *Biochim. Biophys. Acta* **884**, 142–149.

Nimpf, J., Wurm, H., and Kostner, G. M. (1987). β2-Glycoprotein I (apo-H) inhibits the release reaction of human platelets during ADP-induced aggregation. *Atherosclerosis* **63**, 109–114.

Romagnani, S. (1996). Th1 and Th2 in human diseases. *Clin. Immunol. Immunopathol.* **80**, 225–235.

Schousboe, I. (1985). β2-Glycoprotein I: A plasma inhibitor of the contact activation of the intrinsic blood coagulation pathway. *Blood* **66**, 1086–1091.

Shiraishi, N., Hayakawa, S., Chishima, F., and Sato, K. (1996). Tokolysis and herbal medicine [Japanese]. *Recent Prog. Kampo Med. Obstet. Gynecol.* **13**, 40–42.

Sugi, T., and McIntyre, J.A. (1995). Autoantibodies to phosphatidylethanolamine (PE) recognize a kininogen-PE complex. *Blood* **86**, 3083–3089.

Sugi, T., Katsunuma, J., Izumi, S., McIntyre, J. A., and Makino, T. (1999). Prevalence and heterogeneity of antiphosphatidylethanolamine antibodies in patients with habitual early pregnant losses. *Fertil. Steril.* **71,** 1060–1065.

Takakuwa, K., Yasuda, M., Hataya, I., Sekizuka, N., Tamura, M., Arakawa, M., Higashino, M., Hasegawa, I., and Tanaka, K. (1996). Treatment for patients with habitual abortion with positive antiphospholipid antibodies using a traditional Chinese herbal medicine. *J. Perinat. Med.* **24,** 489–494.

Wegmann, T. G., Lin, H., Guilbert, L., and Massmann, T. R. (1993). Bidirectional cytokine interactions in the maternal-fetal relationship: Is successful pregnancy a TH2 phenomenon? *Immunol. Today* **14,** 353–356.

Vitamin D Receptor and Retinoid X Receptor Interactions in Motion

J. Barsony* and K. Prufer

Laboratory of Cell Biochemistry and Biology, National Institutes of Diabetes, Digestive and Kidney Diseases, NIH, Bethesda, Maryland 20892

    I. General Features of VDR and RXR
    II. Regulation of Constitutive Activities of VDR by RXR
    III. Regulation of Calciotropic Effects of Calcitriol by RXR
    IV. Regulation of Antiproliferative Effects of Calcitriol by RXR
    V. RXR as a Dimerization Partner
    VI. RXR Effects on the Physicochemical Properties of VDR
    VII. RXR Effects on DNA Binding and Transcriptional Activities
    VIII. Dimerization Interfaces
    IX. Regulation of VDR/RXR Dimerization
    X. Subcellular Localization of VDR/RXR Heterodimers
    XI. Trafficking of VDR and RXR between Cytoplasm and Nucleus
    XII. Regulation of Unliganded VDR and RXR Shuttling
    XIII. Regulation of Liganded VDR and RXR Shuttling
    XIV. Intranuclear Trafficking and Target Site Binding of VDR and RXR
    XV. Outlook
    References

*Address correspondence to: Julia Barsony, 8 Center Drive, Room 422, Bethesda, MD, 20892, Tel: 301-402-2868; E-mail: jul@helix.nih.gov.

Vitamin D receptor (VDR) and retinoid X receptor (RXR) are members of the nuclear receptor superfamily and they bind target DNA sequences as heterodimers to regulate transcription. This article surveys the latest findings regarding the roles of dimerizing RXR in VDR function and emphasizes potential areas for future developments. We first highlight the importance of dimerization with RXR for both the ligand-independent (hair growth) and ligand-dependent functions of VDR (calcium homeostasis, bone development and mineralization, control of cell growth and differentiation). Emerging information regarding the regulatory control of dimerization based on biochemical, structural, and genetic studies is then presented. Finally, the main focus of this article is a new dynamic perspective of dimerization functions, based on recent research with fluorescent protein chimeras in living cells by microscopy. These studies revealed that both VDR and RXR constantly shuttle between the cytoplasm and the nucleus and between subnuclear compartments, and showed the transient nature of receptor–DNA and receptor–coregulator interactions. Because RXR dimerizes with most of the nuclear receptors, regulation of receptor dynamics by RXR has a broad significance. © 2002, Elsevier Science (USA).

I. GENERAL FEATURES OF VDR AND RXR

Vitamin D receptors (VDRs) and retinoid X receptors (RXRs) belong to the nuclear receptor superfamily. Nuclear receptors are transcriptional regulators that bind to the promoters of their target genes either as homodimers or heterodimers. Type I nuclear receptors include the steroid hormone receptors that regulate the expression of target genes as homodimers. Type II nuclear receptors, including VDR, regulate transcription as heterodimers with RXR. The human genome is reported to contain 48 members of the nuclear receptor family (Maglich *et al.*, 2001), and many of these receptors dimerize with RXR. Type I and type II nuclear receptors regulate transcription of target genes in response to hydrophobic ligands, such as steroid hormones, thyroid hormones, fat-soluble retinoids, and the secosteroid vitamin D. VDR is activated by calcitriol, a hormonal form of vitamin D. RXR is activated by a vitamin A derivative, 9-*cis*-retinoic acid, and by a variety of dietary lipids, including docosahexaenoic acid, phytanic acid, and methoprene acid (Chawla *et al.*, 2001). In addition to ligand-dependent functions, most nuclear receptors also have constitutive activities.

In the 1980s, VDR was cloned along with other nuclear receptors (Hughes *et al.*, 1988). VDR is encoded by a single gene that is located close to the centromere on chromosome 12 (Mathew *et al.*, 1992; Taymans *et al.*, 1999). In addition to the major product of this gene, NH_2-terminal variant VDRs were recently described (Jurutka *et al.*, 2000; Sunn *et al.*, 2001). In the early 1990s, RXRs were also cloned and sequenced (Fleischhauer *et al.*, 1993; Leid *et al.*, 1992). The three RXR subtypes are encoded by separate genes; the RXRα gene is located on the long arm of chromosome 9, RXRβ on the short arm of chromosome 6, and RXRγ on the long arm of chromosome 1 (Almasan *et al.*, 1994). All three subtypes of RXR form dimers with VDR and they bind together to consensus target DNA sequences *in vitro* (Jin *et al.*, 1996). VDR, however, preferentially dimerizes with RXRα and RXRγ to regulate activities of natural promoters (Kephart *et al.*, 1996). The tissue-specific distribution of RXR isotypes contributes to the tissue specificity of VDR functions. RXRα is abundant in visceral tissues, skin, and brain, RXRβ is enriched in the immune system, and RXRγ is mainly expressed in the brain, muscle, and chondrocytes (Almasan *et al.*, 1994; Dolle *et al.*, 1994).

VDR and RXR share a common structural organization with other nuclear receptors. These proteins contain an NH_2-terminal variable region, a DNA-binding region, a hinge region, and a COOH-terminal ligand-binding region. The NH_2-terminal region often contains a ligand-independent transcriptional activation function (AF-1) domain. VDR has a short NH_2-terminal region with no AF-1 function, whereas RXR has an AF-1 domain (Mascrez *et al.*, 2001; Nagpal *et al.*, 1993). The DNA-binding domain (DBD) contains two highly conserved zinc-finger motifs that target each receptor to specific DNA sequences, known as hormone response elements. The hinge region is important for receptor flexibility and may contain interacting sequences for chaperones and proteins of the nuclear import machinery. The COOH-terminal region (ligand-binding domain, LBD) contains the ligand-binding pocket, the ligand-dependent activation function (AF-2) domain, and also homo- and heterodimerization interfaces. The AF-2 domain interacts with coactivators and corepressors. Ligand binding enhances coactivator binding and causes dissociation of corepressors (McKenna *et al.*, 1999).

Further understanding of structural and functional organization has been derived from X-ray crystallography and nuclear magnetic resonance (NMR) studies. The structures of the RXRα DBD (Lee *et al.*, 1993), the unliganded RXRα LBD (Bourguet *et al.*, 1995; Egea *et al.*, 2000), and the liganded RXRα LBD (Bourguet *et al.*, 2000) have been resolved. The three-dimensional structure of the VDR remained elusive until very recently. Removal of a large insertion peptide from the LBD increased its solubility and finally allowed the crystallization and structural determination of the liganded VDR LBD (Rochel *et al.*, 2000).

VDR/RXR heterodimers mediate the effects of calcitriol on calcium homeostasis. Although most mammalian cells express VDR, in many tissues its functional significance remains to be elucidated. Studies on hereditary vitamin D-resistant rickets patients (HVDRR) harboring mutations in the VDR gene and studies on VDR knockout mice revealed that VDR is essential only for intestinal calcium absorption and hair growth, as the bone defects (Amling *et al.*, 1999; Sakati *et al.*, 1986; Tiosano *et al.*, 2001; Van Cromphaut *et al.*, 2001) and the immune defects (Mathieu *et al.*, 2001) were resolved by calcium replacement both in patients and in mice. In contrast, normalization of calcium homeostasis did not restore hair growth in HVDRR patients and in VDR knockout mice. Interestingly, in the VDR/RXRγ double knockout mice, abnormalities of the growth plate did not normalize upon restoration of normal mineral ion homeostasis (Yagishita *et al.*, 2001), suggesting redundancy among RXR dimerization partners and a direct role of VDR in bone.

Moreover, studies in mice revealed that alopecia not only develops as a result of VDR defect (Chen et al., 2001) but also develops as a result of RXRα ablation in skin keratinocytes (Li et al., 2000). These mice models clearly demonstrate that both VDR and RXR functions are essential for a normal hair cycle, whereas the ligand for the VDR is not. Likewise, abnormality in hair growth does not result from vitamin D deficiency of any sort (Panda et al., 2001). Taken together these findings demonstrate that dimerization with RXR is necessary for both ligand-dependent and ligand-independent functions of VDR. Consequently, regulation of dimerization has a strong influence on VDR functions.

II. REGULATION OF CONSTITUTIVE ACTIVITIES OF VDR BY RXR

Until recently, the ligand-independent functions of VDR were not appreciated. Now that the mouse models have confirmed the physiological importance of ligand-independent functions (Sakai et al., 2001), we understand that VDR/RXR heterodimers can be active in the absence of ligand. The regulation of constitutive activities can also be evaluated in reporter assays. Elegant studies showed recently that phosphorylation regulates VDR-mediated transcription by modulating interactions with coactivators such as DRIP205 (Barletta et al., 2002; Matkovits and Christakos, 1995). Another study showed that in the presence of Ets-1, VDR stimulates the prolactin promoter in a ligand-independent manner, and that this effect required an intact AF-2 domain of VDR (Tolon et al., 2000). The interaction with Ets-1 induces a conformational change in VDR that can be detected by increased resistance to proteolytic digestion. Finally, VDR interacts with corepressors such as NcoR and SMRT, as thyroid hormone receptor (TR) and retinoic acid receptor (RAR) does (Tagami et al., 1998). Consistent with this finding, transcriptional silencing by the apo-VDR was observed using the osteopontin response element. In all these instances, dimerization with RXR was essential for the ligand-independent effects of VDR.

III. REGULATION OF CALCIOTROPIC EFFECTS OF CALCITRIOL BY RXR

The clinical significance of RXR in the calcemic effects of calcitriol is manifested in several disease states. For example, in a patient with vitamin D-deficient rickets a point mutation in the VDR gene impaired dimerization with RXR and led to severe hypocalcemia (Whitfield et al., 1996). Animal, human, and *in vitro* data all indicate that excess vitamin A antagonizes the effects of vitamin D on calcium metabolism and bone, and contributes to development of osteoporosis (Binkley and Krueger, 2000; Feskanich et al., 2002; Johansson and Melhus, 2001). Presumably, sequestration of RXR by RAR plays a role in this effect. Another study showed that selective increase of calreticulin expression in the parathyroid gland inhibits the ability of high serum calcitriol to suppress parathyroid hormone (PTH) mRNA levels in chronically hypocalcemic rats by preventing VDR heterodimerization (Sela-Brown et al., 1998). These results all support the importance of RXR in the calciotropic effects of VDR.

IV. REGULATION OF ANTIPROLIFERATIVE EFFECTS OF CALCITRIOL BY RXR

Induction of growth arrest and differentiation by calcitriol and its analogues in certain cancer cells are well documented, and synthetic calcitriol analogues are now used in clinical trials for the treatment of breast and prostate cancer. However, aggressive cancer lines are often resistant to the antiproliferative effects of calcitriol. Some lines even respond to calcitriol with growth stimulation. Vitamin D receptor content and transcriptional activity are not good predictors of the growth inhibitory effects of calcitriol, suggesting that other factors have significant roles (Zhuang et al., 1997). This may be explained by differential calcitriol sensitivity of genes involved in calcemic effect versus those involved in growth regulation (Akutsu et al., 2001).

Several studies demonstrated that RXR plays a role in the control of growth inhibitory effects of calcitriol. Increased and decreased RXRα levels were generated in human monoblastic leukemia cells by stable expression of sense or antisense RXRα mRNA. Studies in these cells showed that increased or decreased sensitivity to the growth inhibitory effect of calcitriol correlates with RXR expression (Brown et al., 1997). Similarly, ovarian tumor cells lost sensitivity to the growth inhibitory effect of retinoic acid after the expression of antisense RXRα mRNA (Wu et al., 1997). We found that accelerated and aberrant proteasomal degradation of RXRα causes resistance to the growth inhibitory effects of calcitriol in rat osteosarcoma cells (ROS) without affecting ligand-induced reporter activity. Either expression of recombinant RXRα or inhibition of proteasomal degradation reversed this resistance. In ROS cells, the aberrant degradation of RXRα led to the accumulation of a 45-kDa RXR fragment, truncated at the C-terminus. When this truncated recombinant RXRα was expressed in CV-1 cells, it had a dominant negative effect on the ability of calcitriol to inhibit cell growth (Prufer et al., 2002). Another study showed that in *ras*-transformed keratinocytes the phosphorylation of S260 in human RXRα caused resistance to the growth inhibitory effects of calcitriol that can be reversed by treatment with mitogen-activated protein (MAP) kinase inhibitor (Solomon et al., 2001). Interestingly, alternative splicing can also generate dominant negative RXR isoforms in cancer cells (Mahajna et al., 1997).

These studies indicate that changes in functional RXR availability have profound influence on VDR-mediated antiproliferative effects. Such changes can occur as a result of RXR transcriptional regulation, degradation, phosphorylation, or sequestration by the ligand-induced activation of other dimerization partners.

V. RXR AS A DIMERIZATION PARTNER

It is well known that dimerization with other partners limits the availability of RXR to dimerize with VDR and thus influences VDR functions. RXRs are promiscuous partners for several nuclear receptors (Table I).

Many of these receptors serve as lipid sensors, regulating the transcription of genes involved in lipid metabolism, storage, transport, and elimination. In addition, RXR heterodimers often regulate cell proliferation and differentiation. This pleiotrophy of functions has an evolutionary origin; RXR and its homologues first regulated transcription of genes involved in morphogenesis, and only later in evolution did RXR recruit diverse receptors, coregulators, and transcription factors to regulate more specialized functions (Laudet, 1997). The complexity of functions that require RXR, taken together with relatively low expression of RXR in many tissues, creates a balance and a competitive environment for multiple signaling pathways.

The question still remains, why do all these receptors require RXR? Why can't VDR, for example, act as a homodimer or monomer? Part of the answer may come from research on the effects of RXR on VDR structure and function.

VI. RXR EFFECTS ON THE PHYSICOCHEMICAL PROPERTIES OF VDR

Although several experimental results indicate that binding of RXR induces conformational changes to VDR, very little is known about these changes. Binding experiments indicated a conformational change, as the ligand-binding capacity of the monomeric VDR was only 10% of that of the VDR/RXR complex (Quack and Carlberg, 2000). Limited protease digestion and gel-shift clipping experiments further demonstrated that binding of RXR induces conformational changes to VDR (Quack and Carlberg, 2000). In addition, conformational changes induced by ligand binding stabilized VDR/RXR dimers (Carlberg et al., 2001). More detailed studies with quantitative dimerization assays and mutational analysis revealed a role for the AF-2 domain in this stabilization (Liu et al., 2001). Studies with RAR and RXR showed that dimerization also increases receptor solubility (Li et al., 1997). Such changes in the solubility of VDR may also take place as a result of dimerization with RXR. Finally, studies with RXR-specific ligands demonstrated that RXR can activate RAR without the presence of an RAR ligand (Jurutka et al., 1996). A similar "phantom ligand" effect of RXR has also been shown for VDR (Kephart et al., 1996), suggesting that the effect of RXR on VDR conformation may be in some ways similar to the effect of ligand binding.

TABLE I Heterodimerization Partners of RXR

Vitamin D receptor (VDR)	Bugge et al., 1992; MacDonald et al., 1993[a]; Zhao et al., 1997[b]; Dong and Noy, 1998[c]; Prufer et al., 2000[d]; Liu et al., 2001[e]
Thyroid receptor (TR)	Bugge et al., 1992; Berrodin et al., 1992[f]; Sugawara et al., 1993[f]; Meier et al., 1993; Lee et al., 1994[a]; Nagaya et al., 1996[f]; Reginato et al., 1996[f]; Collingwood et al., 1997[b]; Huo et al., 1997[f,g]; Kakizawa et al., 1997[b]
Retinoic acid receptor (RAR)	Bugge et al., 1992; Berrodin et al., 1992[f]; Marks et al., 1992[f,g]; Nagpal et al., 1993[a]; Venepally et al., 1997[h]; Depoix et al., 2001[b]
v-ErbA, viral homologue of TRα	Chen and Privalsky, 1993
Peroxisome proliferator-activating receptor (PPAR)	Miyata et al., 1994[b]; Green and Wahli, 1994; Juge-Aubry et al., 1995; Marcus et al., 1995[a]; Huan et al., 1995[a]
NGFI-B	Perlmann and Jansson, 1995; Zetterstrom et al., 1996
Nurr1	Perlmann and Jansson, 1995; Zetterstrom et al., 1996
Benzoate X receptor (BXR)	Blumberg et al., 1998[a,b]; Nishikawa et al., 2000[f]
Orphan receptor 1 (OR1)	Teboul et al., 1995[a]; Wiebel and Gustafsson, 1997[a,f]; Wiebel et al., 1999[a,f]
Constitutive androstane receptor (CAR)	Choi et al., 1997[e]; Forman et al., 1998[a]; Honkakoski et al., 1998[a]
Pregnane X receptor (PXR)	Kliewer et al., 1998[f]; Wan et al., 2000[i]
Farnesoid X receptor (FXR)	Seol et al., 1995[b]; Zavacki et al., 1997[a,f]; Makishima et al., 1999[a]; Wan et al., 2000[i]; Ananthanarayanan et al., 2001[a]
Liver X receptor (LXR)	Willy et al., 1995[a]; Wan et al., 2000[i]
Ecdysone receptor (EcR)	Arbeitman and Hogness, 2000
Cardiac-enriched RXR-interacting protein (CERIP)	Cresci et al., 1999[f]
Short heterodimer partner (SHP)	Seol et al., 1996, 1997[a,b]
Chicken ovalbumin upstream promoter transcription factor (COUP-TF)	Kliewer et al., 1992[a,f]; Cooney et al., 1993

[a] Transactivation assays.
[b] Yeast or mammalian two-hybrid system.
[c] Fluorescence anisotropy titrations.
[d] In vivo, fluorescence resonance energy transfer using green fluorescent protein chimeras of VDR and RXR.
[e] Quantitative dimerization assay with in vitro translated VDR and GST fusion proteins.
[f] Electrophoretic mobility shift assays.
[g] Coimmunoprecipitation.
[h] Size-exclusion fast protein liquid chromatography and sucrose density gradient sedimentation.
[i] In vivo, knockout mouse.

Hopefully, the cocrystallization of VDR and RXR will soon reveal the details of these structural changes.

VII. RXR EFFECTS ON DNA BINDING AND TRANSCRIPTIONAL ACTIVITIES

It is well established that binding of RXR increases the affinity of VDR for DNA and promotes VDR binding specificity. Heterodimers bind to consensus DNA sequences in the promoter regions of their target genes, referred to as vitamin D response elements (VDREs). These VDREs are comprised of 6 base-pair half-elements RGKTCA (R = A or G, K = G or T), which are in most cases arranged in direct repeats separated by three nucleotides (DR3). The DBD of VDR contacts the 3' half-site, whereas the DBD of RXR contacts the 5' half-site. The G at position 3 of the spacer is also conserved in several VDREs (Haussler *et al.*, 1998). There are also examples of other VDRE types: the rat Pit-1, which is a DR spaced by four nucleotides (DR4), the human phospholipase C-γ1, which is spaced by six nucleotides (DR6), and the mouse c-fos, which is an inverted palindrome spaced by nine nucleotides (IP9). Transcription of target genes is either positively or negatively controlled, as summarized in Table II.

For transcriptional repression, unliganded VDR binds to RXR and brings the AF2 domain of RXR into the active orientation. The heterodimer then binds to DNA and recruits corepressors with histone deacetylase activity (Polly *et al.*, 2000), promoting a repressive nucleosome configuration of chromatin.

Calcitriol binding induces a conformational change in VDR, promoting dimerization with RXR and cooperative binding to DNA. This conformational change turns helix 12 of VDR into a closed position and exposes the AF-2 domain, which binds corepressors and coactivators. Agonist binding induces dissociation of corepressors, such as NcoR, SMRT (Tagami *et al.*, 1998), and Alien (Polly *et al.*, 2000) and stimulates the binding of p160 family coactivators such as SRC-1, TIF2, and RAC3. These coactivators then recruit additional CBP/p300 coactivators (Freedman, 1999; Lee and Lee, 2001), which induce chromatin decondensation through histone acetylation/methylation, thereby opening a space in the promoter regions for the DRIP complex binding (Rachez *et al.*, 1999). Either independently or in a step-wise fashion this DRIP complex then recruits or stabilizes the basal transcriptional machinery (Freedman, 1999; Lee and Lee, 2001). Helices 12 of both VDR and RXR (Rachez *et al.*, 2000) and helix 3 of VDR (Kraichely *et al.*, 1999) are involved in this process.

Little is known about the termination of the VDR/RXR interaction with DNA. One possibility is that interaction of activated receptors with SUG1 targets the receptor for ubiquitination and subsequent degradation by the 26 S proteasome complex, and then this degradation could terminate transcription (Masuyama and MacDonald, 1998). However, our photobleaching experiments argue against rapid and permanent destruction of receptors after DNA binding, instead showing that

TABLE II Genes Regulated by VDR/RXR Heterodimers

Gene	Reference
Rat, human osteocalcin (+)[a]	MacDonald et al., 1991; Demay et al., 1990; Markose et al., 1990; Terpening et al., 1991; Ozono et al., 1990; Morrison et al., 1989
Mouse osteopontin (+)	Noda et al., 1990
Avian β_3 integrin (+)	Cao et al., 1993
Rat, 24-OHase (+)	Zierold et al., 1994; Ohyama et al., 1994; Dwivedi et al., 2000
Mouse calbindin-D28$_k$(+)	Gill and Christakos, 1993
NPT2 (+)	Taketani et al., 1997
p21 (+)	M. Liu et al., 1996
Cytochrome P-450 3A (CYP3A4) (+)	Thummel et al., 2001
Fructose-1,6-bisphosphatase (FBPase) (+)	Fujisawa et al., 2000
Transforming growth factor-β (TGF-β) (+)	Wu et al., 1999
Carbonic anhydrase II (CAII) (+)	Quelo et al., 1994
Mouse c-fos (+)	Schrader et al., 1997
Human phospholipase C-γ1 (+)	Xie and Bikle, 1997
Pit-1 (+)	Rhodes et al., 1993
Human atrial natriuretic factor (+)	Kahlen and Carlberg, 1996
Avian, human PTH (−)	S. Liu et al., 1996; Mackey et al., 1996; Demay et al., 1992
Mouse osteocalcin (−)	Lian et al., 1997
Rat bone sialoprotein (BSP) (−) VDR homodimer	Kim et al., 1996
Cyclic AMP-dependent protein kinase inhibitor (PKI) (−)	Rowland-Goldsmith et al., 1999
Rat PTHrP (−)	Kremer et al., 1996; Falzon, 1996
Slow myosin heavy chain 3 (slow MyHC3) (−)	Wang et al., 2001
Interleukin 12 (IL-12) (−)	D'Ambrosio et al., 1998
RA receptor β2 (RARβ2) (−)	Jimenez-Lara and Aranda, 1999
Human atrial natriuretic peptide (hANP) (−)	Wu et al., 1995
Interferon-γ (IFN-γ) (−)	Cippitelli and Santoni, 1998
Nuclear factor of activated T cells (NFAT) (−)	Takeuchi et al., 1998

[a] (+) transactivation; (−) transrepression.

receptors from one target site rapidly move to and bind another site (unpublished observations). Another possibility is the interaction with calreticulin that inhibits DNA binding of VDR and most other nuclear receptors (Holaska et al., 2001; Wheeler et al., 1995). The problem with this hypothesis is that there is not enough calreticulin in the nucleus for all these receptors. The interaction of receptors with heat shock protein (hsp) could also play a role in the termination of signal (Liu and DeFranco, 1999; Freeman and Yamamoto, 2002). If so, the ability of RXR to associate with hsp90, p23, and hsp70 could be important for termination of hormone effects. To date the effect of dimerization on the termination of calcitriol-dependent transcription has yet to be explored.

Without RXR, VDR is unable to interact with the transcription machinery. Consequently any changes in the ability of VDR or RXR to form dimers would prevent both constitutive and ligand-dependent VDR functions. Mutational analysis, structural studies, and cell-free transcription systems revealed the receptor domains that participate in dimerization and promise to elucidate the factors influencing dimerization.

VIII. DIMERIZATION INTERFACES

Large deletion studies, mutational analysis, and structural analysis showed that VDR and RXR dimerize both through their DBD and LBD (Table III). In addition, flexibility of the hinge-domain is also essential for dimerization and DNA binding.

Structural studies indicate that the DNA spacer determines which dimerization surface on RXR is engaged in forming an interface with its partner (Rastinejad, 2001). One dimerization interface is located in the first Zn-finger of the VDR, where amino acid 37 of the VDR contacts residues in the second Zn-finger of the RXR on DR3 (Rastinejad et al., 1995). The second dimerization interface of VDR

TABLE III Heterodimerization Interfaces

	VDR	RXR
DNA binding domain, Zn-fingers and T-box	N37, K91, E92 (Rastinejad et al., 1995)[a]; K91, E92 (Hsieh et al., 1995)[b]; AA 80–124 (Nishikawa et al., 1995)[c]	R182, Q183, R186 (Rastinejad et al., 1995)[a]; AA 176–195 and AA 206–217 (Zechel et al., 1994a,b)[c]
Ligand binding domain, helix 3–4 and 7–10	AA 244–263, D258, I248 (Rosen et al., 1993)[b,c]; F244, L254, Q259, L262 (Whitfield et al., 1995)[b]; AA 239–269 (Jin et al., 1996)[a]; AA 359–402 (Nishikawa et al., 1995)[c]; AA 317–394 (Jin et al., 1996)[c]; I384, Q385 (Chen et al., 1999)[b]	AA 413–443, L418, L419, L422 (Zhang et al., 1994)[b,c]; AA 389–429 (Perlmann et al., 1996); A416, R421 (Lee et al., 1998, 2000)[b]; L419, L420 (Chen et al., 1999)[b]

[a] Shown using structural analysis.
[b] Shown using site-directed mutagenesis.
[c] Shown using deletion analysis.

is situated in the T-box region just C-terminal to the second Zn-finger. Amino acids 91 and 92 mediate heterodimerization between the VDR and RXR DBDs on DR3 (Hsieh et al., 1999). Possibly, the DBDs of the VDR and the RXR do not contact each other when they bind to IP9-type VDRE (Carlberg et al., 2001).

In the LBD of VDR two subdomains corresponding to portions of helices 3–4 and helices 7–10 mediate ligand-dependent dimerization. Consistently, natural mutations have been reported in patients with HVDRR in this region of the VDR that compromise dimerization with RXR, including Q259G, R291C, F251C, I314S, and R391C (Cockerill et al., 1997; Malloy et al., 2001; Whitfield et al., 1996). In the RXR only the C-terminal part of the LBD appears to play a role in dimerization; this part includes portions of helices 7–10.

IX. REGULATION OF VDR/RXR DIMERIZATION

The most important factor that regulates dimerization is the ligand. The role of ligand in regulating the interaction between VDR and RXR has been shown by surface plasmon resonance experiments (Cheskis and Freedman, 1996). Moreover, two-hybrid assays demonstrated that the relative potencies of calcitriol analogues to induce dimerization correlate with their potencies to regulate transcription (Zhao et al., 1997). The effect of ligand on dimerization of RXR with other partners has also been shown. Pull-down and mammalian two-hybrid assays demonstrated that RAR agonists, but not antagonists, increase RXR dimerization with RAR in the absence of DNA (Depoix et al., 2001). Thyroid hormone has been shown to promote dimerization of RXR with TR prior to DNA binding (Collingwood et al., 1997).

Receptor phosphorylation also regulates RXR dimerization with its partners. Phosphorylation of TRβ increased dimerization with RXR and dephosphorylation prevented dimerization and dimer binding to target DNA (Bhat et al., 1994). Conversely, phosphorylation of RAR at serine-157 prevented dimerization with RXRα (Delmotte et al., 1999). VDR is phosphorylated at serine-51 and serine-208, and hyperphosphorylation has been shown to increase transcriptional activity (Hsieh et al., 1991; Jurutka et al., 1996). However, very little is known about the effect of VDR phosphorylation on dimerization function. Results of one study indicated that hyperphosphorylation of serine-208 of VDR has no effect on dimerization with RXR (Jurutka et al., 1996). Phosphorylation of RXRα may have a negative influence on dimerization; in ras-transformed keratinocytes phosphorylation of RXRα disrupted RXR dimerization with VDR and caused insensitivity to the growth inhibitory effects of calcitriol (Solomon et al., 1998, 2001).

Regulation of RXR expression and commitment of RXR to different nuclear receptors could also influence VDR functions by limiting dimerization. Several studies investigated the effect of RXR stoichiometry on VDR functions and showed that sequestration of RXR by the activation of other partner receptors attenuates VDR-mediated responses. For example, transfection with TR repressed VDR-mediated transcription by competing for RXR, as the repression was reversed by cotransfection of RXR (Raval-Pandya *et al.*, 1998). In another study, RXR-specific ligand repressed calcitriol-dependent accumulation of osteocalcin mRNA in ROS 17/2.8 osteosarcoma cells by promoting homodimerization of RXR (MacDonald *et al.*, 1993; Thompson *et al.*, 1998). As we discussed earlier, changes in functional RXR levels are particularly important for the antiproliferative effects of calcitriol.

Molecular chaperones may also influence RXR dimerization and dimer partner functions. This is suggested by data showing that depletion of hsc70 in yeast decreases VDR and RXR concentrations and compromises VDR transcriptional activity (Lutz *et al.*, 2001). Another recent study demonstrated that hsp90 and hsc70 are required for ecdysone receptor dimerization with the *Drosophila* homologue of RXR, EcR (Arbeitman and Hogness, 2000). Clearly, the roles of chaperones in heterodimerization warrant further investigation.

X. SUBCELLULAR LOCALIZATION OF VDR/RXR HETERODIMERS

Regulation of subcellular localization serves as an additional control mechanism for many nuclear receptors and signaling proteins. Well-known examples of transcriptional regulators that move from the cytoplasm into the nucleus under the control of signals from the cell surface include the tubby proteins, NFATs, SMADs, STATs, SREBP, and NF-κB (Cantley, 2001). To explore this type of regulation for VDR and RXR and their dimerizing partners, one must first locate these receptors.

For decades, immunohistological and cell fractionation studies generated controversial results and led to a static model of receptor localization. Immunohistological studies often found RXR (Reichrath *et al.*, 1997; Sugawara *et al.*, 1997) and VDR exclusively in the nucleus, whereas certain immunohistological techniques and cell fractionation experiments showed RXR and VDR also in the cytoplasm (Barsony *et al.*, 1990; Janssen *et al.*, 1999; Prufer *et al.*, 1999). Even more troubling, the resolution of these techniques did not allow localization of VDR/RXR dimer.

The ability to generate multicolor fluorescent protein chimeras of nuclear receptors provided the means to study receptor distribution, shuttling, and protein–protein interactions in living cells by microscopy. From these studies emerged a new concept of nuclear hormone action that accommodated the dynamic nature of these receptors.

First, the steady-state distribution of unliganded and liganded VDR and RXR was clarified by microscopy in living cells. Transcriptionally active fluorescent protein chimeras of VDR were generated first in our laboratory (Racz and Barsony, 1999) and later in other laboratories (Michigami *et al.*, 1999; Sunn *et al.*, 2001). Microscopy in living cells demonstrated that VDR partitions between the cytoplasm and the nucleus (Racz and Barsony, 1999; Sunn *et al.*, 2001). When we compared VDR distribution in living and formaldehyde-fixed cells it became clear that the apparent exclusively nuclear localization of VDR is a fixation artifact (Prufer and Barsony, 2002). In addition, four studies showed in living cells a calcitriol-induced redistribution of VDR from the cytoplasm into the nucleus (Michigami *et al.*, 1999; Prufer *et al.*, 2000; Prufer and Barsony, 2002; Racz and Barsony, 1999). The steady-state distribution of RXR was predominantly nuclear even in living cells (Baumann *et al.*, 2001; Prufer *et al.*, 2000).

More importantly, with a suitable pair of fluorescent protein tags on the VDR and RXR we can detect the location of heterodimers in living cells (Prufer *et al.*, 2000). Fluorescence resonance energy transfer (FRET) detects interaction between two molecules when they are separated by less than 50 nm. In this case, illuminating the blue donor results in emission of light that in turn excites the green acceptor molecule. This emission from the acceptor is detected allowing spatial and temporal determination of binding. We labeled VDR with green fluorescent protein (GFP) at the N-terminus (GFP-VDR) and RXR with blue fluorescent protein (BFP) at the C-terminus (RXR-BFP); only this opposing orientation allowed dimerization between the tagged receptors (Prufer *et al.*, 2000). For control, we created a mutation in the RXR that compromised dimerization (*hd*RXR-BFP). Signal from FRET experiments in COS-7 cells shows steady-state nuclear localization of GFP-VDR/RXR-BFP dimers after calcitriol addition (100 n*M* for 1 h at 37°C) (Fig. 1a; pseudocolor images). For a control, Fig. 1b shows signal from a cell expressing the GFP-VDR and *hd*RXR-BFP. The image from these cells shows the weak nonspecific signal. In other experiments without calcitriol treatment, the FRET signal was detectable both in the nucleus and in the cytoplasm (Prufer *et al.*, 2000). Taken together these results clarify that VDRs and RXRs form dimers without the ligand and that calcitriol binding initiates nuclear accumulation of dimers.

XI. TRAFFICKING OF VDR AND RXR BETWEEN CYTOPLASM AND NUCLEUS

Studies with fluorescent protein chimeras not only clarified the steady-state distribution of VDR and RXR and their dimers, but also led to the realization that VDR and RXR both constantly shuttle between the cytoplasm and the nucleus. Over the past couple of years, the shuttling of several transcription factors has been detected (Table IV). Heterokaryon experiments were used first to demonstrate the relatively slow shuttling of TR (Baumann *et al.*, 2001); later this method was used to examine the shuttling of other nuclear receptors. Recently, microinjection of receptors into *Xenopus laevis* oocytes also demonstrated shuttling of TR (Bunn *et al.*, 2001). We used photobleaching experiments, such as the monitoring of fluorescence recovery after photobleaching (FRAP) and fluorescence loss in photobleaching (FLIP), to demonstrate that both GFP-VDR and YFP-RXR shuttle across the nuclear membrane. This technique is gaining popularity, as evidenced by abstracts presented at the last meeting of the American Society for Cell Biology. These techniques allowed us to characterize the nucleocytoplasmic shuttling

TABLE IV Nucleocytoplasmic Shuttling of Transcription Factors

Androgen receptor (AR)	Georget et al., 1997, 1998; Poukka et al., 2000
Glucocorticoid receptor (GR)	Hache et al., 1999; Liu et al., 1999; Liu and DeFranco, 1999, 2000; Yang et al., 1997; Yang and DeFranco, 1996
Mineralocorticoid receptor (MR)	Fejes-Toth et al., 1998
Estrogen receptor (ER)	Htun et al., 1999; Dauvois et al., 1993
Vitamin D receptor (VDR)	Prufer et al., 2000; Racz and Barsony, 1999
Thyroid hormone receptor (TR)	Baumann et al., 2001; Bunn et al., 2001
Progesterone receptor (PR)	Chandran and DeFranco, 1992; Tyagi et al., 1998; Guiochon-Mantel, 1991, 1994
Aryl hydrocarbon receptor (AhR)	Davarinos and Pollenz, 1999; Richter et al., 2001
IκBα	Huang et al., 2000; Johnson et al., 1999; Rodriguez et al., 1999; Sachdev et al., 2000
NGFI-B	Katagiri et al., 2000
STAT1	McBride et al., 2000
Tomato heat stress transcription factor (HsfA2)	Heerklotz et al., 2001
p53	Akakura et al., 2001; Smart et al., 1999; Zhang and Xiong, 2001
HIV-1 Rev transactivator	Meyer and Malim, 1994; Love et al., 1998
Nuclear factors of activated T cells (NFAT)	Kehlenbach et al., 1998
bZIP transcriptional activator RSG	Igarashi et al., 2001
Cell cycle-regulated transcription factor Ace2p	Jensen et al., 2000
DAX-1	Lalli et al., 2000
Goosecoid-like (gscl)	Galili et al., 2000
Yeast transcription factor Pho4	Kaffman et al., 1998
Class II transactivator (CIITA)	Kretsovali et al., 2001; Cressman et al., 2001
The metal-regulatory transcription factor 1 (MTF-1)	Saydam et al., 2001
COUP-TF	Vlahou and Flytzanis, 2000
Forkhead transcription factors FKHRL1	del Peso et al., 1999

speed; the recovery half-times were 15–30 min for the unliganded and 5–15 min for the liganded VDR, and 20–30 min for the unliganded RXR (Prufer and Barsony, 2002).

We then explored the protein–protein interaction of VDR and RXR that is significant for the regulation of receptor shuttling. Our studies indicated that VDR and RXR shuttling takes place through interactions with the nuclear import and export machinery. The role of import receptors has been extensively studied. Importin

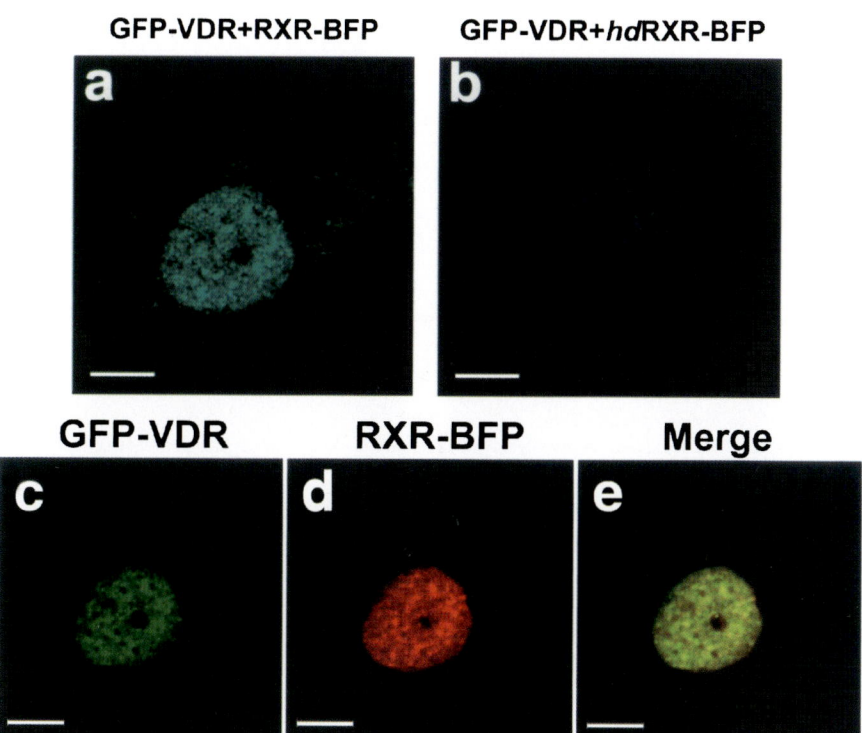

FIGURE 1 VDR and RXR dimers are shown in the nucleus with FRET (a, b) and colocalization (c–e) demonstrates formation of heterodimer foci in the nucleus after calcitrol treatment. Images were taken from living cells by confocal microscopy (Zeiss LSM410). Cells expressed fluorescent protein chimeras of wild-type VDR and either wild-type or heterodimerization incompetent mutant RXR as indicated. Bars = 10 μm.

α binds to lysine/arginine-rich consensus sequences within their cargo proteins, so called nuclear localization sequences (NLS), and then this complex binds to importin β. The latter interacts with the nuclear pore complex in a RanGDP-dependent process and carries the cargo proteins into the nucleus. Within the nucleus, this import complex dissociates and importin α and β return to the cytoplasm. Most nuclear receptors contain one or more classical NLS. In addition to the constitutive NLSs in the DBD, a ligand-regulated NLS was identified in the LBD androgen receptor (AR) (Poukka *et al.*, 2000) and glucocorticoid receptor (GR) (Picard and Yamamoto, 1987). Several nuclear proteins utilize other import pathways; their arginine-rich NLS binds directly to importin β (Palmeri and Malim, 1999). Finally, indirect nuclear import through a piggyback mechanism is also possible. We analyzed the NLS within the DBD of VDR (Prufer *et al.*, 2000) and identified a similar NLS in the RXR (Prufer and Barsony, 2002). Within this NLS of RXR, a lysine or arginine corresponding to R54 of VDR is conserved through all the nuclear receptors that act as heterodimers with RXR, suggesting that this NLS-like region could mediate nuclear import of other RXR partners. Instead of this lysine, the homodimerizing steroid receptors all have a conserved glycine in the same position. The importance of this amino acid in steroid receptor function has not yet been explored. Another NLS has been found in the hinge region of the VDR (Michigami *et al.*, 1999). Because the hinge region is less conserved among nuclear receptors than the DNA-binding domain, this NLS is unique to the VDR. An additional NLS may be located within the second zinc-finger of RXR, as this sequence is homologous to the NLS found in the orphan nuclear receptor TR2 (Yu *et al.*, 1998). The significance of this sequence for RXR import has yet to be determined.

Nuclear export of proteins and RNA is mediated by exportins. These export receptors bind specific sequences, known as nuclear export signals (NES), on their cargos. Exportins also bind RanGTP and interact with the nuclear pore complex. The best characterized export receptor is CRM-1, which binds to leucine-rich NES in cargo proteins. This binding can be inhibited by leptomycin B (LMB), thus inhibiting the export of CRM-1/cargo complexes. Other recently identified exportins are CAS, an importin family member that binds to large NES in the importin family of proteins, and exportin-t, which is responsible for tRNA export. Leucine-rich NES have not been identified in nuclear receptors. However, a recent study demonstrated that the DBD of nuclear receptors contains a different NES. Mutations changing FF to AA within this conserved DBD region prevented nuclear export of the GR and the androgen receptor (AR) (Black *et al.*, 2001). We found that the same mutations within the DBD decrease nuclear export of VDR and RXR (Prufer and Barsony, 2002). Interestingly, the same amino acids are important for calreticulin binding. Because calreticulin content of cells correlated with export efficiency of GR and AR, it was suggested that calreticulin serves as an export receptor for nuclear receptors (Black *et al.*, 2001). However, the interaction of calreticulin with the nuclear pore complex has not been demonstrated, leaving open the possibility that it still needs to utilize an export receptor to cross the nuclear membrane. Thus, the export receptors for nuclear receptors and many other NES-containing proteins (Li and Yen, 2002) remain to be investigated.

Some proteins may be exported by a piggyback mechanism through association with other shuttling proteins or RNA (Segal et al., 2001). Such indirect export has been reported for the RXR via the phosphorylation-dependent export of the orphan nuclear receptor NGFI-B, which contains leucine-rich NESs (Katagiri et al., 2000). Another example is the 14-3-3 protein which interacts with RIP140 corepressor and exports it from the nucleus, thus enhancing transcriptional activity of GR (Zilliacus et al., 2001). Because RIP140 binds to VDR (Masuyama et al., 1997) and RXR (Wiebel et al., 1999) as well, this mechanism may be important for the regulation of calcitriol action. We found that LMB inhibits nuclear export of the unliganded VDR, possibly by inhibiting the export of a VDR-associated protein (Prufer and Barsony, 2002). Calcitriol addition prevented LMB sensitivity of VDR export, suggesting that the protein involved in the export of unliganded VDR may be a corepressor.

Taken together these results show that multiple protein interactions mediate the nuclear import and export of VDR and RXR. Our luciferase reporter assays showed that these interactions are important for transcriptional activities of VDR and RXR (Prufer et al., 2000; Prufer and Barsony, 2002).

XII. REGULATION OF UNLIGANDED VDR AND RXR SHUTTLING

Nucleocytoplasmic shuttling of transcription factors is regulated by several mechanisms, including docking/retention, phosphorylation, and dimerization.

An example for regulation by nuclear retention is β-catenin, which accumulates in the nucleus upon overexpression of TCF/Lef-1 transcription factors and by XRCF-3 binding (Wiechens and Fagotto, 2001). Another example is the small nuclear RING finger protein SNURF, which binds to the AR and increases its nuclear retention (Poukka et al., 2000). We found with permeabilization experiments that nuclear retention/docking and association with nuclear matrix plays a negligible role in the nuclear localization of unliganded VDR and RXR (Prufer and Barsony, 2002). Molecular chaperones such as hsp90 have been implicated in the cytoplasmic docking of unliganded GR. It is generally believed that hsp90 masks the NLS of the GR until the ligand induces a conformational change. Docking of mineralocorticoid receptors (MR) at the endoplasmic reticulum is mediated by binding to 11β-hydroxysteroid dehydrogenase (Odermatt et al., 2001). Cytoplasmic retention of VDR and RXR has not been investigated.

An example for regulation of export by phosphorylation is cyclin B. In interphase, cyclin B resides in the cytoplasm due to its efficient export from the nucleus. At the G_2/M transition, phosphorylation of cyclin B at serine-113 inhibits its binding to the CRM-1 export receptor, thus resulting in nuclear accumulation of cyclin B (Yang et al., 2001). Phosphorylation of all four serines of cyclin B accelerates nuclear import; this too contributes to its nuclear localization. Phosphorylation of

VDR and RXR may have an effect on their subcellular mobility; this has yet to be explored.

Dimerization can also regulate nuclear import of transcription factors and signaling molecules (Lee and Yonehara, 2002). Dimerization with RXR has been shown to increase nuclear retention of TR (Baumann *et al.*, 2001). We have shown that dimerization with RXR regulates nuclear import and export of VDR (Prufer *et al.*, 2000; Prufer and Barsony, 2002). First we found that coexpression of RXR-BFP with GFP-VDR in HEK293 cells resulted in nuclear accumulation of unliganded GFP-VDR, and a heterodimerization mutant of RXR-BFP failed to alter GFP-VDR distribution (Prufer *et al.*, 2000). We then used coexpression experiments with wild-type and NLS mutant GFP-VDR and RXR-BFP to study the effect of dimerization on nuclear import. The *nls*GFP-VDR itself accumulated in the cytoplasm, but coexpression of RXR-BFP brought it into the nucleus. Coexpression of RXR-BFP also restored the ability of *nls*GFP-VDR to induce reporter activity both constitutively and in a calcitriol-dependent manner (Prufer *et al.*, 2000). The *nls*RXR-BFP itself was exclusively in the cytoplasm, and it prevented nuclear import of coexpressed GFP-VDR. Coexpression of *nls*RXR-BFP also inhibited baseline transcriptional activity of GFP-VDR (Prufer and Barsony, 2002). These experiments demonstrated that RXR dominates the import of unliganded VDR.

The first indication that dimerization influences VDR export came from experiments that monitored temperature-sensitive export of GFP-VDR. Lowering the temperature from 37°C to 22°C induced relocalization of GFP-VDR from the nucleus into the cytoplasm. Coexpression of RXR-BFP prevented this GFP-VDR redistribution (Prufer *et al.*, 2000). More detailed studies using permeabilization and photobleaching methods in COS-7 cells showed that although VDR and RXR monomers export at about the same rate (~50% loss from the nucleus after 1 h), dimers are exported at a slower rate. When RXR-BFP was coexpressed, only $38 \pm 13\%$ of GFP-VDR fluorescence was lost during 1 h after permeabilization, whereas $63 \pm 12\%$ was lost when the dimerization-deficient *hd*RXR-BFP was coexpressed ($p < 0.001$). Another effect of dimerization was the loss of LMB sensitivity of VDR export (Prufer and Barsony, 2002). This finding may be explained by a masking of NES by the dimerizing RXR (Prufer and Barsony, 2002). Taken together these data support the notion that dimerization regulates both nuclear import and export of the unliganded VDR. Taken together with the effect of RXR coexpression on TR import and export, our findings suggest that dimerization with RXR could influence the subcellular trafficking of other partner receptors.

XIII. REGULATION OF LIGANDED VDR AND RXR SHUTTLING

It is well known that ligand binding induces a rapid redistribution of several nuclear receptors [GR, MR, AR, TR, and progesterone receptor (PR)] from the cytoplasm into the nucleus. In view of the fact that these receptors constantly shuttle between the nucleus and cytoplasm, we have to reinterpret this redistribution as a change in the balance between nuclear import and export.

We found that calcitriol binding induces a similar redistribution of GFP-VDR within 5–30 min (Racz and Barsony, 1999). To understand these changes, we studied nuclear import and export of the liganded GFP-VDR. Mutating the NLS in the DNA-binding region substantially decreased this hormone-dependent nuclear accumulation, suggesting that interaction with importin α is also involved in the import of the liganded VDR. We also found that deletion of the C-terminal AF-2 region of VDR prevented the hormone-induced change in VDR distribution (Racz and Barsony, 1999). This finding suggests that either another NLS is located in the LBD of the VDR or the binding of a coactivator indirectly facilitates VDR shuttling. Dimerization had a different effect on the import of liganded and unliganded VDR. Coexpression of *nls*RXR-BFP with the wild-type GFP-VDR did not prevent calcitriol-induced nuclear accumulation; instead the liganded GFP-VDR brought *nls*RXR-BFP into the nucleus.

Calcitriol also altered certain VDR export characteristics. Photobleaching and permeabilization assays demonstrated that the export of the liganded VDR is three times faster than the export of the unliganded VDR. In spite of the faster export, liganded VDR accumulated in the nucleus under steady-state conditions, because the import rate increased even more than the export rate. Unlike the unliganded VDR export, the liganded VDR export was not sensitive to LMB, although both unliganded and liganded VDR export was inhibited by mutations within the calreticulin-binding region (Prufer and Barsony, 2002). Mutational analysis showed that the hormone-induced changes in nuclear import and export correlate with VDR transcriptional activities (Prufer *et al.*, 2000; Racz and Barsony, 1999) corroborating the functional significance of receptor trafficking.

Our understanding of the protein–protein interactions that mediate liganded VDR/RXR shuttling is rapidly growing, along with the rapid progress in the general understanding of nuclear protein import and export. This progress will reveal regulatory checkpoints in this phase of receptor activation.

XIV. INTRANUCLEAR TRAFFICKING AND TARGET SITE BINDING OF VDR AND RXR

Recent research demonstrated that steroid receptors are in constant rapid motion within the nucleus and that they accumulate in discrete nuclear foci after hormone binding (McNally *et al.*, 2000; Stenoien *et al.*, 2001; Tyagi *et al.*, 2000). Photobleaching experiments indicated that liganded GFP-GR transiently interacts with an MMTV array; GFP-GR was replaced at this DNA target site within seconds (Barsony *et al.*, 1997; McNally *et al.*, 2000). Furthermore, recent FRET studies in living cells with GFP fusion proteins revealed that the ligand-dependent coactivator binding with RAR and ER lasts just seconds (Llopis *et al.*, 2000). For the VDR and RXR, photobleaching studies with fluorescent protein chimeras revealed that they also move rapidly within the nucleus and accumulate in foci upon agonist treatment (Prufer *et al.*, 2000; Prufer and Barsony, 2002; Racz and Barsony, 1999). Mutations that compromised DNA binding of GFP-VDR similarly compromised its ligand-induced accumulation in intranuclear foci (Prufer *et al.*, 2000). In cells that coexpressed GFP-VDR and RXR-BFP, calcitriol induced RXR-BFP accumulation in foci that also colocalized with GFP-VDR foci (Fig. 1c–e). This colocalization is visualized by merging digital images from GFP-VDR and from RXR-BFP. FRET experiments confirmed that GFP-VDR and RXR-BFP bind to foci as heterodimers (Fig. 1a). Taken together these results demonstrate VDR/RXR heterodimer binding to target DNA sequences in living cells. Understanding the influence of RXR on the intranuclear mobility and target site binding dynamics of VDR still awaits further studies.

XV. OUTLOOK

From these studies a new concept of nuclear receptor function is emerging. This concept is considerably more dynamic than previous models. We now understand that VDR, RXR, and all other nuclear receptors are rapidly moving between the cytoplasm and the nucleus and between intranuclear compartments. We no longer view receptor binding to DNA and to other proteins as permanent, but rather as a transient interaction that takes place successively. This constantly changing environment, and these constantly changing partners, provide a versatile and flexible regulatory network, and superior adaptability.

Dimerization of type II nuclear receptors with RXR, including dimerization of VDR, is a central protein–protein interaction. It influences receptor conformation and nearly all known stages of receptor activation. These include but are not limited to VDR ligand binding, shuttling, interaction with DNA, coregulators, and proteins of the transcription machinery, and VDR degradation. Emerging data point to several factors that regulate VDR dimerization, such as cross-talk between nuclear receptors and other signaling pathways, phosphorylation events, competition for nuclear import and export receptors, heat shock proteins, and the generation of dominant negative RXR forms. Attempts to use the advantages of dimerization go beyond receptor biology. At the MIT Artificial Intelligence Laboratory, a group of researchers harness dimerization of nuclear receptors to build programmable bioorganisms with sensors and actuators precisely controlled by logic circuitry. To aid in this biocircuit design process, simulation and verification tools are developed, which may become instrumental in modeling nuclear receptor actions (Weiss and Knight, 2000).

We have just begun to explore the significance and regulation of receptor dimerization. The insights gained from VDR and RXR knockout mice models clearly demonstrate the existence and importance of ligand-independent VDR function. The prospect of using calcitriol analogues for cancer treatment is more promising now that we understand the importance of increasing RXR expression at the same time. Finally, genetic investigations, including linkage analysis, could eventually lead to the discovery of natural genetic mutations and alleles of RXR and could elucidate the molecular basis for a number of diseases.

REFERENCES

Akakura, S., Yoshida, M., Yoneda, Y., and Horinouchi, S. (2001). A role for Hsc70 in regulating nucleocytoplasmic transport of a temperature-sensitive p53 (p53Val-135). *J. Biol. Chem.* **276**, 14649–14657.

Akutsu, N., Lin, R., Bastien, Y., Bestawros, A., Enepekides, D. J., Black, M. J., and White, J. H. (2001). Regulation of gene expression by 1alpha,25-dihydroxyvitamin D3 and its analog EB1089 under growth-inhibitory conditions in squamous carcinoma cells. *Mol. Endocrinol.* **15**, 1127–1139.

Almasan, A., Mangelsdorf, D. J., Ong, E. S., Wahl, G. M., and Evans, R. M. (1994). Chromosomal localization of the human retinoid X receptors. *Genomics* **20**, 397–403.

Amling, M., Priemel, M., Holzmann, T., Chapin, K., Rueger, J. M., Baron, R., and Demay, M. B. (1999). Rescue of the skeletal phenotype of vitamin D receptor-ablated mice in the setting of normal mineral ion homeostasis: Formal histomorphometric and biomechanical analyses. *Endocrinology* **140**, 4982–4987.

Ananthanarayanan, M., Balasubramanian, N., Makishima, M., Mangelsdorf, D. J., and Suchy, F. J. (2001). Human bile salt export pump promoter is transactivated by the farnesoid X receptor/bile acid receptor. *J. Biol. Chem.* **276**, 28857–28865.

Arbeitman, M. N., and Hogness, D. S. (2000). Molecular chaperones activate the Drosophila ecdysone receptor, an RXR heterodimer. *Cell* **101**, 67–77.

Barletta, F., Freedman, L. P., and Christakos, S. (2002). Enhancement of VDR-mediated transcription by phosphorylation: Correlation with increased interaction between the VDR and DRIP205, a subunit of the VDR-interacting protein coactivator complex. *Mol. Endocrinol.* **16**, 301–314.

Barsony, J., Pike, J. W., DeLuca, H. F., and Marx, S. J. (1990). Immunocytology with microwave-fixed fibroblasts shows 1 alpha,25-dihydroxyvitamin D3-dependent rapid and estrogen-dependent slow reorganization of vitamin D receptors. *J. Cell Biol.* **111**, 2385–2395.

Barsony, J, Carroll, J, McKoy, W., Renyi, I., Gould, D. L., Htun, H., and Hager, G. L. (1997). Intracellular traffick of glucocorticoid receptors: Studies with green fluorescent protein chimeras in living cells. In "Microscopy and Microanalysis" (G. W. Bailey, R. V. W. Dimlich, J. J. McCarthy, and T. P. Pretlov, Eds.), pp. 131–132. Springer-Verlag, New York.

Baumann, C. T., Maruvada, P., Hager, G. L., and Yen, P. M. (2001). Nuclear cytoplasmic shuttling by thyroid hormone receptors. Multiple protein interactions are required for nuclear retention. *J. Biol. Chem.* **276**, 11237–11245.

Berrodin, T. J., Marks, M. S., Ozato, K., Linney, E., and Lazar, M. A. (1992). Heterodimerization among thyroid hormone receptor, retinoic acid receptor, retinoid X receptor, chicken ovalbumin upstream promoter transcription factor, and an endogenous liver protein. *Mol. Endocrinol.* **6**, 1468–1478.

Bhat, M. K., Ashizawa, K., and Cheng, S. Y. (1994). Phosphorylation enhances the target gene sequence-dependent dimerization of thyroid hormone receptor with retinoid X receptor. *Proc. Natl. Acad. Sci. USA* **91**, 7927–7931.

Binkley, N., and Krueger, D. (2000). Hypervitaminosis A and bone. *Nutr. Rev.* **58**, 138–144.

Black, B. E., Holaska, J. M., Rastinejad, F., and Paschal, B. M. (2001). DNA binding domains in diverse nuclear receptors function as nuclear export signals. *Curr. Biol* **11**, 1749–1758.

Blumberg, B., Kang, H., Bolado, J., Jr., Chen, H., Craig, A. G., Moreno, T. A., Umesono, K., Perlmann, T., De Robertis, E. M., and Evans, R. M. (1998). BXR, an embryonic orphan nuclear receptor activated by a novel class of endogenous benzoate metabolites. *Genes Dev.* **12**, 1269–1277.

Bourguet, W., Ruff, M., Chambon, P., Gronemeyer, H., and Moras, D. (1995). Crystal structure of the ligand-binding domain of the human nuclear receptor RXR-alpha [see comments]. *Nature* **375**, 377–382.

Bourguet, W., Vivat, V., Wurtz, J. M., Chambon, P., Gronemeyer, H., and Moras, D. (2000). Crystal structure of a heterodimeric complex of RAR and RXR ligand-binding domains. *Mol. Cell* **5**, 289–298.

Brown, T. R., Stonehouse, T. J., Branch, J. S., Brickell, P. M., and Katz, D. R. (1997). Stable transfection of U937 cells with sense or antisense RXR-alpha cDNA suggests a role for RXR-alpha in the control of monoblastic differentiation induced by retinoic acid and vitamin D. *Exp. Cell Res.* **236**, 94–102.

Bugge, T. H., Pohl, J., Lonnoy, O., and Stunnenberg, H. G. (1992). RXR alpha, a promiscuous partner of retinoic acid and thyroid hormone receptors. *EMBO J.* **11**, 1409–1418.

Bunn, C. F., Neidig, J. A., Freidinger, K. E., Stankiewicz, T. A., Weaver, B. S., McGrew, J., and Allison, L. A. (2001). Nucleocytoplasmic shuttling of the thyroid hormone receptor alpha. *Mol. Endocrinol.* **15**, 512–533.

Cantley, L. C. (2001). Transcription Translocating tubby. *Science* **292**, 2019–2021.

Cao, X., Ross, F. P., Zhang, L., MacDonald, P. N., Chappel, J., and Teitelbaum, S. L. (1993). Cloning of the promoter for the avian integrin beta 3 subunit gene and its regulation by 1,25-dihydroxyvitamin D3. *J. Biol. Chem.* **268**, 27371–27380.

Carlberg, C., Quack, M., Herdick, M., Bury, Y., Polly, P., and Toell, A. (2001). Central role of VDR conformations for understanding selective actions of vitamin D(3) analogues. *Steroids* **66**, 213–221.

Chandran, U. R., and DeFranco, D. B. (1992). Internuclear migration of chicken progesterone receptor, but not simian virus-40 large tumor antigen, in transient heterokaryons. *Mol. Endocrinol.* **6**, 837–844.

Chawla, A., Repa, J. J., Evans, R. M., and Mangelsdorf, D. J. (2001). Nuclear receptors and lipid physiology: Opening the X-files. *Science* **294**, 1866–1870.

Chen, C. H., Sakai, Y., and Demay, M. B. (2001). Targeting expression of the human vitamin D receptor to the keratinocytes of vitamin D receptor null mice prevents alopecia. *Endocrinology* **142**, 5386–5389.

Chen, H. W., and Privalsky, M. L. (1993). The erbA oncogene represses the actions of both retinoid X and retinoid A receptors but does so by distinct mechanisms. *Mol. Cell. Biol.* **13**, 5970–5980.

Chen, S., Costa, C. H., Nakamura, K., Ribeiro, R. C., and Gardner, D. G. (1999). Vitamin D-dependent suppression of human atrial natriuretic peptide gene promoter activity requires heterodimer assembly. *J Biol. Chem.* **274**, 11260–11266.

Cheskis, B., and Freedman, L. P. (1996). Modulation of nuclear receptor interactions by ligands: Kinetic analysis using surface plasmon resonance. *Biochemistry* **35**, 3309–3318.

Choi, H. S., Chung, M., Tzameli, I., Simha, D., Lee, Y. K., Seol, W., and Moore, D. D. (1997). Differential transactivation by two isoforms of the orphan nuclear hormone receptor CAR. *J. Biol. Chem.* **272**, 23565–23571.

Cippitelli, M., and Santoni, A. (1998). Vitamin D3: A transcriptional modulator of the interferon-gamma gene. *Eur. J. Immunol.* **28**, 3017–3030.

Cockerill, F. J., Hawa, N. S., Yousaf, N., Hewison, M., O'Riordan, J. L., and Farrow, S. M. (1997). Mutations in the vitamin D receptor gene in three kindreds associated with hereditary vitamin D resistant rickets. *J. Clin. Endocrinol. Metab.* **82**, 3156–3160.

Collingwood, T. N., Butler, A., Tone, Y., Clifton-Bligh, R. J., Parker, M. G., and Chatterjee, V. K. (1997). Thyroid hormone-mediated enhancement of heterodimer formation between thyroid hormone receptor beta and retinoid X receptor. *J. Biol. Chem.* **272**, 13060–13065.

Cooney, A. J., Leng, X., Tsai, S. Y., O'Malley, B. W., and Tsai, M. J. (1993). Multiple mechanisms of chicken ovalbumin upstream promoter transcription factor-dependent repression of transactivation by the vitamin D, thyroid hormone, and retinoic acid receptors. *J. Biol. Chem.* **268**, 4152–4160.

Cresci, S., Clabby, M. L., and Kelly, D. P. (1999). Evidence for a novel cardiac-enriched retinoid X receptor partner. *J. Biol. Chem.* **274**, 25668–25674.

Cressman, D. E., O'Connor, W. J., Greer, S. F., Zhu, X. S., and Ting, J. P. (2001). Mechanisms of nuclear import and export that control the subcellular localization of class II transactivator. *J. Immunol.* **167**, 3626–3634.

D'Ambrosio, D., Cippitelli, M., Cocciolo, M. G., Mazzeo, D., Di Lucia, P., Lang, R., Sinigaglia, F., and Panina-Bordignon, P. (1998). Inhibition of IL-12 production by 1,25-dihydroxyvitamin D3. Involvement of NF-kappaB downregulation in transcriptional repression of the p40 gene. *J. Clin. Invest.* **101,** 252–262.

Dauvois, S., White, R., and Parker, M. G. (1993). The antiestrogen ICI 182780 disrupts estrogen receptor nucleocytoplasmic shuttling. *J. Cell Sci.* **106**(Pt. 4), 1377–1388.

Davarinos, N. A., and Pollenz, R. S. (1999). Aryl hydrocarbon receptor imported into the nucleus following ligand binding is rapidly degraded via the cytosplasmic proteasome following nuclear export. *J. Biol. Chem.* **274,** 28708–28715.

Delmotte, M. H., Tahayato, A., Formstecher, P., and Lefebvre, P. (1999). Serine 157, a retinoic acid receptor alpha residue phosphorylated by protein kinase C in vitro, is involved in RXR. RARalpha heterodimerization and transcriptional activity. *J. Biol. Chem.* **274,** 38225–38231.

del Peso, L., Gonzalez, V. M., Hernandez, R., Barr, F. G., and Nunez, G. (1999). Regulation of the forkhead transcription factor FKHR, but not the PAX3-FKHR fusion protein, by the serine/threonine kinase Akt. *Oncogene* **18,** 7328–7333.

Demay, M. B., Gerardi, J. M., DeLuca, H. F., and Kronenberg, H. M. (1990). DNA sequences in the rat osteocalcin gene that bind the 1,25-dihydroxyvitamin D3 receptor and confer responsiveness to 1,25-dihydroxyvitamin D3. *Proc. Natl. Acad. Sci. USA* **87,** 369–373.

Demay, M. B., Kiernan, M. S., DeLuca, H. F., and Kronenberg, H. M. (1992). Sequences in the human parathyroid hormone gene that bind the 1,25-dihydroxyvitamin D3 receptor and mediate transcriptional repression in response to 1,25-dihydroxyvitamin D3. *Proc. Natl. Acad. Sci. USA* **89,** 8097–8101.

Depoix, C., Delmotte, M. H., Formstecher, P., and Lefebvre, P. (2001). Control of retinoic acid receptor heterodimerization by ligand-induced structural transitions. A novel mechanism of action for retinoid antagonists. *J. Biol. Chem.* **276,** 9452–9459.

Dolle, P., Fraulob, V., Kastner, P., and Chambon, P. (1994). Developmental expression of murine retinoid X receptor (RXR) genes. *Mech. Dev.* **45,** 91–104.

Dong, D., and Noy, N. (1998). Heterodimer formation by retinoid X receptor: Regulation by ligands and by the receptor's self-association properties. *Biochemistry* **37,** 10691–10700.

Dwivedi, P. P., Omdahl, J. L., Kola, I., Hume, D. A., and May, B. K. (2000). Regulation of rat cytochrome P450C24 (CYP24) gene expression. Evidence for functional cooperation of Ras-activated Ets transcription factors with the vitamin D receptor in 1,25-dihydroxyvitamin D(3)-mediated induction. *J. Biol. Chem.* **275,** 47–55.

Egea, P. F., Mitschler, A., Rochel, N., Ruff, M., Chambon, P., and Moras, D. (2000). Crystal structure of the human RXRalpha ligand-binding domain bound to its natural ligand: 9-cis retinoic acid. *EMBO J.* **19,** 2592–2601.

Falzon, M. (1996). DNA sequences in the rat parathyroid hormone-related peptide gene responsible for 1,25-dihydroxyvitamin D3-mediated transcriptional repression. *Mol. Endocrinol.* **10,** 672–681.

Fejes-Toth, G., Pearce, D., and Naray, F. T. (1998). Subcellular localization of mineralocorticoid receptors in living cells: Effects of receptor agonists and antagonists. *Proc. Natl. Acad. Sci. USA* **95,** 2973–2978.

Feskanich, D., Singh, V., Willett, W. C., and Colditz, G. A. (2002). Vitamin A intake and hip fractures among postmenopausal women. *JAMA* **287,** 47–54.

Fleischhauer, K., McBride, O. W., DiSanto, J. P., Ozato, K., and Yang, S. Y. (1993). Cloning and chromosome mapping of human retinoid X receptor beta: Selective amino acid sequence conservation of a nuclear hormone receptor in mammals. *Hum. Genet.* **90,** 505–510.

Forman, B. M., Tzameli, I., Choi, H. S., Chen, J., Simha, D., Seol, W., Evans, R. M., and Moore, D. D. (1998). Androstane metabolites bind to and deactivate the nuclear receptor CAR-beta. *Nature* **395,** 612–615.

Freeman, B. C., and Yamamoto, K. R. (2002). Disassembly of transcriptional regulatory complexes by molecular chaperones. *Science* **296,** 2232–2235.

Freedman, L. P. (1999). Increasing the complexity of coactivation in nuclear receptor signaling. *Cell* **97,** 5–8.

Fujisawa, K., Umesono, K., Kikawa, Y., Shigematsu, Y., Taketo, A., Mayumi, M., and Inuzuka, M. (2000). Identification of a response element for vitamin D3 and retinoic acid in the promoter region of the human fructose-1,6-bisphosphatase gene. *J. Biochem. (Tokyo)* **127**, 373–382.

Galili, N., Nayak, S., Epstein, J. A., and Buck, C. A. (2000). Rnf4, a RING protein expressed in the developing nervous and reproductive systems, interacts with Gscl, a gene within the DiGeorge critical region. *Dev. Dyn.* **218**, 102–111.

Georget, V., Lobaccaro, J. M., Terouanne, B., Mangeat, P., Nicolas, J. C., and Sultan, C. (1997). Trafficking of the androgen receptor in living cells with fused green fluorescent protein-androgen receptor. *Mol. Cell. Endocrinol.* **129**, 17–26.

Georget, V., Terouanne, B., Lumbroso, S., Nicolas, J. C., and Sultan, C. (1998). Trafficking of androgen receptor mutants fused to green fluorescent protein: A new investigation of partial androgen insensitivity syndrome. *J Clin. Endocrinol. Metab.* **83**, 3597–3603.

Gill, R. K., and Christakos, S. (1993). Identification of sequence elements in mouse calbindin-D28k gene that confer 1,25-dihydroxyvitamin D3- and butyrate-inducible responses. *Proc. Natl. Acad. Sci. USA* **90**, 2984–2988.

Green, S., and Wahli, W. (1994). Peroxisome proliferator-activated receptors: Finding the orphan a home. *Mol. Cell. Endocrinol.* **100**, 149–153.

Guiochon-Mantel, A., Lescop, P., Christin-Maitre, S., Loosfelt, H., Derrot-Applanat, M., and Milgrom, E. (1991). Nucleocytoplasmic shuttling of the progesterone receptor. *Embo. J.* **10**, 3851–3859.

Guiochon-Mantel, A., Delabre, K., Lescop, P., and Milgrom, E. (1994). Nuclear localization signals also mediate the outward movement of proteins from the nucleus. *Proc. Natl. Acad. Sci. USA* **91**, 7179–7183.

Hache, R. J., Tse, R., Reich, T., Savory, J. G., and Lefebvre, Y. A. (1999). Nucleocytoplasmic trafficking of steroid-free glucocorticoid receptor. *J. Biol. Chem.* **274**, 1432–1439.

Haussler, M. R., Whitfield, G. K., Haussler, C. A., Hsieh, J. C., Thompson, P. D., Selznick, S. H., Dominguez, C. E., and Jurutka, P. W. (1998). The nuclear vitamin D receptor: Biological and molecular regulatory properties revealed. *J. Bone Miner. Res.* **13**, 325–349.

Heerklotz, D., Doring, P., Bonzelius, F., Winkelhaus, S., and Nover, L. (2001). The balance of nuclear import and export determines the intracellular distribution and function of tomato heat stress transcription factor HsfA2. *Mol. Cell. Biol.* **21**, 1759–1768.

Holaska, J. M., Black, B. E., Love, D. C., Hanover, J. A., Leszyk, J., and Paschal, B. M. (2001). Calreticulin is a receptor for nuclear export. *J. Cell Biol.* **152**, 127–140.

Honkakoski, P., Zelko, I., Sueyoshi, T., and Negishi, M. (1998). The nuclear orphan receptor CAR-retinoid X receptor heterodimer activates the phenobarbital-responsive enhancer module of the CYP2B gene. *Mol. Cell Biol.* **18**, 5652–5658.

Hsieh, J. C., Jurutka, P. W., Galligan, M. A., Terpening, C. M., Haussler, C. A., Samuels, D. S., Shimizu, Y., Shimizu, N., and Haussler, M. R. (1991). Human vitamin D receptor is selectively phosphorylated by protein kinase C on serine 51, a residue crucial to its trans-activation function. *Proc. Natl. Acad. Sci. USA* **88**, 9315–9319.

Hsieh, J. C., Jurutka, P. W., Selznick, S. H., Reeder, M. C., Haussler, C. A., Whitfield, G. K., and Haussler, M. R. (1995). The T-box near the zinc fingers of the human vitamin D receptor is required for heterodimeric DNA binding and transactivation. *Biochem. Biophys. Res. Commun.* **215**, 1–7.

Hsieh, J. C., Whitfield, G. K., Oza, A. K., Dang, H. T., Price, J. N., Galligan, M. A., Jurutka, P. W., Thompson, P. D., Haussler, C. A., and Haussler, M. R. (1999). Characterization of unique DNA-binding and transcriptional-activation functions in the carboxyl-terminal extension of the zinc finger region in the human vitamin D receptor. *Biochemistry* **38**, 16347–16358.

Htun, H., Holth, L. T., Walker, D., Davie, J. R., and Hager, G. L. (1999). Direct visualization of the human estrogen receptor alpha reveals a role for ligand in the nuclear distribution of the receptor. *Mol. Biol. Cell* **10**, 471–486.

Huan, B., Kosovsky, M. J., and Siddiqui, A. (1995). Retinoid X receptor alpha transactivates the hepatitis B virus enhancer 1 element by forming a heterodimeric complex with the peroxisome proliferator-activated receptor. *J. Virol.* **69**, 547–551.

Huang, T. T., Kudo, N., Yoshida, M., and Miyamoto, S. (2000). A nuclear export signal in the N-terminal regulatory domain of IkappaBalpha controls cytoplasmic localization of inactive NF-kappaB/IkappaBalpha complexes. *Proc. Natl. Acad. Sci. USA* **97,** 1014–1019.

Hughes, M. R., Malloy, P. J., Kieback, D. G., Kesterson, R. A., Pike, J. W., Feldman, D., and O'Malley, B. W. (1988). Point mutations in the human vitamin D receptor gene associated with hypocalcemic rickets. *Science* **242,** 1702–1705.

Huo, B., Dozin, B., and Nikodem, V. M. (1997). Identification of a nuclear protein from rat developing brain as heterodimerization partner with thyroid hormone receptor-beta. *Endocrinology* **138,** 3283–3289.

Igarashi, D., Ishida, S., Fukazawa, J., and Takahashi, Y. (2001). 14-3-3 proteins regulate intracellular localization of the bZIP transcriptional activator RSG. *Plant Cell* **13,** 2483–2497.

Janssen, J. J., Kuhlmann, E. D., Van Vugt, A. H., Winkens, H. J., Janssen, B. P., Deutman, A. F., and Driessen, C. A. (1999). Retinoic acid receptors and retinoid X receptors in the mature retina: Subtype determination and cellular distribution. *Curr. Eye Res.* **19,** 338–347.

Jensen, T. H., Neville, M., Rain, J. C., McCarthy, T., Legrain, P., and Rosbash, M. (2000). Identification of novel Saccharomyces cerevisiae proteins with nuclear export activity: Cell cycle-regulated transcription factor ace2p shows cell cycle-independent nucleocytoplasmic shuttling. *Mol. Cell. Biol.* **20,** 8047–8058.

Jimenez-Lara, A. M., and Aranda, A. (1999). Vitamin D represses retinoic acid-dependent transactivation of the retinoic acid receptor-beta2 promoter: The AF-2 domain of the vitamin D receptor is required for transrepression. *Endocrinology* **140,** 2898–2907.

Jin, C. H., Kerner, S. A., Hong, M. H., and Pike, J. W. (1996). Transcriptional activation and dimerization functions in the human vitamin D receptor. *Mol. Endocrinol.* **10,** 945–957.

Johansson, S., and Melhus, H. (2001). Vitamin A antagonizes calcium response to vitamin D in man. *J. Bone Miner. Res.* **16,** 1899–1905.

Johnson, C., Van Antwerp, D., and Hope, T. J. (1999). An N-terminal nuclear export signal is required for the nucleocytoplasmic shuttling of IkappaBalpha. *EMBO J.* **18,** 6682–6693.

Juge-Aubry, C. E., Gorla-Bajszczak, A., Pernin, A., Lemberger, T., Wahli, W., Burger, A. G., and Meier, C. A. (1995). Peroxisome proliferator-activated receptor mediates cross-talk with thyroid hormone receptor by competition for retinoid X receptor. Possible role of a leucine zipper-like heptad repeat. *J. Biol. Chem.* **270,** 18117–18122.

Jurutka, P. W., Hsieh, J. C., Nakajima, S., Haussler, C. A., Whitfield, G. K., and Haussler, M. R. (1996). Human vitamin D receptor phosphorylation by casein kinase II at Ser-208 potentiates transcriptional activation. *Proc. Natl. Acad. Sci. USA* **93,** 3519–3524.

Jurutka, P. W., Remus, L. S., Whitfield, G. K., Thompson, P. D., Hsieh, J. C., Zitzer, H., Tavakkoli, P., Galligan, M. A., Dang, H. T., Haussler, C. A., and Haussler, M. R. (2000). The polymorphic N terminus in human vitamin D receptor isoforms influences transcriptional activity by modulating interaction with transcription factor IIB. *Mol. Endocrinol.* **14,** 401–420.

Kaffman, A., Rank, N. M., O'Neill, E. M., Huang, L. S., and O'Shea, E. K. (1998). The receptor Msn5 exports the phosphorylated transcription factor Pho4 out of the nucleus. *Nature* **396,** 482–486.

Kahlen, J. P., and Carlberg, C. (1996). Functional characterization of a 1,25-dihydroxyvitamin D3 receptor binding site found in the rat atrial natriuretic factor promoter. *Biochem. Biophys. Res. Commun.* **218,** 882–886.

Kakizawa, T., Miyamoto, T., Kaneko, A., Yajima, H., Ichikawa, K., and Hashizume, K. (1997). Ligand-dependent heterodimerization of thyroid hormone receptor and retinoid X receptor. *J. Biol. Chem.* **272,** 23799–23804.

Katagiri, Y., Takeda, K., Yu, Z. X., Ferrans, V. J., Ozato, K., and Guroff, G. (2000). Modulation of retinoid signalling through NGF-induced nuclear export of NGFI-B. *Nat. Cell Biol.* **2,** 435–440.

Kehlenbach, R. H., Dickmanns, A., and Gerace, L. (1998). Nucleocytoplasmic shuttling factors including Ran and CRM1 mediate nuclear export of NFAT in vitro. *J. Cell Biol.* **141,** 863–874.

Kephart, D. D., Walfish, P. G., DeLuca, H., and Butt, T. R. (1996). Retinoid X receptor isotype identity directs human vitamin D receptor heterodimer transactivation from the 24-hydroxylase vitamin D response elements in yeast. *Mol. Endocrinol.* **10,** 408–419.

Kim, R. H., Li, J. J., Ogata, Y., Yamauchi, M., Freedman, L. P., and Sodek, J. (1996). Identification of a vitamin D3-response element that overlaps a unique inverted TATA box in the rat bone sialoprotein gene. *Biochem. J.* **318**(Pt. 1), 219–226.

Kliewer, S. A., Umesono, K., Heyman, R. A., Mangelsdorf, D. J., Dyck, J. A., and Evans, R. M. (1992). Retinoid X receptor-COUP-TF interactions modulate retinoic acid signaling. *Proc. Natl. Acad. Sci. USA* **89**, 1448–1452.

Kliewer, S. A., Moore, J. T., Wade, L., Staudinger, J. L., Watson, M. A., Jones, S. A., McKee, D. D., Oliver, B. B., Willson, T. M., Zetterstrom, R. H., Perlmann, T., and Lehmann, J. M. (1998). An orphan nuclear receptor activated by pregnanes defines a novel steroid signaling pathway. *Cell* **92**, 73–82.

Kraichely, D. M., Collins, J. J., DeLisle, R. K., and MacDonald, P. N. (1999). The autonomous transactivation domain in helix H3 of the vitamin D receptor is required for transactivation and coactivator interaction. *J. Biol. Chem.* **274**, 14352–14358.

Kremer, R., Sebag, M., Champigny, C., Meerovitch, K., Hendy, G. N., White, J., and Goltzman, D. (1996). Identification and characterization of 1,25-dihydroxyvitamin D3- responsive repressor sequences in the rat parathyroid hormone-related peptide gene. *J. Biol. Chem.* **271**, 16310–16316.

Kretsovali, A., Spilianakis, C., Dimakopoulos, A., Makatounakis, T., and Papamatheakis, J. (2001). Self-association of class II transactivator correlates with its intracellular localization and transactivation. *J. Biol. Chem.* **276**, 32191–32197.

Lalli, E., Ohe, K., Hindelang, C., and Sassone-Corsi, P. (2000). Orphan receptor DAX-1 is a shuttling RNA binding protein associated with polyribosomes via mRNA. *Mol. Cell. Biol.* **20**, 4910–4921.

Laudet, V. (1997). Evolution of the nuclear receptor superfamily: Early diversification from an ancestral orphan receptor. *J. Mol. Endocrinol.* **19**, 207–226.

Lee, J. W., Moore, D. D., and Heyman, R. A. (1994). A chimeric thyroid hormone receptor constitutively bound to DNA requires retinoid X receptor for hormone-dependent transcriptional activation in yeast. *Mol. Endocrinol.* **8**, 1245–1252.

Lee, K. C., and Lee, K. W. (2001). Nuclear receptors, coactivators and chromatin: New approaches, new insights. *Trends Endocrinol. Metab.* **12**, 191–197.

Lee, K. K., and Yonehara, S. (2002). Phosphorylation and dimerization regulate nucleocytoplasmic shuttling of mammalian STE20-like kinase (MST). *J. Biol. Chem.* **277**, 12351–12358.

Lee, M. S., Kliewer, S. A., Provencal, J., Wright, P. E., and Evans, R. M. (1993). Structure of the retinoid X receptor alpha DNA binding domain: A helix required for homodimeric DNA binding. *Science* **260**, 1117–1121.

Lee, S. K., Na, S. Y., Kim, H. J., Soh, J., Choi, H. S., and Lee, J. W. (1998). Identification of critical residues for heterodimerization within the ligand-binding domain of retinoid X receptor. *Mol. Endocrinol.* **12**, 325–332.

Lee, S. K., Lee, B., and Lee, J. W. (2000). Mutations in retinoid X receptor that impair heterodimerization with specific nuclear hormone receptor. *J. Biol. Chem.* **275**, 33522–33526.

Leid, M., Kastner, P., Lyons, R., Nakshatri, H., Saunders, M., Zacharewski, T., Chen, J. Y., Staub, A., Garnier, J. M., and Mader, S. (1992). Purification, cloning, and RXR identity of the HeLa cell factor with which RAR or TR heterodimerizes to bind target sequences efficiently. *Cell* **68**, 377–395.

Li, B., and Yen, T. S. (2002). Characterization of the nuclear export signal of polypyrimidine tract binding protein. *J. Biol. Chem.* **277**, 10306–10314.

Li, C., Schwabe, J. W., Banayo, E., and Evans, R. M. (1997). Coexpression of nuclear receptor partners increases their solubility and biological activities. *Proc. Natl. Acad. Sci. USA* **94**, 2278–2283.

Li, M., Indra, A. K., Warot, X., Brocard, J., Messaddeq, N., Kato, S., Metzger, D., and Chambon, P. (2000). Skin abnormalities generated by temporally controlled RXRalpha mutations in mouse epidermis. *Nature* **407**, 633–636.

Lian, J. B., Shalhoub, V., Aslam, F., Frenkel, B., Green, J., Hamrah, M., Stein, G. S., and Stein, J. L. (1997). Species-specific glucocorticoid and 1,25-dihydroxyvitamin D responsiveness in mouse MC3T3-E1 osteoblasts: Dexamethasone inhibits osteoblast differentiation and vitamin D downregulates osteocalcin gene expression. *Endocrinology* **138**, 2117–2127.

Liu, J., and DeFranco, D. B. (1999). Chromatin recycling of glucocorticoid receptors: Implications for multiple roles of heat shock protein 90. *Mol. Endocrinol.* **13,** 355–365.

Liu, J., and DeFranco, D. B. (2000). Protracted nuclear export of glucocorticoid receptor limits its turnover and does not require the exportin 1/CRM1-directed nuclear export pathway. *Mol. Endocrinol.* **14,** 40–51.

Liu, J., Xiao, N., and DeFranco, D. B. (1999). Use of digitonin-permeabilized cells in studies of steroid receptor subnuclear trafficking. *Methods* **19,** 403–409.

Liu, M., Iavarone, A., and Freedman, L. P. (1996). Transcriptional activation of the human p21(WAF1/CIP1) gene by retinoic acid receptor. Correlation with retinoid induction of U937 cell differentiation. *J. Biol. Chem.* **271,** 31723–31728.

Liu, S. M., Koszewski, N., Lupez, M., Malluche, H. H., Olivera, A., and Russell, J. (1996). Characterization of a response element in the 5′-flanking region of the avian (chicken) PTH gene that mediates negative regulation of gene transcription by 1,25-dihydroxyvitamin D3 and binds the vitamin D3 receptor. *Mol. Endocrinol.* **10,** 206–215.

Liu, Y. Y., Nguyen, C., Ali Gradezi, S. A., Schnirer, I., and Peleg, S. (2001). Differential regulation of heterodimerization by 1alpha, 25-dihydroxyvitamin D(3) and its 20-epi analog. *Steroids* **66,** 203–212.

Llopis, J., Westin, S., Ricote, M., Wang, Z., Cho, C. Y., Kurokawa, R., Mullen, T. M., Rose, D. W., Rosenfeld, M. G., Tsien, R. Y., Glass, C. K., and Wang, J. (2000). Ligand-dependent interactions of coactivators steroid receptor coactivator-1 and peroxisome proliferator-activated receptor binding protein with nuclear hormone receptors can be imaged in live cells and are required for transcription. *Proc. Natl. Acad. Sci. USA* **97,** 4363–4368.

Love, D. C., Sweitzer, T. D., and Hanover, J. A. (1998). Reconstitution of HIV-1 rev nuclear export: Independent requirements for nuclear import and export. *Proc. Natl. Acad. Sci. USA* **95,** 10608–10613.

Lutz, W., Kohno, K., and Kumar, R. (2001). The role of heat shock protein 70 in vitamin D receptor function. *Biochem. Biophys. Res. Commun.* **282,** 1211–1219.

MacDonald, P. N., Haussler, C. A., Terpening, C. M., Galligan, M. A., Reeder, M. C., Whitfield, G. K., and Haussler, M. R. (1991). Baculovirus-mediated expression of the human vitamin D receptor. Functional characterization, vitamin D response element interactions, and evidence for a receptor auxiliary factor *J. Biol. Chem.* **266,** 18808–18813.

MacDonald, P. N., Dowd, D. R., Nakajima, S., Galligan, M. A., Reeder, M. C., Haussler, C. A., Ozato, K., and Haussler, M. R. (1993). Retinoid X receptors stimulate and 9-cis retinoic acid inhibits 1,25-dihydroxyvitamin D3-activated expression of the rat osteocalcin gene. *Mol. Cell. Biol.* **13,** 5907–5917.

Mackey, S. L., Heymont, J. L., Kronenberg, H. M., and Demay, M. B. (1996). Vitamin D receptor binding to the negative human parathyroid hormone vitamin D response element does not require the retinoid x receptor. *Mol. Endocrinol.* **10,** 298–305.

Maglich, J. M., Sluder, A., Guan, X., Shi, Y., McKee, D. D., Carrick, K., Kamdar, K., Willson, T. M., and Moore, J. T. (2001). Comparison of complete nuclear receptor sets from the human, Caenorhabditis elegans and Drosophila genomes. *Genome Biol. Res.* **2,** 0029.1–0029.7.

Mahajna, J., Shi, B., and Bruskin, A. (1997). A four-amino-acid insertion in the ligand-binding domain inactivates hRXRbeta and renders dominant negative activity. *DNA Cell Biol.* **16,** 463–476.

Makishima, M., Okamoto, A. Y., Repa, J. J., Tu, H., Learned, R. M., Luk, A., Hull, M. V., Lustig, K. D., Mangelsdorf, D. J., and Shan, B. (1999). Identification of a nuclear receptor for bile acids. *Science* **284,** 1362–1365.

Malloy, P. J., Zhu, W., Zhao, X. Y., Pehling, G. B., and Feldman, D. (2001). A novel inborn error in the ligand-binding domain of the vitamin D receptor causes hereditary vitamin D-resistant rickets. *Mol. Genet. Metab.* **73,** 138–148.

Marcus, S. L., Miyata, K. S., Rachubinski, R. A., and Capone, J. P. (1995). Transactivation by PPAR/RXR heterodimers in yeast is potentiated by exogenous fatty acid via a pathway requiring intact peroxisomes. *Gene Expr.* **4,** 227–239.

Markose, E. R., Stein, J. L., Stein, G. S., and Lian, J. B. (1990). Vitamin D-mediated modifications in protein-DNA interactions at two promoter elements of the osteocalcin gene. *Proc. Natl. Acad. Sci. USA* **87,** 1701–1705.

Marks, M. S., Hallenbeck, P. L., Nagata, T., Segars, J. H., Appella, E., Nikodem, V. M., and Ozato, K. (1992). H-2RIIBP (RXR beta) heterodimerization provides a mechanism for combinatorial diversity in the regulation of retinoic acid and thyroid hormone responsive genes. *EMBO J.* **11,** 1419–1435.

Mascrez, B., Mark, M., Krezel, W., Dupe, V., LeMeur, M., Ghyselinck, N. B., and Chambon, P. (2001). Differential contributions of AF-1 and AF-2 activities to the developmental functions of RXR alpha. *Development* **128,** 2049–2062.

Masuyama, H., and MacDonald, P. N. (1998). Proteasome-mediated degradation of the vitamin D receptor (VDR) and a putative role for SUG1 interaction with the AF-2 domain of VDR. *J. Cell. Biochem.* **71,** 429–440.

Masuyama, H., Jefcoat, S. C. J., and MacDonald, P. N. (1997). The N-terminal domain of transcription factor IIB is required for direct interaction with the vitamin D receptor and participates in vitamin D-mediated transcription. *Mol. Endocrinol.* **11,** 218–228.

Mathew, S., Murty, V. V., Hunziker, W., and Chaganti, R. S. (1992). Subregional mapping of 13 single-copy genes on the long arm of chromosome 12 by fluorescence in situ hybridization. *Genomics* **14,** 775–779.

Mathieu, C., Van Etten, E., Gysemans, C., Decallonne, B., Kato, S., Laureys, J., Depovere, J., Valckx, D., Verstuyf, A., and Bouillon, R. (2001). In vitro and in vivo analysis of the immune system of vitamin D receptor knockout mice. *J. Bone Miner. Res.* **16,** 2057–2065.

Matkovits, T., and Christakos, S. (1995). Ligand occupancy is not required for vitamin D receptor and retinoid receptor-mediated transcriptional activation. *Mol. Endocrinol.* **9,** 232–242.

McBride, K. M., McDonald, C., and Reich, N. C. (2000). Nuclear export signal located within the DNA-binding domain of the STAT1transcription factor. *EMBO J.* **19,** 6196–6206.

McKenna, N. J., Xu, J., Nawaz, Z., Tsai, S. Y., Tsai, M. J., and O'Malley, B. W. (1999). Nuclear receptor coactivators: Multiple enzymes, multiple complexes, multiple functions. *J. Steroid Biochem. Mol. Biol.* **69,** 3–12.

McNally, J. G., Muller, W. G., Walker, D., Wolford, R., and Hager, G. L. (2000). The glucocorticoid receptor: Rapid exchange with regulatory sites in living cells. *Science* **287,** 1262–1265.

Meier, C. A., Parkison, C., Chen, A., Ashizawa, K., Meier-Heusler, S. C., Muchmore, P., Cheng, S. Y., and Weintraub, B. D. (1993). Interaction of human beta 1 thyroid hormone receptor and its mutants with DNA and retinoid X receptor beta. T3 response element-dependent dominant negative potency. *J. Clin. Invest* **92,** 1986–1993.

Meyer, B. E., and Malim, M. H. (1994). The HIV-1 Rev trans-activator shuttles between the nucleus and the cytoplasm. *Genes Dev.* **8,** 1538–1547.

Michigami, T., Suga, A., Yamazaki, M., Shimizu, C., Cai, G., Okada, S., and Ozono, K. (1999). Identification of amino acid sequence in the hinge region of human vitamin D receptor that transfers a cytosolic protein to the nucleus. *J. Biol. Chem.* **274,** 33531–33538.

Miyata, K. S., McCaw, S. E., Marcus, S., and Capone, J. P. (1994). The peroxisome proliferator-activated receptor interacts with the retinoid X receptor in vivo. *Gene* **148,** 327–330.

Morrison, N. A., Shine, J., Fragonas, J. C., Verkest, V., McMenemy, M. L., and Eisman, J. A. (1989). 1,25-Dihydroxyvitamin D-responsive element and glucocorticoid repression in the osteocalcin gene. *Science* **246,** 1158–1161.

Nagaya, T., Nomura, Y., Fujieda, M., and Seo, H. (1996). Heterodimerization preferences of thyroid hormone receptor alpha isoforms. *Biochem. Biophys. Res. Commun.* **226,** 426–430.

Nagpal, S., Friant, S., Nakshatri, H., and Chambon, P. (1993). RARs and RXRs: Evidence for two autonomous transactivation functions (AF-1 and AF-2) and heterodimerization in vivo. *EMBO J.* **12,** 2349–2360.

Nishikawa, J., Kitaura, M., Imagawa, M., and Nishihara, T. (1995). Vitamin D receptor contains multiple dimerization interfaces that are functionally different. *Nucleic Acids Res.* **23,** 606–611.

Nishikawa, J., Saito, K., Sasaki, M., Tomigahara, Y., and Nishihara, T. (2000). Molecular cloning and functional characterization of a novel nuclear receptor similar to an embryonic benzoate receptor BXR. *Biochem. Biophys. Res. Commun.* **277,** 209–215.

Noda, M., Vogel, R. L., Craig, A. M., Prahl, J., DeLuca, H. F., and Denhardt, D. T. (1990). Identification of a DNA sequence responsible for binding of the 1,25-dihydroxyvitamin D3 receptor and 1,25-dihydroxyvitamin D3 enhancement of mouse secreted phosphoprotein 1 (SPP-1 or osteopontin) gene expression. *Proc. Natl. Acad. Sci. USA* **87,** 9995–9999.

Odermatt, A., Arnold, P., and Frey, F. J. (2001). The intracellular localization of the mineralocorticoid receptor is regulated by 11beta-hydroxysteroid dehydrogenase type 2. *J. Biol. Chem.* **276,** 28484–28492.

Ohyama, Y., Ozono, K., Uchida, M., Shinki, T., Kato, S., Suda, T., Yamamoto, O., Noshiro, M., and Kato, Y. (1994). Identification of a vitamin D-responsive element in the 5′-flanking region of the rat 25-hydroxyvitamin D3 24-hydroxylase gene. *J. Biol. Chem.* **269,** 10545–10550.

Ozono, K., Liao, J., Kerner, S. A., Scott, R. A., and Pike, J. W. (1990). The vitamin D-responsive element in the human osteocalcin gene. Association with a nuclear proto-oncogene enhancer. *J. Biol. Chem.* **265,** 21881–21888.

Palmeri, D., and Malim, M. H. (1999). Importin beta can mediate the nuclear import of an arginine-rich nuclear localization signal in the absence of importin alpha. *Mol. Cell. Biol.* **19,** 1218–1225.

Panda, D. K., Miao, D., Tremblay, M. L., Sirois, J., Farookhi, R., Hendy, G. N., and Goltzman, D. (2001). Targeted ablation of the 25-hydroxyvitamin D 1alpha-hydroxylase enzyme: Evidence for skeletal, reproductive, and immune dysfunction. *Proc. Natl. Acad. Sci. USA* **98,** 7498–7503.

Perlmann, T., and Jansson, L. (1995). A novel pathway for vitamin A signaling mediated by RXR heterodimerization with NGFI-B and NURR1. *Genes Dev.* **9,** 769–782.

Perlmann, T., Umesono, K., Rangarajan, P. N., Forman, B. M., and Evans, R. M. (1996). Two distinct dimerization interfaces differentially modulate target gene specificity of nuclear hormone receptors. *Mol. Endocrinol.* **10,** 958–966.

Picard, D., and Yamamoto, K. R. (1987). Two signals mediate hormone-dependent nuclear localization of the glucocorticoid receptor. *EMBO J.* **6,** 3333–3340.

Polly, P., Herdick, M., Moehren, U., Baniahmad, A., Heinzel, T., and Carlberg, C. (2000). VDR-Alien: A novel, DNA-selective vitamin D(3) receptor-corepressor partnership. *FASEB J.* **14,** 1455–1463.

Poukka, H., Karvonen, U., Yoshikawa, N., Tanaka, H., Palvimo, J. J., and Janne, O. A. (2000). The RING finger protein SNURF modulates nuclear trafficking of the androgen receptor. *J. Cell Sci* **113**(Pt. 17), 2991–3001.

Prufer, K., and Barsony, J. (2002). Retinoid X receptor dominates the nuclear import and export of the unliganded vitamin D receptor. *Mol. Endocrinol.* **16,** 1738–1751.

Prufer, K., Veenstra, T. D., Jirikowski, G. F., and Kumar, R. (1999). Distribution of 1,25-dihydroxyvitamin D3 receptor immunoreactivity in the rat brain and spinal cord. *J. Chem. Neuroanat.* **16,** 135–145.

Prufer, K., Racz, A., Lin, G. C., and Barsony, J. (2000). Dimerization with retinoid X receptors promotes nuclear localization and subnuclear targeting of vitamin D receptors. *J. Biol. Chem.* **275,** 41114–41123.

Prufer, K., Schroder, C., Hegyi, K., and Barsony, J. (2002). Degradation of retinoid X receptors influences sensitivity of rat osteosarcoma cells to the antiproliferative effects of calcitriol. *Mol. Endocrinol.* **16,** 961–976.

Quack, M., and Carlberg, C. (2000). The impact of functional vitamin D(3) receptor conformations on DNA-dependent vitamin D(3) signaling. *Mol. Pharmacol.* **57,** 375–384.

Quelo, I., Kahlen, J. P., Rascle, A., Jurdic, P., and Carlberg, C. (1994). Identification and characterization of a vitamin D3 response element of chicken carbonic anhydrase-II. *DNA Cell Biol.* **13,** 1181–1187.

Rachez, C., Lemon, B. D., Suldan, Z., Bromleigh, V., Gamble, M., Naar, A. M., Erdjument-Bromage, H., Tempst, P., and Freedman, L. P. (1999). Ligand-dependent transcription activation by nuclear receptors requires the DRIP complex. *Nature* **398,** 824–828.

Rachez, C., Gamble, M., Chang, C. P., Atkins, G. B., Lazar, M. A., and Freedman, L. P. (2000). The DRIP complex and SRC-1/p160 coactivators share similar nuclear receptor binding determinants but constitute functionally distinct complexes. *Mol. Cell. Biol.* **20,** 2718–2726.

Racz, A., and Barsony, J. (1999). Hormone-dependent translocation of vitamin D receptors is linked to transactivation. *J. Biol. Chem.* **274,** 19352–19360.

Rastinejad, F. (2001). Retinoid X receptor and its partners in the nuclear receptor family. *Curr. Opin. Struct. Biol.* **11,** 33–38.

Rastinejad, F., Perlmann, T., Evans, R. M., and Sigler, P. B. (1995). Structural determinants of nuclear receptor assembly on DNA direct repeats [see comments]. *Nature* **375,** 203–211.

Raval-Pandya, M., Freedman, L. P., Li, H., and Christakos, S. (1998). Thyroid hormone receptor does not heterodimerize with the vitamin D receptor but represses vitamin D receptor-mediated transactivation. *Mol. Endocrinol.* **12,** 1367–1379.

Reginato, M. J., Zhang, J., and Lazar, M. A. (1996). DNA-independent and DNA-dependent mechanisms regulate the differential heterodimerization of the isoforms of the thyroid hormone receptor with retinoid X receptor. *J. Biol. Chem.* **271,** 28199–28205.

Reichrath, J., Mittmann, M., Kamradt, J., and Muller, S. M. (1997). Expression of retinoid-X receptors (-alpha, -beta, -gamma) and retinoic acid receptors (-alpha,-beta,-gamma) in normal human skin: An immunohistological evaluation. *Histochem. J.* **29,** 127–133.

Rhodes, S. J., Chen, R., DiMattia, G. E., Scully, K. M., Kalla, K. A., Lin, S. C., Yu, V. C., and Rosenfeld, M. G. (1993). A tissue-specific enhancer confers Pit-1-dependent morphogen inducibility and autoregulation on the pit-1 gene. *Genes Dev.* **7,** 913–932.

Richter, C. A., Tillitt, D. E., and Hannink, M. (2001). Regulation of subcellular localization of the aryl hydrocarbon receptor (AhR). *Arch. Biochem. Biophys.* **389,** 207–217.

Rochel, N., Wurtz, J. M., Mitschler, A., Klaholz, B., and Moras, D. (2000). The crystal structure of the nuclear receptor for vitamin D bound to its natural ligand. *Mol. Cell* **5,** 173–179.

Rodriguez, M. S., Thompson, J., Hay, R. T., and Dargemont, C. (1999). Nuclear retention of IkappaBalpha protects it from signal-induced degradation and inhibits nuclear factor kappaB transcriptional activation. *J. Biol. Chem.* **274,** 9108–9115.

Rosen, E. D., Beninghof, E. G., and Koenig, R. J. (1993). Dimerization interfaces of thyroid hormone, retinoic acid, vitamin D, and retinoid X receptors. *J. Biol. Chem.* **268,** 11534–11541.

Rowland-Goldsmith, M. A., Holmquist, B., and Henry, H. L. (1999). Genomic cloning, structure, and regulatory elements of the 1 alpha, 25(OH)2D3 down-regulated gene for cyclic AMP-dependent protein kinase inhibitor. *Biochim. Biophys. Acta* **1446,** 414–418.

Sachdev, S., Bagchi, S., Zhang, D. D., Mings, A. C., and Hannink, M. (2000). Nuclear import of IkappaBalpha is accomplished by a ran-independent transport pathway. *Mol. Cell. Biol.* **20,** 1571–1582.

Sakai, Y., Kishimoto, J., and Demay, M. B. (2001). Metabolic and cellular analysis of alopecia in vitamin D receptor knockout mice. *J. Clin. Invest* **107,** 961–966.

Sakati, N., Woodhouse, N. J., Niles, N., Harfi, H., de Grange, D. A., and Marx, S. (1986). Hereditary resistance to 1,25-dihydroxyvitamin D: Clinical and radiological improvement during high-dose oral calcium therapy. *Horm. Res.* **24,** 280–287.

Saydam, N., Georgiev, O., Nakano, M. Y., Greber, U. F., and Schaffner, W. (2001). Nucleo-cytoplasmic trafficking of metal-regulatory transcription factor 1 is regulated by diverse stress signals. *J. Biol. Chem.* **276,** 25487–25495.

Schrader, M., Kahlen, J. P., and Carlberg, C. (1997). Functional characterization of a novel type of 1 alpha,25-dihydroxyvitamin D3 response element identified in the mouse c-fos promoter. *Biochem. Biophys. Res. Commun.* **230,** 646–651.

Segal, S. P., Graves, L. E., Verheyden, J., and Goodwin, E. B. (2001). RNA-regulated TRA-1 nuclear export controls sexual fate. *Dev. Cell* **1,** 539–551.

Sela-Brown, A., Russell, J., Koszewski, N. J., Michalak, M., Naveh-Many, T., and Silver, J. (1998). Calreticulin inhibits vitamin D's action on the PTH gene in vitro and may prevent vitamin D's effect in vivo in hypocalcemic rats. *Mol. Endocrinol.* **12,** 1193–1200.

Seol, W., Choi, H. S., and Moore, D. D. (1995). Isolation of proteins that interact specifically with the retinoid X receptor: two novel orphan receptors. *Mol. Endocrinol.* **9,** 72–85.

Seol, W., Choi, H. S., and Moore, D. D. (1996). An orphan nuclear hormone receptor that lacks a DNA binding domain and heterodimerizes with other receptors. *Science* **272,** 1336–1339.

Seol, W., Chung, M., and Moore, D. D. (1997). Novel receptor interaction and repression domains in the orphan receptor SHP. *Mol. Cell. Biol.* **17,** 7126–7131.

Smart, P., Lane, E. B., Lane, D. P., Midgley, C., Vojtesek, B., and Lain, S. (1999). Effects on normal fibroblasts and neuroblastoma cells of the activation of the p53 response by the nuclear export inhibitor leptomycin B. *Oncogene* **18,** 7378–7386.

Solomon, C., Sebag, M., White, J. H., Rhim, J., and Kremer, R. (1998). Disruption of vitamin D receptor-retinoid X receptor heterodimer formation following ras transformation of human keratinocytes. *J. Biol. Chem.* **273,** 17573–17578.

Solomon, C., Kremer, R., White, J. H., and Rhim, J. S. (2001). Vitamin D resistance in RAS-transformed keratinocytes: Mechanism and reversal strategies. *Radiat. Res.* **155,** 156–162.

Stenoien, D. L., Patel, K., Mancini, M. G., Dutertre, M., Smith, C. L., O'Malley, B. W., and Mancini, M. A. (2001). FRAP reveals that mobility of oestrogen receptor-alpha is ligand- and proteasome-dependent. *Nat. Cell Biol.* **3,** 15–23.

Sugawara, A., Yen, P. M., Darling, D. S., and Chin, W. W. (1993). Characterization and tissue expression of multiple triiodothyronine receptor-auxiliary proteins and their relationship to the retinoid X-receptors. *Endocrinology* **133,** 965–971.

Sugawara, A., Sanno, N., Takahashi, N., Osamura, R. Y., and Abe, K. (1997). Retinoid X receptors in the kidney: Their protein expression and functional significance. *Endocrinology* **138,** 3175–3180.

Sunn, K. L., Cock, T. A., Crofts, L. A., Eisman, J. A., and Gardiner, E. M. (2001). Novel N-terminal variant of human VDR. *Mol. Endocrinol.* **15,** 1599–1609.

Tagami, T., Lutz, W. H., Kumar, R., and Jameson, J. L. (1998). The interaction of the vitamin D receptor with nuclear receptor corepressors and coactivators. *Biochem. Biophys. Res. Commun.* **253,** 358–363.

Taketani, Y., Miyamoto, K., Tanaka, K., Katai, K., Chikamori, M., Tatsumi, S., Segawa, H., Yamamoto, H., Morita, K., and Takeda, E. (1997). Gene structure and functional analysis of the human Na+/phosphate co-transporter. *Biochem. J.* **324**(Pt. 3), 927–934.

Takeuchi, A., Reddy, G. S., Kobayashi, T., Okano, T., Park, J., and Sharma, S. (1998). Nuclear factor of activated T cells (NFAT) as a molecular target for 1alpha,25-dihydroxyvitamin D3-mediated effects. *J. Immunol.* **160,** 209–218.

Taymans, S. E., Pack, S., Pak, E., Orban, Z., Barsony, J., Zhuang, Z., and Stratakis, C. A. (1999). The human vitamin D receptor gene (VDR) is localized to region 12cen-q12 by fluorescent in situ hybridization and radiation hybrid mapping: Genetic and physical VDR map. *J. Bone Miner. Res.* **14,** 1163–1166.

Teboul, M., Enmark, E., Li, Q., Wikstrom, A. C., Pelto-Huikko, M., and Gustafsson, J. A. (1995). OR-1, a member of the nuclear receptor superfamily that interacts with the 9-cis-retinoic acid receptor. *Proc. Natl. Acad. Sci. USA* **92,** 2096–2100.

Terpening, C. M., Haussler, C. A., Jurutka, P. W., Galligan, M. A., Komm, B. S., and Haussler, M. R. (1991). The vitamin D-responsive element in the rat bone Gla protein gene is an imperfect direct repeat that cooperates with other cis-elements in 1,25-dihydroxyvitamin D3-mediated transcriptional activation. *Mol. Endocrinol.* **5,** 373–385.

Thompson, P. D., Jurutka, P. W., Haussler, C. A., Whitfield, G. K., and Haussler, M. R. (1998). Heterodimeric DNA binding by the vitamin D receptor and retinoid X receptors is enhanced by 1,25-dihydroxyvitamin D3 and inhibited by 9-cis-retinoic acid. Evidence for allosteric receptor interactions. *J. Biol. Chem.* **273,** 8483–8491.

Thummel, K. E., Brimer, C., Yasuda, K., Thottassery, J., Senn, T., Lin, Y., Ishizuka, H., Kharasch, E., Schuetz, J., and Schuetz, E. (2001). Transcriptional control of intestinal cytochrome P-4503A by 1alpha,25-dihydroxy vitamin D3. *Mol. Pharmacol.* **60,** 1399–1406.

Tiosano, D., Weisman, Y., and Hochberg, Z. (2001). The role of the vitamin D receptor in regulating vitamin D metabolism: A study of vitamin D-dependent rickets, type II. *J. Clin. Endocrinol. Metab.* **86,** 1908–1912.

Tolon, R. M., Castillo, A. I., Jimenez-Lara, A. M., and Aranda, A. (2000). Association with Ets-1 causes ligand- and AF2-independent activation of nuclear receptors. *Mol. Cell. Biol.* **20,** 8793–8802.

Tyagi, R. K., Amazit, L., Lescop, P., Milgrom, E., and Guiochon-Mantel, A. (1998). Mechanisms of progesterone receptor export from nuclei: Role of nuclear localization signal, nuclear export signal, and ran guanosine triphosphate. *Mol. Endocrinol.* **12,** 1684–1695.

Tyagi, R. K., Lavrovsky, Y., Ahn, S. C., Song, C. S., Chatterjee, B., and Roy, A. K. (2000). Dynamics of intracellular movement and nucleocytoplasmic recycling of the ligand-activated androgen receptor in living cells. *Mol. Endocrinol.* **14,** 1162–1174.

Van Cromphaut, S. J., Dewerchin, M., Hoenderop, J. G., Stockmans, I., Van Herck, E., Kato, S., Bindels, R. J., Collen, D., Carmeliet, P., Bouillon, R., and Carmeliet, G. (2001). Duodenal calcium absorption in vitamin D receptor-knockout mice: Functional and molecular aspects. *Proc. Natl. Acad. Sci. USA* **98,** 13324–13329.

Venepally, P., Reddy, L. G., and Sani, B. P. (1997). Analysis of homo- and heterodimerization of retinoid receptors in solution. *Arch. Biochem. Biophys.* **343,** 234–242.

Vlahou, A., and Flytzanis, C. N. (2000). Subcellular trafficking of the nuclear receptor COUP-TF in the early embryonic cell cycle. *Dev. Biol.* **218,** 284–298.

Wan, Y. J., An, D., Cai, Y., Repa, J. J., Hung-Po, C. T., Flores, M., Postic, C., Magnuson, M. A., Chen, J., Chien, K. R., French, S., Mangelsdorf, D. J., and Sucov, H. M. (2000). Hepatocyte-specific mutation establishes retinoid X receptor alpha as a heterodimeric integrator of multiple physiological processes in the liver. *Mol. Cell. Biol.* **20,** 4436–4444.

Wang, G. F., Nikovits, W., Jr., Bao, Z. Z., and Stockdale, F. E. (2001). Irx4 forms an inhibitory complex with the vitamin D and retinoic X receptors to regulate cardiac chamber-specific slow MyHC3 expression. *J. Biol. Chem.* **276,** 28835–28841.

Weiss, R., and Knight, T. F. (2000). Engineered Communications for Microbial Robotics. 6th International Workshop on Computational Robotics. http://www.swiss.ai.mit.edu/~rweiss/bioprogramming/

Wheeler, D. G., Horsford, J., Michalak, M., White, J. H., and Hendy, G. N. (1995). Calreticulin inhibits vitamin D3 signal transduction. *Nucleic Acids Res.* **23,** 3268–3274.

Whitfield, G. K., Hsieh, J. C., Nakajima, S., MacDonald, P. N., Thompson, P. D., Jurutka, P. W., Haussler, C. A., and Haussler, M. R. (1995). A highly conserved region in the hormone-binding domain of the human vitamin D receptor contains residues vital for heterodimerization with retinoid X receptor and for transcriptional activation. *Mol. Endocrinol.* **9,** 1166–1179. Erratum: *Mol. Endocrinol.* 1995, **9**(11), 1509.

Whitfield, G. K., Selznick, S. H., Haussler, C. A., Hsieh, J. C., Galligan, M. A., Jurutka, P. W., Thompson, P. D., Lee, S. M., Zerwekh, J. E., and Haussler, M. R. (1996). Vitamin D receptors from patients with resistance to 1,25-dihydroxyvitamin D3: Point mutations confer reduced transactivation in response to ligand and impaired interaction with the retinoid X receptor heterodimeric partner. *Mol. Endocrinol.* **10,** 1617–1631.

Wiebel, F. F., and Gustafsson, J. A. (1997). Heterodimeric interaction between retinoid X receptor alpha and orphan nuclear receptor OR1 reveals dimerization-induced activation as a novel mechanism of nuclear receptor activation. *Mol. Cell. Biol.* **17,** 3977–3986.

Wiebel, F. F., Steffensen, K. R., Treuter, E., Feltkamp, D., and Gustafsson, J. A. (1999). Ligand-independent coregulator recruitment by the triply activatable OR1/retinoid X receptor-alpha nuclear receptor heterodimer. *Mol. Endocrinol.* **13,** 1105–1118.

Wiechens, N., and Fagotto, F. (2001). CRM-1- and Ran-independent nuclear export of β-catenin. *Curr. Biol.* **11,** 18–27.

Willy, P. J., Umesono, K., Ong, E. S., Evans, R. M., Heyman, R. A., and Mangelsdorf, D. J. (1995). LXR, a nuclear receptor that defines a distinct retinoid response pathway. *Genes Dev.* **9,** 1033–1045.

Wu, J., Garami, M., Cao, L., Li, Q., and Gardner, D. G. (1995). 1,25(OH)2D3 suppresses expression and secretion of atrial natriuretic peptide from cardiac myocytes. *Am. J. Physiol.* **268,** E1108–E1113.

Wu, S., Zhang, Z. P., Zhang, D., Soprano, D. R., and Soprano, K. J. (1997). Reduction of both RAR and RXR levels is required to maximally alter sensitivity of CA-OV3 ovarian tumor cells to growth suppression by all-trans-retinoic acid. *Exp. Cell Res.* **237,** 118–126.

Wu, Y., Craig, T. A., Lutz, W. H., and Kumar, R. (1999). Identification of 1 alpha,25-dihydroxyvitamin D3 response elements in the human transforming growth factor beta 2 gene. *Biochemistry* **38,** 2654–2660.

Xie, Z., and Bikle, D. D. (1997). Cloning of the human phospholipase C-gamma1 promoter and identification of a DR6-type vitamin D-responsive element. *J. Biol. Chem.* **272,** 6573–6577.

Yagishita, N., Yamamoto, Y., Yoshizawa, T., Sekine, K., Uematsu, Y., Murayama, H., Nagai, Y., Krezel, W., Chambon, P., Matsumoto, T., and Kato, S. (2001). Aberrant growth plate development in VDR/RXR gamma double null mutant mice. *Endocrinology* **142,** 5332–5341.

Yang, J., and DeFranco, D. B. (1996). Assessment of glucocorticoid receptor-heat shock protein 90 interactions in vivo during nucleocytoplasmic trafficking. *Mol. Endocrinol.* **10,** 3–13.

Yang, J., Liu, J., and DeFranco, D. B. (1997). Subnuclear trafficking of glucocorticoid receptors in vitro: Chromatin recycling and nuclear export. *J. Cell Biol.* **137,** 523–538.

Yang, J., Song, H., Walsh, S., Bardes, E. S., and Kornbluth, S. (2001). Combinatorial control of cyclin B1 nuclear trafficking through phosphorylation at multiple sites. *J. Biol. Chem.* **276,** 3604–3609.

Yu, Z., Lee, C. H., Chinpaisal, C., and Wei, L. N. (1998). A constitutive nuclear localization signal from the second zinc-finger of orphan nuclear receptor TR2. *J. Endocrinol.* **159,** 53–60.

Zavacki, A. M., Lehmann, J. M., Seol, W., Willson, T. M., Kliewer, S. A., and Moore, D. D. (1997). Activation of the orphan receptor RIP14 by retinoids. *Proc. Natl. Acad. Sci USA* **94,** 7909–7914.

Zechel, C., Shen, X. Q., Chambon, P., and Gronemeyer, H. (1994a). Dimerization interfaces formed between the DNA binding domains determine the cooperative binding of RXR/RAR and RXR/TR heterodimers to DR5 and DR4 elements. *EMBO J.* **13,** 1414–1424.

Zechel, C., Shen, X. Q., Chen, J. Y., Chen, Z. P., Chambon, P., and Gronemeyer, H. (1994b). The dimerization interfaces formed between the DNA binding domains of RXR, RAR and TR determine the binding specificity and polarity of the full-length receptors to direct repeats. *EMBO J.* **13,** 1425–1433.

Zetterstrom, R. H., Solomin, L., Mitsiadis, T., Olson, L., and Perlmann, T. (1996). Retinoid X receptor heterodimerization and developmental expression distinguish the orphan nuclear receptors NGFI-B, Nurr1, and Nor1. *Mol. Endocrinol.* **10,** 1656–1666.

Zhang, X. K., Salbert, G., Lee, M. O., and Pfahl, M. (1994). Mutations that alter ligand-induced switches and dimerization activities in the retinoid X receptor. *Mol. Cell. Biol.* **14,** 4311–4323.

Zhang, Y., and Xiong, Y. (2001). A p53 amino-terminal nuclear export signal inhibited by DNA damage-induced phosphorylation. *Science* **292,** 1910–1915.

Zhao, X. Y., Eccleshall, T. R., Krishnan, A. V., Gross, C., and Feldman, D. (1997). Analysis of vitamin D analog-induced heterodimerization of vitamin D receptor with retinoid X receptor using the yeast two-hybrid system. *Mol. Endocrinol.* **11,** 366–378.

Zhuang, S. H., Schwartz, G. G., Cameron, D., and Burnstein, K. L. (1997). Vitamin D receptor content and transcriptional activity do not fully predict antiproliferative effects of vitamin D in human prostate cancer cell lines. *Mol. Cell. Endocrinol.* **126,** 83–90.

Zierold, C., Darwish, H. M., and DeLuca, H. F. (1994). Identification of a vitamin D-response element in the rat calcidiol (25-hydroxyvitamin D3) 24-hydroxylase gene. *Proc. Natl. Acad. Sci. USA* **91,** 900–902.

Zilliacus, J., Holter, E., Wakui, H., Tazawa, H., Treuter, E., and Gustafsson, J. (2001). Regulation of glucocorticoid receptor activity by 14-3-3-dependent intracellular relocalization of the corepressor rip140. *Mol. Endocrinol.* **15,** 501–511.

INDEX

A

Acetylcholine
  endomorphins, 266
Acetylcoenzyme A (acyl-CoA), 109–110, 198
Acipimox, 51
Adenine nucleotide translocase (ANT), 99
  role, 100–101
Adenosine diphosphate, 101
Adenylyl cyclase, 38
Adipogenesis
  SREBP-1c, 181
AEA. *See* Anandamide
Aequorea victoria
  green fluorescent protein, 82
Agouti-related peptide, 297–298
AIB1, 137
AIF, 10
Alendronate, 53, 57
Alkaline phosphatase (ALP), 14, 41
ALP, 14, 41
Alzheimer's dementia, 51
α-melanocyte-stimulating hormone (α-MSH), 283
Amplified in breast cancer 1 (AIB1), 137
α-MSH, 283
Anabolic parathyroid hormone
  bone, 20–50
Anandamide (AEA), 110, 111, 228, 242–248. *See also* N-acylethanolamines (NAE)
  biosynthesis, 231
  blood pressure, 244
  heart rate, 244
  peripheral actions, 243
  platelet activation, 245–246
  smooth muscle relaxation, 244
  structure, 227
Androgen receptor, 130–131
  conformation, 133–138
  diagram, 130
  ligand binding, 131–132
  mRNA
    levels, 132–133
  nervous system, 131
  transactivation capacity, 131–132
Androgen Receptor Gene Mutations Database
  World Wide Web Server, 134
Androgens, 196
  insensitivity, 128
Animals
  PTH-induced growth and fracture mending, 21–26
Animal studies
  statins, 51
Anoikis, 9
ANT, 99
  role, 100–101
Antiphosphatidylethanolamine antibody (aPE), 337
APE, 337

Aplastic anemia
  osteopetrosis-induced, 26
Apoptosis
  homelessness-induced, 9
Apoptosis-inducing factor (AIF), 10
Aquaporins, 208
Arabidopsis, 198, 200–201, 203–205
  seedlings, 205
Arabidopsis thaliana, 196
Arachidonic acid, 114, 115
Aromatase, 3
Artificial chromosomes, 139
Atherogenic remnant lipoproteins
  SREBP-1c, 182–183
Atorvastatin, 57
Autoimmunity-related habitual abortion
  herbal factors, 333–342

B

Basic helix-loop-helix leucine zipper
  (bHLH-zip), 169
Basic multicellular units (BMUs), 4
  efficiency, 11–12
BDNF, 9
Benzafibrate, 51
17 β-estradiol, 18
BFGF, 44
BGFP-5, 7
BHLH-zip, 169
Bisphosphonates, 19–20, 23
BMP, 49
BMP-2, 7
  gene promoters, 50–55
BMP-4
  promoter, 50–55
BMUs, 4
  efficiency, 11–12
Bone, 12–15
  anabolic parathyroid hormone, 20–50
  remodeling, 4
Bone growth
  leptin, 13
  stimulators, 1–58
Bone loss
  menopause, 12–20
  stopping, 19–20
Bone morphogenic protein (BMP), 49
Bone morphogenic protein (BMP)-2, 7
  gene promoters, 50–55
Bone morphogenic protein (BMP)-4
  promoter, 50–55

Bone-remodeling compartment (BRC), 7
Bone sialoprotein (BSP), 41
Brain, 12–15
Brain-derived neurotrophic factor (BDNF), 9
Brassica napus, 196
Brassinolide
  campesterol, 203–206
  cell division, 208–209
  cell expansion, 207–208
  future prospects, 217–218
  physiological responses, 207–216
  receptor, 210–215
  signal transduction, 210–217
    downstream construction, 215–217
Brassinosteroid insensitive 1 (BRII), 210–214
Brassinosteroids, 195–218
  mevalonate, 198–206
  structure, 197
BRC, 7
BRII, 210–214
BSP, 41

C

CAAX (CSCL) motif, 155
Caenorhabditis elegans, 321
Calcitonin, 23
Calcitriol
  RXR, 348–350
Calmodulin, 53
Campesterol, 201, 205
  brassinolide, 203–206
  cycloartenol, 200–203
Cannabis sativa, 226
Carbonyl cyanide m-chlorophenylhydrazone
  (CCCP), 106
L-carnitine, 110
Caspase, 20
Catharanthus roseus, 200
Caveolae, 53
CBP, 134, 140
CCCP, 106
CCT. *See* Cytosolic chaperonin-containing
  t-complex polypeptide 1
Cell binding
  visual monitoring, 91–93
Cell death
  PTP, 107
  saturated fatty acids, 114
  unsaturated fatty acids, 115–116
Cell differentiation
  organ morphogenesis, 209–210

Cellular fluorescence, 86–87
Central melanocortin signaling system, 282
Central nervous system
  endocannabinoids, 240–242
Cerivastatin, 57
CFU-S, 38
Channel concept, 103
Chaperonin family, 323
Cholestyramine, 51
Chromosomes
  artificial, 139
Chronic relapsing experimental allergic
    encephalomyelitis (CREAE), 240–241
Ciprofibrate, 51
Clodronate, 20
Clofibrate, 51
CoA, 109–110, 198
Colestipol, 51, 176
Collagen I, 14
Colony-forming unit stem cells (CFU-S),
    38
Competitive inhibition, 88
Corticosteroids, 196
COX, 99, 150
cPLAs, 115
CREAE, 240–241
CREB-binding protein (CBP), 134, 140
CSCL motif, 155
Cyclic AMP, 34, 44
Cycloartenol, 199
  biosynthetic pathways, 202
  campesterol, 200–203
Cyclooxygenase (COX), 99, 150
Cyclosporin A, 110
Cynomolgus monkeys, 30
Cytosolic chaperonin-containing t-complex
    polypeptide (CCT), 313–326
  characteristics, 319
  evolution, 320–321
  expression, 321–324
  function, 315–319
  structure, 319–320
  subunit roles, 324–326
Cytosolic phospholipase A2 (cPLAs), 115

D

Dentatorubral-pallidoluysian atrophy (DRPLA),
    129
DHT, 136–137
Dictyostelium discoideum, 322
Dihydrotestosterone (DHT), 136–137

DNA binding
  RXR, 352–354
Dopamine, 18
Dose curve, 85–86
Drosophila runt gene, 41
DRPLA, 129

E

EAAT1 glutamate transporters, 6
Ecdysteroids, 196
Eicosanoid family, 149
Eicosanoids, 158–159
Endocannabinoids, 225–247
  central nervous system, 240–242
  degradation, 233–236
  fertility, 246–247
  molecular targets, 236–239
  multiple sclerosis, 240–241
  Parkinson's' disease, 242
  periphery, 243–244, 243–247
  platelet activation, 245–246
  synthesis, 230–231
Endomorphin, 257–271
  acetylcholine, 266
  action, 261–262
  discovery, 258–267
  future directions, 270–271
  localization, 259–261
  memory, 266
  structure-activity relationship, 267–270
  vascular systems, 265
Endomorphin-1
  activities, 265–267
  analogues
    bioactivity, 269
  antinociceptive activity, 262–264
  binding stimulation, 261–262
  derivatives
    β-amino acid, 270
  localization, 259–261
Endomorphin-2
  activities, 265–267
  antinociceptive activity, 262–264
  binding stimulation, 261–262
  Dmt analogues
    binding activity, 271
  localization, 259–261
  stereoisometric derivatives
    receptor binding, 268
Endothelin-1, 8
Enkephalins, 258

Epinephrine, 18
Epoxomicin, 52
ERE, 15
Escherichia coli, 85, 102
Estradiol, 18
Estrogen, 3, 12–15, 15, 23, 196
 monocytes, 16
 osteoclast population explosion, 15–19
Estrogen response element (ERE), 15
Etidronate, 20

F

FACKEL gene, 201
Farnesyl pryophosphate synthase (FPP), 19–20, 174
Fat, 12–15
Fatty acids
 cellular effects, 112–116
 cycling, 100
 induction, 105–106
 mitochondrial membrane potential
  in situ, 112–114
 mitochondrial permeability transition pore, 104–109
 mitochondrial respiration, 108
 proton conductance
  static models, 102
 protonophores, 99–104
 PTP, 108–109
 structural requirements, 103–104
Fatty acylethanolamides, 110–111
Fatty livers, 183
 SREBP-1c, 182
Fenofibrate, 51
Fertility, 128
FGF-2, 7, 49
Fibroblast growth factor (FGF)-2, 7, 49
FLIP, 357
Flip-flop, 100
Fluorescence loss in photobleaching (FLIP), 357
Fluorescence recovery after photobleaching (FRAP), 357
Fluvastatin, 57
Formalin
 nociceptive response, 264
Forteo, 49
Fosamax, 19–20
FPP, 19–20, 174
Fracture reduction
 statins, 56

FRAP, 357
Fructose, 178

G

GABA, 242
GAM, 25
γ-aminobutyric acid (GABA), 242
Gating potential, 105
Gemfibrozil, 51
Gene-activated collagen matrix (GAM), 25
Gene expression modification
 SREBPs, 173–174
General Practice Research Database, 56
Geranylgeraniol, 20
GFP. *See* Green fluorescent protein
GLAST (EAAT1) glutamate transporters, 6
Glucocorticoid therapy, 31
Glutamate, 5
 osteotransmitter, 5–6
Glutamate secretors, 6
GnRH, 12
Gonadotropin-releasing hormone (GnRH), 12
GPCR, 158–159, 293–296
 type II, 32
GPCR kinase (GRK), 153
G-protein-coupled receptors (GPCR), 158–159, 293–296
Green fluorescent protein (GFP)
 Aequorea victoria, 82
 applications, 85–93
 defined, 83–85
 leucine zipper domain, 83
 ligand fusions
  receptor-mediated endocytosis, 81–93
GRK, 153
Gynecomastia, 128

H

Habitual abortion
 autoimmunity-related
  herbal factors, 333–342
 causes, 335–337
 immunological disorders, 336
 Sairei-to, 337
 Tokishakuyaku-san, 337
HBM, 12
Helianthus tuberosus, 208
Hemorphin-4, 267

Herbal factors
  autoimmunity-related habitual abortion, 333–342
Herbal medicine
  obstetrics and gynecology, 335
Heregulin-α, 84
Heterodimerization interfaces, 354–355
High bone mass (HBM), 12
HMG-CoA
  reductase, 51–52
HMG-CoA synthase genes, 169
HOBIF, 13–14
Homelessness-induced apoptosis, 9
Hormone replacement therapy (HRT), 27
Hormone response elements (HRE), 130
HPTH, 37
HRE, 130
HRT, 27
Human prostacyclin receptor, 149–159
  knockout and transgene studies, 157–158
  ligand binding, 152–153
  mutation studies, 155–156
  signal transduction, 152–153
Huntington's disease, 129
Hydroxymethylglutaryl-coenzyme A (HMG-CoA)
  reductase, 51–52
Hydroxymethylglutaryl-coenzyme A (HMG-CoA) synthase genes, 169
Hypothalamic gene regulation
  leptin, 288–290
Hypothalamic osteoblast inhibitory factor (HOBIF), 13–14
Hypothalamic signal transduction
  leptin receptors, 286–288
Hypothalamus
  leptin, 281–301
Hypoxia, 151

I

IGF-binding protein (BGFP)-5, 7
IGFs-I, 7, 44, 49
IL-4, 341
Immune reaction
  maternal, 338–341
Immunological disorders
  habitual abortion, 336
Insulin/glucose
  SREBP-1c expression, 177–178
Insulin-like growth factors (IGFs)-I, 7, 44, 49

Insulin resistance
  SREBP-1c, 181–183
Interacting proteins, 137
Interleukin (IL)-4, 341
Internalization
  schematic, 91
  visual monitoring, 91–93
IP receptor
  deficiency, 158
  distribution, 151–152
  expression, 151–152
  ligation, 153
  regulation, 153–154

J

Japanese herbal medicine
  obstetrics and gynecology, 335

K

Kennedy's disease, 128
Knob protein, 84
Koric bones, 17
K receptors, 259

L

Lactacystin, 52
Lanosterol
  biosynthetic pathways, 202
LDL receptor, 169
Leptin, 14
  bone growth, 13
  future directions, 301
  hypothalamic gene regulation, 288–290
  hypothalamus, 281–301
  receptors, 284–285
    hypothalamic signal transduction, 286–288
  signaling system
    human mutations, 291–292
    system, 283–284
Leptomycin B (LMB), 359
Lidocaine
  nociceptive response, 264
Ligands, 84
  binding
    androgen receptor, 131–132
    SMNA, 136–137
Ligand vitamin D receptors (VDR)
  RXR shuttling, 361–362

Linolenic acid, 114
Lipid metabolism
  cross-talk
    transcription factors, 183–186
Lipid synthesis
  transcriptional regulation, 178–179
Lipid synthetic genes
  SREBP
    energy metabolism, 167–186
Lipogenic enzymes, 177
Lipooxygenase (LOX), 99, 115
Liver X receptor/retinoid X receptor (LXR/RXR), 180
  SREBP-1c, 180
LMB, 359
Lovastatin, 52, 57, 176
Low-density lipoprotein (LDL) receptor, 169
LOX, 99, 115
Luciferase gene, 50–55
Luciferase reporter gene, 180
LXR/RXR, 180
  SREBP-1c, 180

M

Macaca fascicularis, 23, 30
Mammalian endogenous opioid peptides
  amino acid sequences, 260
MAP kinase pathway, 151–152
Maternal immune reaction
  Sairei-to, 338–341
  Tokishakuyaku-san, 338–341
Mechanosensitive calcium channels (MSCCs), 45
Melanocortin
  coreceptors, 298–299
  future directions, 301
  receptors, 293–296
  signaling
    targets, 299–300
    system, 292–293
      human mutations, 300–301
Memory
  endomorphins, 266
Menopause
  bone loss, 12–20
  stopping, 19–20
Methoxyprogesterone, 28
Mevalonate
  brassinosteroids, 198–206
Microcracks, 18

Mitochondrial energy dissipation
  fatty acids, 97–117
Mitochondrial membrane potential
  fatty acids
    in situ, 112–114
Mitochondrial permeability transition pore (PTP)
  fatty acids, 104–109
Mitochondrial respiration
  fatty acids, 108
Mitogen-activated protein (MAP) kinase
  pathway, 151–152
Monocytes
  estrogen, 16
Morphiceptin, 267
Morphine antinociception, 264
mRNA
  androgen receptor
    levels, 132–133
MSCCs, 44
Multiple PGI2 receptors, 158–159
Multiple sclerosis
  endocannabinoids, 240–241
Mutant ataxin-1, 135

N

N-acylethanolamines (NAE), 110–111, 228–229. *See also* Anandamide (AEA)
N-arachidonoyl-phosphatidylethanolamine (NArPE), 230
N-containing bisphosphonate (N-BP), 19–20, 52
Nervous system
  androgen receptor, 131
NES, 359
Neurofilament light chain (NFL), 138
Neuronal dysfunction
  polyglutamine expansion, 140
Neuronal intranuclear inclusions, 134–135
Neuron-specific enolase (NSE), 138
Neuropeptide Y (NPY), 288
NFL, 138
Niacin/nicotinic acid, 51
Nitrogen-containing bisphosphonates (N-BPs), 19–20, 52
Norepinephrine, 18
NPY, 288
NSE, 138
Nuclear export signals (NES), 359
Nucleocytoplasmic shuttling
  transcription factors, 358

INDEX

O

Obesity, 300–301
Obstetrics and gynecology
  Japanese herbal medicine, 335
OPG gene, 25, 47, 50
Opioid peptides, 258
  binding activities, 260
Opioid tetrapeptide
  locus coeruleus neuronal activity, 268
Organ morphogenesis
  cell differentiation, 209–210
Ostabolin C, 37, 38
Osteoblast, 18
Osteoblastic stromal cells
  estrogen, 16
Osteocalcin, 14, 30, 41, 57
Osteocytes, 5
Osteogenic mechanism, 51–55
Osteoid, 9
Osteopetrosis-induced aplastic anemia, 26
Osteopontin, 41
Osteoporosis
  defined, 3–4
Osteoprotegerin, 4
Ovariectomy, 15

P

Palmitoylcarnitine, 99
Parathyroid hormone (PTH), 4, 23
  anabolic action, 31–50
  bone, 23
  bone formation
    PTH-induced bone loss, 46–48
  clinical prospects, 48–50
  induced bone growth
    humans, 26–31
  induced bone loss
    PTH-induced bone formation, 46–48
  induced growth and fracture mending
    animals, 21–26
  pill, 49
  subcutaneous injection, 49–50
Parkinson's' disease
  endocannabinoids, 242
Permeability transition pore (PTP), 98
  cell death, 107
  fatty acids, 104–109
  mitochondrial
    fatty acids, 104–109
  opening, 106–107

PFA, 10
Phantom ligand, 350
Phosphatidylcholine (PtdCh), 34
Phosphoenolpyruvate carboxykinase, 181
Phospholipase A2 (PLAs)
  cytosolic, 115
Phospholipase C, 10
Phospholipase D (PLD), 34
Phosphonoformic acid (PFA), 10
Pit1, 45
Pit2 transporter, 45
PKA, 34
PKC, 153
Plant steroids, 196
Platelet activation
  AEA, 245–246
  endocannabinoids, 245–246
PLD, 34
Polar zipper, 133
Polyglutamine expansion
  neuronal dysfunction, 140
Polyglutamine repeat expansion, 129, 137
POMC, 296–297
  precursor, 282–283
  producing neurons, 289
Posttranslational modifications, 135
Pravastatin, 56, 57
Premarin, 28
Preosteoclasts, 6
Prion protein (PrP) promoter, 139
Procollagen, 41
Promoter analysis
  SREBP-1c, 179–180
Proopiomelanocortin (POMC), 296–297
  precursor, 282–283
  producing neurons, 289
Prostaglandin PGI2, 150
Prostaglandins, 17
Protease, 10
Protein
  production, 84–85
Protein kinase C (PKC), 153
Protein-protein interactions, 133
Proteolysis, 135–136
PrP promoter, 139
PtdCh, 34
PTF. *See* Parathyroid hormone
PTHR1 receptor
  signaling, 31–38
PTP. *See* Permeability transition pore

Q

Quinones, 111

R

RAFS, 336
Raloxifene, 19–20, 23
RANKL, 39
Receptor binding
 detection, 85–87
Receptor-binding mutant
 analysis, 88–89
Receptor mediated endocytosis, 82, 90–91
 GFP-ligand fusions, 81–93
Receptor tyrosine kinase (RTK), 36
Relaxant receptors, 155
Remodeling space, 9
Reproductive autoimmune failure syndrome
 (RAFS), 336
Retinoid X receptor (RXR), 180
 calcitriol, 348–350
 dimerization
  partner, 350
  regulation, 355–356
 DNA binding, 352–354
 general features, 346–348
 heterodimerization partners, 351
 heterodimers
  genes regulated by, 353
  subcellular localization, 356–357
 shuttling
  ligand VDR, 361–362
 transcription, 352–354
 VDR, 345–364
  cytoplasm/nucleus trafficking, 357–360
  physiochemical properties, 350–352
  target site binding, 362–363
Rho–kinase, 55
Rodents
 statins, 52
Rough endoplasmic reticulum (rER), 169
RTK, 36
RXR. *See* Retinoid X receptor

S

Saccharomyces cerevisiae, 200
Sairei–to, 334–335
 habitual abortion, 337
 maternal immune reaction, 338–341
Saturated fatty acids
 cell death, 114

SBMA, 128
 clinical view, 128–129
 mouse models, 138–139
 X-linked, 128
Selective estrogen receptor modulator
 (SERMs), 2, 19–20, 23
SERMs, 2, 19–20, 23
Sitosterol, 199
SMNA
 ligand, 136–137
SNURF, 360
Speech, 128
Spinal and bulbar muscular atrophy (SBMA),
 128
 clinical view, 128–129
 mouse models, 138–139
 X-linked, 128
Spinocerebellar ataxias, 129
Squalene, 199
SRC-1, 134, 137
SRE, 169
SREBP. *See* Sterol regulatory element-binding
 protein
SREBP-1
 lipogenic enzyme genes
  nutritional regulation, 176–177
SREBP-2, 174–175
 lipid synthesis
  transcriptional regulation, 178–179
SREBP-1a, 174–175
SREBP-1c. *See* Sterol regulatory
 element-binding protein (SREBP)-1c
Starvation
 thyroid hormone, 289
Statins, 50–55
 animal studies, 51
 clinical prospects, 57–58
 fracture reduction, 56
 humans, 55–57
 rodents, 52
Steroid receptor coactivator 1 (SRC-1), 134,
 137
Sterol regulatory element-binding protein
 (SREBP)
 cleavage, 170–171
 clinical aspects, 181–183
 DNA-binding sites, 171–172
 future aspects, 186
 gene expression modification, 173–174
 isoforms
  target DNA specificity, 175
  knockout mice, 176

lipid synthetic genes
   energy metabolism, 167–186
   members, 174–175
   structure, 169–174
   target genes, 172–173
   transgenic mice, 175–176
   in vivo functions, 175–178
Sterol regulatory element-binding protein (SREBP)-1
   lipogenic enzyme genes
      nutritional regulation, 176–177
Sterol regulatory element-binding protein (SREBP)-2, 174–175
   lipid synthesis
      transcriptional regulation, 178–179
Sterol regulatory element-binding protein (SREBP)-1a, 174–175
Sterol regulatory element-binding protein (SREBP)-1c, 174–175
   adipogenesis, 181
   atherogenic remnant lipoproteins, 182–183
   expression
      insulin/glucose, 177–178
   fatty livers, 182
   insulin resistance, 181–183
   lipid synthesis
      transcriptional regulation, 178–179
   promoter
      oxysterol-inducible region, 179–180
      regulation, 179
      SRE, 179–180
   promoter analysis, 179–180
Sterol-regulatory element (SRE), 169
Stigmasterol, 199
Swallowing, 128

T

Target DNA specificity
   SREBP isoforms, 175
Target genes
   SREBPs, 172–173
T cells
   estrogen, 16
Testicular atrophy, 128
Testosterone, 3, 132, 136–137
Tetrahydrocannabinol, 226
Tetrahymena pyriformis, 321
TGF-β, 7
Thiosters, 98
Thrombotic response
   vascular injury, 157–158

Thyroid hormone
   starvation, 289
Thyroparathyroidectomy, 46
TIF2, 137
Tiludronate, 20
TNF-α, 17, 339
TNF-β, 17
Tokishakuyaku-san, 334–335
   habitual abortion, 337
   maternal immune reaction, 338–341
Tomato
   brassinolide biosynthesis, 206
Transactivation capacity
   androgen receptor, 131–132
Transcription
   RXR, 352–354
Transcription factors
   cross-talk
      lipid metabolism, 183–186
      nucleocytoplasmic shuttling, 358
Transcription intermediary factor 2 (TIF2), 137
Transforming growth factors-β (TGF-β), 7
Transglutaminase, 133
TRiC. *See* Cytosolic chaperonin-containing t-complex polypeptide 1
Trinucleotide repeat disease, 127–140
Tumor necrosis factor-α (TNF-α), 17, 339
Tumor necrosis factor-β (TNF-β), 17
Type II G-protein-coupled receptor (GPCR), 32

U

Uncoupling proteins (UCPs)
   role, 101–103
Unsaturated fatty acids
   cell death, 115–116

V

Vascular systems
   endomorphins, 265
VDR
   general features, 346–348
   retinoid X receptors, 345–364
   RXR
      cytoplasm/nucleus trafficking, 357–360
      shuttling, 361–362
      target site binding, 362–363

VDREs, 352
Vitamin D receptors (VDR)
 dimerization
  regulation, 355–356
 general features, 346–348
 heterodimers
  genes regulated by, 353
  subcellular localization, 356–357
 retinoid X receptors, 345–364
 RXR
  cytoplasm/nucleus trafficking, 357–360
  target site binding, 362–363
Vitamin D response elements (VDREs), 352
Voltage dependence, 105

W

Women's Health Initiative Observational Study (WHI–OS), 57

X

Xenoestrogens, 18
Xenopus laevis, 357
X-linked spinal and bulbar muscular atrophy (SBMA), 128

Z

Zinnia elegans, 209